ANALYSIS OF LINEAR DYNAMIC SYSTEMS

ANALYSIS OF LINEAR DYNAMIC SYSTEMS

A Unified Treatment for Continuous and Discrete Time and Deterministic and Stochastic Signals

John B. Lewis

Professor of Electrical Engineering
The Pennsylvania State University

MATRIX PUBLISHERS, INCORPORATED
CHAMPAIGN, ILLINOIS • USA

10 9 8 7 6 5 4 3 2 1

Library of Congress catalog card number: 77-79249

Matrix Publishers, Inc.
Champaign, IL 61820

ISBN: 0-916460-20-7

This book was prepared by typewriter composition.
Illustrations by Scientific Illustrators.
In-house Editor was Merl K. Miller.
Series Editor is Andrew P. Sage.

CONTENTS

PREFACE

Since most modern signal and systems analyses and designs involve the use of continuous-time and discrete-time signals and systems, as well as deterministic and stochastic signals, a unified and balanced treatment of these topics, using a common terminology and notation, and with extensive cross references, relates the various concepts to one another in a larger framework.

This book serves as a single comprehensive reference for review and study for those who have a basic knowledge of engineering and who have found in their work in communications systems and control systems as well as in bioengineering, economics, transportation systems, instrumentation, and computer simulation the need for a broader knowledge of signals and systems.

The many small examples in Chapter 2 through 5 illustrate the concepts with a minimum of manipulation and follow directly from the material presented. The emphasis is on the topics under discussion. The problems at the end of each chapter suggest applications of the topics, but the complete development of applications that require the background of Chapters 2 through 5 is given in Chapter 6. Here the emphasis is on what is being modelled, on what is being analyzed, and on how various assumptions and approximations are required in order to use the preceding theoretical results. Sufficient material is presented so that the transition from theory to application is demonstrated without any need to digress from the main concern of the chapter, applications.

The mathematical subjects that have direct bearing on the material are reviewed in three appendixes. This should be an adequate review for those familiar with these topics. For those who are not, the appendixes can serve as a guide, but the more complete expositions referenced should be used for study. In either case, we start in Chapter 2 with the assumption that the basic mathematical background is satisfactory.

The use of simulation diagrams, algorithmic descriptions and standard mathematical operations indicate that extensive analog and digital computations are used, especially when problems of higher dimension are encountered. Appendix D outlines a general-purpose digital program which, if available, can be used in many of the problems.

Although the language of the book is mathematical, it is not intended as a book on applied mathematics. The presentations of theorems as formal statements is used where this format makes the hypothesis and consequent obvious, as in the part of Chapter 5 dealing with stability. No proofs of theorems are given, for they are well known. References which provide proofs are given for the reader whose interests lie in this aspect of the subject.

Extensive use of summary tables is made to place each result in proper context; related equation numbers can be seen at a glance. The underlying unity of the study is clearly shown.

At least three <u>study plans</u> are possible when the book is used as a text. For 3-credit lecture courses, the following are suggested:

1. A one-semester course where a basic background in first and second year mathematics and mechanics is assumed. The course then covers:
   Mathematical Topics - Appendixes A,B,C.
   Signal Models - Chapter 2 (except 2.9)
   System Models - Chapter 3
   System Response - Chapter 4 (except 4.4 and 4.5)

or

2. A one-semester course where no mathematics review is required. Then cover:
   Signal Models - Chapter 2 (except 2.9)
   System Models - Chapter 3
   System Response - Chapter 4 (except 4.4)
   System Stability - Chapter 5, Section 5.1

or

3. A two-semester course where no (or only a brief) mathematics review is required. For the first semester:
   Signal Models - Chapter 2
   System Models - Chapter 3
   System Response - Chapter 4
   and for the second semester:
   System Performance - Chapter 5
   Applications - Chapter 6
   It is desirable that the second semester then include a considerable amount of project work on analog and digital computers, e.g. system response simulation (both deterministic and stochastic cases) sensitivity, signal processing, identification, and model evaluation.

If more applications are wanted in the first two courses, the material is written so that by eliminating certain sections of special interest (e.g., 2.6, 2.9, 3.5, 3.6, 3.7, 4.4, 4.5), parts of Chapter 6 can be used. Section 6.2 can be used after completing Chapter 2 and Section 6.3 can be used after Chapter 4.

University Park, PA
June, 1977

John B. Lewis

ACKNOWLEDGEMENTS

Any author of a new book on signals and linear systems owes much to earlier writers and researchers. The literature on these subjects is large and goes back almost two centuries. Some new results are given, but I also have drawn extensively from material that has evolved over the past four decades. I have given specific sources whenever I was aware of published material, and to those whose work I have missed and not cited, I sincerely regret the omission.

Special credit is due Lee Lewis whose steadfast demand for good writing, regardless of the specialized nature of the material, has made this final version far better than earlier ones. The suggestions for the right word or phrase, for the simple, direct statement, or for the positive form of expression have saved me from many a dull or ambiguous sentence. The careful reading of much of the manuscript as well as a very considerable amount of editorial work are gratefully acknowledged.

The material in the book has been used in lecture courses for several years. To the students who have listened and questioned and to the colleagues who have taught, I express my appreciation. Profs. John L. Brown, Jr. and James F. Delansky were particularly helpful in their discussions with me. My work with graduate students also has led to a number of additions and I am indebted to Drs. G. Bucek, J. Keenan, F. Symons, Jr., Y.-S. Jan, and M. Bloom and to B. Kuhn and B. Towe. The theses of R. Huber and W. L. Miller provided material in Chapter 6.

Mrs. Brenda Wagner skillfully typed the early drafts as well as the final version of the book. Her conscientious and cheerful participation in this project was a great help.

MATHEMATICAL SYMBOLS AND NOTATION

$=$	equal
$\equiv$	equal by definition
$\neq$	not equal
$\cong$	approximately equal
$\geq$	greater than or equal
$\leq$	less than or equal
$\forall$	for all
$\Rightarrow$	implies
$\Leftrightarrow$	if, and only if
$\rightarrow$	goes to, approaches
∞	unbounded, infinite
$\lvert\alpha\rvert$	magnitude or absolute value of α
ε	belongs to
$\exists$	there exists
$\ni$	such that
$f(\cdot)$	function f
f^*	complex conjugate of f
lim	limit

$f(0+)$	right hand limit at 0
$f(0-)$	left hand limit at 0
$\dot{f}(t)$	derivative of f with respect to t
t	continuous-time variable
n	discrete-time variable
$[a,b]$	closed interval
D	derivative operator, e.g., $Df(t) = \frac{df}{dt}$
E	shift operator, e.g., $Ef(n)=f(n+1)$
$f^{(k)}$	kth derivative
Δ_b	first backward difference operator
$\sum_i$	summation over i
$\prod_i$	product over i
R_r	r-dimension vector space
$\underline{a}$	column vector
A	matrix
A^T	matrix transpose
A^{-1}	matrix inverse
$\det A$	determinant of A
$\lVert\underline{a}\rVert$	norm of $\underline{a}$
$\lVert A\rVert$	norm of A
Θ	null matrix
I	unit matrix
$Pr(\ )$	probability of
$E[\ ]$	expected value
$\langle\ \rangle$	time average (also inner product in App. B)
$(\mid)$	given
$\{\ \}$	set or sequence
$\oplus$	or
$\otimes$	and
$\cup$	union
$\cap$	intersection
$A \subset B$	A is a proper subset of B
$\frac{\partial f}{\partial x}$	partial derivative of f with respect to x
$\mathrm{Sa}x$	$\frac{\mathrm{Sin}x}{x}$
$u_{(M)}$	M-term approximation

Chapter 1

INTRODUCTION

1.1 SIGNALS, SYSTEMS, AND MODELS

The history of signal and system theory can be traced through the development of the calculus, differential and difference equations, probability and statistics, linear algebra and vector spaces and transform theory. Recent signal and system theory is the result of the work of Wiener, Kalman, Bellman and others who contributed both new theories and equally important applications of those theories to a great array of problems in communications, control, and computation. In such disciplines as physics, chemistry, biology, economics, sociology, etc., the analysis and design of signals and systems can solve many problems. The common ground is mathematical abstraction.

Information or energy (or both) flows through the system from an input terminal to an output terminal as shown in Figure 1.1. An operation on (or transformation of) the input

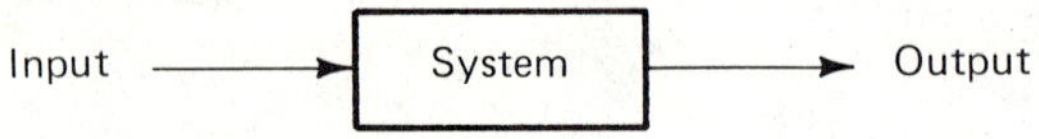

Figure 1.1

is performed to produce the output. The inputs and outputs are signals; the operation is the system. The signals contain information, and, in cases where the signals are physical quantities (e.g., voltage, pressure, light intensity), they also contain energy. Systems, which occur in limitless variety, consist of a number of interacting elements, and the ability to deal with the complex interactions of those elements makes the study of signals and systems both interesting and useful.

In this book, the language used to describe signals and systems is mathematical, and it is necessary that models, mathematical representations, be used. Modelling is central to much of the use of signal and system theories. In modelling, we are concerned with making compromises between descriptions which involve so much detail that the complexity makes its use difficult, if not impossible, and descriptions that are so simple that the phenomena of interest are lost. An elementary example is shown in Figure 1.2 where an iron core audio transformer is to be modelled. A "complete" model would include iron saturation, eddy current and hysteresis loss in the core, winding resistance variation with frequency (skin effect), magnetic flux leakage, capacitance between the turns of the windings, capacitance between the two windings and between the windings and the core, etc. However, if only small currents and certain frequencies are of interest, various linear, lumped-parameter models can be used [1.]. Thus, the mathematical descriptions for the low and high frequency linear equivalents can be obtained by the application of basic lumped circuit theory --

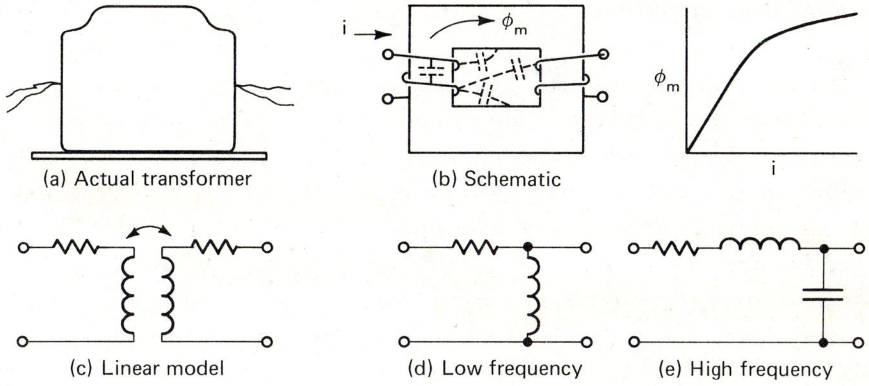

Figure 1.2

the Kirchoff current and voltage laws. Clearly, the development of a model depends on the intended use.

Modelling also must bring together the mathematical forms and the data obtained from measurements on actual signals and systems. In this book this procedure is referred to as identification. Generally a verification of the form of the model is needed as well as a set of numerical values for the parameters (e.g., the resistances, inductances, and capacitances in Figure 1.2). The use of least squares methods -- methods which minimize a sum or integral of the square of the error between model and experimental values -- is introduced briefly in Chapter 2 on signal models and in Chapter 4 on system models.

Although no model ever corresponds exactly to reality, it must be remembered that no amount of correct mathematical procedure or accurate computation can offset a poor choice of model. One must distinguish at all times between the model and the real world, and a model must be permitted to evolve as conditions change or as additional data become available [2.]. In practical problems, experience and new information are essential to the evolution of the model. A useful model must serve practical ends, and good models are useful for analyzing signals and systems and for predicting the behavior of signals and systems.

1.2 ANALYSIS AND DESIGN

Analysis of signals and systems is concerned with discovering their properties. The information in a signal may be related to its presence or absence or to some complex code. Common properties of interest are periodicities, peak values, zero crossings, spectral content, and correlation time. For systems, properties such as stability, linearity, and response modes are of interest. Most of these can be inferred from the models; this is the approach taken in this book. Obviously, it is important then that a close relationship to actual signals and systems be maintained in the model development. Tests are needed which compare the models with the signals and systems they represent. One of the advantages of using models is that many tests can be done on a model that cannot be done on a real system, and the wide availability of general-purpose analog and digital computers has accelerated the development of methods of signal and system analysis by computer simulation.

In a more restricted technological view, interest in signals and systems stems from the desire to design and synthesize. Problems such as the design of signals to minimize the effects of noisy transmission channels, signals which afford a maximum probability of detection (e.g., in a radar system), signals which do not exceed dynamic ranges, and signals which have restricted bandwidths (e.g., in a multiplex system) are common. Designs of systems to perform a great variety of functions are needed: signal transmission, detection, and filtering, control of complex physical processes, analog and digital computation, and monitoring and checking other systems are examples. The ability to model and to analyze is essential to achieving these designs, and much of the material in this book is aimed at these goals. At a more advanced level, we are interested in the idea of a "best" design, and this leads to the need for optimization -- an important aspect of the study of signals and systems. There is a large amount of literature on optimal estimation and control, operations research (scheduling, routing, inventory control, etc.) as well as on the mathematical background of mathematical programming, the calculus of variations, and related numerical methods.

1.3 GENERAL SYSTEMS

Although Figure 1.1 gives a simple view of a meaning of signals and systems, there is a much more general interpretation, particularly for systems [17.]. Associated signals or variables are part of the general system. Some examples are a number of particles interacting, a society of animals, an industrial complex, and a digital computer [4.]. A system in this broad sense need not involve physical objects, e.g., we have algorithms for computation and supervisory programs of a time-shared computer. Recently much interest has developed in using computer simulation and system modelling to study the entire world as a dynamical system [5.] as well as industrial and urban systems [6.][7.]. In fact, this broad interpretation of systems can be found in statements describing the areas of interest of two journals:

"Foundations of general systems theory and methodology in various branches of science, technology, humanities, and the arts, general systems philosophy, and general systems education. Examples of topics within these areas are: principles of modelling, and simulation, systems analysis and synthesis, optimization, identification, general principles of experimentation, problems associated with extremely complex systems, studies concerned with various classes of systems such as self-organizing, adaptive, self-producing, fuzzy, hierarchical, etc." [8.];

"Large-scale systems, theory and applications; optimization, decision analysis, problem definition, modelling, simulation, test, and evaluation. Foundations of cybernetics, pattern recognition, adaptive and learning systems, biocybernetics; man-machine systems. Representative applications include complex hardware, behavioral, biological, ecological,

educational, environmental, health care, management, socio-economic, transportation, and urban systems" [9.].
The distinction between theory and application exemplified in the preceding statements will be followed in subsequent chapters.

System and signal <u>theory</u> is concerned with general results -- the more general the better -- and consequently they are largely mathematical. <u>Applications</u>, on the other hand, involve specifics and the use of appropriate theories. Typical references on signal theory (which are more advanced than the following material) are [10.][11.][12.] and typical references on system theory (also advanced) are [13.][14.][15.][16.]. These references provide theoretical material related to Chapters 2 through 4 -- a coverage of signals and systems far less general than that implied by the two statements quoted above. Applications are discussed in both of the journals to which we have referred, and many additional references are given in Chapter 6 which is entirely on applications.

1.4 CONTENT

As is apparent in the previous section, no single book can cover more than a small part of the general area of signal and system theory and applications. However, if only certain classes of signals and systems are chosen for study, an introduction to a great many ideas which provide a point of departure for more advanced and specialized study can be developed. The aim of this book is to provide a background in signal and system models, system response, and system performance under the following assumptions:

1. The signals are either <u>deterministic</u> (functions of continuous or discrete time) or are stationary <u>stochastic</u> processes or sequences.
2. The systems are described by ordinary <u>linear</u> differential or difference equations.

Within this framework a parallel can be developed between continuous-time and discrete-time results. This parallel development reenforces many ideas. Although there are many similarities, enough exceptions occur to warrant careful analysis in each case. In addition, a number of useful similarities between deterministic and stochastic signal models exist, and parallel developments are carried out wherever possible.

The scope of the book is summarized in Figure 1.3. Briefly, the plan is to develop models in Chapters 2 and 3 and to develop the response results, i.e., input/output forms, in Chapter 4 for all the cases implied by Figure 1.3. Chapter 5 is then concerned primarily with measures of system performance -- accuracy, sensitivity, and reliability. Chapter 6, as pointed out above, provides some

SIGNALS

Continuous-Time	*Discrete-Time*
Deterministic	Deterministic
Stochastic processes	Stochastic sequences

SYSTEMS

Continuous-Time	*Discrete-Time*
Lumped, Linear	Lumped, Linear

Figure 1.3

examples of applications in digital signal processing (spectral estimation), R-L-C network analysis, aircraft control, and power system analysis.

In choosing this particular range of topics for study in this book, we also note some topics that are not covered. Examples are complex signals, nonlinear systems, distributed parameter systems, and stochastic (i.e., random parameter) systems. Much research is still being done on all of these. Furthermore, new ways of dealing with systems problems are being developed which involve major advances over the traditional methods. For example, there has been, in recent years, a growing interest in algebraic methods [18.], and an excellent introduction is given in [19.].

Summary tables show the various results in proper relationship and give a compact presentation in the form of equation numbers. The main results are shown in the following tables:

Signal Models	Tables 2.6, 2.7, 2.8
System Models	Table 3.4
System Response	Tables 4.2, 4.3
Stability	Table 5.1

Frequent reference to these should be made so that each new development can be seen in perspective.

1.5 TERMINOLOGY

We shall, wherever possible, use terms on which there is general agreement in the literature of signals and systems. Generally, when new terms are first introduced, they are italicized to emphasize that a defining statement is being given. Most of the needed mathematical terms are provided in the appendixes. The terms continuous-time and discrete-time as well as deterministic and stochastic are introduced in Chapter 2. We shall also make frequent reference to the "time domain" and the "frequency domain". To make clear the sense in which these terms are used, we can refer to Figure 1.4. The figure also shows the symbols used for the various transforms. Although the Fourier ($\mathfrak{F}$) and Laplace ($\mathfrak{L}$) Transforms and Fourier Series ($\mathfrak{F}$ -Series) are well known, the transforms on discrete-time data may not be. These are the Z-Transform ($\mathfrak{Z}$) and the Fourier Transform ($\mathfrak{F}_d$). We also have a Fourier Series ($\mathfrak{F}_d$-Series) on periodic discrete-time data which is closely related to the Discrete Fourier Transform (DFT). All four transforms are reviewed in Appendix A, the two Fourier Series forms are reviewed in Chapter 2, and the DFT is developed in Chapter 6, Section 6.2.2

1.6 BACKGROUND

The background subjects for the study of signals and systems depend on the goal. Since the goal here is to provide a broad but not advanced coverage of signals and systems, there are some basic areas of which knowledge is assumed. In mathematics, an understanding of the calculus, linear differential and difference equations (reviewed briefly in Chapters 3 and 4), transform theory, matrices and vector spaces, and probability (reviewed in Appendixes A, B, and C, respectively) is assumed. In physics and engineering, the basic methods in mechanics and electric circuits are needed. Finally, some knowledge of analog and digital

FREQUENCY DOMAIN*

			Continuous		Discrete
			s	ω	$k\omega_o$
TIME DOMAIN	Continuous	t	Laplace Transform $\mathcal{L}\, f(t)=F(s)$	Fourier Transform $\mathfrak{F}\, f(t)=F(j\omega)$	Fourier Series $\mathfrak{F}$-Series
	Discrete	n	Z-Transform $\mathfrak{Z}\, f(n)=F(z)$	Fourier Transform $\mathfrak{F}_d f(n)=F(\varepsilon^{j\theta})$	Fourier# Series $\mathfrak{F}_d$-Series DFT
			z	θ	$k\lambda_o$

Figure 1.4

*The transform domain is called the "frequency domain" for both continuous-time and discrete-time. Although s, ω, $k\omega_o$ have the dimension inverse seconds (sec^{-1}), z, θ, and $k\lambda_o$ are dimensionless and the "frequency" terminology is only by analogy.

#The $\mathfrak{F}_d$-Series (Fourier Series for discrete-time functions) and the Discrete Fourier Series (DFT) have the same mathematical form. The DFT usually appears as a numerical approximation to the $\mathfrak{F}$-Transform where sampling in both time ($t \to nT_s$) and frequency ($\omega \to k\Omega$) are used. The $\mathfrak{F}$-Series and $\mathfrak{F}_d$-Series are expansions of periodic functions.

computation is helpful since many of the models in the book are illustrated with the use of simulation diagrams. Related block diagrams and signal flow graphs [3.] are omitted although these methods are sometimes used in systems analysis. Beyond this background knowledge, it is assumed that the reader brings his own special interest such as communications, control, economics, experimental design, reliability, or human factors to this study, and, using this methodology, will be able to find solutions to his own particular problems.

REFERENCES

1. Terman, F., "Radio Engineering," New York: McGraw-Hill Book Co., 3rd Ed., 1947.
2. Golomb, S., "Mathematical Models: Uses and Limitations," IEEE Trans. on Reliability, v. R-20, n. 3, Aug. 1971, pp. 130-131.
3. Mason, S., "Feedback Theory - Some Properties of Signal Flow Graphs," Proc. IRE, v. 41, n. 9, Sept. 1953, pp. 1144-1156.
4. Zadeh, L. "From Circuit Theory to System Theory," Proc. IRE, v. 50, n. 5, May 1962, pp. 856-865.
5. Forrester, J., "World Dynamics," Cambridge, Mass.: Wright-Allan Press, 1971.
6. Forrester, J., "Industrial Dynamics," Cambridge, Mass.: M.I.T. Press, 1961.
7. Forrester, J., "Urban Dynamics," Cambridge, Mass.: M.I.T. Press, 1969.
8. ____________, International Journal of General Systems, New York: Gordan and Breach.
9. ____________, IEEE Transactions on Systems, Man, and Cybernetics, New York: IEEE.
10. Franks, L., "Signal Theory," Englewood Cliffs, N.J., 1969.
11. Thomas, J., "Introduction to Statistical Communication Theory," New York: Wiley, 1972.

12. Gallager, R., "Information Theory and Reliable Communication," New York: Wiley, 1968.
13. Zadeh, L. and Desoer, C., "Linear System Theory," New York: McGraw-Hill, 1963.
14. Wymore, A. W., "A Mathematical Theory of Systems Engineering," New York: Wiley, 1967.
15. Padulo, L. and Arbib, M., "System Theory," Philadelphia: W. B. Saunders, 1974.
16. Brockett, R., "Finite Dimensional Linear Systems," New York: Wiley, 1970.
17. von Bertalanffy, L., "General System Theory," New York: Braziller, 1968.
18. Kalman, R., et al, "Topics in Mathematical System Theory," New York: McGraw-Hill, 1969.
19. Sain, M., "The Growing Algebraic Presence in Systems Engineering: An Introduction," IEEE Proc., v. 64, n. 1, Jan. 1976, pp. 96-111.

Chapter 2

SIGNAL MODELS

2.1 DEFINITIONS

In engineering, the word signal is used in many ways and it is reasonable to expect that we have some intuitive concept of its meaning. However, a more explicit definition is needed so that a distinction can be drawn between actual signals and signal models. Usually we assume that an actual signal carries information, particularly where it is an output or observation or response associated with a physical system. Terms such as input or actuating or excitation signals occur too. These also may contain energy. Basically, a signal model is a function, a single-valued mathematical form. This mathematical signal model may be used to represent the evolution in time of physical variables such as voltage, current, displacement, pressure, and force. In all cases discussed in this chapter, models of signals are needed so that a mathematical analysis can be done to discover signal properties that are of interest.

2.1.1 Signal Models

Although in some instances we may be given a signal model or we may have sufficient evidence to justify assuming one, the more interesting possibilities occur when we are required to obtain a signal model based on an analysis of experimental data. In all three cases we must be familiar with a great variety of signal models, and this chapter is concerned with the classification and description of signal models. The term signal generally will refer to a signal model; an actual signal or experimental data will be specified as it is used.

Figure 2.1 illustrates, in a general way, the nature of our goals in this chapter. We assume that a true signal is being generated, but we frequently cannot make direct measurements on it. Therefore, we try to find a model which

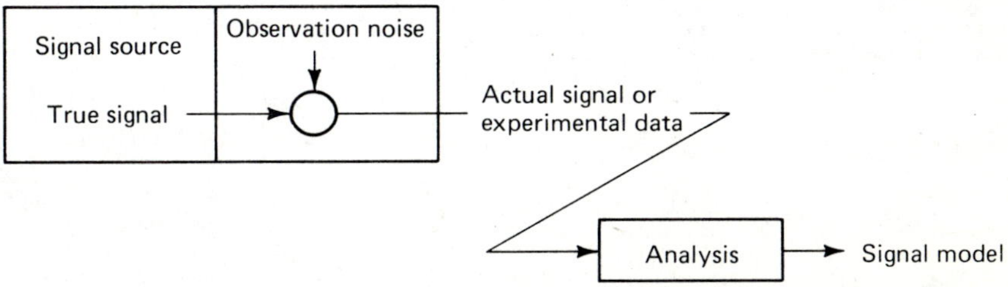

Figure 2.1

is in some way a good approximation to the true signal. Once we have a model -- obtained from experimental data, from a specification, or by assumption -- we can proceed to study the signal, i.e., we try to discover signal characteristics such as the nature of its variation in time and its frequency content. We also can check our model against experimental evidence to verify it, and if it is satisfactory, we can use our model to predict future behavior of the signal. Furthermore, as we shall see in Chapter 3, the signal models become the inputs and outputs of the systems we wish to study. Many system properties may be inferred from the system output signals when specific input signal models are used. In communications and information systems, it is the "signal content" itself that is of interest. It should be clear then that a major component of the general study of signals and systems outlined in Chapter 1 is an understanding of signal models.

2.1.2 Domains and Ranges, Discrete and Continuous

The signal u(t) is a real-valued function of t, and t usually represents time. The real variable t takes on all values on an interval, the domain of u. In set notation, two domains frequently used in this book are:

$$T_{c1} = \{t: t \geq 0\}$$

or

$$T_{c2} = \{t: -\infty < t < \infty\} .$$

In both cases, u(t) is called a <u>continuous-time signal</u>.* For T_{c1}, u(t) is <u>one-sided</u> and for T_{c2} it is <u>two-sided</u>. Similarly, the signal v(n) is a real-valued function of n, an integer argument. The variable n may or may not be related to certain instants of time, t_n. Two domains for v(n) are:

$$T_{d1} = \{n: n \geq 0, n \text{ integer}\}$$

or

$$T_{d2} = \{n: -\infty < n < \infty, n \text{ integer}\}$$

In both cases, v(n) is called a <u>discrete-time signal</u> and the terms one-sided and two-sided apply as before. In many situations, a subset of T_{c1} or T_{c2} containing elements t_n is generated by <u>sampling</u>. An appropriate change of variable can then be used to define the corresponding T_{d1} or T_{d2}. For example, if the adjacent instants at which the samples are

---

*Note that a continuous-time signal is not necessarily a continuous function.

taken are T_s seconds apart (uniform sampling),

$$T_s = t_{n+1} - t_n \ , \quad \forall n$$

and

$$t_n = nT_s$$

then $u(t_n) = u(nT_s)$ can be related directly to a v(n). In other cases discrete-time functions occur as a natural way to describe certain signals, particularly where digital computer processing is anticipated. Two typical signals are shown in Figure 2.2.

The amplitude values or range of u(t) or v(n) may be the real numbers or a subset of real numbers. Two sets which will be used are:

$$U_1 = \{u(t)\colon u(t) \in R_1\}$$

or

$$U_2 = \{u(t)\colon u(t) \in A \subset R_1\}$$

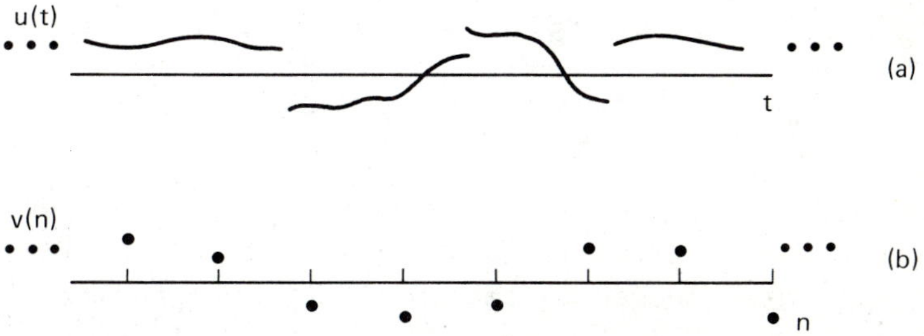

Figure 2.2 (a) Continuous-time signal, (b) Discrete-time signal

with similar expressions for v(n) in terms of V_1 and V_2. The set A is an amplitude quantization set and is one of the two sets which describes a process of quantization. These sets are defined in Section 2.9. Signals may occur originally as quantized signals as described by U_2 and V_2, or by a conversion of u(t) or v(n) in a quantizer. The quantized signals are $u_q(t)$ and $v_q(n)$. The process of quantization is discussed in detail in 2.9, but it is important to observe here that $u_q(t)$ and $v_q(n)$ must be single-valued. Two typical quantized signals are shown in Figure 2.3.

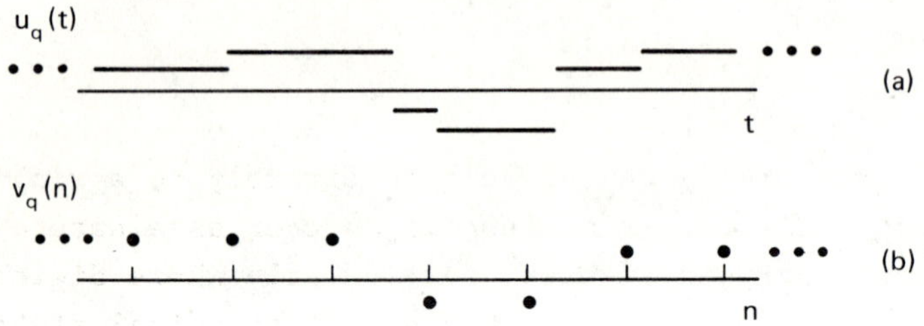

Figure 2.3 (a) Quantized continuous-time signal, (b) Quantized discrete-time signal

2.1.3 Signal Conversions

The processes of sampling and quantizing are signal conversions. It is possible, for example, to start with u(t), sample it to produce $u(nT_s)$, and then to quantize this to get $u_q(nT_s)$. This is illustrated in Figure 2.4. A typical application is analog to digital signal conversion.

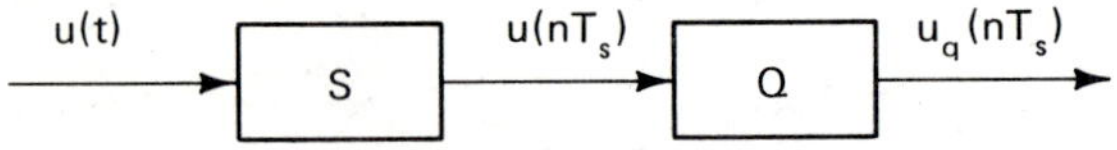

Figure 2.4 Sampling and Quantizing

S is a uniform sampler and Q is a quantizer. One question that occurs immediately is that of interchangeability. If S and Q are interchanged so that u(t) is converted to a $u_q(t)$ which is then converted to a $u_q'(nT_s)$, it is quite possible that $u_q(nT_s) \neq u_q'(nT_s)$. This situation is considered in Section 2.9 where conditions under which the interchange is valid are given. The well known "sample and hold and then quantize" methods used in some digital voltmeters and analog-to-digital converters are examples of the situation illustrated in Figure 2.4. Two other signal conversions related to S and Q are reconstruction, R, and dequantization, F, a particular kind of filter, R converts a discrete-time signal, e.g., $u(nT_s)$ to $u_r(t)$ or $u_q(nT_s)$ to $u_{qr}(t)$. F converts a quantized signal to one with a continuum of amplitude values, e.g., $u_q(t)$ to $u_{qf}(t)$ or $v_q(n)$ to $v_{qf}(n)$. These are discussed in Section 2.9 and examples of F and R are shown in Figure 2.5.

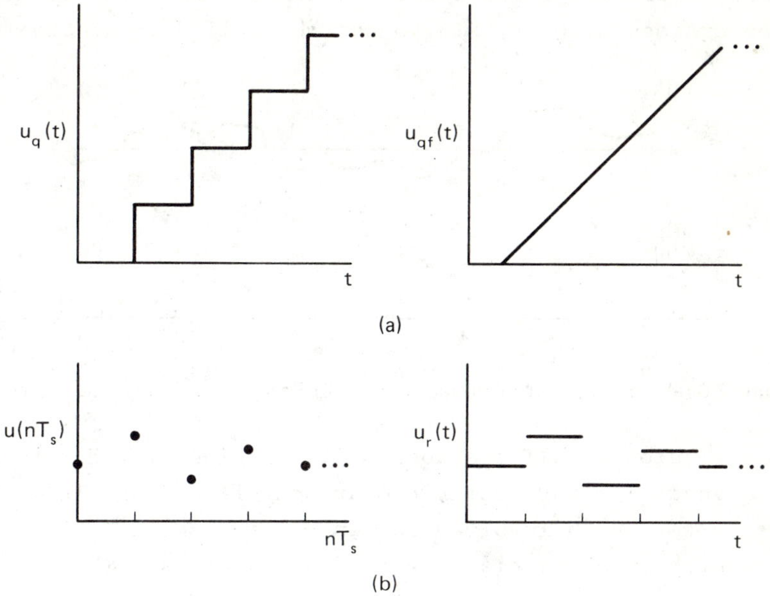

Figure 2.5 (a) Dequantization, (b) Reconstruction

2.1.4 Deterministic and Stochastic Signals

The signals u(t) and v(n), and signals derived from them by the S, Q, F, R conversions, are all <u>deterministic</u> signals since for any $t \in T_{c1}$ (or $t \in T_{c2}$) or $n \in T_{d1}$ (or $n \in T_{d2}$) there is exactly one value of the signal. Furthermore, when t or n varies over the entire domain of the signal, the signal is completely determined. A signal is <u>stochastic</u> if it is defined on a sample space $\mathcal{S}$ as well as on T_{c1}, T_{c2}, T_{d1}, or T_{d2}. Thus the stochastic signal notation $U(t,\sigma)$ or $V(n,\sigma)$ (see Appendix C) is implied although the σ frequently is dropped. Sets of stochastic signals defined on T_{c1} or T_{c2} will be called <u>stochastic processes</u> and those defined on T_{d1} or T_{d2} will be called <u>stochastic sequences</u>. These will be discussed in Section 2.7 where the concept of an ensemble of sample functions or realizations (for fixed σ) of the stochastic process or stochastic sequence is introduced. The two forms for the realizations are $u^i(t)$ or $v^i(n)$, i=1,2,3, ..., and Figure 2.6 shows two examples of these realizations. The conversions S, Q. F, R also can be applied to stochastic

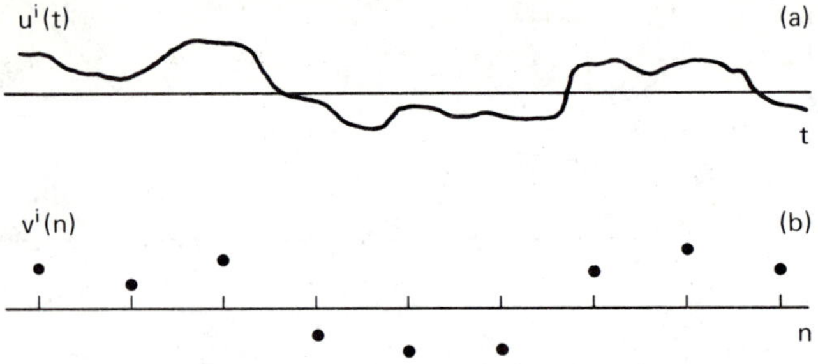

Figure 2.6 (a) Realization of stochastic process, (b) Realization of stochastic sequence

signals to produce all the possibilities given for deterministic signals. A particular comparison of interest is between U(t), which involves continuous random variables, and $U_q(t)$, which involves discrete random variables. (See Appendix C). A similar comparison can be made between V(n) and $V_q(n)$.

2.1.5 Vector Signals

Vector signals can be formed by using a finite number of scalar signals as components (see Appendix B). The notation is then $\underline{u}(t)$ and $\underline{v}(n)$ for deterministic vector signals and $\underline{U}(t)$ and $\underline{V}(n)$ for stochastic vector signals. Appropriate forms for the four signal conversions also follow immediately.

2.1.6 Energy and Power Signals

Another signal classification concerns the energy and average power in a signal where the terms energy and power are used in a general sense, i.e., they usually are not joules and watts. Starting with instantaneous power as the square of the signal, for <u>energy signals</u>

$$E_c = \int_{-\infty}^{\infty} u^2(\lambda)\,d\lambda < \infty$$

or

$$E_d = \sum_{k=-\infty}^{\infty} v^2(k) < \infty$$

Thus energy signals have finite energy. Similarly, for <u>power signals</u>

$$0 < P_c = \lim_{T\to\infty} \frac{1}{2T} \int_{-T}^{T} u^2(\lambda)\,d\lambda < \infty$$

or

$$0 < P_d = \lim_{N\to\infty} \frac{1}{2N+1} \sum_{k=-N}^{N} v^2(k) < \infty \ .$$

Thus power signals have finite average power as it is defined. It also can be seen that energy signals have zero average power and power signals have infinite energy. These two classifications are important with regard to the Fourier Transforms of u(t) or v(n) (see Appendix A). Generally the transforms exist for energy signals and for some power signals.

2.2 ELEMENTARY FORMS AND LINEAR COMBINATIONS

The first step in developing a signal model is to define a set of elementary forms. In some cases these elementary forms are used directly as models for test signals in systems analysis. In other cases the elementary forms serve as a basis for more general signal models -- particularly where linear combinations of the elementary forms are used. Four common forms in wide use are singularity functions, sinusoidal functions, real exponential functions, and complex exponential functions. These are developed first for both continuous-time and discrete-time signal models and in each case the simple forms are then used in linear combinations to model classes of functions such as polynomials and periodic functions. These ideas precede the extension to more general forms including the use of orthogonal basis functions.

2.2.1 Singularity Functions

The function

$$u(t) = \frac{1}{k!} t^k \qquad t \geq 0 \tag{2.1}$$
$$= 0 \qquad t < 0$$

for* k=0,1,2,... has, for a given finite k, a finite number of derivatives at t=0. (Note that for k=0, $\lim_{t \to 0+} u(t) = 1$)
For example, for k=2

$$u(t) = \frac{1}{2} t^2 \qquad t \geq 0$$

and

$$\dot{u}(t) = Du(t) = t \qquad t \geq 0$$

but

$$\ddot{u}(t) = D^2u(t) = 1 \qquad t > 0$$

and the second derivative does not exist at t=0. Functions which do not have finite derivatives of all orders (for all t) are frequently called <u>singularity</u> functions. Using 2.1, the following are defined: For

k=0, $u(t) = u_1(t)$, a unit step*

k=1, $u(t) = u_2(t)$, a unit ramp

k=2, $u(t) = u_3(t)$, a unit square ,

etc. If k = j-1 is used in 2.1, the general form is

$$u_j(t) = \frac{1}{(j-1)!} t^{j-1}, \quad t \geq 0 \tag{2.2}$$
$$= 0 \quad , \quad t < 0$$

---

* 0! = 1 and 0° is defined as 1.

for $j \geq 1$. Next, consider the first derivative of $u_j(t)$,

$$Du_j(t) = \frac{1}{(j-2)!} t^{j-2}$$

or (2.3)

$$Du_j(t) = u_{j-1}(t)$$

for $j \geq 2$. By integrating 2.3,

$$\int_{-\infty}^{t} u_{j-1}(\lambda)d\lambda = u_j(t) \qquad (2.4)$$

and Equations 2.3 and 2.4 then serve as a pair for going from one singularity function to another for $j \geq 2$.

To define the $u_j(t)$, $j < 1$, we could try to use Equation 2.4. For $j = 1$,

$$\int_{-\infty}^{t} u_o(\lambda)d\lambda = u_1(t), \quad t \neq 0 \qquad (2.5)$$

which gives a property of the unit impulse $u_o(t)$ for $t \neq 0$. Also, differentiating 2.5 for $t \neq 0$ gives another property

$$u_o(t) = 0, \quad t \neq 0 . \qquad (2.6)$$

However, these properties do not define $u_o(t)$ satisfactorily. An intuitive definition that frequently is used is the limit of a rectangular pulse $s_o(t,\Delta)$ of duration Δ and amplitude $1/\Delta$ which starts at $t = -\Delta/2$. We then let

$$u_o(t) = \lim_{\Delta \to 0} s_o(t,\Delta) . \qquad (2.7)$$

Other pulse shapes with unit area also can be used. Using 2.7,

$$\int_{-\infty}^{t} u_o(\lambda)\,d\lambda = \int_{-\infty}^{t} \lim_{\Delta\to 0} s_o(\lambda,\Delta)\,d\lambda$$

and assuming an interchange of integration and limit is valid,

$$\int_{-\infty}^{t} u_o(\lambda)\,d\lambda = \lim_{\Delta\to 0} \int_{-\infty}^{t} s_o(\lambda,\Delta)\,d\lambda = \lim_{\Delta\to 0} s_1(t,\Delta) = u_1(t)$$

where s_1 is defined as the integral of s_o. This result is consistent with 2.5. Also, differentiating 2.5 gives

$$u_o(t) = \frac{d}{dt} u_1(t) = \frac{d}{dt} \lim_{\Delta\to 0} s_1(t,\Delta)$$

and assuming an interchange of differentiation and limit is valid,

$$u_o(t) = \lim_{\Delta\to 0} \frac{d}{dt} s_1(t,\Delta) = \lim_{\Delta\to 0} s_o(t,\Delta)$$

and

$$u_o(t) = 0, \quad t \neq 0$$

which is consistent with 2.6.

A much more satisfactory way to define the $u_j(t)$, $j < 1$, is to use the concept of <u>generalized</u> <u>functions</u> or specifically, distributions [1.]. We use the form

$$\int_{-\infty}^{\infty} f(\lambda)\phi(\lambda)d\lambda \tag{2.7a}$$

where f is a distribution and ϕ is an ordinary function called a testing function. The distribution f is defined in terms of the value of the integral for a class of functions ϕ. The functions $\phi(\lambda)$ that we use here possess all derivatives and become zero for $|\lambda|$ sufficiently large.

Using the previous notation, we let $u_o(\cdot)$ be a distribution and $u(\cdot)$ be a testing function and define the unit impulse u_o such that

$$\int_{-\infty}^{\infty} u(t-\lambda)u_o(\lambda)d\lambda = u(t) \tag{2.8}$$

This is called a sifting property. Other equivalent forms are

$$\int_{-\infty}^{\infty} u(\lambda)u_o(t-\lambda)d\lambda = u(t) \tag{2.9}$$

and

$$\int_{-\infty}^{\infty} u(\lambda)u_o(\lambda-t)d\lambda = u(t) \quad . \tag{2.10}$$

Having defined u_o as a distribution, we also need extensions of 2.3 and 2.4 for $j = 1$. To do this we consider a generalized derivative of the distribution f in 2.7a, i.e.,

$$\int_{-\infty}^{\infty} \frac{df(\lambda)}{d\lambda}\,\phi(\lambda)d\lambda$$

Integrating this by parts and using the restrictions on ϕ,

$$\int_{-\infty}^{\infty} \frac{df(\lambda)}{d\lambda}\,\phi(\lambda)d\lambda = -\int_{-\infty}^{\infty} f(\lambda)\,\frac{d\phi(\lambda)}{d\lambda}\,d\lambda$$

Using the derivative of u_1 as a distribution in this result

we obtain

$$\int_{-\infty}^{\infty} \frac{du_1}{d\lambda} u(\lambda)d\lambda = -\int_{-\infty}^{\infty} u_1(\lambda) \frac{du}{d\lambda} d\lambda$$

$$= -\int_{0}^{\infty} \frac{du}{d\lambda} d\lambda = u(o)-u(\infty) = u(o).$$

Comparing this to 2.10 with t = 0, and using the definition of equality of distributions,

$$\frac{du_1}{d\lambda} = u_o(\lambda)$$

which is 2.3 for j = 1. Integrating this (in the sense of distributions), we obtain 2.4 for j = 1. By repeating the sequence of steps that starts with 2.8, we can define other distributions u_{-k}, k=1,2,..., with

$$\int_{-\infty}^{\infty} u(t-\lambda)u_{-k}(\lambda)d\lambda = \frac{d^k}{dt^k} u(t) \tag{2.11}$$

or an equivalent form

$$\int_{-\infty}^{\infty} u_{-k}(\lambda-t)u(\lambda)d\lambda = (-1)^k \frac{d^k}{dt^k} u(t) \tag{2.12}$$

Also, using generalized derivatives as in the preceding steps we can show that

$$Du_o(t) = u_{-1}(t)$$

$$Du_{-1}(t) = u_{-2}(t)$$

$$\vdots$$

$$Du_{-k}(t) = u_{-k-1}(t)$$

or for j = -k,

$$Du_j(t) = u_{j-1}(t)$$

as the extension of Equation 2.3. Integrating this we obtain the extension of Equation 2.4. In fact, with the understanding that the $u_j(t)$ for $j \leq 0$ are defined as distributions by Equation 2.11, Equations 2.3 and 2.4 are used for all j. The complete set of $u_j(t)$ is then written generally as

$$u_j(t-t_o) = \frac{1}{(j-1)!}(t-t_o)^{j-1}, \quad t \geq t_o \qquad (2.13)$$
$$= 0 \qquad t < t_o$$

for $j \geq 1$, and with

$$Du_j(t-t_o) = u_{j-1}(t-t_o) \qquad (2.14)$$

and

$$\int_{-\infty}^{t} u_{j-1}(\lambda-t_o)\,d\lambda = u_j(t-t_o) \qquad (2.15)$$

for all j. These three equations summarize our results for the set of continuous-time singularity functions $u_j(t)$, $\forall$ j. Equation 2.9 is useful for signal resolution in terms of generalized singularity functions as described later in this chapter.

Figure 2.7 shows the singularity functions most frequently used in developing signal models.

Next we use the singularity functions $u_j(t)$ to develop models for a class of signals. Polynomial signals easily can be written as a sum of $u_j(t)$. For N+1 terms,

$$u_{(N+1)}(t) = a_o + a_1t + a_2t^2 + \ldots + a_Nt^N, \quad t \geq 0$$
$$= 0 \qquad , \quad t < 0$$

where the subscript in parentheses, (N+1), is used to specify the number of terms in the sum. It also is the number of coefficients a_i, i=0,1, ..., N. An equivalent form is then

$$u_{(N+1)}(t) = a_ou_1(t) + a_1u_2(t) + 2!a_2u_3(t) + \ldots$$
$$+ N!a_Nu_{N+1}(t)$$

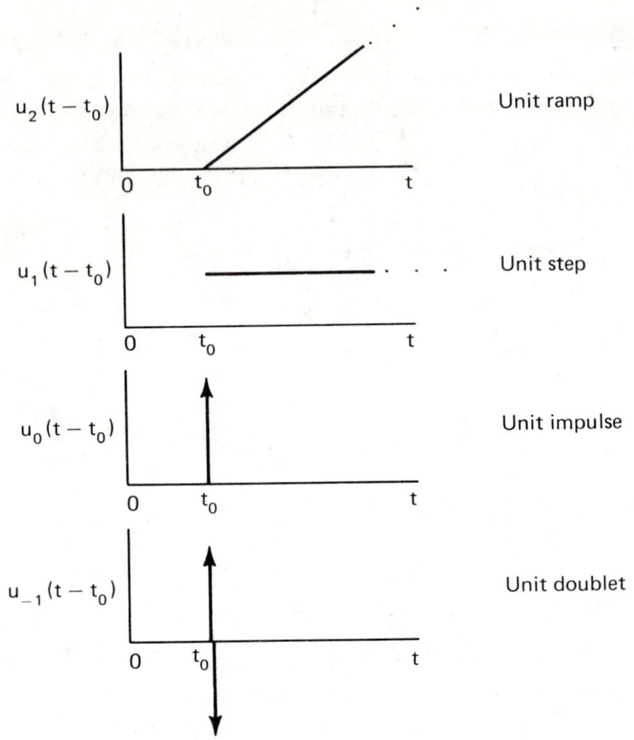

Figure 2.7

or

$$u_{(N+1)}(t) = \sum_{j=1}^{N+1} (j-1)!a_{j-1}u_j(t) \quad . \tag{2.16}$$

The model $u_{(N+1)}(t)$ is a linear combination of elementary functions which in this case are the singularity functions $u_j(t)$, $j=1,2,\ldots,N+1$.

We now can consider a specific problem of the type suggested in Figure 2.1. Let $u(t)$ be the true or ideal signal and $u_x(t)$ be a record of experimental observations, $t \geq 0$. If it is known that $u(t)$ is such that a polynomial model like 2.16 can be used as a good approximation, the problem (for given N+1) is to find coefficients a_{j-1}, $j=1,2,\ldots,N+1$ which make some nonnegative function of error as small as possible. The error is the difference between $u(t)$ and $u_{(N+1)}(t)$, i.e., $e(t) = u(t)-u_{(N+1)}(t)$. For example, we could choose the a_{j-1} to make the error zero at N+1 values of t, an "exact fit".

This usually is not a good procedure, however, since our measurements are not accurate and in using $u_x(t)$ rather $u(t)$ (which is not available) we would be obtaining an "exact fit" to inaccurate data. A better procedure, the well-known method of least squares [2.], is to choose the a_{j-1} such that for M points $u_x(t_k)$, $M>N+1$, the function

$$S = \sum_{k=1}^{M} (u_x(t_k) - u_{(N+1)}(t_k))^2$$

is minimal. Note that u_x has replaced u in the error since u is not measured. By taking

$$\frac{\partial S}{\partial a_{j-1}} = 0, \quad j = 1,2,\ldots,N+1$$

(after substituting 2.16), a necessary and sufficient condition for minimal S, we obtain

$$\underline{a} = G_s^{-1} \underline{u}_x$$

where (2.17)

$$\underline{a}^T = (a_o \; a_1 \; 2!a_2 \; \ldots \; N!a_N)$$

and the elements of G_s are

$$g_{ij} = \sum_{k=1}^{M} u_i(t_k)u_j(t_k), \quad i,j = 1,2,\ldots,N+1$$

and the elements of $\underline{u}_x$ are

$$u_{xj} = \sum_{k=1}^{M} u_x(t_k)u_j(t_k).$$

Equation 2.17 is the solution for the required coefficients in terms of the singularity functions (u_i and u_j) and the experimental data (u_x) at times t_k. The required inverse exists since for $M>N+1$, the $u_j(t)$, $j \geq 1$, are linearly independent for $t \in [0,T]$, $T < \infty$, i.e., the identity

$$\alpha_1 u_1(t) + \alpha_2 u_2(t) + \ldots + \alpha_M u_M(t) = 0$$

for all $t \in [0,T]$ implies $\alpha_1 = \alpha_2 = \ldots = \alpha_M = 0$. Example 2.1 illustrates the method for N=2, M=4.

If a record $u_x(t)$, $0 \le t \le T$, is given, the a_{j-1} are chosen such that

$$I = \int_0^T (u_x(\lambda) - u_{(N+1)}(\lambda))^2 d\lambda$$

is minimal. The result is similar to that above with

$$\underline{a} = G_I^{-1}\, \underline{u}'_x \tag{2.18}$$

where the elements of G_I are

$$g_{ij} = \int_0^T u_i(t)u_j(t)dt\ , \quad i,j = 1,2,\ldots,N+1$$

and the elements of $\underline{u}'_x$ are

$$u'_{xj} = \int_0^T u_x(t)u_j(t)dt\ .$$

Example 2.1

For the points $u_x(t_1)$, $u_x(t_2)$, $u_x(t_3)$, $u_x(t_4)$ known, find the a_{j-1}, j=1,2,3, such that $u_{(3)}(t)$ as given by Equation 2.16 is a least squares fit. Assume $0<t_1<t_2<t_3<t_4 \le T<\infty$. Now for N=2 and M=4,

$$u_{(3)}(t_k) = a_0 u_1(t_k) + a_1 u_2(t_k) + 2a_2 u_3(t_k)$$

for k=1,2,3,4 and the error

$$e(t_k) = u_x(t_k) - u_{(3)}(t_k)$$
$$= u_x(t_k) - \sum_{j=1}^{3} (j-1)!\,a_{j-1}u_j(t_k).$$

Let

$$S = \sum_{k=1}^{4} e(t_k)^2\ .$$

A necessary and sufficient condition for minimal S with respect to the a_{j-1} is

$$\frac{\partial S}{\partial a_{j-1}} = 0, \quad j = 1,2,3 .$$

Differentiating and setting the partial derivatives equal to zero leads to

$$u_x(t_k)u_1(t_k) = a_o\sum u_1(t_k)^2 + a_1\sum u_1(t_k)u_2(t_k) + 2a_2\sum u_1(t_k)u_3(t_k)$$

$$u_x(t_k)u_2(t_k) = a_o\sum u_2(t_k)u_1(t_k) + a_1\sum u_2(t_k)^2 + 2a_2\sum u_2(t_k)u_3(t_k)$$

$$u_x(t_k)u_3(t_k) = a_o\sum u_3(t_k)u_1(t_k) + a_1\sum u_3(t_k)u_2(t_k) + 2a_2\sum u_3{}^2(t_k)$$

where all the sums are $\sum_{k=1}^{4}$.

In matrix form,

$$\underline{u}_x = G_S\underline{a}$$

where $\underline{u}_x$ is a column vector with elements as shown on the left side of the equations, and

$$\underline{a}^T = [a_o \; a_1 \; 2a_2],$$

and G_S is a Gram matrix whose ijth element is

$$g_{ij} = \sum u_i(t_k)u_j(t_k) .$$

The linear independence of the singularity functions $u_j(t)$,

$j \geq 1$, means that G_S is nonsingular, therefore,

$$\underline{a} = G_S^{-1}\underline{u}_x .$$

It is obvious that Equation 2.17 is well-suited to digital computation where Equation 2.13 (for $t_o=0$) gives the $u_j(t_k)$ and where $u_x(t_k)$ is the input data to a program. It should be emphasized that we must decide whether Equation 2.16 is a suitable model, and if so, what N+1 should be used. Furthermore, for a large number of points, i.e., large M, there is greater redundancy, and if all readings are reliable, we expect a better model in the sense of a smaller S. We note also that this method does not use a statistical model for the source of errors; this problem is discussed in Chapter 4.

A generalization of 2.16 would allow arbitrary starting times for the $u_j(t)$, thus we have

$$u_{(N+1)}(t) = \sum_{j=1}^{N+1} (j-1)!a_{j-1}u_j(t-\tau_j)$$

or using a sum of such terms,

$$u_{(J)}(t) = \sum_{j=1}^{N+1}(j-1)!a_{j-1}u_j(t-\tau_j) + \sum_{k=1}^{P+1}(k-1)!b_{k-1}u_k(t-\sigma_k) + \dots + \sum_{\ell=1}^{R+1}(\ell-1)!c_{\ell-1}u_\ell(t-\lambda_\ell) \qquad (2.19)$$

for $J = N+1 + P+1 +\dots+ R+1$. Many signal models of great practical interest can be written in this form. This represents the class of functions that the elementary functions $u_j(t)$, $j \geq 1$, can model. For signals which have impulses, doublets, etc., extra terms of the form $\alpha_j u_j(t-\tau_j)$, $j \leq 0$, can be added to the summations.

Example 2.2

For u(t) as shown,

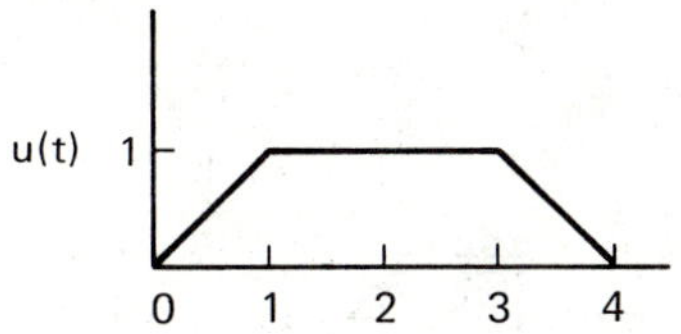

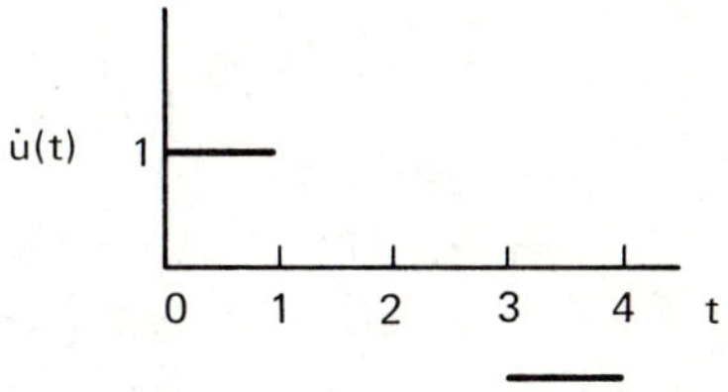

$$u(t) = u_2(t) - u_2(t-1) - u_2(t-3) + u_2(t-4) \ .$$

Also, using Equation 2.14,

$$\dot{u}(t) = u_1(t) - u_1(t-1) - u_1(t-3) + u_1(t-4)$$

as shown.

A set of discrete-time functions analogous to the singularity functions $u_j(t)$ also can be developed. Starting with

$$\begin{aligned} v(n) &= \frac{1}{k!} n^k \qquad n \geq 0 \\ &= 0 \qquad\qquad n < 0 \end{aligned}$$

for k = 0,1,2, ..., the following are defined: For

$k = 0, \quad v(n) = v_1(n)$, a unit step

$k = 1, \quad v(n) = v_2(n)$, a unit ramp

$k = 2, \quad v(n) = v_3(n)$, a unit square ,

etc. (Note that 0^0 is defined as 1 here.)

If k = j-1 is used, the general form is

$$\begin{aligned} v_j(n) &= \frac{1}{(j-1)!} n^{j-1} \qquad n \geq 0 \\ &= 0 \qquad\qquad\qquad n < 0 \end{aligned} \tag{2.20}$$

for $j \geq 1$. Next, consider the <u>first backward difference</u> of $v_j(n)$, $j > 2$

$$\Delta_b v_j(n) = v_j(n) - v_j(n-1)$$

or

$$\Delta_b v_j(n) = \frac{n^{j-1}}{(j-1)!} - \frac{(n-1)^{j-1}}{(j-1)!} \qquad n \geq 1$$
$$= 0 \qquad n \leq 0 \qquad (2.21)$$

A difference relationship analogous to Eq. 2.3 can be developed from Eq. 2.21. However, the result is not as simple as that for the singularity functions because of the use of the first backward difference operator. Higher order operators could be used, but the simplicity of the first-order operator has a computational advantage. Also, the resulting forms for j=1 and j=0, the forms of greatest use in analysis, turn out to be directly analogous to 2.3. From 2.21, for $j>2$ and $n\geq 1$, using the Binomial Theorem on the second term,

$$\Delta_b v_j(n) = \frac{1}{(j-1)!}\{n^{j-1} - [n^{j-1} - (j-1)n^{j-2} + \frac{(j-1)(j-2)}{2!}n^{j-3} - \ldots + (j-1)n^{j-2} + (-1)^{j-1}]\}$$
$$= \frac{n^{j-2}}{(j-2)!} - \frac{n^{j-3}}{2!(j-3)} + \ldots - \frac{(-1)^{j-2}n}{(j-2)!} - \frac{(-1)^{j-1}}{(j-1)!} \qquad (2.22)$$

and

$$\Delta_b v_j(n) = 0, \qquad n \leq 0 .$$

The first term on the right of 2.22 is in the form of $v_{j-1}(n)$ and there are j-2 additional terms. For j=2,

$$\Delta_b v_2(n) = v_2(n) - v_2(n-1)$$

or

$$\Delta_b v_2(n) = 1 \qquad n \geq 1$$
$$= 0 \qquad n \leq 0 \qquad (2.23)$$

For j=1,

$$\Delta_b v_1(n) = 1 \qquad n = 0$$
$$= 0 \qquad n \neq 0 \qquad (2.24)$$

Equation 2.24 is now used to define $v_o(n)$ as

$$v_o(n) = 1 \qquad n = 0$$
$$= 0 \qquad n \neq 0$$

and this is called the unit pulse function. Using $v_o(n)$, 2.22, 2.23, and 2.24 can be written as

$$\Delta_b v_j(n) = v_{j-1}(n) - \frac{1}{2!} v_{j-2}(n) + \dots - \frac{(-1)^{j-2}}{(j-2)!} v_2(n) - \frac{(-1)^{j-1}}{(j-1)!} v_1(n) + \frac{(-1)^{j-1}}{(j-1)!} v_o(n) \qquad (2.25)$$

for $j > 2$,

and
$$\Delta_b v_2(n) = v_1(n) - v_o(n) \qquad (2.26)$$

and
$$\Delta_b v_1(n) = v_o(n) \ . \qquad (2.27)$$

The functions $v_j(n)$ for $j < 0$ can now be defined by using a summation in a way which is analogous to the use of 2.4. Therefore, let

$$\sum_{i=-\infty}^{n} v_{j-1}(i) = v_j(n), \qquad j \le 0$$

and note that this is true for j=1 also. For notational convenience below we, therefore, write this as

$$\sum_{i=-\infty}^{n} v_{j-1}(i) = v_j(n), \qquad j < 2 \ . \qquad (2.28)$$

Now taking the first backward difference of 2.28, we have

$$\sum_{i=-\infty}^{n} v_{j-1}(i) - \sum_{i=-\infty}^{n-1} v_{j-1}(i) = \Delta_b v_j(n)$$

or

$$v_{j-1}(n) = \Delta_b v_j(n), \qquad j < 2 \qquad (2.29)$$

Equations 2.25, 2.26, and 2.29 now relate the functions $v_j(n)$ using the first backward difference operator for $j > 2$, $j = 2$, and $j < 2$, respectively. To complete a set of equations relating the $v_j(n)$ using a summation, Equations 2.25 and 2.26 are summed. In general

$$\sum_{i=-\infty}^{n} \Delta_b v_j(i) = \sum_{i=-\infty}^{n} v_j(i) - \sum_{i=-\infty}^{n} v_j(i-1)$$

$$= \sum_{i=-\infty}^{r} v_j(i) - \sum_{i=-\infty}^{n-1} v_j(i) = v_j(n)$$

and using 2.25, for $j > 2$,

$$v_j(n) = \sum_{i=-\infty}^{n} [v_{j-1}(i) - \frac{1}{2} v_{j-2}(i) + \ldots + \frac{(-1)^{j-1}}{(j-1)!} v_o(i)] \tag{2.30}$$

and using 2.26, for $j = 2$,

$$v_2(n) = \sum_{i=-\infty}^{n} (v_1(i) - v_o(i)) . \tag{2.31}$$

Equations 2.30, 2.31, and 2.28 then relate the functions $v_j(n)$ using the summation operation for $j > 2$, $j = 2$, and $j < 2$, respectively. Thus for the functions $v_j(n)$ defined here, there are three sets of equations that relate them by differencing and summing, but for the singularity functions $u_j(t)$ only Equations 2.3 and 2.4 are needed when the functions for $j \leq 0$ are distributions.

The more general form for the set of discrete-time analogs to the singularity functions is

$$v_j(n-n_o) = \frac{1}{(j-1)!} (n-n_o)^{j-1}, \quad n \geq n_o \tag{2.32}$$

$$= 0 \qquad , \quad n < n_o$$

for $j \geq 1$, and with

$$\Delta_b v_j(n-n_o) = v_{j-1}(n-n_o) - \frac{1}{2!} v_{j-2}(n-n_o) + \ldots + \frac{(-1)^{j-1}}{(j-1)!} v_o(n-n_o) \tag{2.33a}$$

from 2.25 for $j > 2$, and

$$\Delta_b v_j(n-n_o) = v_1(n-n_o) - v_o(n-n_o) \tag{2.33b}$$

from 2.26 for j = 2, and

$$\Delta_b v_j(n-n_o) = v_{j-1}(n-n_o) \tag{2.33c}$$

from 2.29 for j < 2. Also,

$$\sum_{i=-\infty}^{n}[v_{j-1}(i-n_o) - \frac{1}{2} v_{j-2}(i-n_o) + \ldots + \frac{(-1)^{j-1}}{(j-1)!} v_o(i-n_o)] = v_j(n-n_o) \tag{2.34a}$$

from 2.30 for j > 2, and

$$\sum_{i=-\infty}^{n}[v_1(i-n_o) - v_o(i-n_o)] = v_2(n-n_o) \tag{2.34b}$$

from 2.31 for j = 2, and

$$\sum_{i=-\infty}^{n} v_{j-1}(i-n_o) = v_j(n-n_o) \tag{2.34c}$$

from 2.28 for j < 2. Equations 2.32, 2.33, and 2.34 summarize our definitions of the set of discrete-time functions v_j, $\forall$ j. They are analogous to Equations 2.13, 2.14, and 2.15.

The function $v_o(n)$ also has a sifting property analogous to 2.8 for a given signal v(n):

$$\sum_{i=-\infty}^{\infty} v(n-i)v_o(i) = \sum_{i=-\infty}^{\infty} v(i)v_o(n-i) = v(n) \tag{2.35}$$

Figure 2.8 shows some functions frequently used in developing models of discrete-time signal models.

The functions $v_j(n)$ can be used in linear combinations to develop models of discrete-time polynomials. For

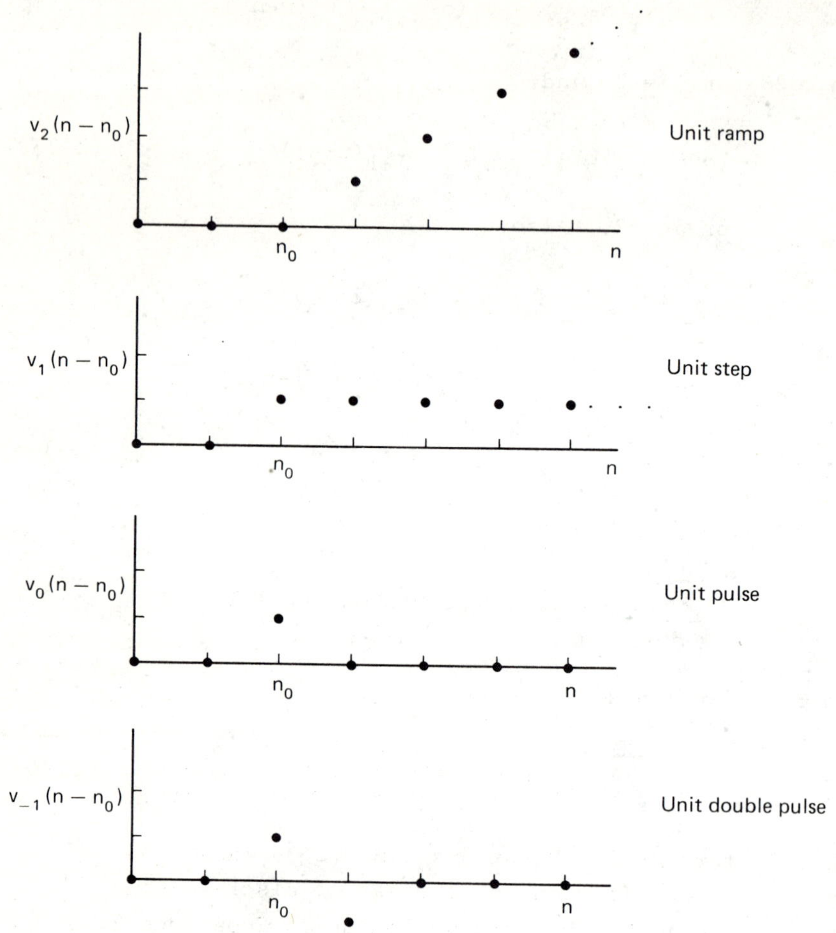

Figure 2.8

$$v_{(N+1)}(n) = a_o + a_1 n + a_2 n^2 + \ldots + a_N n^N, \qquad n \geq 0$$

$$= 0 \qquad , \qquad n < 0$$

an equivalent form is

$$v_{(N+1)}(n) = a_o v_1(n) + a_1 v_2(n) + 2!a_2 v_3(n) + \ldots$$

$$+ N!a_N v_{N+1}(n)$$

or

$$v_{(N+1)}(n) = \sum_{j=1}^{N+1} (j-1)!\,a_{j-1} v_j(n) \qquad (2.36)$$

This model can be used as a least squares approximation to a signal v(n). The data are $v_x(n)$, $n = n_1, n_2, \ldots, n_M$ and $M > N+1$ as before. Usually $n_1, n_2, \ldots, n_M$ is a sequence of integers and the data are then equally-spaced. Problem 7 at the end of this chapter is an exercise in finding a result analogous to Equation 2.17. We note that the $v_j(n)$, $j \geq 1$, are linearly independent on $n \in [o, n_M]$.

Generalizations of 2.36 are

$$v_{(N+1)}(n) = \sum_{j=1}^{N+1} (j-1)!\,a_{j-1} v_j(n-n_j)$$

or using a sum of these terms

$$v_{(J)}(n) = \sum_{j=1}^{N+1} (j-1)!\,a_{j-1} v_j(n-n_j) + \sum_{k=1}^{P+1} (k-1)!\,b_{k-1} v_k(n-m_k) + \ldots + \sum_{\ell=1}^{R+1} (\ell-1)!\,c_{\ell-1} v_\ell(n-p_\ell) \qquad (2.37)$$

where $J = N+1 + P+1 + \ldots + R+1$.

<u>Example 2.3</u>

For v(n) as shown,

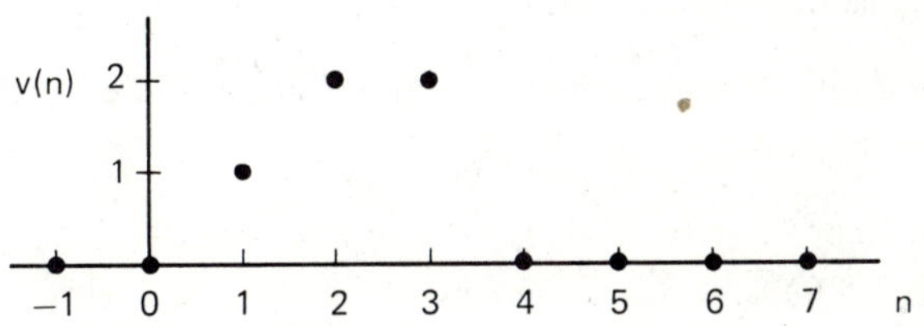

$$v(n) = v_2(n) - v_2(n-2) - 2v_1(n-4) .$$

Also,

$$\Delta_b v(n) = (v_1(n)-v_0(n)) - (v_1(n-2)-v_0(n-2)) - 2v_0(n-4)$$

$$= v_1(n-1) - v_1(n-3) - 2v_0(n-4)$$

as shown.

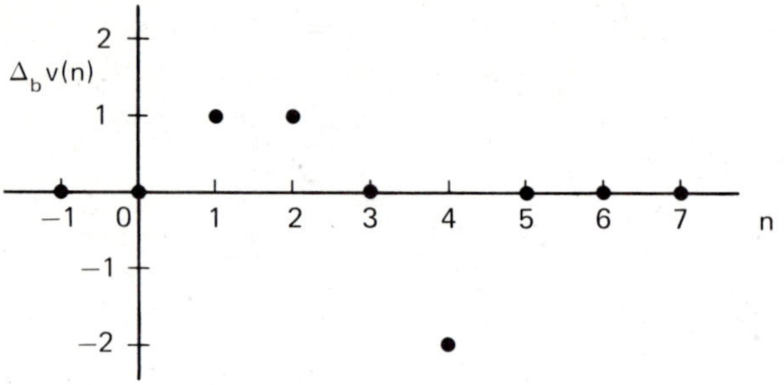

In reviewing this rather long Section 2.2.1, we find that we now have completed one cycle of a pattern we shall repeat four more times in subsequent sections. The steps are listed here with the corresponding equation numbers for our results so far.

1. Define a set of continuous-time elementary functions (u_j of 2.13, 2.14, 2.15)

2. Define a model using a linear combination of the elementary functions ($u_{(N+1)}(t)$ of 2.16)

3. Give a method of finding the parameters of the model (2.17 and 2.18)

4. Repeat 1. for a set of analogous discrete-time functions (v_j of 2.32, 2.33, 2.34)

5. Repeat 2. for a discrete-time model ($v_{(N+1)}(n)$ of 2.36)

6. Repeat 3. for a discrete-time case (Problem 7 of Chapter 2).

2.2.2 Real Exponentials

The second elementary form of interest is the real exponential function

$$u_{ei}(t) = \varepsilon^{-a_i t}, \quad t \geq 0, \quad a_i > 0,$$

for i = 1,2,3, ... Linear combinations of such functions are then used to model a signal u(t) as

$$u(t) = \sum_{i=1}^{\infty} A_i \varepsilon^{-a_i t}, \quad t \geq 0 . \tag{2.38}$$

Although 2.38 is a useful form when the parameters (the A_i and a_i) are known, the determination of these parameters for a given set of data, using N terms, generally is difficult. If at least 2N points are given, an iterative least squares method can be used [3.]. The signal u(t) must then have the properties

$$1. \quad \int_0^{\infty} u(t)^2 dt < \infty .$$

$$2. \quad \lim_{t\to\infty} u(t) = 0 .$$

The determination of the parameters of a signal model of this type is greatly simplified if instead of using the elementary form $u_{ei}(t)$, an orthonormal set of functions $\psi_i(t)$, which are themselves sums of real exponentials, is used [4.]. An orthonormal set of functions $f_i(t)$, i = 1,2,3, ... on an interval $[t_1, t_2]$ is such that

$$\int_{t_1}^{t_2} f_i(t) f_j(t) dt = 1, \quad i = j$$
$$= 0, \quad i \neq j .$$

Using the functions $\psi_i(t)$,

$$u(t) = \sum_{i=1}^{N} B_i \psi_i(t) + e(t), \quad t \geq 0$$
$$= u_{(N)}(t) + e(t) , \tag{2.39}$$

and the B_i coefficients which minimize

$$\int_0^{\infty} e^2(t)\,dt$$

are

$$B_i = \int_0^{\infty} u(t)\psi_i(t)\,dt, \qquad i = 1, \ldots, N.$$

As in the previous section, u_x replaces u when experimental data are used. The functions $\psi_i(t)$, which are orthonormal on $[0,\infty]$, are tabulated in [4.] for i = 1,2,3,4,5. For example

$$\psi_1(t) = \sqrt{2c}\ \varepsilon^{-ct}, \quad t \geq 0$$

$$\psi_2(t) = \sqrt{c}\ [6\varepsilon^{-2ct} - 4\varepsilon^{-ct}]; \quad t \geq 0$$

and c is a real parameter chosen to make ε^{-ct} a "good" approximation to u(t) for large t. The more general case for different exponents as well as different coefficients in the form 2.38 also has been developed [5.].

A two-sided form for 2.38 also is useful, particularly where u(t) is an even function such as the autocorrelation function discussed later. We let

$$u(t) = \sum_{i=1}^{\infty} A_i \varepsilon^{a_i|t|}, \quad -\infty < t < \infty, \quad a_i > 0$$

and the extension of 2.39 also follows.

The discrete-time elementary function analogous to the real exponential is

$$v_{ei}(n) = (b_i)^n, \quad n \geq 0, \quad 0 < b_i < 1,$$

for i = 1,2,3, ... and a linear combination to model a v(n) is

$$v(n) = \sum_{i=1}^{\infty} D_i (b_i)^n, \quad n \geq 0 . \qquad (2.40)$$

An analogous set of orthonormal discrete-time functions $\phi_i(n)$ also can be developed when using equally-spaced data. For

$$v(n) = \sum_{i=1}^{N} E_i \phi_i(n) + e(n) , \qquad (2.41)$$
$$= v_{(N)}(n) + e(n)$$

the E_i coefficients which minimize

$$\sum_{n=0}^{\infty} e(n)^2$$

are

$$E_i = \sum_{n=0}^{\infty} v(n)\phi_i(n) \qquad i = 1, 2, \ldots, N.$$

It is necessary that

1. $\sum_{n=0}^{\infty} v(n)^2 < \infty$

2. $v(n) \to 0$ for $n \to \infty$.

The functions $\phi_1(n)$ and $\phi_2(n)$ are, for example,

$$\phi_1(n) = \sqrt{1-b^2}\,(b^n)$$

$$\phi_2(n) = \frac{1}{b}\sqrt{1-b^4}\,[(1+b+b^2)(b^{2n}) - (1+b)(b^n)]$$

where b is a parameter chosen to make b^n a "good" approximation to v(n) for large n. The functions $\phi_i(n)$ for i = 3,4, ... can be generated by continuing a discrete-time version of the method in [4.]. The $\phi_i(n)$ are such that

$$\sum_{n=0}^{\infty} \phi_i(n)\phi_j(n) = 1, \qquad i = j$$
$$= 0, \qquad i \neq j$$

Example 2.4

For the data $v_x(n)$ shown, find the parameters of a two-term approximation, i.e., find $v_{(N)}(n)$ for N=2.

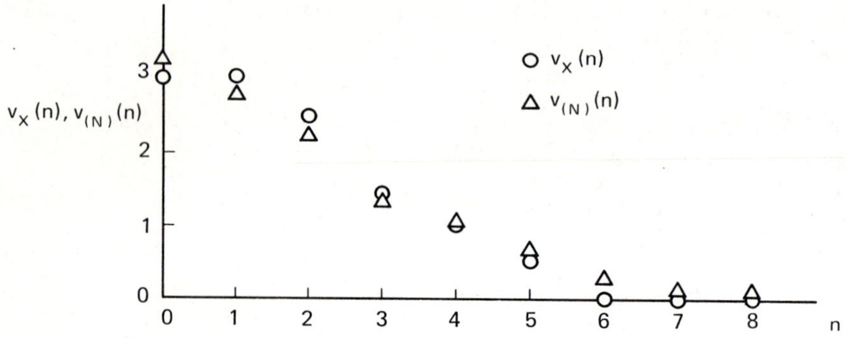

Using $v_x(0) = 3$ and $v_x(5) = 1/2$, we have

$$v_x(5)/v_x(0) = b^5/b^0 = b^5 = 1/6 \ ,$$

we then let the trial value of b = 0.7. Using this

$$\phi_1(n) = 0.714(0.7)^n, \qquad n \geq 0$$

$$\phi_2(n) = 2.73[(0.7)^{2n} - 0.776(0.7)^n], \qquad n \geq 0$$

Using these,

$$E_1 = \sum_{n=0}^{\infty} v_x(n)\phi_1(n) = 5.11$$

$$E_2 = \sum_{n=0}^{\infty} v_x(n)\phi_2(n) = -0.646$$

and the approximation function is

$$v_{(2)}(n) = 5.01(0.7)^n - 1.762(0.7)^{2n} \qquad n \geq 0$$

which also is plotted. The trial value of b resulted in a maximum error of about 0.57 and a summed square error of about 1.14. A series of computations using $0.05 \leq b \leq 0.95$ shows that for b approximately 0.55, the minimal summed square error is about 0.25.

Having completed this second cycle (outlined at the end of

the previous section), we note an important difference from the first result. In the case of the $u_j(t)$ and $v_j(n)$ for modelling signals, it is necessary to solve simultaneous equations to find the parameters a_{j-1}, e.g., Equation 2.17. However, when we use the orthonormal functions $\psi_i(t)$ and $\phi_i(t)$, the parameters B_i and E_i can be found <u>one at a time</u>. This has an obvious computational advantage, and we are, therefore, led to seek other sets of orthonormal functions.

2.2.3 Sinusoids

The third elementary form which is useful for developing signal models is the set of sinusoidal functions. Two common elementary functions for i = 1,2,3, ... are

$$u_{ai}(t) = \sin(i\omega_o t + \alpha_i), \qquad t \in T_{c2}$$

or

$$u_{bi}(t) = \cos(i\omega_o t + \beta_i), \qquad t \in T_{c2}$$

and since $(j = \sqrt{-1})$

$$\varepsilon^{j(\omega_o t+\phi)} = \cos(\omega_o t+\phi) + j\,\sin(\omega_o t+\phi) ,$$

we have

$$u_{ai}(t) = \mathrm{Im}\ \varepsilon^{j(i\omega_o t+\alpha_i)}$$

and

$$u_{bi}(t) = \mathrm{Re}\ \varepsilon^{j(i\omega_o t+\beta_i)} .$$

The simple forms $u_{ai}(t)$ and $u_{bi}(t)$ also combine easily to give new functions of the same form. Thus, for the sum

$$\begin{aligned} u_{ci}(t) &= u_{ai}(t) + u_{bi}(t) \\ &= \sin(i\omega_o t+\alpha_i) + \cos(i\omega_o t+\beta_i) \\ &= C_i\cos(i\omega_o t+\gamma_i) \end{aligned}$$

where

$$C_i = \{(\cos\,\alpha_i-\sin\,\beta_i)^2 + (\sin\,\alpha_i+\cos\,\beta_i)^2\}^{\frac{1}{2}}$$

and

$$\gamma_i = -\tan^{-1} \frac{[\cos\alpha_i - \sin\beta_i]}{[\sin\alpha_i + \cos\beta_i]} \quad .$$

The elementary forms $u_{ai}(t)$ and $u_{bi}(t)$ can be used in linear combinations to model a class of signals as was done previously. In this case, we consider a periodic signal $u(t)$ with period $T_o = 2\pi/\omega_o$ satisfying the Dirichlet conditions $(-T_o/2 \leq t \leq T_o/2)$:

1. $u(t)$ has a finite number of maxima and minima.
2. $u(t)$ has a countable number of finite discontinuities.
3. $u(t)$ has a finite number of infinite discontinuities.
4. $\int_{-T_o/2}^{T_o/2} |u(t)|dt < \infty$

The trigonometric form of the Fourier Series expansion of $u(t)$ is then

$$u(t) = \frac{b_o}{2} + \sum_{k=1}^{\infty} b_k \cos k\,\omega_o t + \sum_{k=1}^{\infty} a_k \sin k\,\omega_o t \tag{2.42}$$

with

$$b_o = \frac{2}{T_o}\int_{-T_o/2}^{T_o/2} u(t)dt$$

$$b_k = \frac{2}{T_o}\int_{-T_o/2}^{T_o/2} u(t)\cos k\,\omega_o t\,dt$$

$$a_k = \frac{2}{T_o}\int_{-T_o/2}^{T_o/2} u(t)\sin k\,\omega_o t\,dt \;.$$

For a finite number of terms in the summations in 2.42, the coefficients give the best approximation to $u(t)$ in the least squares sense. Thus for M terms in each sum, the approximating model is $u_{(N)}(t)$ and b_o, b_k, and a_k are such that

$$\int_{-T_o/2}^{T_o/2} [u(t) - u_{(N)}(t)]^2\,dt, \qquad N = 2M+1$$

is minimal. The simple forms for the coefficients b_o, b_k, and a_k occur because $u_{ai}(t)$ and $u_{bi}(t)$, i = 1,2,3, ..., form orthonormal sets of functions on $[-T_o/2, T_o/2]$. Equation 2.42 also can be put in the form

$$u(t) = \frac{b_o}{2} + \sum_{k=1}^{\infty} c_k \cos(k\omega_o t + \gamma_k)$$

with

$$c_k = \sqrt{a_k^2 + b_k^2} \quad \text{and} \quad \gamma_k = -\tan^{-1} a_k/b_k \; .$$

<u>Example 2.5</u>

For u(t) as shown

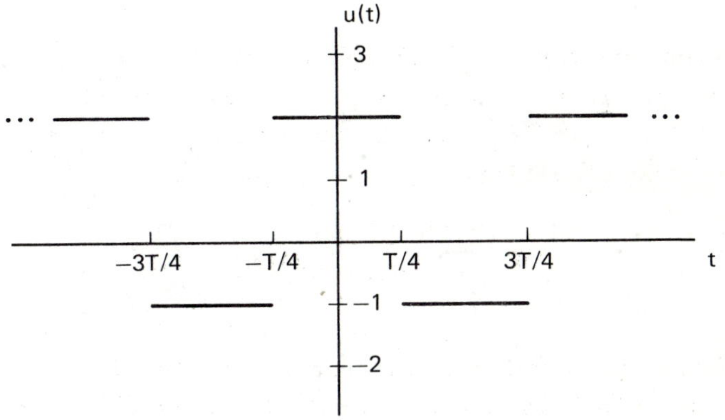

$$b_o = \frac{2}{T}\int_{-T/2}^{T/2} u(t)\,dt = \frac{2}{T} \cdot \frac{T}{2} = 1$$

and for $k \geq 1$,

$$a_k = 0$$

$$b_k = \frac{2}{T}\int_{-T/2}^{T/2} u(t) \cos k\,\omega_o t \; dt, \quad \omega_o = \frac{2\pi}{T}$$

$$= \frac{6}{k\pi}(-1)^{(k-1)/2} \quad k \text{ odd}$$

$$= 0 \qquad k \text{ even}$$

Thus

$$u(t) = \frac{1}{2} + \frac{6}{\pi} \cos \omega_o t - \frac{2}{\pi} \cos 3\, \omega_o t + \ldots .$$

The discrete-time sinusoidal forms are

$$v_{ai}(n) = \sin(i\lambda_o n + \alpha_i), \quad n \in T_{d2}$$

or

$$v_{bi}(n) = \cos(i\lambda_o n + \gamma_i). \quad n \in T_{d2}$$

As before, $v_{ai}(n)$ and $v_{bi}(n)$ combine to give the form

$$v_{ci}(n) = C_i \cos(i\lambda_o n + \beta_i),$$

The elementary forms $v_{ai}(n)$ and $v_{bi}(n)$ can be used in linear combinations to model periodic discrete-time signals. We consider a periodic v(n) with period N_o and $\lambda_o = 2\pi/N_o$, satisfying the condition

$$1. \sum_{i=0}^{N_o-1} |v(i)| < \infty$$

Trigonometric polynomial forms (or a "finite" Fourier Series") used in harmonic analysis [9.] follow. We note that the sets of discrete-time functions $v_{ai}(n)$ and $v_{bi}(n)$, $i = 1,2,3 \ldots$, are orthonormal on $[0, N_o-1]$. There are now two cases:

$$v(n) = v_{(N)} + e(n), \quad N_o \text{ odd}, \; N = 2K+1 = N_o$$

or

$$v(n) = v_{(N)} + e(n), \quad N_o \text{ even}, \; N = 2K'+2 = N_o$$

which define K and K´. By minimizing

$$\sum_{n=0}^{N_o-1} [v(n) - v_{(N)}(n)]^2$$

we obtain two forms:

<u>N_o odd</u>

$$v_{(N)}(n) = \frac{b_o}{2} + \sum_{k=1}^{K} b_k \cos k\lambda_o n + \sum_{k=1}^{K} a_k \sin k\lambda_o n \qquad (2.43)$$

with $K = (N_o-1)/2$ and with

$$b_o = \frac{2}{N_o} \sum_{n=0}^{N_o-1} v(n)$$

$$b_k = \frac{2}{N_o} \sum_{n=0}^{N_o-1} v(n) \cos k \lambda_o n$$

$$a_k = \frac{2}{N_o} \sum_{n=0}^{N_o-1} v(n) \sin k \lambda_o n$$

<u>N_o even</u>

$$v_{(N)}(n) = \frac{b_o}{2} + \sum_{k=1}^{K'} b_k \cos k\lambda_o n + \sum_{k=1}^{K'} a_k \sin k\lambda_o n + \frac{b_{N_o/2}}{2} \cos n\pi \qquad (2.44)$$

with b_o, b_k, a_k as given for Eq. 2.43, $k = 1,2, \ldots, K'$ and

$$b_{N_o/2} = \frac{2}{N_o} \sum_{n=0}^{N_o-1} (-1)^n v(n) .$$

with $K' = (N_o-2)/2$.

<u>Example 2.6</u>

For v(n) as shown, find a Fourier Series.

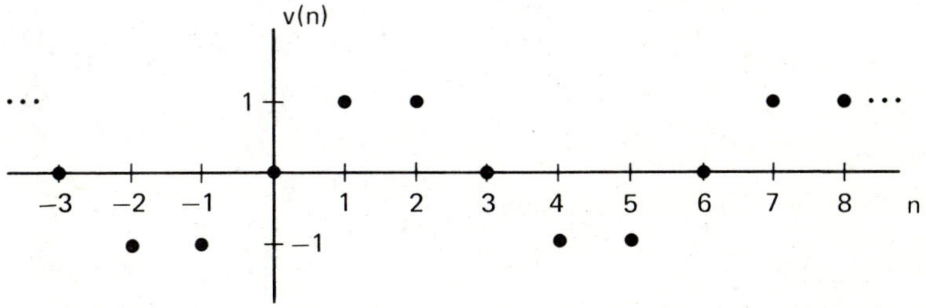

$N_0 = 6$ and $\lambda_o = 2\pi/6$, therefore, use Eq. 2.44

$$b_0 = \frac{2}{6}\sum_{n=0}^{5} v(n) = 0$$

$$b_k = \frac{1}{3}\sum_{n=0}^{5} v(n) \cos k\,\frac{\pi}{3}\,n = 0, \qquad k = 1,2$$

$$a_k = \frac{1}{3}\sum_{n=0}^{5} v(n) \sin k\,\frac{\pi}{3}\,n$$

$$= \frac{2}{\sqrt{3}}, \quad k = 1$$

$$= 0, \qquad k = 2$$

$$b_3 = \frac{2}{6}\sum_{n=0}^{5} (-1)^n v(n) = 0 \ .$$

Thus $v(n) = \dfrac{2}{\sqrt{3}} \sin \dfrac{\pi}{3}\, n$.

2.2.4 Imaginary Exponents

A fourth elementary form is the exponential with imaginary exponent

$$u_{gi}(t) = \varepsilon^{ji\omega_o t}, \quad i = 1,2,3 \ldots, \quad j = \sqrt{-1}$$

$$t \in T_{c2}$$

A linear combination of this form results in the exponential Fourier Series expansion of a periodic function $u(t)$, $T_o = 2\pi/\omega_o$, satisfying the Dirichlet conditions stated earlier. Similar to the previous section we have

$$u(t) = \sum_{k=-\infty}^{\infty} U_k \varepsilon^{jk\omega_o t} \tag{2.45}$$

with

$$U_k = \frac{1}{T_o}\int_{-T_o/2}^{T_o/2} u(t)\varepsilon^{-jk\omega_o t}\, dt \ .$$

This is referred to as an $\mathfrak{F}$-Series in other sections.

Similarly, for discrete-time functions,

$$v_{gi}(n) = \varepsilon^{ji\lambda_o n}, \quad i = 1,2,3 \ldots, \; j = \sqrt{-1}$$

$$n \; \varepsilon \; T_{d2}$$

A linear combination results in a "finite" Fourier Series expansion of a periodic v(n), $N_o = 2\pi/\lambda_o$. Thus, the two cases are:

$\underline{N_o \text{ odd}}$

$$v_{(N)}(n) = \sum_{k=-K}^{K} V_k \varepsilon^{jk\lambda_o n}, \quad N = 2K+1 = N_o \qquad (2.46)$$

with

$$V_k = \frac{1}{N_o} \sum_{n=0}^{N_o-1} v(n) \varepsilon^{-jk\lambda_o n}$$

for $K = (N_o-1)/2$.

$\underline{N_o \text{ even}}$

$$v_{(N)}(n) = \sum_{k=-K'}^{K'} V'_k \; \varepsilon^{jk\lambda_o n} + \frac{b_{N_o/2}}{2} \cos n\pi$$

$$= \sum_{k=-K'}^{K'+1} V'_k \; \varepsilon^{jk\lambda_o n}, \quad N = 2K'+2 = N_o \qquad (2.47)$$

for $K' = (N_o-2)/2$ and where the expression for V'_k is the same as that for V_k, and $b_{N_o/2}$ is given after Eq. 2.44. These are called $\mathfrak{F}_d$-Series in other sections.

We observe that the set of $u_{gi}(t)$ is orthonormal on $[-T_o/2, T_o/2]$ and the set of $v_{gi}(n)$ is orthonormal on $[0, N_o-1]$.

2.2.5 Complex Exponents

The cases using real and imaginary exponents can be combined to give complex exponents. Thus for continuous-time functions, the elementary form is

$$u_{di}(t) = \varepsilon^{(\sigma_{oi}+ji\omega_o)t} = \varepsilon^{s_{oi}t}, \quad t \geq 0$$

with $\sigma_{oi} \leq 0$, $\forall$ i. A linear combination is

$$u(t) = \sum_{i=1}^{\infty} A_i \varepsilon^{s_{oi}t} \tag{2.48}$$

This form also has been found useful and it is possible, using a finite number of terms, to obtain the parameters of the expansion using an iterative technique [5.]. The A_i and s_{oi} may be real or complex and when complex, must occur in conjugate pairs. The discrete-time form is

$$v_{di}(n) = (\alpha_{oi}+j\beta_{oi})^n = (z_{oi})^n, \quad n \geq 0$$

with $|z_{oi}| \leq 1$, $\forall$ i. A linear combination is

$$v(n) = \sum_{i=1}^{\infty} B_i (z_{oi})^n \tag{2.49}$$

The damped oscillatory forms used in 2.48 and 2.49 provide a better approximation to some experimental data than the previous exponential forms. The coefficients are found, using a finite number of terms, by starting with the minimization of an integral square or summed square error as was shown in previous cases. Unfortunately, the resulting equations are nonlinear in the s_{oi} or z_{oi}; the iterative method referred to previously is a way of solving for the required parameters. Another pair of forms for these decaying or damped exponentials is

$$u_{fi}(t) = \varepsilon^{\sigma_{oi}t} \cos(i\omega_o t+\phi_i), \quad t \geq 0, \quad \sigma_{oi} \geq 0 \tag{2.48a}$$

and

$$u_{fi}(n) = (\rho_{oi})^n \cos(i\lambda_o n+\psi_i), \quad \rho_{oi} \leq 1, \; n \geq 0, \tag{2.49a}$$

and these are real. Linear combinations of these two forms also can be used to approximate u(t) or v(n). A least squares approximation again requires the solution of non-linear equations.

The material of Section 2.2 is summarized in Table 2.1. Results for both continuous-time and discrete-time signals are given.

2.3 GENERAL MODELS

2.3.1 Elementary Functions and Spectral Functions

An examination of the equations listed in Table 2.1 shows that there are two basic forms depending on whether the signal model is for continuous-time or discrete-time. In all cases the equation has two factors under the summation signs. One depends on the discrete summation variable only and the

TABLE 2.1

Elementary Function	Signal Model
Singularity $u_j(t)$ $v_j(n)$	Polynomial Eq. 2.16 Eq. 2.36
Real Exponent $u_{ei}(t)$ $v_{ei}(n)$	Exponential Sum Eq. 2.38, 2.39 Eq. 2.40, 2.41
Sine/Cosine $u_{ai}(t)$, $u_{bi}(t)$ $u_{ai}(n)$, $v_{bi}(n)$	Periodic Eq. 2.42 Eq. 2.43, 2.44
Imaginary Exponent $u_{gi}(t)$ $u_{gi}(n)$	Periodic Eq. 2.45 Eq. 2.46, 2.47
Complex Exponent $u_{di}(t)$ $u_{di}(n)$	Damped Oscillation Eq. 2.48 Eq. 2.49

other depends on the summation variable and the time variable. To generalize these forms, the coefficients, which depend on the single discrete argument, will be called <u>real spectral functions</u>. Also, the second factors, which depend on two arguments, are the <u>elementary functions</u>. Then, for a continuous-time signal u(t), the general model is

$$u(t) = \sum_{k} \ell(k) q_2(t,k) \tag{2.50}$$

where $\ell(k)$ is the real spectral function and $q_2(t,k)$ is the elementary function. For example, for Equation 2.38, with $k \to i$, we get

$$\ell(k) \to A_i$$

$$q_2(t,k) \to \varepsilon^{-a_i t}, \quad t \geq 0$$

and similar substitutions are used for the other continuous-time signal models. For a discrete-time signal v(n), the general model is

$$v(n) = \sum_{k} m(k) q_4(n,k) \tag{2.51}$$

with m(k) the real spectral function and $q_4(n,k)$ the elementary function. For example, for Equation 2.40, with $k \to i$, we get

$$m(k) \to D_i$$

$$q_4(n,k) \to (b_i)^n$$

and with similar substitutions for the other discrete-time signal models.

A change of the discrete argument k in Equations 2.50 and 2.51 to a continuous argument λ results in two more forms. Thus, for u(t),

$$u(t) = \int f(\lambda) q_1(t,\lambda) d\lambda \tag{2.52}$$

and for v(n),

$$v(n) = \int g(\lambda) q_3(n,\lambda) d\lambda \quad . \tag{2.53}$$

As before, $f(\lambda)$ and $g(\lambda)$ are real spectral functions, and $q_1(t,\lambda)$ and $q_3(n,\lambda)$ are elementary functions.

Taken together, these four forms apply to many signal models of general interest. Three classes of signal models used in this book which are examples of these forms are: (1) signal resolution in terms of the $u_j(t)$ or $v_j(n)$, (2) the representation of signal conversions, and (3) the description of signals in a function space which is analogous to a finite dimensional vector space (see Appendix B). Furthermore, the introduction of complex as well as real arguments brings the integral transforms of Appendix A into this framework as discussed in the next section. For the four forms with real arguments, Table 2.2 shows their relationship.

TABLE 2.2
ELEMENTARY FUNCTION TIME ARGUMENT

		Continuous (t)	Discrete (n)
Spectral Function	Continuous Argument	Eq. 2.52	Eq. 2.53
	Discrete Argument	Eq. 2.50	Eq. 2.51

[Signal Resolution in u_j, v_j]
[Sampling]
[Reconstruction]
[Signal Space Models]

One particular occurrence of the form 2.52 is the Hilbert transform [6.] where

$$u(t) = -\frac{1}{\pi} \int_{-\infty}^{\infty} \frac{\hat{u}(\lambda)}{t-\lambda} d\lambda$$

with

$$\hat{u}(t) = \frac{1}{\pi}\int_{-\infty}^{\infty}\frac{u(\lambda)}{t-\lambda}\,d\lambda$$

The elementary function $-1/\pi(t-\lambda)$ results from taking the inverse Fourier transform of a frequency domain description of an ideal device which shifts the phase of all signal frequencies by $-\pi/2$. Such a device has a transfer function (discussed in Chapter 3) which is

$$H(\omega) = -j \text{ sgn } \omega$$

(where sgn x = 1 for x>0 and sgn x = -1 for x<0) and

$$\mathcal{F}^{-1}H(\omega) = h(t)$$

is its unit impulse response. Using the symmetry property of the Fourier Transform (see Section A.1.3 of Appendix A) and entry 8. of Table A.2,

$$h(t) = \frac{1}{\pi t}$$

For u(t) the input and $\hat{u}(t)$ the output of this device, the output is found by using real convolution (see Section A.1.4 and Chapter 4). We obtain

$$\hat{u}(t) = u(t)*h(t) = \frac{1}{\pi}\int_{-\infty}^{\infty}\frac{u(\lambda)}{t-\lambda}\,d\lambda \tag{2.54a}$$

and since passing the signal through two such devices produces $-\pi$ radians phase shift, the second output is -u(t). Therefore, we obtain [41.]

$$u(t) = -\hat{u}(t)*h(t) = -\frac{1}{\pi}\int_{-\infty}^{\infty}\frac{\hat{u}(\lambda)}{t-\lambda}\,d\lambda \tag{2.54b}$$

The Hilbert Transform pair, a pair of real signal transforms, occurs frequently in signal processing. Discrete-time forms (as in 2.51 and 2.53) have become important in digital signal processing [42.], e.g., for representing band pass signals in terms of low pass signals.

2.3.2 Signal Resolution

Signal resolution in terms of the elementary forms $u_j(t)$ and $v_j(n)$ follows from the sifting properties expressed by Equations 2.9 and 2.35 where $u_o(t)$ and $v_o(n)$ are used as the elementary forms. Thus for $t_o \leq t < \infty$,

$$u(t) = \int_{t_o}^{\infty} u(\lambda) u_o(t-\lambda) d\lambda \qquad (2.55)$$

is in the form of 2.52 with $u(\lambda)$ the real spectral function and $u_o(t-\lambda)$ the elementary function, i.e., the spectral function is the same as the signal itself in this case. Integration of 2.55 by parts results in the form

$$u(t) = \int_{t_o}^{\infty} \dot{u}(\lambda) u_1(t-\lambda) d\lambda \qquad (2.55a)$$

provided that $u(t_{o-}) = 0$ and the integral includes any impulse at t_o due to a discontinuity* of $u(t)$ at t_o. Continued integration by parts results in similar forms of signal resolution in terms of the $u_j(t)$, $j \geq 1$ where the spectral functions are $u^{(j)}(t)$. Similarly, for $n_o \leq n < \infty$, with $v(n_o-1) = 0$,

$$v(n) = \sum_{i=n_o}^{\infty} v(i) v_o(n-i) \qquad (2.56)$$

is in the form of 2.51 with $v(i)$ the real spectral function and $v_o(n-i)$ the elementary function. Summation of 2.56 by parts leads to

$$v(n) = v(n_o) v_1(n-n_o) + \sum_{i=n_o+1}^{\infty} v_1(n-i)\ {}_b v(i) \ . \qquad (2.56a)$$

Repeated summation by parts leads to similar forms of signal

---

*Strictly speaking, this would violate the conditions on testing functions in Eq. 2.8.

resolution in terms of the $v_j(n)$, $j \geq 1$. The resulting forms, however, are increasingly complicated because of the extra terms in 2.31 and 2.30.

Signal resolution in terms of other elementary functions such as real exponentials also can be developed (see Problem 34 at the end of this chapter).

Example 2.7

For the signal

$$u(t) = \begin{cases} 0 & t < 0 \\ t & 0 \leq t \leq 2 \\ 2 & t > 2 \end{cases}$$

find resolutions of u(t) in terms of $u_0(t)$ and $u_1(t)$.

Using Eq. 2.19,

$$u(t) = u_2(t) - u_2(t-2)$$

and

$$\dot{u}(t) = u_1(t) - u_1(t-2) \ ,$$

thus using 2.55,

$$u(t) = \int_0^\infty [u_2(\lambda)-u_2(\lambda-2)]u_0(t-\lambda)d\lambda$$

and using 2.55a,

$$u(t) = \int_0^\infty [u_1(\lambda)-u_1(\lambda-2)]u_1(t-\lambda)d\lambda \ .$$

2.3.3 Sampling and Reconstruction

The representation of uniform impulse sampling of u(t) (see Appendix A and Section 2.9) and reconstruction of a $u(nT_s)$ using a zero-order hold can be put in the forms of 2.53 and 2.50, respectively. For a continuous signal u(t) and a uniform sampling function $u_{T_s}(t)$, it is shown in Section 2.9 that

$$u(nT_s) = \int_{-\infty}^\infty u(\lambda)u_0(\lambda-nT_s)d\lambda \qquad (2.57)$$

for all integer n. The spectral function is $u(\lambda)$ and the elementary function is $u_o(\lambda - nT_s)$.

For a reconstruction of $u(nT_s)$ such that

$$u_r(t) = u(nT_s), \qquad nT_s \le t < (n+1)T_s$$

we define

$$u_h(t,n) = u_1(t-nT_s) - u_1(t-(n+1)T_s) \; .$$

The reconstructed signal is then

$$u_r(t) = \sum_{n=-\infty}^{\infty} u(nT_s)u_h(t,n) \tag{2.58}$$

which is in the form of 2.50.

2.3.4 Signal Spaces

A signal space model can be developed in analogy to the finite-dimensional vector spaces of Appendix B. For energy signals u(t) and v(n) (see Section 2.1), an N-term model of the form

$$u_{(N)}(t) = \sum_{k=1}^{N} \ell(k)q_2(t,k) \tag{2.59}$$

or

$$v_{(N)}(n) = \sum_{k=1}^{N} m(k)q_4(n,k) \tag{2.60}$$

is used. The elementary or basis functions $q_2(t,k)$ and $q_4(n,k)$, k=1,2, ..., N, must be orthonormal on some interval $[t_1,t_2]$ or $[n_1,n_2]$. The spectral functions are then given by

$$\ell(k) = \int_{t_1}^{t_2} u(t)q_2(t,k)\,dt \tag{2.61}$$

or

$$m(k) = \sum_{n_1}^{n_2} v(n)q_4(n,k) \tag{2.62}$$

where u(t) and v(n) are the signals being approximated by 2.59 and 2.60, respectively, in a least squares sense. Examples of these forms appeared in the previous section in Equations 2.39 and 2.41, where the elementary functions were $\psi_i(t)$ and $\phi_i(n)$. Equation 2.42, when used with N terms, also is in this form since the sets of cos $k\omega_o t$ and sin $k\omega_o t$ are orthonormal on $[-T_o/2, T_o/2]$. Eq. 2.43 and 2.44 also are in this form since the sets of cos $k\lambda_o n$ and sin $k\lambda_o n$ are orthonormal on $[0, N_o-1]$.

The models 2.59 and 2.60 are linear combinations of the elementary or basis functions, and the geometrical interpretation is that $\ell(k)$ and m(k) are the components of the signals in an N-dimensional signal space. In Example 2.4, the model for $v_x(n)$ is in a 2-dimensional signal space and the components are $E_1=5.11$ and $E_2=-0.646$.

2.3.5 Complete Orthonormal Sets

One further remark on the properties of $q_2(t,k)$ and $q_4(n,k)$ in the previous section must be made. The choice of N in most applications is a compromise between accuracy and complexity. It is important that as N increases, the approximation improves. In fact, as $N \to \infty$, the sequence $\{u_{(N)}(t)\}$ must converge to u(t) or the sequence $\{v_{(N)}(n)\}$ must converge to v(n). In this case, the sets of orthonormal functions form <u>complete</u> orthonormal sets [6.]. In the two examples given above, the $\psi_i(t)$ and $\phi_i(t)$ as well as the sinusoids do form complete orthonormal sets. In addition, Legendre polynomials, Laguerre functions, and Walsh functions are commonly used [6.][7.]. The <u>Laguerre functions</u> are of interest because they can be generated using the impulse response of an electrical network -- a cascade of one single-pole and N all-pass networks. They are orthonormal on $[0,\infty]$ and the first three are:

$$f_o(t) = \varepsilon^{-t/2} \qquad t \in T_{cl} \tag{2.63a}$$

$$f_1(t) = \varepsilon^{-t/2}(1-t) \qquad t \in T_{cl} \tag{2.63b}$$

$$f_2(t) = \frac{1}{2}\varepsilon^{-t/2}(2-t)^2 \quad t \in T_{c1} \tag{2.63c}$$

These could be used to model signals which are similar to those for which the $\psi_i(t)$ of Section 2.2.2 were suitable. The Legendre functions are orthonormal on [-1,1] and with scale changing they can be used to model polynomial-like signals instead of the $u_j(t)$ of Section 2.2.1 (which are not orthonormal). The first three are:

$$g_o(t) = \frac{1}{\sqrt{2}} \qquad -1 \leq t \leq 1 \tag{2.64a}$$

$$g_1(t) = \sqrt{\frac{3}{2}}\, t \quad -1 \leq t \leq 1 \tag{2.64b}$$

$$g_2(t) = \sqrt{\frac{5}{2}}\left(\frac{3}{2}t^2 - \frac{1}{2}\right) \qquad -1 \leq t \leq 1 \tag{2.64c}$$

The Walsh Functions are useful in digital representations of signals since they have only two amplitudes, +1 and -1. They are orthonormal on [0,1]. Example 2.8 illustrates their use. Obviously we also can develop discrete-time counterparts to these in the same way the $\phi_i(n)$ were suggested by the $\psi_i(t)$ of Section 2.2.2.

Example 2.8

Find an approximation to $u_s(\tau)=3 \sin \omega_o\tau$, $0 \leq \tau \leq 2\pi/\omega_o$, using the first six Walsh functions $W_o, W_1, \ldots, W_5$. Thus, using Equation 2.59,

$$u_{(N)}(t) = \sum_{k=0}^{5} \ell(k) W_k(t)$$

with N=6. The Walsh functions are orthonormal on $t \in [0,1]$, and the signal $u_s(\tau)$ is scaled by letting

$$\tau = \frac{2\pi}{\omega_0} t ,$$

thus

$$u(t) = 3 \sin 2\pi t, \qquad 0 \leq t \leq 1$$

The functions $W_0, W_1, \ldots, W_5$ are shown below. The index i on W_i tells the number of zero crossings in the open interval (0,1).

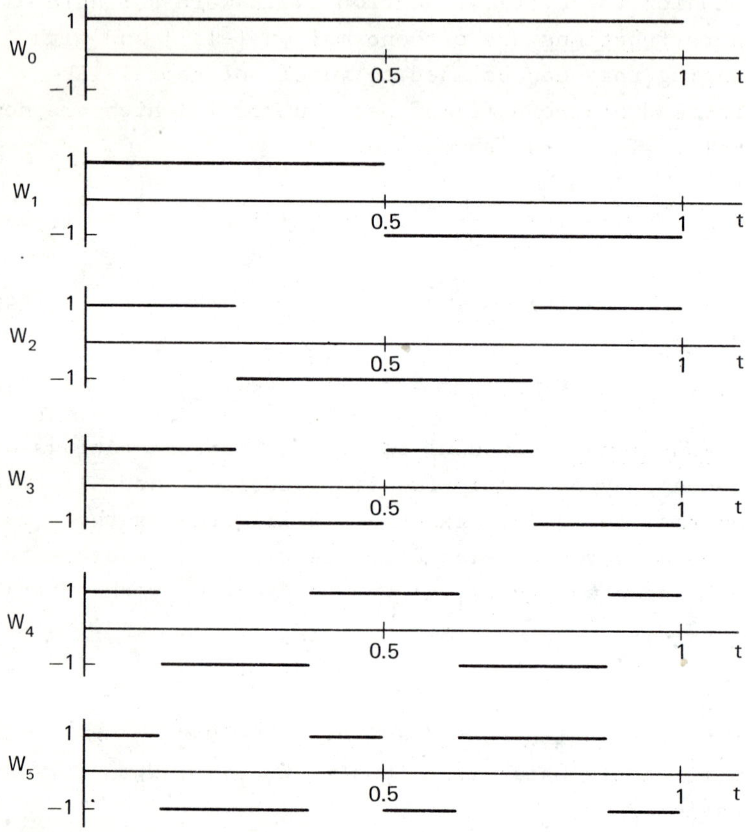

Now

$$\ell(k) = \int_0^1 u(t) W_k \, dt$$

and the following spectral coefficients are found:

$$\ell(0) = 0$$

$$\ell(1) = 6/\pi$$

$$\ell(2) = 0$$

$$\ell(3) = 0$$

$$\ell(4) = 0$$

$$\ell(5) = -(\sqrt{2}-1)6/\pi$$

Thus

$$u_{(6)}(t) = \frac{6}{\pi}[W_1(t) - 0.414W_5(t)]$$

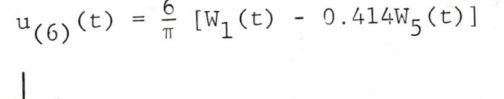

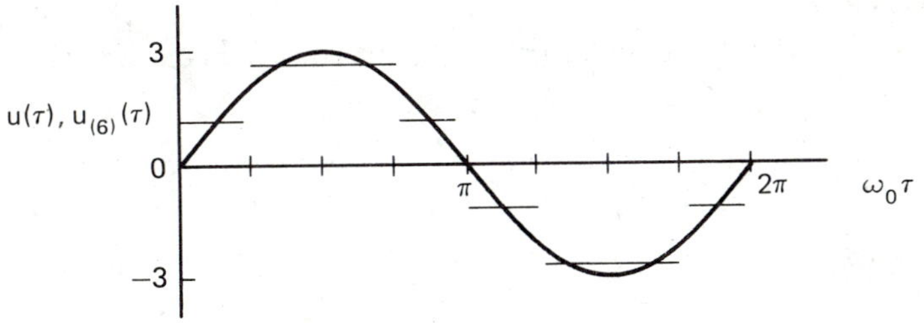

2.4 FREQUENCY DOMAIN MODELS

The signal models of the two previous sections required only real arguments as shown in Table 2.2. However, the general concept of an integral transform includes elementary functions (or kernels) with both real and complex arguments. The Hilbert transform is an example where real arguments occur; the spectral function as well as the elementary function has a real argument. The transformed signal description remains in the "time domain." By contrast, the transforms of Appendix A -- the Fourier, Laplace and Z-transforms -- have elementary functions with imaginary or complex arguments and the transformed signal description is in the "frequency domain."* It happens that it is sometimes more convenient to study a signal in terms of its frequency (or complex argument) domain description. The long history of

---

*Recall the discussion in Chapter 1 on "frequency" for discrete-time signals.

using sinusoidal signals has resulted in a wide familiarity with this point of view. Many other integral transforms have been defined (e.g., the Mellin transform), but those above which use a Fourier kernel have proved to be of great value in signal analysis and in studying the transmission of signals through linear systems. Thus, the frequency domain signal models given here are obtained by use of the transforms of Appendix A on the time domain signal models of the previous two sections.

2.4.1 Poles and Zeros

For the (one-sided) singularity functions, from Eq. 2.2,

$$\mathcal{L} u_j(t) = \mathcal{L} \frac{t^{j-1}}{(j-1)!}, \qquad j \geq 1,\ \sigma > 0,$$

or

$$= s^{-j} .$$

With the extensions of 2.3 to 2.14 (for $t_o=0$) and property 2. of Sect. A.2.3, App. A,

$$u_j(t) = s^{-j}, \qquad \forall_j, \quad \sigma > 0 . \tag{2.65}$$

From 2.20 or 2.32 with $n_o=0$,

$$\mathfrak{z}\, v_j(n) = \mathfrak{z} \frac{1}{(j-1)!} n^{j-1}, \qquad j \geq 1,\ n \geq 0 .$$

A general form like 2.65 does not follow although there is some similarity. We use 2.32 and 2.33c and the following are found:

$$\mathfrak{z}\, v_3(n) = \frac{z(z+1)}{2(z-1)^3}, \qquad \rho > 1 \tag{2.66a}$$

$$\mathfrak{z}\, v_2(n) = \frac{z}{(z-1)^2}, \qquad \rho > 1 \tag{2.66b}$$

$$\mathfrak{z}\, v_1(n) = \frac{z}{(z-1)}, \qquad \rho > 1 \tag{2.66c}$$

$$\mathcal{Z}\, v_o(n) = 1 \tag{2.66d}$$

$$\mathcal{Z}\, v_{-1}(n) = \frac{(z-1)}{z}, \quad \rho > 0 \tag{2.66e}$$

$$\mathcal{Z}\, v_{-2}(n) = \frac{(z-1)^2}{z^2}, \quad \rho > 0 \tag{2.66f}$$

$$\mathcal{Z}\, v_{-3}(n) = \frac{(z-1)^3}{z^3}, \quad \rho > 0 \tag{2.66g}$$

We observe that there is a jth order pole at z=1 for $j \geq 1$ and a jth order zero at z=1 for $j \leq -1$, and these are analogous to the poles and zeros at s=0 for $\mathcal{L}\, u_j(t)$.

A summary for the singularity functions used most often is given in Table 2.3.

For the (one-sided) real exponential functions,

$$\mathcal{L}\, u_{ei}(t) = \mathcal{L}\, \varepsilon^{-a_i t}, \qquad i \geq 1,\ t \geq 0$$

$$= \frac{1}{s+a_i}$$

and

$$\mathcal{Z}\, v_{ei}(n) = \mathcal{Z}\, (b_i)^n, \qquad i \geq 1,\ n \geq 0$$

$$= \frac{z}{z-b_i}\ .$$

TABLE 2.3

$u_j(t)$	$U_j(s)$	$v_j(n)$	$V_j(z)$
$u_{-2}(t)$	s^2	$v_{-2}(n)$	$(z-1)^2/z^2$
$u_{-1}(t)$	s	$v_{-1}(n)$	$(z-1)/z$
$u_o(t)$	1	$v_o(n)$	1
$u_1(t)$	$1/s$	$v_1(n)$	$z/(z-1)$
$u_2(t)$	$1/s^2$	$v_2(n)$	$z/(z-1)^2$

Thus the poles of the frequency domain models are at $s=-a_i$ and $z=b_i$.

For the one-sided sinusoidal functions with $\alpha_i=0$ and $\beta_i=0$,

$$\mathcal{L}\, u_{ai}(t)u_1(t) = \mathcal{L}\sin(i\omega_o t), \qquad i \geq 1,\ t \geq 0$$

$$= \frac{i\omega_o}{s^2+(i\omega_o)^2} \qquad \alpha > 0$$

and

$$\mathcal{L}\, u_{bi}(t)u_1(t) = \mathcal{L}\cos(i\omega_o t), \qquad i \geq 1,\ t \geq 0$$

$$= \frac{s}{s^2+(i\omega_o)^2}, \qquad \sigma > 0 .$$

Also,

$$\mathcal{Z}\, v_{ai}(n)v_1(n) = \mathcal{Z}\sin(i\lambda_o n), \qquad i \geq 1,\ n \geq 0$$

$$= \frac{z\sin(i\lambda_o)}{z^2-2z\cos(i\lambda_o)+1}, \qquad \rho > 1$$

and

$$\mathcal{Z}\, v_{bi}(n)v_1(n) = \mathcal{Z}\cos(i\lambda_o n), \qquad i \geq 1,\ n \geq 0$$

$$= \frac{z(z-\cos(i\lambda_o))}{z^2-2z\cos(i\lambda_o)+1}, \qquad \rho > 1$$

Thus the poles of the frequency domain models are at $s = \pm j(i\omega_o)$ or at $z = \cos(i\lambda_o)\pm j\sin(i\lambda_o)$. The forms of the transforms of the corresponding two-sided sinusoidal functions are quite different. For example,

$$\mathcal{F}\, B\cos\omega_o t = \pi B[u_o(\omega-\omega_o)+u_o(\omega+\omega_o)]$$

and

$$\mathcal{F}_d B\cos\lambda_o n = \pi B[u_o(\gamma-\gamma_o)+u_o(\gamma-\gamma_o^*)]$$

as shown in Tables A.2 and A.7 of Appendix A.

Finally, for the (one-sided) damped oscillatory forms of

2.48a and 2.49a, for $\phi_i=0$ and $\psi_i=0$, we get

$$\mathcal{L}\, u_{fi}(t) = \frac{(s+\sigma_{oi})}{(s+\sigma_{oi})^2+(i\omega_o)^2}$$

and

$$\mathcal{Z}\, v_{fi}(n) = \frac{z(z-\Theta_{oi})}{z^2-2z\Theta_{oi}+\rho_{oi}^2}$$

where $\Theta_{oi}=\rho_{oi}\cos i\lambda_o$. Thus the poles of these frequency domain models are at $s=-\sigma_{oi}\pm j(i\omega_o)$ and at $z=\rho_{oi}(\cos i\lambda_o \pm j\sin i\lambda_o)$.

Figures 2.9 and 2.10 show typical pole-zero plots for the frequency domain models of the one-sided continuous-time and discrete-time elementary functions. It is obvious that in all cases, sums of the frequency domain forms can be used as was done previously with the time domain models. For example, using u_{ei},

$$U(s) = \sum_i \frac{A_i}{s+a_i}$$

and using v_{ei},

$$V(z) = \sum_i \frac{D_i z}{z-b_i} .$$

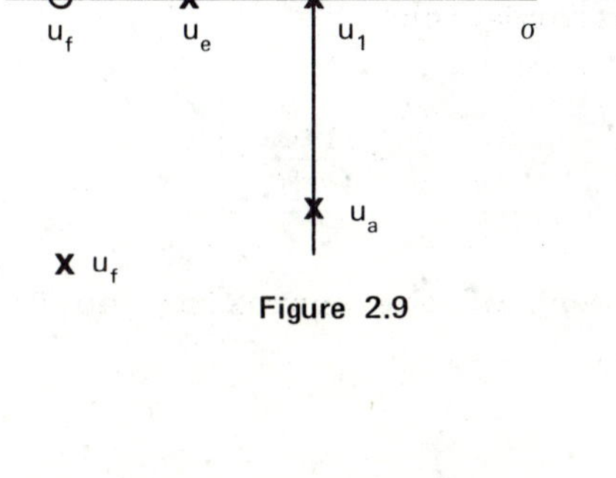

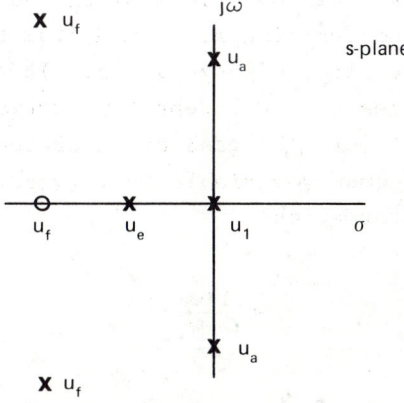

Figure 2.9

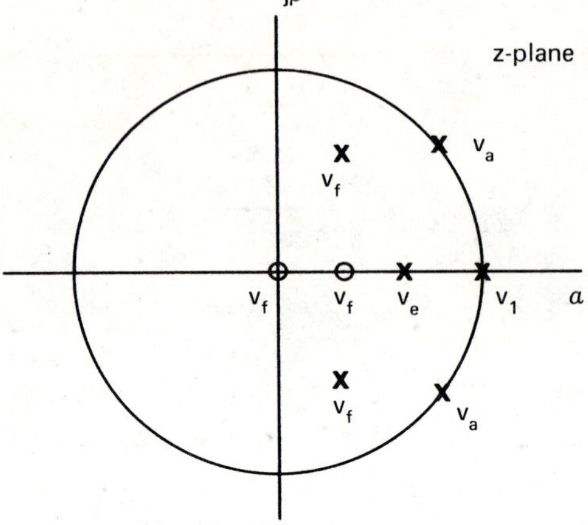

Figure 2.10

2.4.2 Frequency Domain Approximations

Inspection of all the one-sided models shows that in general, U(s) will be a rational function of s and V(z) will be a rational function of z. This means that methods that are widely used in network synthesis and control system design for approximating a magnitude or phase function of frequency can be used. One well known method [8.] uses logarithmic plots so that positions of the poles and zeros of the approximating function are found from the intersections (break frequencies) of asymptotes. In the case of continuous-time signals, the independent frequency variable is ω (where $s \to j\omega$). In the case of discrete-time signals, the independent frequency variable is v (where $w \to jv$) and the bilinear transformation

$$z = \frac{1+w}{1-w}$$

is used. Improvement of the approximation is made by

adjusting the positions of the poles and zeros. In analogy to the time-domain approximation models of Section 2.2,

$$|U_x(j\omega)|,\ \phi_x(j\omega),\ |V_x(jv)|,\ \psi_x(jv)$$

are the experimental magnitude and phase data, and

$$|U_{(M)}(j\omega)|,\ \phi_{(M)}(j\omega),\ |V_{(M)}(jv)|,\ \psi_{(M)}(jv)$$

are the models which use a finite number of poles and zeros.

Another method [17.][18.] of approximation uses the Integral Square and Sum Square Theorems (see Appendix A) and makes a best approximation in the least integral square or least sum square sense (in the time domain). It is necessary that the integral square and sum square errors be finite, and therefore, that the poles of $E(s) = U_x(s) - U_{(M)}(s)$ lie in the left half s-plane or the poles of $E(z) = V_x(z) - V_{(M)}(z)$ lie inside the unit circle of the z-plane. Thus the class of models $U_{(M)}(s)$ and $V_{(M)}(z)$ is correspondingly restricted and the experimental data are assumed to satisfy these conditions too. In general, the equations which must be solved for the parameters of the $U_{(M)}(s)$ and $V_{(M)}(z)$ models are nonlinear, and iterative methods of solution on a computer are required. This was true, for example, for Eq. 2.38 also. The use of frequency domain error criteria (e.g., functions of $E(s)$ or $E(z)$) is considered in Chapter 5, Section 5.2.

Example 2.9

a) For a signal model

$$u_{(2)}(t) = a_o u_1(t) + a_1 u_2(t) \qquad a_o,\ a_1 > 0$$

$$\mathcal{L}\, u_{(2)}(t) = U_{(2)}(s) = \frac{a_o}{s} + \frac{a_1}{s^2}, \qquad \sigma > 0$$

$$= \frac{a_o s + a_1}{s^2} = a_o\left(\frac{s + a_1/a_o}{s^2}\right)$$

and $U_{(2)}(s)$ is a rational function of s with a second-order pole at the origin and a zero $-a_1/a_0$. For $s = j\omega$, i.e., $\sigma \to 0$,

$$U_{(2)}(j\omega) = a_0\left(\frac{j\omega + a_1/a_0}{(j\omega)^2}\right)$$

and

$$|U_{(2)}(j\omega)| = \frac{a_0}{\omega^2}\sqrt{\left(\frac{a_1}{a_0}\right)^2 + \omega^2}\ .$$

This can be used to approximate the data $|U_x(j\omega)|$. If logarithmic coordinates are used, the break frequency is $\omega_1 = a_1/a_0$ and the asymptote slopes are -2 for $\omega < \omega_1$ and -1 for $\omega > \omega_1$. The gain a_0 can be found by using one particular value of ω, e.g., for $\omega = 1$,

$$|U_{(2)}| = a_0\sqrt{\left(\frac{a_1}{a_0}\right)^2 + 1}$$

thus both a_0 and a_1 are determined by choosing ω_1 and $|U_{(2)}|$ at $\omega = 1$. This method does not give an optimal model in the sense that a nonnegative function of error is minimal. Note that $U_{(2)}(j\omega)$ is <u>not</u> the Fourier Transform of $u_{(2)}(t)$. The Fourier Transform of $u_{(2)}(t)$ is

$$\begin{aligned}\mathcal{F}\, u_{(2)}(t) &= a_0\, \mathcal{F}\, u_1(t) + a_1\, \mathcal{F}\, u_2(t)\\ &= a_0[\pi u_0(\omega) + \frac{1}{j\omega}]\\ &\qquad + a_1[j\pi u_{-1}(t) - \frac{1}{\omega^2}]\end{aligned}$$

b) For a signal model

$$u_{(2)}(t) = A_1\varepsilon^{-a_1 t} + A_2\varepsilon^{-a_2 t} \qquad t \geq 0$$

$$a_1,\ a_2 > 0,\ a_1 \neq a_2.$$

$$\mathcal{L}\, u_{(2)}(t) = U_{(2)}(s) = \frac{A_1}{s+a_1} + \frac{A_2}{s+a_2}, \qquad \sigma > \mathrm{Max}(-a_1, -a_2)$$

$$= \frac{(A_1+A_2)s+(A_1a_2+A_2a_1)}{(s+a_1)(s+a_2)} .$$

For $K = (A_1+A_2)$ and $b = (A_1a_2+A_2a_1)/K$,

$$U_{(2)}(s) = \frac{K(s+b)}{(s+a_1)(s+a_2)}$$

and

$$|U_{(2)}(j\omega)| = \frac{K\sqrt{\omega^2+b^2}}{\sqrt{(\omega^2+a_1^2)(\omega^2+a_2^2)}}$$

which is used to approximate $|U_x(j\omega)|$. If logarithmic coordinates are used, and for example, $a_1 < b < a_2$, the slopes of the asymptotes are

$$0 \text{ for } \omega < a_1$$

$$-1 \text{ for } a_1<\omega<b$$

$$0 \text{ for } b < \omega < a_2$$

$$-1 \text{ for } \omega > a_2$$

As before, K can be fixed for $\omega = 1$ where

$$|U_{(2)}| = \frac{K\sqrt{1+b^2}}{\sqrt{(1+a_1^2)(1+a_2^2)}} .$$

and three break frequencies corresponding to a_1, b, and a_2 must be chosen. Reference [8.] gives a method for adjusting the constants of the model in a sequential way which makes successive improvements in the approximation. Reference [18.] gives an iterative method for finding the constants that make the integral of $e(t)^2 = (u_x(t) - u_{(2)}(t))^2$ minimal.

c) For a signal model

$$v_{(2)}(n) = a_o v_1(n) + a_1 v_2(n), \qquad 0 < a_1/a_o < 1$$

$$a_o, a_1 > 0$$

$$\mathcal{Z}\, v_{(2)}(n) = V_{(2)}(z) = \frac{a_o z}{z-1} + \frac{a_1 z}{(z-1)^2}, \qquad \rho > 1$$

$$= \frac{a_o z(z + \frac{a_1 - a_o}{a_o})}{(z-1)^2}$$

and $V_{(2)}(z)$ is a rational function of z with a second-order pole at z=1 and zeros at $1-a_1/a_o$ and zero. For $z=\gamma$, i.e., $\rho \to 1$,

$$V_{(2)}(\gamma) = \frac{a_o \gamma(\gamma+b)}{(\gamma-1)^2}$$

where $b = (a_1-a_o)/a_o$, and $\gamma = \varepsilon^{j\theta}$, and

$$|V_{(2)}(\varepsilon^{j\theta})| = \left| \frac{a_o \varepsilon^{j\theta}(\varepsilon^{j\theta}+b)}{(\varepsilon^{j\theta}-1)^2} \right|$$

$$= \frac{a_o \sqrt{b^2 + 1 + 2b\cos\theta}}{2(1-\cos\theta)}$$

This is periodic with singularities at $\theta=0, \pm 2\pi, \ldots,$ and could be used to approximate a $|V_x(\gamma)|$ for $-\pi<\theta\leq\pi$. A more convenient method is to use the bilinear transformation and then use the same asymptote method described in a) of this example. Thus for

$$z = \frac{1+w}{1-w}$$

$$V_{(2)} = \frac{a_0(\frac{1+w}{1-w})[(\frac{1+w}{1-w}) + b]}{(\frac{1+w}{1-w} - 1)^2}$$

or

$$V_{(2)} = \frac{a_o(1-b)}{4} \frac{w^2 + \frac{2}{1-b} w + (\frac{1+b}{1-b})}{w^2}$$

and for w=jv, and for the constant multiplier $K = \frac{a_o(1-b)}{4}$

$$V_{(2)}(jv) = K(\frac{(jv)^2 + \frac{2}{1-b}(jv) + (\frac{1+b}{1-b})}{(jv)^2})$$

and we get

$$|V_{(2)}(jv)| = K \frac{[(\frac{1+b}{1-b} - v^2)^2 + \frac{4}{(1-b)^2} v^2]^{\frac{1}{2}}}{v^2}$$

which can be used to approximate a $|V_x(jv)|$ found from the original $|V_x(\gamma)|$. Note that $V_{(2)}$ is rational in jv (as $U_{(2)}$ in a) is rational in $j\omega$) and logarithmic coordinates and asymptote approximations can be used. The break frequency is $v_1 = (1+b/1-b)^{\frac{1}{2}}$ and the slopes are -2 for $v < v_1$ and 0 for $v > v_1$.

d) For a signal model

$$v_{(2)}(n) = B_1(b_1)^n + B_2(b_2)^n \qquad n \geq 0$$

$$0 < b_1,\ b_2 < 1,\ b_1 \neq b_2$$

$$\mathfrak{z}\, v_{(2)}(n) = V_{(2)}(z) = \frac{B_1 z}{z-b_1} + \frac{B_2 z}{z-b_2}, \qquad \rho > \mathrm{Max}(b_1,b_2)$$

$$= \frac{Kz(z-b)}{(z-b_1)(z-b_2)}$$

where $K = (B_1+B_2)$ and $b = (B_1b_2+B_2b_1)/K$. For $z = \gamma$,

$$V_{(2)}(\gamma) = \frac{K\gamma(\gamma-b)}{(\gamma-b_1)(\gamma-b_2)} .$$

Also,

$$|V_{(2)}(\varepsilon^{j\theta})| = K \left\{\frac{(1+b^2)-2b\cos\theta}{[(1+b_1^2)-2b_1\cos\theta][(1+b_2^2)-2b_2\cos\theta]}\right\}^{\frac{1}{2}}$$

and this is periodic. Using the bilinear transformation, again with the usual change in notation,

$$V_{(2)}(jv) = \frac{K_1[(jv)^2 + \frac{2}{1+b}(jv) + b']}{(jv+b_1')(jv+b_2')}$$

where

$$K_1 = \frac{K(1+b)}{(1+b_1)(1+b_2)}, \quad b' = \frac{1-b}{1+b}, \quad b_1' = \frac{1-b_1}{1+b_1}, \quad b_2' = \frac{1-b_2}{1+b_2}$$

and

$$|V_{(2)}(jv)| = \frac{K_1[(b'-v^2)^2 + \frac{4}{(1+b)^2}v^2]^{\frac{1}{2}}}{[(v^2+b_1'^2)(v^2+b_2'^2)]^{\frac{1}{2}}} .$$

In analogy to b) of this example, the method of [18.] now can be used. This method of [18.], when changed to discrete-time models, can be used to find the parameters of $V_{(N)}(z)$ such that the sum of $e(n)^2 = (v_x(n) - v_{(N)}(n))^2$ is minimal.

2.5 SIGNAL SPECTRA

2.5.1 Real and Complex Spectra

In Section 2.3 the real spectral functions $\ell(k)$, $m(k)$, $f(\lambda)$ and $g(\lambda)$ were introduced in Equations 2.50, 2.51, 2.52 and 2.53. Frequency domain signal models were introduced in Section 2.4 by using Fourier, Laplace, and Z-Transforms. These led to spectral functions and elementary functions (or kernels) with complex or imaginary arguments, that is s, z, $j\omega$ or γ. For periodic continuous-time signals, the exponential Fourier Series of Eq. 2.45 is similar in form to Eq. 2.50 with

U_k equivalent to $\ell(k)$

$\varepsilon^{jk\omega_o t}$ equivalent to $q_2(t,k)$

and for periodic discrete-time signals, the polynomials of Equations 2.46 and 2.47 are similar to Equation 2.51 with

$$V_k \text{ equivalent to } m(k)$$

$$\varepsilon^{jk\lambda_o n} \text{ equivalent to } q_4(n,k) \; .$$

Similar comparisons using the inverse $\mathfrak{F}$ and $\mathfrak{F}_d$ transforms of Equations A.4 and A.14 of Appendix A with obvious changes in notation yield the following relationships with Equations 2.52 and 2.53:

$$U(j\omega) \text{ equivalent to } f(\lambda)$$

$$\varepsilon^{j\omega t} \text{ equivalent to } q_1(t,\lambda)$$

and

$$V(\gamma) \text{ equivalent to } g(\lambda)$$

$$\gamma^n \text{ equivalent to } q_3(n,\lambda) \; .$$

The corresponding forms using the Laplace and Z-Transforms also can be used to find the spectral functions $U(j\omega)$ and $V(\gamma)$ by using the relationships between transforms given in Section A.4 of Appendix A.

2.5.2 Amplitude and Phase Spectra

The eight elementary functions and their corresponding spectral functions provide another way of developing signal models. In fact, it is common practice to measure and plot the spectral functions of signals as alternatives to the explicit function of t or n in Section 2.2. The commonest forms in use are U_k, the discrete spectral function of a periodic signal u(t), or $|U(j\omega)|$ and Angle of $U(j\omega)$, the continuous-ω amplitude and phase spectra of a signal u(t). (In many digital computer analyses of signals, a discretized version of $|U(j\omega)|$ is used; this should not be confused with

U_k or $|V(\gamma)|$. An introduction to these methods is given in Chapter 6 where the Discrete Fourier Transform (DFT) is used.) Table 2.4 summarizes the eight forms which include continuous-time and discrete-time signals and spectral functions resulting from elementary functions with both real and complex arguments. The table includes all the forms established for Table 2.2 plus those discussed in this section. Therefore, it covers a wide choice of signal models, and many other possible models result from other choices of elementary functions or transforms. In each case, an expression for finding the spectral function must be included with the choice of an elementary function.

TABLE 2.4

Spectral Function Argument			Elementary Function Time Argument: Continuous (t)	Elementary Function Time Argument: Discrete (n)
	Continuous	Real	Eq. 2.52 t, λ	Eq. 2.53 n, λ
	Continuous	Complex	Eq. A.4* $t, j\omega$	Eq. A.14* n, γ
	Discrete	Real	Eq. 2.50 t, k	Eq. 2.51 n, k
	Discrete	Complex	Eq. 2.45 $t, jk\omega_o$	Eq. 2.46, 2.47 $n, jk\lambda_o$

*For cases of correspondence with Laplace and Z-Transforms see Appendix A, Section A.4.

Several simple examples of signal spectra are given below to illustrate the variety that is possible. In many practical situations it is easier to compare signals and to detect certain phenomena by inspection of the spectra of u(t) or v(n) rather than u(t) or v(n) themselves.

<u>Example 2.10</u>

a) For Example 2.2, u(t) and $\dot{u}(t)$ are given, and using Eq. 2.55a, a special form of Eq. 2.52,

$$u(t) = \int_0^\infty \dot{u}(\lambda)u_1(t-\lambda)d\lambda \ ,$$

where the spectral function is

$$\dot{u}(\lambda) = u_1(\lambda)-u_1(\lambda-1)-u_1(\lambda-3)+u_1(\lambda-4)$$

and the elementary function is

$$u_1(t-\lambda) \ .$$

b) From Example 2.3, v(n) and $\Delta_b(n)$ are given, and using Eq. 2.56a, a special form of Eq. 2.51,

$$v(n) = \sum_{k=1}^{\infty} v_1(n-k)\Delta_b v(k)$$

where the spectral function is

$$\Delta_b v(k) = v_1(k-1)-v_1(k-3)-2v_0(k-4)$$

and the elementary function is

$$v_1(n-k) \ .$$

c) Another example of Eq. 2.51 is from Example 2.4 where the spectral function is E_k, k=1,2 and the elementary functions are $\phi_k(n)$ as given after Eq. 2.41.

d) Using the results of Example 2.8, an example of Eq. 2.50 is shown where the spectral function is

$$\ell(k), \quad k=0,1,\ \dots,\ 5$$

and the elementary functions are the Walsh functions

$$W_k(t), \quad k=0,1,\ \dots,\ 5, \quad 0 \le t \le 1\ .$$

e) A signal $u(t) = \varepsilon^{-at}u_1(t)$ is uniformly impulse sampled (see Appendix A and Sect. 2.9). Using Equation 2.57, a special form of Eq. 2.53,

$$u(nT_s) = \varepsilon^{-anT_s}u_1(nT_s) = \int_0^{\infty} \varepsilon^{-a\lambda}u_1(\lambda)u_o(\lambda-nT_s)d\lambda\ .$$

The spectral function is

$$\varepsilon^{-a\lambda}u_1(\lambda)\ ,$$

the signal itself, and the elementary function is

$$u_o(\lambda-nT_s)\ .$$

f) Similar to the above, a reconstructed signal as given by Eq. 2.58, a form of Eq. 2.50, has the spectral function

$$\varepsilon^{-anT_s}u_1(nT_s)\ ,$$

the samples. The elementary function is

$$u_h(t,n)\ .$$

g) For the signal u(t) shown, the amplitude spectrum U_k is required.

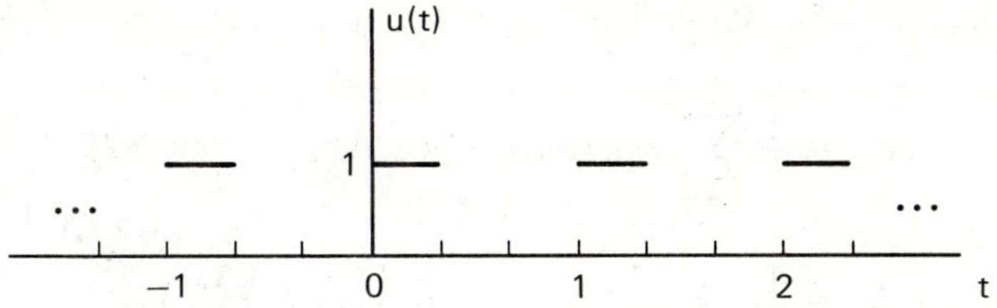

Using Eq. 2.45,

$$U_k = \int_0^{1/3} \exp(-jk\omega_o t)dt, \quad \omega_o = 2\pi, \; T_o = 1$$

$$= \frac{1}{k\pi} \exp(-jk\pi/3) \sin(k\ \pi/3) \ .$$

Thus,

$$U_k = \frac{1}{3} \left| \frac{\sin(k\ \pi/3)}{(k\ \pi/3)} \right| = \frac{1}{3} \left| \frac{\sin(k\ \frac{\omega_o}{6})}{(k\ \frac{\omega_o}{6})} \right|$$

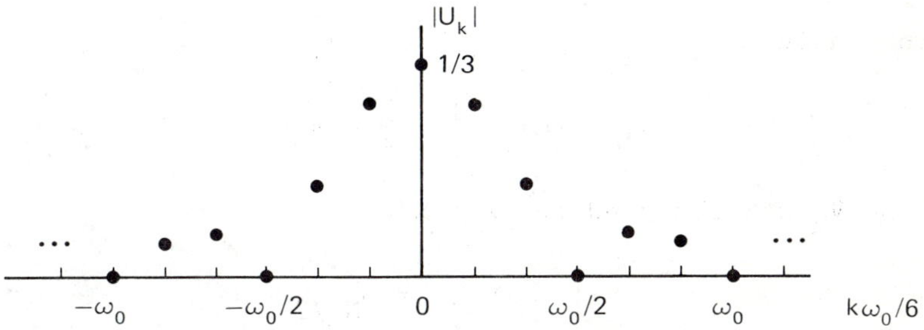

h) For the signal v(n) shown, the amplitude spectrum V_k is required.

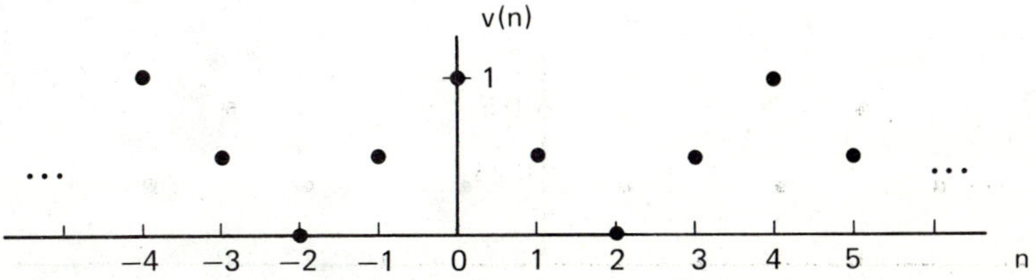

Using Eq. 2.47, $K' = 1$

$$V_k = \frac{1}{4}\sum_{n=0}^{3} v(n)\exp(-jk\lambda_o n), \quad \lambda_o = \pi/2$$

$$V_k = \frac{1}{4}\left(1+\frac{1}{2}\cos k\pi + \frac{1}{2}\cos\frac{k3\pi}{2}\right)$$

$$- j\frac{1}{4}\left(\frac{1}{2}\sin k\pi + \frac{1}{2}\sin\frac{k3\pi}{2}\right) .$$

Thus

$$|V_o| = 0.5$$

$$|V_1| = |V_{-1}| = 0.25$$

and

$$b_{N_o/2} = 0.$$

Using these,

$$v(n) = \frac{1}{2} + \frac{1}{2}\cos\frac{n\pi}{2} .$$

If the V_k are computed for $k > K'$,

$$|V_2| = |V_{-2}| = b_{N_o/2} = 0$$

$$|V_3| = |V_{-3}| = 0.25 = |V_1|$$

$$|V_4| = |V_{-4}| = 0.5 = |V_o|, \text{ etc.}$$

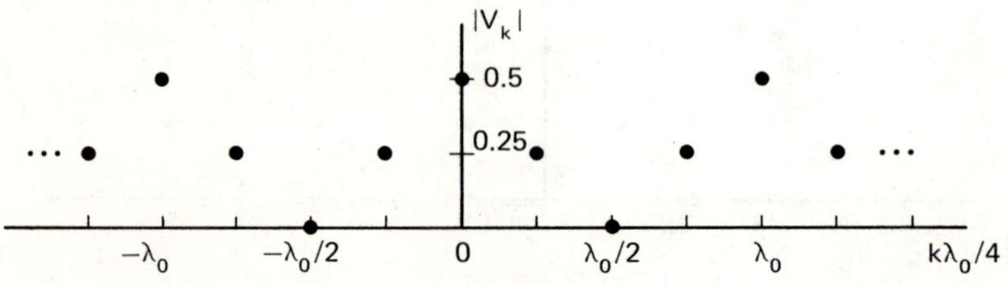

i) For

$$u(t) = \varepsilon^{-at}u_1(t), \quad a > 0,$$

find the amplitude spectrum $|U(j\omega)|$. Now, using Eq. A.4,

$$\mathfrak{F}\, u(t) = \frac{1}{a+j\omega} = U(j\omega) \quad ,$$

thus

$$|U(j\omega)| = \frac{1}{\sqrt{a^2+\omega^2}} \quad ,$$

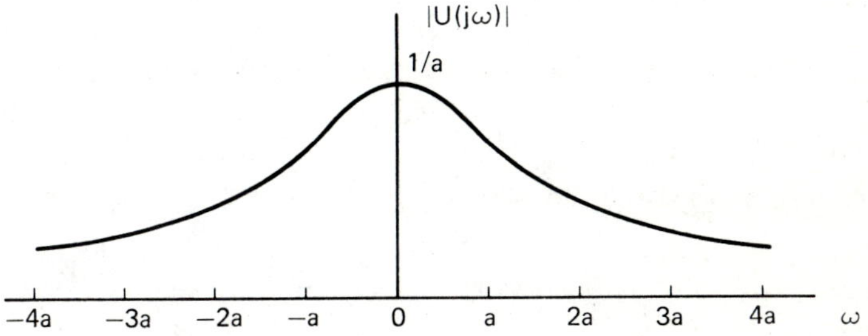

j) For

$$v(n) = b^n v_1(n), \qquad 0 < b < 1 \ ,$$

find the amplitude spectrum $|V(\gamma)| = |V(\varepsilon^{j\theta})|$. Now using Eq. A.14,

$$\mathfrak{F}_d v(n) = \frac{\gamma}{\gamma - b} = V(\gamma) \ .$$

Then

$$V(\varepsilon^{j\theta}) = \frac{\varepsilon^{j\theta}}{\varepsilon^{j\theta}-b}$$

and

$$|V(\varepsilon^{j\theta})| = \frac{1}{\sqrt{1-2b\cos\theta+b^2}}$$

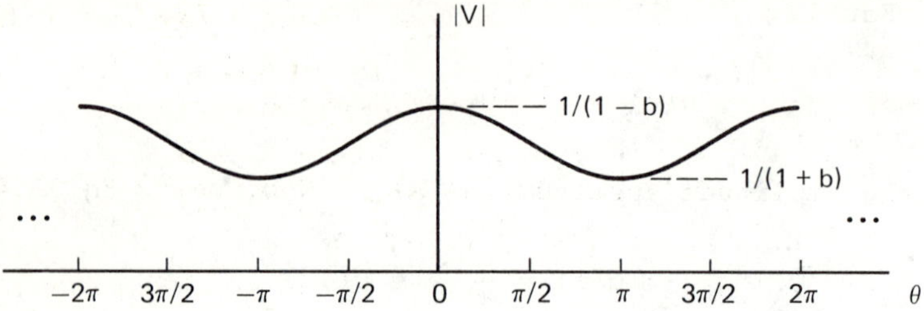

k) For Example 2.9b where

$$U_{(2)}(j\omega) = \frac{A_1}{j\omega+a_1} + \frac{A_2}{j\omega+a_2}$$

the amplitude spectrum is

$$|U_{(2)}(j\omega)| = \frac{K\sqrt{\omega^2+b^2}}{\sqrt{(\omega^2+a_1^2)(\omega^2+a_2^2)}}$$

l) From Example 2.9d where

$$V_{(2)}(\gamma) = \frac{K\gamma(\gamma-b)}{(\gamma-b_1)(\gamma-b_2)}$$

the amplitude spectrum is

$$|V_{(2)}(\varepsilon^{j\theta})| = K\left\{\frac{(1+b^2) - 2b\cos\theta}{[(1+b_1^2) - 2b_1\cos\theta][(1+b_2^2) - 2b_2\cos\theta]}\right\}^{\frac{1}{2}}$$

Note that the amplitude spectra of parts h and j of this example are periodic. This is true in general for the sum 2.46 and the inverse transform A.14 since both concern discrete-time functions where the function is defined on equally-spaced points in the time domain. It also should be noted that only for parts g and h are the amplitude spectra discrete (in the frequency domain). This is true in general for signals u(t) or v(n) that are periodic in the time domain. For the spectral functions associated with imaginary arguments, the phase spectra also can be found.

m) From part i,

$$U(j\omega) = \frac{1}{a+j\omega} ,$$

thus

$$\text{Ang}U(j\omega) = \theta(j\omega) = -\tan^{-1}\frac{\omega}{a}$$

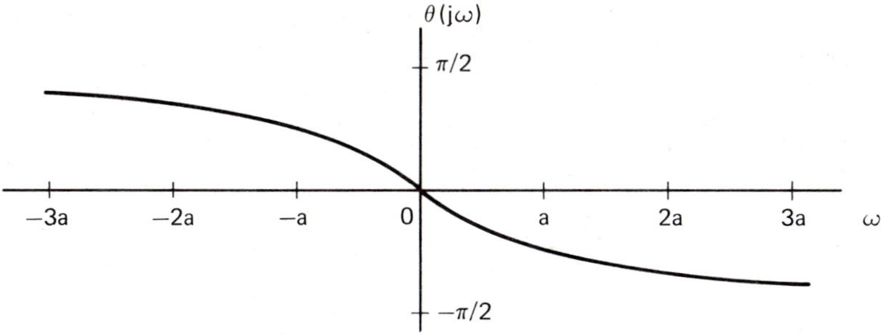

2.5.3 Energy Spectra

By using the concept of signal energies E_c and E_d introduced in Section 2.1, energy spectra also can be defined. Thus for an energy signal u(t), use of the Parseval Theorem (Appendix A, Section A.1.4) leads to

$$E_c = \int_{-\infty}^{\infty} u^2(\lambda)\,d\lambda = \frac{1}{2\pi}\int_{-\infty}^{\infty} |U(j\omega)|^2 d\omega$$

and $|U(j\omega)|^2$ is called the <u>energy spectrum</u> of u(t). It has the dimension energy per Hz, and consequently is also sometimes called energy density. The signal energy in an interval $[\omega_1,\omega_2]$ is then

$$\frac{1}{\pi}\int_{\omega_1}^{\omega_2} |U(j\omega)|^2 d\omega, \qquad 0 \le \omega_1 < \omega_2 .$$

Similarly for an energy signal v(n),

$$E_d = \sum_{k=-\infty}^{\infty} v^2(k) = \frac{1}{2\pi}\int_{-\pi}^{\pi} |V(\varepsilon^{j\theta})|^2 d\theta$$

which follows from the sum square form of Appendix A, Section A.3 for $z \to \gamma = \varepsilon^{j\theta}$. Thus $|V(\varepsilon^{j\theta})|^2$ is the energy spectrum of v(n). In Example 2.10, squaring the amplitude spectra of parts i and j produces the corresponding energy spectra. The energy in an interval $[\theta_1, \theta_2]$ is

$$\frac{1}{\pi}\int_{\theta_1}^{\theta_2} |V(\varepsilon^{j\theta})|^2 d\theta, \qquad 0 \leq \theta_1 < \theta$$

2.5.4 Power Spectra

For signals that are not energy signals, a power spectrum can be defined. Thus for

$$\begin{aligned} u_T(t) &= u(t), && |t| \leq T \\ &= 0 && |t| > T \end{aligned}$$

and

$$\mathfrak{F}\, u_T(t) = U_T(j\omega) ,$$

the power spectrum is

$$P_{cs}(\omega) = \lim_{T\to\infty} \frac{|U_T(j\omega)|^2}{2T} \tag{2.67}$$

and the average power is

$$P_c = \frac{1}{2\pi}\int_{-\infty}^{\infty} P_{cs}(\omega)\, d\omega .$$

For a periodic signal u(t), with U_k the coefficients of a Fourier Series expansion,

$$P_{cs}(\omega) = 2\pi \sum_{k=-\infty}^{\infty} |U_k|^2 u_o(\omega - k\omega_o)$$

and

$$P_c = \frac{1}{T_o}\int_{-T_o/2}^{T_o/2} u^2(\lambda)\, d\lambda = \sum_{k=-\infty}^{\infty} |U_k|^2 .$$

This can be used to compute P_c for Example 2.10g.

Similarly, let

$$v_N(n) = v(n), \qquad |n| \leq N$$
$$= 0 \quad , \qquad |n| > N$$

and

$$\mathfrak{F}_d v_N(n) = V_N(\gamma) = V_N(\varepsilon^{j\theta})$$

the power spectrum is

$$P_{ds}(\theta) = \lim_{N\to\infty} \frac{|V_N(\varepsilon^{j\theta})|^2}{2N+1} \tag{2.68}$$

and the average power is

$$P_d = \frac{1}{2\pi}\int_{-\pi}^{\pi} P_{ds}(\theta)\,d\theta$$

For a <u>periodic</u> <u>signal</u> v(n), with V_k the coefficients of a (discrete) Fourier Series expansion and N_o <u>odd</u>,

$$P_{ds}(\theta) = 2\pi \sum_{k=-K}^{K} |V_k|^2 \frac{u_o(\theta - k\lambda_o)}{Sa^2(\frac{\theta - k\lambda_o}{2})}$$

$$= 2\pi \sum_{k=-K}^{K} |V_k|^2 \, u_o(\theta - k\lambda_o)$$

where $Sa(x) = \frac{\sin x}{x}$ and $K = \frac{N_o - 1}{2}$. Then

$$P_d = \frac{1}{N_o}\sum_{k=0}^{N_o-1} v^2(k) = \sum_{k=-K}^{K} |V_k|^2 .$$

A similar development for N_o <u>even</u> leads to

$$P_d = \sum_{k=-K'}^{K'} |V_k'|^2 + \frac{1}{4}(b_{N_o/2})^2 = \sum_{k=-K'}^{k=K'+1} |V_k'|^2$$

where $K' = (N_o-2)/2$ as in Eq. 2.47. This can be used to compute P_d for Example 2.10h.

In summarizing the results on frequency domain spectra, we divide the signals into five categories:

a) Those that satisfy the sufficiency conditions for the $\mathfrak{F}$ and $\mathfrak{F}_d$ transforms. The corresponding Laplace or Z-Transforms have no poles on the imaginary axis of the s-plane or the unit circle of the z-plane. This generally is equivalent to the class of energy signals or signals that are square integrable or square summable. See Examples 2.9b and 2.9d.

b) Those that do not satisfy the sufficiency conditions, but do have $\mathfrak{F}$ or $\mathfrak{F}_d$ transforms as well Laplace and Z-Transforms. However, the corresponding transforms differ in form -- usually in the occurrence of impulses (or higher order derivatives) in the Fourier forms. See Example 2.9a.

c) Those that have Laplace or Z-Transforms, but do not have $\mathfrak{F}$ or $\mathfrak{F}_d$ Transforms. Consider Example 2.9b with $a_1, a_2 < 0$.

d) Those that have $\mathfrak{F}$ and $\mathfrak{F}_d$ Transforms, but do not have Laplace or Z-Transforms. Consider periodic functions (two-sided); regions of convergence do not exist. This generally is the class of power signals.

e) Those that do not have $\mathfrak{F}$ and Laplace Transforms or $\mathfrak{F}_d$ and Z-Transforms. Consider the function $\varepsilon^{t^2}u_1(t)$.

Obviously those in a) are "well-behaved" and many useful frequency domain results are available. Unfortunately, many signals of wide interest in engineering applications are in b) and d), and care must be taken in the meaning of the spectrum, e.g., in characterizing signal energy. In b) and d), the impulses in the Fourier spectra must be included. These occur when specific frequency components are present, e.g., a term $\sin \omega_o t$. In c), where the <u>Fourier Transforms do not exist</u> and E_c and E_d are not finite, it sometimes is useful to use an integral of $|U|$ over $\omega\varepsilon[\omega_1,\omega_2]$ or of $|V|$ over $\theta\varepsilon[\theta_1,\theta_2]$ as suggested in Section 2.5.3. However, we find $|U|$ and $|V|$ in this situation by substituting $j\omega$ for s in $U(s) = \mathcal{L}\, u(t)$ or $\varepsilon^{j\theta}$ for z in $V(z) = \mathfrak{z}\, v(n)$. Power spectra are often useful for signals in d).

2.6 AUXILIARY FUNCTIONS

The previous sections of this chapter have discussed several ways of representing u(t) or v(n): as functions of t or n using linear combinations of elementary functions, as functions of t or n using real spectral functions and orthonormal sets of functions, as functions of t or n by resolution in terms of an elementary function, and as functions of s or z (or $j\omega$ and γ) and the pole/zero patterns and spectral functions associated with these. Another possibility is to define an auxiliary function of a time shift variable. This is useful as a measure of how strongly signal fluctuations that are separated in time are related. Detection of signals is frequently made simpler by designs that produce narrow auxiliary functions. For deterministic signals of finite energy as defined in Section 2.1, the auxiliary function is called a translation function [10.] or time ambiguity function [6.]; for the stochastic signals discussed in the next section, they are correlation functions. Translation functions are also useful in the design of certain Wiener filters [10.].

2.6.1 Translation Functions

The function

$$w_u(\tau) = \int_{-\infty}^{\infty} u(t)u(t+\tau)dt, \qquad -\infty < \tau < \infty \tag{2.69}$$

is the autotranslation function of u(t). Similarly,

$$w_v(i) = \sum_{n=-\infty}^{\infty} v(n)v(n+i), \qquad -\infty < i < \infty \tag{2.70}$$

is the autotranslation function of v(n). By taking the Two-Sided Laplace Transform of $w_u(\tau)$, a corresponding frequency representation is obtained:

$$\mathcal{L}_2 w_u(\tau) = W_u(s)$$

or

$$W_u(s) = \int_{-\infty}^{\infty} \varepsilon^{-s\tau} \left[\int_{-\infty}^{\infty} u(t)u(t+\tau)dt \right] d\tau \ .$$

Interchanging the order of integration,

$$W_u(s) = \int_{-\infty}^{\infty} dt \ \varepsilon^{st} u(t) \int_{-\infty}^{\infty} d\tau \ \varepsilon^{-s(t+\tau)} u(t+\tau)$$

or

$$W_u(s) = U(-s)U(s) \tag{2.71}$$

with a region of convergence $-c < \sigma < c$, $c > 0$. Similarly, taking the Two-Sided Z-Transform of $w_v(i)$,

$$\mathfrak{z}_2 w_v(i) = W_v(z)$$

or

$$W_v(z) = \sum_{n=-\infty}^{\infty} z^{-i} \sum_{n=-\infty}^{\infty} v(n)v(n+i)$$
$$= \sum_{i=-\infty}^{\infty} z^{n} v(n) \left[\sum_{i=-\infty}^{\infty} z^{-(n+i)} v(n+i) \right]$$

or

$$W_v(z) = V(z^{-1})V(z) \tag{2.72}$$

with a region of convergence $r < \rho < r^{-1}$, $0 < r < 1$. From the definitions of $w_u(\tau)$ and $w_v(i)$, it is clear that

$$w_u(\tau) = w_u(-\tau)$$

and

$$w_v(i) = w_v(-i) \ ,$$

that is, the autotranslation functions are even. Also for $\tau=0$ and $i=0$, the integral square and summed square values of u(t) and v(i) are found. These results can be obtained in the frequency domain as

$$w_u(0) = \mathcal{L}_2^{-1} W_u(s)\big|_{\tau=0} = \mathcal{L}_2^{-1} U(-s)U(s)\big|_{\tau=0}$$

$$= \frac{1}{2\pi j}\int_{-j\infty}^{j\infty} U(-s)U(s)ds$$

which is the Integral Square Theorem of Appendix A. Similarly,

$$w_v(0) = \mathcal{Z}_2^{-1} W_v(z)\big|_{i=0} = \mathcal{Z}_2^{-1} V(z^{-1})V(z)\big|_{i=0}$$

$$= \frac{1}{2\pi j}\oint_\Gamma V(z^{-1})V(z)z^{-1}dz$$

which is the Sum Square Theorem of Appendix A. Referring to the earlier definitions of signal energy and energy spectra, it is clear that $w_u(0)$ and $w_v(0)$ are signal energies, and that $W_u(j\omega)$ and $W_v(\gamma)$ are energy spectra. Similar forms can be defined for two signals $u_p(t)$ and $u_q(t)$ (or $v_p(n)$ and $v_q(n)$) to produce crosstranslation functions [10.],

$$w_{u_p u_q}(\tau) \text{ and } w_{v_p v_q}(i) \ .$$

The definitions also can be extended for vector signals $\underline{u}(t)$ or $\underline{v}(n)$. Thus the autotranslation matrices are

$$W_{\underline{u}}(\tau) = \int_{-\infty}^{\infty} \underline{u}(t)\underline{u}^T(t+\tau)dt \qquad (2.73)$$

and

$$W_{\underline{v}}(i) = \sum \underline{v}(n)\underline{v}^T(n+i) \ , \tag{2.74}$$

and the corresponding frequency domain matrices are

$$W_{\underline{u}}(s) = \mathcal{L}_2 W_{\underline{u}}(\tau) = \underline{U}(-s)\underline{U}^T(s) \tag{2.75}$$

and

$$W_{\underline{v}}(z) = \mathfrak{z}_2 W_{\underline{v}}(i) = \underline{V}(z^{-1})\underline{V}^T(z) \ . \tag{2.76}$$

The elements of $W_{\underline{u}}(\tau)$ and $W_{\underline{v}}(i)$ are autotranslation functions on the main diagonals and crosstranslation functions elsewhere. These forms will be useful in Chapter 5 where we derive expressions for system accuracy or fidelity.

Example 2.11

For the signal $u(t) = \varepsilon^{-at}u_1(t)$, $a > 0$, find $w_u(\tau)$, $W_u(s)$, and the signal energy,

$$w_u(\tau) = \int_{-\infty}^{\infty} \varepsilon^{-at}u_1(t)\exp(-a(t+\tau))u_1(t+\tau)dt \ .$$

For $\tau < 0$,

$$w_u(\tau) = \int_{-\tau}^{\infty} \exp(-2at)\varepsilon^{-a\tau}dt = \frac{1}{2a}\varepsilon^{a\tau}.$$

For $\tau \geq 0$,

$$w_u(\tau) = \int_{0}^{\infty} \exp(-2at)\varepsilon^{-a\tau}dt = \frac{1}{2a}\varepsilon^{a\tau}.$$

Therefore,

$$w_u(\tau) = \frac{1}{2a}\exp(-a|\tau|) \ .$$

Also, since $w_u(\tau)$ is even (see Appendix A),

$$W_u(s) = \frac{1}{2a}\left[\frac{1}{s+a} + \frac{1}{-s+a}\right] = \frac{1}{a^2-s^2} \ .$$

As a check,

$$\mathcal{L}_2 u(t) = \frac{1}{s+a} = U(s)$$

and

$$W_u(s) = U(-s)U(s) = \frac{1}{-s+a} \cdot \frac{1}{s+a} = \frac{1}{a^2-s^2} .$$

Also,

$$w_u(0) = \int_{-\infty}^{\infty} u^2(t)dt = 1/2a = E_c .$$

As a check, using a contour integral,

$$w_u(0) = \frac{1}{2\pi j} \int_{-j\infty}^{j\infty} \frac{-ds}{(s-a)(s+a)} = 1/2a$$

which also can be checked by direct integration of $u^2(t)$.

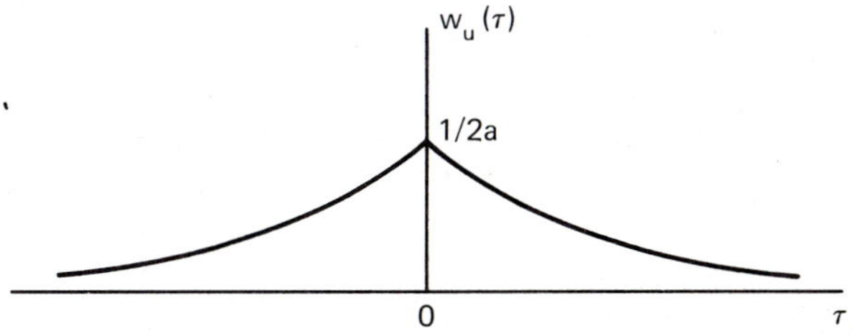

2.7 STOCHASTIC SIGNALS

The signals considered up to this point are deterministic, that is for every t or n there is one corresponding value of the signal u(t) or v(n). By contrast, the signal models of this section are concerned with signals whose values at any t or n are known only in a probabilistic sense. Signals described in this way are stochastic processes, when defined on $t \varepsilon T_{c2}$ or stochastic sequences, when defined on $n \varepsilon T_{d2}$. Thus, in place of u(t) and v(n), the random variables $U(t;\sigma)$ and $V(n;\sigma)$ are the starting point. These random variables are functions of a time variable as well as of points in a sample space, and for any fixed t_1 or n_1, the random variables $U(t_1;\sigma)$ and $V(n_1;\sigma)$ have the properties discussed in Appendix C. The dependence of these random variables on σ is not usually noted explicitly, although the concepts of sample spaces and probabilities associated with events for subsets of the sample spaces still underlie the development of stochastic signal models. Sets of the random variables $U(t)$ or $V(n)$ are denoted $\{U(t)\}$ or $\{V(n)\}$, a standard notation for stochastic processes and stochastic sequences, respectively. The sets also include the necessary probability density functions described in the following paragraphs.

In addition to the behavior of the random variables $U(t_1)$ and $V(n_1)$ for fixed t_1 or n_1 where σ varies over a sample space $\mathcal{S}$, it also is necessary to consider the case of a fixed point σ_i in a sample space with t or n varying over an interval. In general, $t \varepsilon T_{c2}$ and $n \varepsilon T_{d2}$, and for a given σ_i, the corresponding ith sample functions (or realizations) of $\{U(t)\}$ or $\{V(n)\}$ are $u^i(t)$ or $v^i(n)$ as was shown in Figure 2.6. The concept of stochastic processes and sequences, therefore, brings together previous ideas of random variables (fixed t or n) and signal models that are functions of t or n (fixed σ). The notation $U(t_1) = U_1$ and $V(n_1) = V_1$ further simplifies the following discussions; it is important to keep in mind that the first numbered subscript refers to fixed times.

2.7.1 Density and Distribution Functions

The distribution function for U_1 is $P_{U_1}(u_1;t_1)$ and

$$\Pr(U_1 \leq u_1) = P_{U_1}(u_1;t_1) .$$

The corresponding density function is related by

$$P_{U_1}(u_1;t_1) = \int_{-\infty}^{u_1} p_{U_1}(\lambda;t_1)\,d\lambda$$

where it is assumed that U_1 is a continuous random variable.

If U_1 is a discrete random variable,

$$Pr(U_1 \leq u_{1k}) = P_{U_1}(u_{1k};t_1)$$

and

$$P_{U_1}(u_{1k};t_1) = \sum_{u_{1j} \leq u_{1k}} P_{U_1}(u_{1j},t_1) \ .$$

The corresponding notation for V_1 follows similarly. Thus

$$Pr(V_1 \leq v_1) = P_{V_1}(v_1;n_1)$$

with

$$P_{V_1}(v_1;n_1) = \int_{-\infty}^{v_1} P_{V_1}(\lambda;n_1)d\lambda$$

where V_1 is a continuous random variable and

$$Pr(V_1 \leq v_{1k}) = P_{V_1}(v_{1k};n_1)$$

with

$$P_{V_1}(v_{1k};n_1) = \sum_{v_{1j} \leq v_{1k}} p_{V_1}(v_{1j};n_1)$$

where V_1 is a discrete random variable. We see that there are four cases of interest according to whether the time variable is continuous or discrete and whether the random variable is continuous or discrete. The four distribution and density functions are sometimes called <u>first-order functions</u> since they are associated with one particular time. However, these functions may change as different times are chosen. The time variable is written explicitly if this is the case, otherwise it frequently is suppressed.

The next step is to consider two time instants t_1, t_2 or n_1, n_2. The random variables are then U_1, U_2, and V_1, V_2, and these pairs have joint density functions. Thus for the four possibilities above there are*

---

*Note that U_1, U_2, v_1, v_2 etc. are <u>not</u> singularity functions in this context.

$$p_{U_1U_2}(u_1, u_2; t_1, t_2) ,$$

$$p_{U_1U_2}(u_{1j}, u_{1k}; t_1, t_2),$$

$$p_{V_1V_2}(v_1, v_2; n_1\ n_2),$$

and

$$p_{V_1V_2}(v_{1j}, v_{2k}, n_1, n_2) .$$

These are called <u>second density functions</u>, and using the corresponding distribution functions, the probabilities of joint events can be stated. Thus

$$\Pr(U_1 \leq u_1, U_2 \leq u_2) = P_{U_1U_2}(u_1, u_2; t_1, t_2)$$

and similar forms follow for the other three cases. This procedure is then continued for t_1, t_2, t_3 or n_1, n_2, n_3 for third density functions, and in fact for any finite integer m, the time values $t_1, t_2, \ldots, t_m$ or $n_1, n_2, \ldots, n_m$ are associated with m^{th} density functions. As m becomes large, an increasingly complete probabilistic description of the process $\{U(t)\}$ or the sequence $\{V(n)\}$ is developed. These families of joint density functions are the required probabilistic characterization of $\{U(t)\}$ or $\{V(n)\}$. In almost all applications, however, only the first and second density functions are used to develop useful statistical parameters. Also, in most experimental work, estimation of density functions for $m > 2$ is a formidable task. It is necessary that these joint density functions for finite m satisfy a <u>consistency condition</u> such that for $q < m$, it is possible to calculate the qth density function from the mth density function. Finally, there are some important mathematical questions when m becomes infinite. For the stochastic sequence, a countably infinite set of random variables with corresponding infinite dimensional density functions results as a logical extension of the case for finite m. For a stochastic process, however, the extension to continuous time leads to difficulties with the corresponding density functions [11.]. This question will not be discussed here since only distribution and density functions for finite m will be used.

Example 2.12

A stochastic sequence $\{V(n)\}$ has, for each n

$$V(n) = \pm 1 \text{ with } \begin{cases} Pr(V=1) = 1/2 \\ Pr(V=-1) = 1/2 \end{cases}$$

and $V(n_1)=V_1$ is independent of $V(n_2)=V_2$, $n_1 \neq n_2$. Find $p_{V_1}(v_{1j};n_1)$ and $p_{V_1V_2}(v_{1j},v_{2k};n_1,n_2)$. A sample sequence $v^i(n)$ is as shown, and $\{V(n)\}$ is called a Bernoulli sequence.

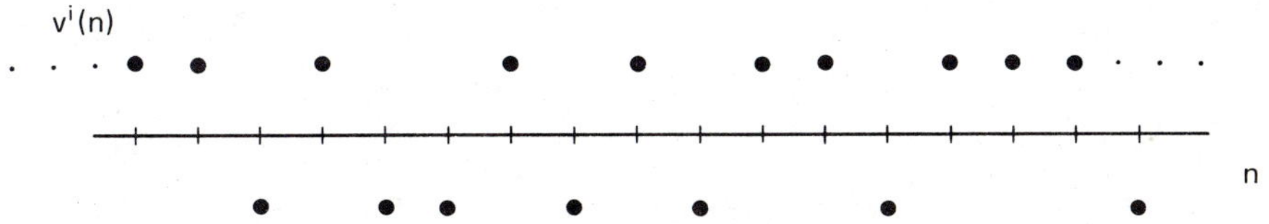

For any n_1,

$$p_{V_1}(v_{1j};n_1) = \tfrac{1}{2}v_o(v_{1j}+1) + \tfrac{1}{2}v_o(v_{1j}-1).$$

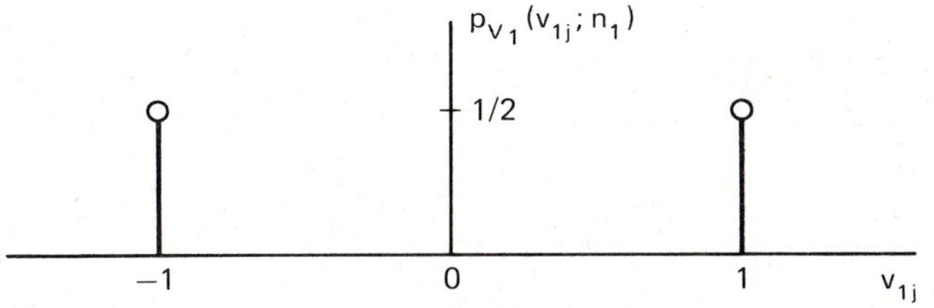

For any n_1, n_2 with $n_1 \neq n_2$, $V(n_1)$ is independent of $V(n_2)$, thus

$$p_{V_1V_2}(v_{1j},v_{2k};n_1,n_2) = p_{V_1}(v_{1j};n_1) \cdot p_{V_2}(v_{2k};n_2)$$

and

$$p_{V_1V_2}(v_{1j},v_{2k};n_1,n_2) = [\tfrac{1}{2}v_o(v_{1j}+1) + \tfrac{1}{2}v_o(v_{1j}-1)]\cdot$$
$$\cdot[\tfrac{1}{2}v_o(v_{2k}+1) + \tfrac{1}{2}v_o(v_{2k}-1)]$$
$$= \tfrac{1}{4}v_o(v_{1j}+1)v_o(v_{2k}+1)$$
$$+ \tfrac{1}{4}v_o(v_{1j}+1)v_o(v_{2k}-1)$$
$$+ \tfrac{1}{4}v_o(v_{1j}-1)v_o(v_{2k}+1)$$
$$+ \tfrac{1}{4}v_o(v_{1j}-1)v_o(v_{2k}-1)$$

where $v_o(\cdot)$ is the unit pulse function of Section 2.2.1.

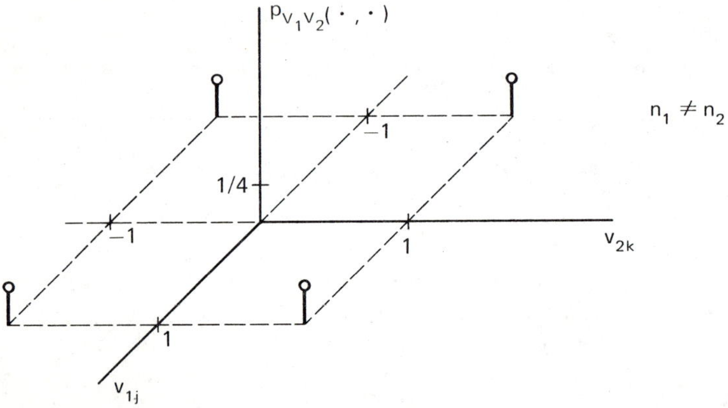

Note the consistency of $p_{V_1V_2}(\ )$ and $p_{V_1}(\ )$ or $p_{V_2}(\ )$; thus for example,

$$\sum_{v_{2k}} p_{V_1V_2}(\ ) = p_{V_1}(\ ).$$

For $n_1 = n_2$, there are two points in the sample space: $v_{1j}=v_{2k}=1$ and $v_{1j}=v_{2k}=-1$. Therefore,

$$p_{V_1V_2}(v_{1j},v_{1j};n_1,n_1) = \tfrac{1}{2}v_o(v_{1j}+1) + \tfrac{1}{2}v_o(v_{1j}-1).$$

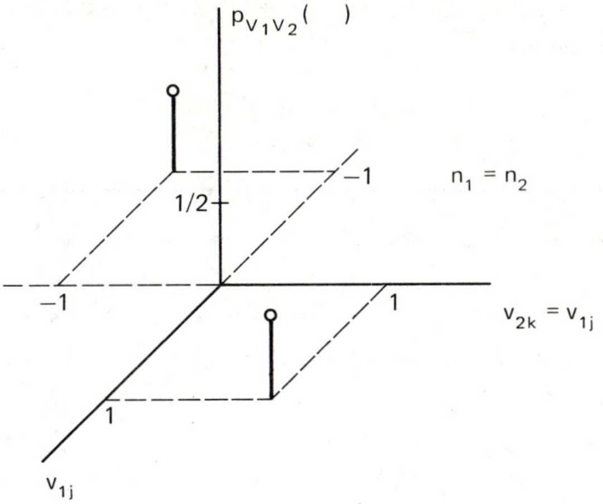

Example 2.13

A stochastic sequence {V(n)} has

$$V(n) = 1 \text{ or } V(n) = -1$$

with transition probabilities (where $Pr(V=\alpha|V=\beta)=Pr(\alpha|\beta)$)

$$Pr(1|1) = Pr(-1|-1) = 3/4$$

$$Pr(1|-1) = Pr(-1|1) = 1/4$$

This is an example of a Markov chain, and a transition diagram is:

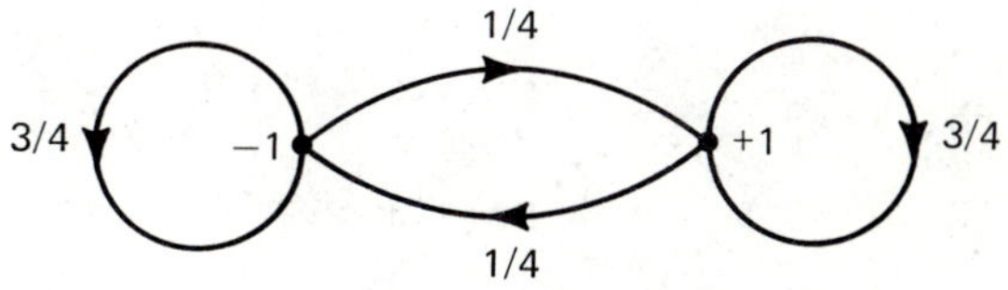

Since $Pr(1) + Pr(-1) = 1$, and

$$Pr(1) = Pr(1|1)Pr(1) + Pr(1|-1)Pr(-1)$$

$$Pr(-1) = Pr(-1|-1)Pr(-1) + Pr(-1|1)Pr(1) \quad ,$$

it follows that Pr(1) = Pr(-1) = 1/2. A sample sequence $v^i(n)$ is as shown.

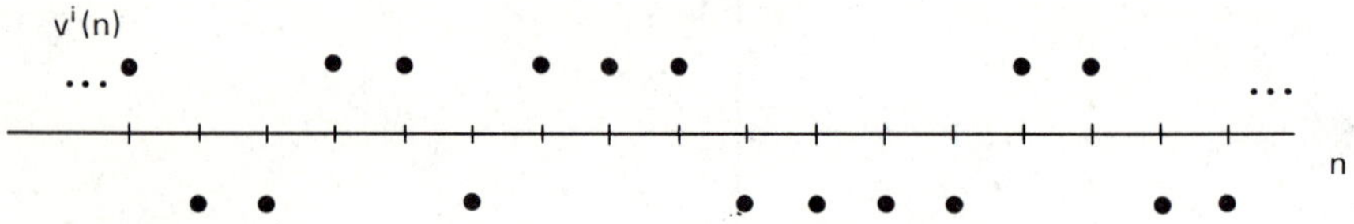

The runs of 1 or -1 tend to be longer in this example compared to the previous one since $V(n_1)$ and $V(n_2)$ are not independent here. For any n_1,

$$p_{V_1}(v_{1j};n_1) = \tfrac{1}{2}v_0(v_{1j}+1) + \tfrac{1}{2}v_0(v_{1j}-1)$$

as in Example 2.12.

To find the second density functions $p_{V_1V_2}(\;)$, we consider cases for $n_2=n_1$, $n_2=n_1+1$, $n_2=n_1+2$, ..., for any n_1. For $n_1=n_2$, the result is the same as in Example 2.12. For $n_2=n_1+k$, $k\geq 1$, we use the transition probability matrix

$$P = \begin{bmatrix} 3/4 & 1/4 \\ 1/4 & 3/4 \end{bmatrix}$$

where the elements p_{ij} are the <u>one-step</u> conditional probabilities that are given. Also we define

$$P_k = P^k .$$

Using this, for k=1,

$$Pr(V_1=1,V_2=1) = Pr(1|1)Pr(1) = 3/8$$

$$Pr(V_1=1,V_2=-1) = Pr(-1|1)Pr(1) = 1/8$$

$$Pr(V_1=-1,V_2=1) = Pr(1|-1)Pr(-1) = 1/8$$

$$Pr(V_1=-1,V_2=-1) = Pr(-1|-1)Pr(-1) = 3/8 .$$

These are simply the elements of P multiplied by Pr(1) = Pr(-1) = 1/2. From these for $n_2 = n_1+1$,

$$p_{V_1V_2}(v_{1j}, v_{2k}; n_1, n_2) = 3/8 v_o(v_{1j}+1)v_o(v_{2k}+1) + 3/8 v_o(v_{1j}-1)v_o(v_{2k}-1) + 1/8 v_o(v_{1j}+1)v_o(v_{2k}-1) + 1/8 v_o(v_{1j}-1)v_o(v_{2k}+1).$$

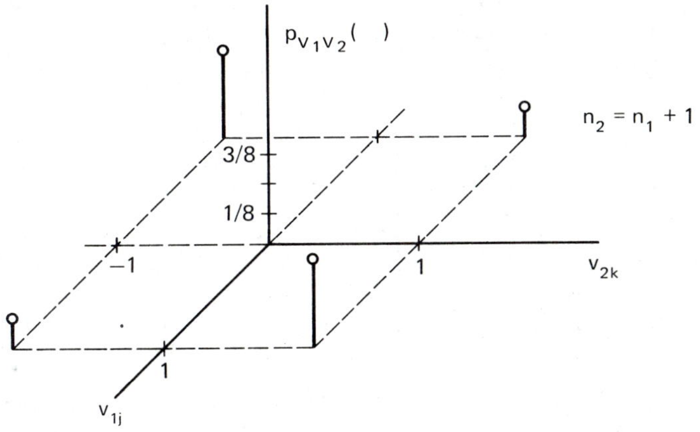

For k = 2, i.e., for <u>two steps</u>,

$$P_2 = P^2 = \begin{bmatrix} 5/8 & 3/8 \\ 3/8 & 5/8 \end{bmatrix},$$

and using Pr(1) = Pr(-1) = 1/2 as before,

$$Pr(1,1) = Pr(-1,-1) = 5/16$$

$$Pr(-1,1) = Pr(1,-1) = 3/16$$

and the form of $p_{V_1V_2}(v_{1j}, v_{2k}; n_1, n_2)$ for $n_2 = n_1+2$ is the same as the previous one with obvious changes in the numerical coefficients.

This procedure is repeated for k = 3, 4, 5, ... It is of interest to consider the limiting case for $k \to \infty$, thus we wish to find

$$\lim_{k\to\infty} P^k .$$

Using Equation B.17 of Appendix B,

$$P^k = \sum_{j=1}^{s} \lambda_j^k C_j$$

where the C_j are the constituent idempotents of P. The characteristic values λ_j or P are found from

$$\lambda I - P = \begin{bmatrix} \lambda - 3/4 & -1/4 \\ -1/4 & \lambda - 3/4 \end{bmatrix}$$

and $p(\lambda) = (\lambda - 3/4)^2 - 1/16 = \lambda^2 - 3/2\lambda + 1/2$ to give $\lambda_1, \lambda_2 = 1, 1/2$. Thus,

$$P^k = 1^k \begin{bmatrix} 1/2 & 1/2 \\ 1/2 & 1/2 \end{bmatrix} + (1/2)^k \begin{bmatrix} 1/2 & -1/2 \\ -1/2 & 1/2 \end{bmatrix}$$

$$= \begin{bmatrix} 1/2(1+1/2^k) & 1/2(1-1/2^k) \\ 1/2(1-1/2^k) & 1/2(1+1/2^k) \end{bmatrix}$$

and

$$\lim_{k\to\infty} P^k = \begin{bmatrix} 1/2 & 1/2 \\ 1/2 & 1/2 \end{bmatrix}$$

Using Pr(1) = Pr(-1) = 1/2, the four values (nonzero) of $p_{V_1V_2}(\ )$, for $n_2 = n_1+k$ with $k \to \infty$, are all 1/4. This agrees with Example 2.12 where independent $V(n_1)$ and $V(n_2)$ correspond to the case of $V(n_1)$ and $V(n_2)$ for $k \to \infty$ in this example, i.e., $V(n_1)$ and $V(n_2)$ tend toward independence as n_1 and n_2 become widely separated.

2.7.2 Stochastic Process and Sequences

The stochastic processes or sequences discussed up to this point have been described by a set of joint density functions. There are alternative ways of describing a process or sequence, and one that is particularly useful is to specify a function of t or n where the parameters of the function are random variables with known density functions. These are sometimes called deterministic processes or deterministic sequences. An example of the former is a process $\{U(t)\}$ with

$$U(t) = A \sin \omega_o t + B \cos \omega_o t, \quad \forall t$$

where A and B are random variables. An example of the latter is a sequence $\{V(n)\}$ with

$$V(n) = An + B, \quad n \geq 0$$

where again A and B are random variables. It is possible to derive the first and second density functions for U and V in such cases although it usually is difficult. Fortunately, the expected values that are needed can be found directly from the known characteristics of the random variables in these cases.

2.7.3 First and Second-order Statistics

Having established the meaning of a stochastic process or sequence, it is natural to ask what signal models of deterministic signals have analogous forms for these signals. The question is of interest (1) with regard to the time domain and frequency domain models, (2) with regard to the transforms of Table 2.4, and (3) with regard to the auxiliary functions called translation functions. It happens that there are several comparable forms for stochastic signal models. A convenient starting place is with the moments of the process or sequence and the auxiliary functions called correlation functions. These are used to establish certain properties of the stochastic signals which are then useful in establishing conditions on other model forms -- for example, Fourier forms.

The <u>mean value</u> of a stochastic process $\{U(t)\}$ is

$$m_U(t_1) = E[U(t_1)] = E[U_1]$$

and using $p_{U_1}(u_1)$,

$$m_U(t_1) = \int_{-\infty}^{\infty} u_1 p_{U_1}(u_1)\,du_1$$

or using $p_{U_1}(u_{1j})$,

$$m_U(t_1) = \sum_{u_{1j}} u_{1j} p_{U_1}(u_{1j}) \ .$$

Similarly, for a stochastic sequence $\{V(n)\}$,

$$m_V(n_1) = E[V(n_1)] = E[V_1]$$

and using $p_{V_1}(v_1)$,

$$m_V(n_1) = \int_{-\infty}^{\infty} v_1 p_{V_1}(v_1)\,dv_1$$

or using $p_{V_1}(v_{1j})$,

$$m_V(n_1) = \sum_{v_{1j}} v_{1j} p_{V_1}(v_{1j}) \ .$$

Similarly, the <u>mean square values</u> of $\{U(t)\}$ and $\{V(n)\}$ are

$$s_U^2(t_1) = E[U^2(t_1)] = E[U_1^2]$$

and

$$s_V^2(n_1) = E[V^2(n_1)] = E[V_1^2],$$

and these are evaluated using the same four density functions used for m_U and m_V. Finally, the variances of $\{U(t)\}$ and $\{V(n)\}$ are

$$\sigma_U^2(t_1) = E[(U(t_1)-m_U(t_1))^2]$$

and

$$\sigma_V^2(n_1) = E[(V(n_1)-m_V(n_1))^2]$$

which also are evaluated using the first density functions. The mean, mean square, and variance are called first-order statistics of $\{U(t)\}$ or $\{V(n)\}$.

The second density functions of $\{U(t)\}$ or $\{V(n)\}$ are used to define a measure of the nature of the fluctuations from one time to another. These are called second-order statistics. The autocorrelation function of the process $\{U(t)\}$ is

$$r_U(t_1,t_2) = E[U(t_1)U(t_2)] = E[U_1U_2] \tag{2.77}$$

$$= \int_{-\infty}^{\infty}\int_{-\infty}^{\infty} u_1u_2p_{U_1U_2}(u_1,u_2;t_1,t_2)du_1du_2$$

when U_1 and U_2 are continuous random variables, or

$$r_U(t_1,t_2) = \sum_{u_{2k}}\sum_{u_{1j}} u_{1j}u_{2k}p_{U_1U_2}(u_{1j},u_{2k};t_1t_2)$$

when U_1 and U_2 are discrete random variables. Similar forms for a sequence $\{V(n)\}$ follow from the definition

$$r_V(n_1,n_2) = E[V(n_1)V(n_2)] = E[V_1V_2]. \tag{2.78}$$

Closely related to the autocorrelation functions are the covariance functions

$$c_U(t_1,t_2) = E[(U(t_1)-m_U(t_1))(U(t_2)-m_U(t_2))]$$

and

$$c_V(n_1,n_2) = E[(V(n_1)-m_V(n_1))(V(n_2)-m_V(n_2))].$$

These are simply normalized forms of $r_U(t_1,t_2)$ and $r_V(n_1,n_2)$ and sometimes are preferred when the mean values are not zero so that certain time averages will remain finite. Also, the autocorrelation and covariance functions are related by

$$c_U(t_1,t_2) = r_U(t_1,t_2) - m_U(t_1)m_U(t_2)$$

and

$$c_V(n_1,n_2) = r_V(n_1,n_2) - m_V(n_1)m_V(n_2).$$

Example 2.14

Consider the stochastic sequence $\{V(n)\}$ of Example 2.12 where $p_{V_1}(v_{1j};n_1)$ and $p_{V_1V_2}(v_{1j},v_{2k};n_1n_2)$ were found. These are now used to find m_V, S_V^2, σ_V^2, and $r_V(n_1,n_2)$.

$$m_V(n_1) = E[V_1] = \sum_{v_{1j}} v_{1j}p_{V_1}(v_{1j};n_1)$$
$$= (-1)(1/2) + (1)(1/2) = 0$$

$$S_V^2(n_1) = E[V_1^2] = \sum_{v_{1j}} v_{1j}^2 p_{V_1}(v_{1j};n_1)$$
$$= (-1)^2(1/2) + (1)^2(1/2) = 1$$

and since $m_V(n_1) = 0$, $\sigma_V^2(n_1) = S_V^2(n_1) = 1$, and these are true for all n_1.

$$r_V(n_1,n_2) = E[V_1V_2]$$
$$= \sum_{v_{2k}}\sum_{v_{1j}} v_{1j}v_{2k}p_{V_1V_2}(v_{1j},v_{2k};n_1n_2).$$

For $n_1 \neq n_2$,

$$r_V(n_1,n_2) = (1)(1)(1/4)+(1)(-1)(1/4)+(-1)(1)(1/4) + (-1)(-1)(1/4)$$
$$= 0$$

and for $n_1 = n_2$

$$r_V(n_1,n_2) = (1)(1)(1/2)+(-1)(-1)(1/2) = 1$$

Thus,

$$r_V(n_1,n_2) = v_o(n_2-n_1).$$

Example 2.15

Consider the stochastic sequence $\{V(n)\}$ of Example 2.13. Since the first density function is the same as in the previous example,

$$m_V(n_1) = 0, \; S_V^2(n_1) = \sigma_V^2(n_1) = 1, \; \forall n_1.$$

For $n_1 = n_2$,

$$r_V(n_1,n_2) = 1$$

as in the previous example. However, for $n_2 = n_1 + 1$,

$$\begin{aligned} r_V(n_1,n_2) &= (1)(1)(3/8)+(-1)(1)(1/8)+(1)(-1)(1/8) \\ &\qquad +(-1)(-1)(3/8) \\ &= 1/2 \; . \end{aligned}$$

Similarly, for $n_2 = n_1 + 2$

$$\begin{aligned} r_V(n_1,n_2) &= (1)(1)(5/16)+(-1)(1)(3/16)+(1)(-1)(3/16) \\ &\qquad +(-1)(-1)(5/16) \\ &= 1/4 \end{aligned}$$

and for $n_2 = n_1 + k$,

$$r_V(n_1,n_2) = \frac{1}{2^k}, \qquad k \geq 0$$

which follows from p^k found in Example 2.13. From the symmetry of this example in n_1 and n_2,

$$r_V(n_1,n_2) = \frac{1}{2^{-k}}, \quad k \leq 0.$$

<u>Example 2.16</u>

A stochastic process $\{U(t)\}$ is given by

$$U(t) = A \sin t + B \cos t$$

with A and B independent random variables. $m_A = m_B = 0$, and $\sigma_A^2 = \sigma_B^2 = \sigma^2$.

$$m_U(t_1) = E[U(t_1)]$$

$$= E[A \sin t_1 + B \cos t_1]$$

and using the linearity of the expectation operation,

$$m_U(t_1) = E[A] \sin t_1 + E[B] \cos t_1 = 0, \quad \forall\, t_1 .$$

Also,

$$r_U(t_1,t_2) = E[U(t_1)U(t_2)]$$

$$= E[(A \sin t_1 + B \cos t_1)\cdot$$
$$\cdot (A \sin t_2 + B \cos t_2)]$$

$$= E[A^2] \sin t_1 \sin t_2$$
$$+ E[AB] \sin t_1 \cos t_2$$
$$+ E[AB] \sin t_2 \cos t_1$$
$$+ E[B^2] \cos t_1 \cos t_2$$

and since $E[AB] = E[A]\,E[B]$, for A and B independent, and $m_A = m_B = 0$,

$$r_U(t_1,t_2) = \sigma^2(\cos t_1 \cos t_2 + \sin t_1 \sin t_2)$$
$$= \sigma^2 \cos (t_2 - t_1) \ .$$

This is an example of a deterministic process where the first and second moments of the random variables A and B are used directly for finding the required expected value.

2.7.4 Stationarity

An important class of stochastic processes and sequences is the class of stationary processes or sequences. Processes or sequences are <u>strictly stationary</u> if their first, second, third, ..., m-th, ..., distribution and density functions remain unchanged for a translation in time. Another way of expressing this is that

$p_{U_1}(.;.)$ is not a function of t_1

$p_{U_1U_2}(..;..)$ is only a function of $\tau = t_2-t_1$

$p_{U_1U_2U_3}(...;...)$ is only a function of $\tau_1 = t_2-t_1$ and $\tau_2 = t_3-t_2$

etc., with similar statements for the distribution functions and for stochastic sequences. An example of a strictly stationary sequence $\{V(n)\}$ is given in Example 2.12 where the independence of $V(n_1)$ and $V(n_2)$, $n_1 \neq n_2$, can be used. To prove strict stationarity, it generally is necessary to examine the density functions of all orders, and as stated previously, these usually are not known. Except in special cases, their measurement also is out of the question.

A much more useful definition is that of wide-sense stationarity. A stochastic process $\{U(t)\}$ is <u>wide-sense stationary</u> if

1. m_U is constant
2. r_U is a function of $\tau = t_2-t_1$ only,

and similarly, for $\{V(n)\}$, if

1. m_V is constant
2. r_V is a function of $k = n_2 - n_1$ only.

Thus, proof of wide-sense stationarity depends only on the first and second density functions.

Useful properties of the autocorrelation functions of real wide-sense stationary processes or sequences are:

1. $r_U(\tau) = r_U(-\tau)$, even function (2.79a)
2. $r_U(o) = E[U^2] = S_U^2 \geq |r_U(\tau)|, \quad \forall \tau$, i.e.,

   $r_U(\tau)$ has a maximum at $\tau = 0$.

and using the convention $t_2 \geq t_1$, we have for 2.77,

3. $r_U(\tau) = E[U(t)U(t+\tau)]$,

by letting $t_1 = t$ and $t_2 = t_1 + \tau = t + \tau$. Similarly

1. $r_V(k) = r_V(-k)$, even function (2.79b)
2. $r_V(o) = E[V^2] = S_V^2 \geq |r_V(k)|, \quad \forall k$

and using the convention $n_2 \geq n_1$, we have for 2.78,

3. $r_V(k) = E[V(n)V(n+k)]$

by letting $n_1 = n$ and $n_2 = n_1 + k = n + k$.

<u>Example 2.17</u>

a) From Example 2.14

$$m_V(n_1) = 0, \qquad \forall n_1$$

and

$$r_V(n_1, n_2) = v_o(n_2 - n_1)$$

or

$$r_V(k) = v_o(k), \qquad k = n_2 - n_1.$$

Therefore, the sequence {V(n)} is wide-sense stationary.

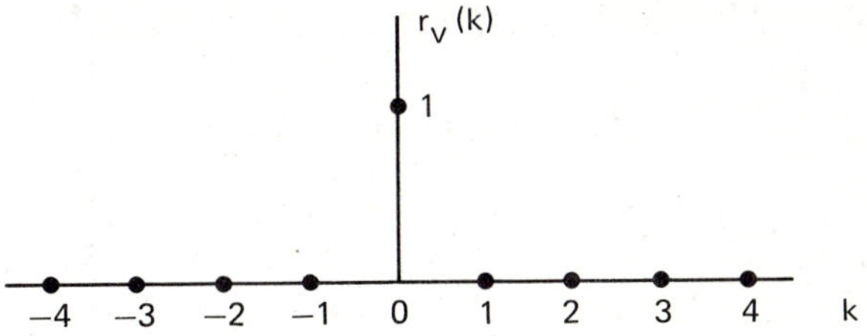

b) From Example 2.15

$$m_V(n_1) = 0, \quad \forall n_1$$

$$r_V(n_1,n_2) = \frac{1}{2^k}, \qquad k \geq 0$$

$$= \frac{1}{2^{-k}}, \qquad k \leq 0$$

for

$$k = n_2 - n_1,$$

or

$$r_V(k) = (\tfrac{1}{2})^{|k|}, \qquad \forall k .$$

Therefore, the sequence {V(n)} is wide-sense stationary.

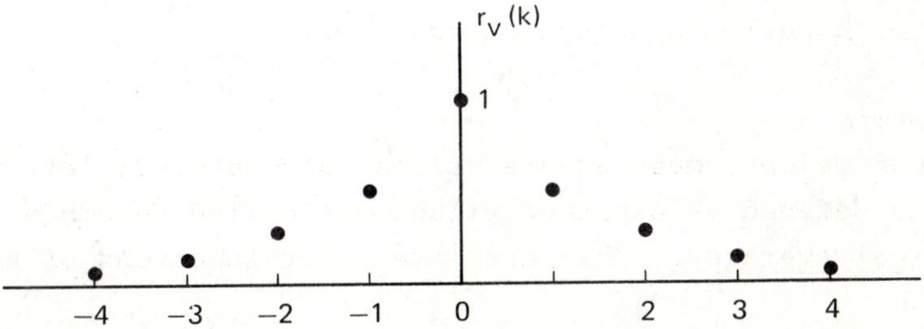

Comparison of the autocorrelation functions in a) and b) shows how the difference in the interdependence of the V(n) changes the character of the autocorrelation function. In a) the V(n) are independent; in b) the V(n) are related through the given transition probabilities.

c) From Example 2.16

$$m_U(t_1) = 0, \quad \forall t_1$$

and

$$r_U(t_1,t_2) = \sigma^2 \cos(t_2-t_1)$$

or

$$r_U(\tau) = \sigma^2 \cos \tau, \qquad \tau = t_2-t_1 .$$

Therefore, the process {U(t)} is wide-sense stationary.

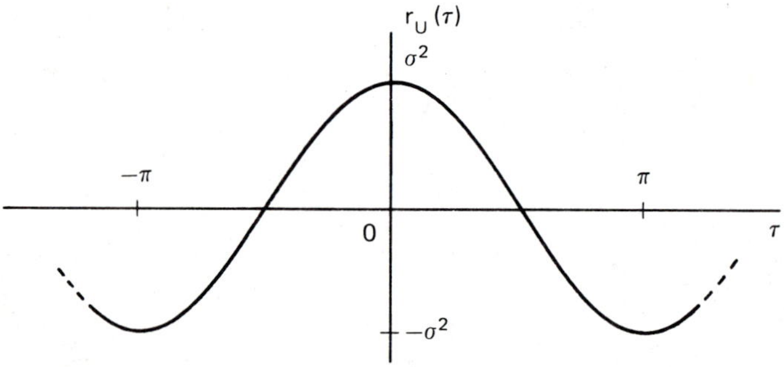

Note that the three autocorrelation functions are even and satisfy the inequalities of Equations 2.79. Part c) is an example of a periodic stochastic process.

2.7.5 Ergodicity

The mean values, mean square values, and autocorrelation functions defined as expected values are called ensemble or <u>statistical averages</u>. They are taken over functions of the random variables U_i and V_i, i = 1, 2, for fixed times. By contrast, time averages may be used over a sample function $u^i(t)$ or a sample sequence $v^i(n)$ for fixed σ. Thus, we let

$$<u^i(t)> = \lim_{T\to\infty} \frac{1}{2T} \int_{-T}^{T} u^i(\lambda)\,d\lambda$$

and

$$<u^i(t)u^i(t+\tau)> = \lim_{T\to\infty} \frac{1}{2T} \int_{-T}^{T} u^i(\lambda)u^i(\lambda+\tau)\,d\lambda \qquad (2.80)$$

be first and second order time averages of $u^i(t)$ from the process $\{U(t)\}$. Similarly,

$$<v^i(n)> = \lim_{N\to\infty} \frac{1}{2N+1} \sum_{j=-N}^{N} v^i(j)$$

and

$$<v^i(n)v^i(n+k)> = \lim_{N\to\infty} \frac{1}{2N+1} \sum v^i(j)v^i(j+k) \qquad (2.81)$$

are time averages of $v^i(n)$. A question of great importance in the analysis of random processes and sequences (particularly in experimental analysis) concerns the equality of statistical and time averages. In fact, the property of ergodicity is defined as follows for both $\{U(t)\}$ and $\{V(n)\}$: If

1. $\{U(t)\}$ is wide-sense stationary

2. $E[U(t_1)] = m_U = <u^i(t)>$

3. $E[U(t_1)U(t_2)] = r_U(\tau) = <u^i(t)u^i(t+\tau)>$,
   $\tau = t_2 - t_1$

then $\{U(t)\}$ is an ergodic process, and the converse is true also. If

1. $\{V(n)\}$ is wide-sense stationary

2. $E[V(n_1)] = m_V = \langle v^i(n) \rangle$

3. $E[V(n_1)V(n_2)] = r_V(k) = \langle v^i(n)v^i(n+k) \rangle,$

$k = n_2 - n_1$

then $\{V(n)\}$ is an ergodic sequence, and the converse is true also.

The equalities must hold for all t_1, t_2, τ or n_1, n_2, k and for all i. Furthermore, the right hand equailities in 2. and 3. are understood to be in the sense of "with probability equal 1".

In many experimental studies, only one or a few sample functions are available for analysis. In these cases, if the ergodicity property can be assumed (which implies wide-sense stationarity), time averages of single, very long records can be used to estimate the means and autocorrelation functions as shown in Chapter 6, Section 6.2.

Example 2.18

Consider the wide-sense stationary process of Example 2.16, where

$$U(t) = A \sin t + B \cos t.$$

A sample function is

$$u^i(t) = a \sin t + b \cos t$$

where a and b are realizations of the random variables A and B. Then

$$\langle u^i(t) \rangle = \lim_{T\to\infty} \frac{1}{2T} \int_{-T}^{T} (a \sin \lambda + b \cos \lambda) d\lambda = 0$$

which is the same as $m_U = 0$. However,

$$\langle u^i(t)u^i(t+\tau) \rangle = \lim_{T\to\infty} \int_{-T}^{T} [a \sin \lambda + b \cos \lambda] \cdot$$

$$\cdot [a \sin (\lambda+\tau) + b \cos (\lambda+\tau] d\lambda$$

$$= \frac{a^2+b^2}{2} \cos \tau .$$

This is not equal to $E[U(t_1)U(t_2)] = r_U(\tau) = \sigma^2 \cos \tau$ found previously. Thus the process $\{U(t)\}$ is not ergodic. This is expected since $\{U(t)\}$ is a deterministic process and in general, a time average over a sample function for a particular a and b would not equal a statistical average for fixed t. There are cases, however, where a deterministic process is ergodic, e.g., a case of random phase only where $U(t) = a \sin(\omega t + \Theta)$ with Θ uniformly distributed on $[0,2\pi]$. See Problem 35 at the end of this chapter.

<u>Example 2.19</u>

Consider the wide-sense stationary sequence of Example 2.12 and Example 2.14. An argument based on the nature of the sample functions can be used to justify the ergodicity assumption in this case. Thus for

$$<v^i(n)> = \lim_{N\to\infty} \frac{1}{2N+1} \sum_{j=-N}^{N} v^i(j)$$

with values of +1 and -1 equally likely. As N becomes very large the sum should approach zero with probability equal one and we have

$$<v^i(n)> = 0 = m_V$$

found previously. Also, for

$$<v^i(n)v^i(n+k)> = \lim_{N\to\infty} \frac{1}{2N+1} \sum_{j=-N}^{N} v^i(j)v^i(j+k)$$

with $k \neq 0$, the sum should approach zero for large N since positive and negative products are equally likely. For k=0, however, all terms are positive and the average is 1. Thus

$$<v^i(n)v^i(n+k)> = v_o(k) = r_V(k)$$

found previously.

2.7.6 Spectral Density Functions

The autocorrelation functions $r_U(\tau)$ and $r_V(k)$ of the wide-sense stationary process $\{U(t)\}$ or sequence $\{V(n)\}$ are "time domain" auxiliary function stochastic signal models. Frequency domain characterization of stochastic signals can be defined by using transforms of $u^i(t)$ and $v^i(n)$ or transforms of $r_U(\tau)$ and $r_V(k)$. The usual approach is to define the <u>spectral density function</u>, or power spectral density for a <u>wide-sense stationary process</u>, as

$$\mathfrak{F}\, r_U(\tau) = R_U(j\omega) = \int_{-\infty}^{\infty} r_U(\tau)\exp(-j\omega\tau)\,d\tau\ , \qquad (2.82)$$

or

$$\mathcal{L}_2 r_U(\tau) = R_U(s).$$

Similarly,

$$\mathfrak{F}_d r_V(k) = R_V(\gamma) = \sum_{k=-\infty}^{\infty} r_V(k)\gamma^{-k},\ \gamma = \varepsilon^{j\theta} \qquad (2.83)$$

or

$$\mathfrak{z}_2 r_V(k) = R_V(z)\ .$$

(See note after Eq. 2.87 concerning the dimension of $R_V(z)$.) Useful properties of spectral density functions of real stochastic processes or sequences are:

1. $R_U(\omega) = R_U(-\omega),\ R_U(s) = R_U(-s)$ (2.84)

2. $R_V(\gamma) = R_V(\gamma^{-1}),\ R_V(z) = R_V(z^{-1})$

and

3. $R_U(\omega) \geq 0$

4. $R_V(\gamma) \geq 0$

and the pole/zero pattern of $R_U(s)$ has a symmetry about the imaginary axis of the s-plane while the pole/zero pattern of $R_V(z)$ has a "symmetry" about the unit circle of the z-plane. Both, of course, are symmetric about the real axis. Properties 3. and 4. mean that spectral density functions are nonnegative definite (see Appendix B). Also,

5. $$R_U(0) = \int_{-\infty}^{\infty} r_U(\tau)\,d\tau$$

6. $$R_V(1) = \sum_{k=-\infty}^{\infty} r_V(k)$$

and

7. $$r_U(0) = \frac{1}{2\pi}\int_{-\infty}^{\infty} R_U(j\omega)\,d\omega$$

8. $$r_V(0) = \frac{1}{2\pi}\int_{-\pi}^{\pi} R_V(\varepsilon^{j\theta})\,d\theta$$

Also, using Properties 1. and 2. of 2.84,

9. $$r_U(\tau) = \frac{1}{\pi}\int_{0}^{\infty} R_U(j\omega)\cos\omega\tau\,d\omega$$

10. $$R_U(j\omega) = 2\int_{0}^{\infty} r_U(\tau)\cos\omega\tau\,d\tau$$

and

11. $$r_V(k) = \frac{1}{\pi}\int_{0}^{\pi} R_V(\varepsilon^{j\theta})\cos k\theta\,d\theta$$

12. $$R_V(\varepsilon^{j\theta}) = 2\sum_{k=0}^{\infty} r_V(k)\cos k\theta - r_V(0)$$

In the definitions of $R_U(j\omega)$ and $R_V(\gamma)$, the averaging was done first to obtain the correlation functions and these were then transformed. As an alternative, for an ergodic process $\{U(t)\}$, let the random variable

$$U_T(t) = U(t), \quad |t| \le T$$
$$= 0 \quad , \quad |t| > T$$

and

$$\mathfrak{F}\, U_T(t) = U_T(j\omega)$$

or

$$U_T(j\omega) = \int_{-\infty}^{\infty} U(t) f_T(t) \exp(-j\omega t)\, dt$$

where $f_T(t) = 1$, $|t| \leq T$ and $f_T(t) = 0$, $|t| > T$. Now the average power over $[-T,T]$ is

$$\frac{1}{2T}\int_{-\infty}^{\infty} U_T^2(t)\, dt = \frac{1}{4\pi T}\int_{-\infty}^{\infty} |U_T(j\omega)|^2 d\omega \qquad (2.85)$$

by the Parseval Theorem (Appendix A). Taking the limit as $T\to\infty$ of the expected value of both sides,

$$\lim_{T\to\infty} E[\frac{1}{2T}\int_{-\infty}^{\infty} U_T^2(t)\, dt] = \lim_{T\to\infty} E[\frac{1}{4\pi T}\int_{-\infty}^{\infty} |U_T(j\omega)|^2 d\omega].$$

On the left,

$$\lim_{T\to\infty} E[\frac{1}{2T}\int_{-\infty}^{\infty} U_T^2(t)\, dt] = S_U^2 = r_U(0) = \frac{1}{2\pi}\int_{-\infty}^{\infty} R_U(j\omega)\, d\omega.$$

On the right, assuming the limit exists, and interchanging expectation and integration,

$$\lim_{T\to\infty} E[\frac{1}{4\pi T}\int_{-\infty}^{\infty} |U_T(j\omega)|^2 d\omega] = \frac{1}{2\pi}\int_{-\infty}^{\infty}\{\lim_{T\to\infty} E[\frac{|U_T(j\omega)|^2}{2T}]\}d\omega$$

Thus, we have

$$\frac{1}{2\pi}\int_{-\infty}^{\infty} R_U(j\omega)\, d\omega = \frac{1}{2\pi}\int_{-\infty}^{\infty}\{\lim_{T\to\infty} E[\frac{|U_T(j\omega)|^2}{2T}]\}d\omega.$$

This result leads us to ask how the two integrands are related. Now direct calculation of

$$\begin{aligned} U_T(-j\omega)U_T(j\omega) &= |U_T(j\omega)|^2 \\ &= \iint_{-\infty}^{\infty} U(\sigma)U(\lambda) f_T(\sigma) f_T(\lambda) \exp(-j\omega(\lambda-\sigma))\, d\lambda\, d\sigma \end{aligned}$$

and

$$E[|U_T(j\omega)|^2] = \iint_{-\infty}^{\infty} r_U(\lambda-\sigma) f_T(\sigma) f_T(\lambda) \exp(-j\omega(\lambda-\sigma)) d\lambda d\sigma .$$

For $\alpha = \lambda-\sigma$,

$$E[|U_T(j\omega)|^2] = \int_{-\infty}^{\infty} d\alpha r_U(\alpha) \exp(-j\omega\alpha) \int_{-\infty}^{\infty} d\sigma \; f_T(\sigma) f_T(\alpha+\sigma)$$

$$= 2T \int_{-\infty}^{\infty} (1 - \frac{|\alpha|}{2T}) r_U(\alpha) \exp(-j\omega\alpha) d\alpha \; .$$

Then, dividing by 2T and taking the limit as $T\to\infty$,

$$\lim_{T\to\infty} E[\frac{|U_T(j\omega)|^2}{2T}] = \int_{-\infty}^{\infty} r_U(\alpha) \exp(-j\omega\alpha) d\alpha = R_U(j\omega) \quad (2.86)$$

provided that

$$\int_{-\infty}^{\infty} |\alpha r_U(\alpha)| d\alpha < \infty \; ,$$

and the right hand equality follows from the Fourier Transform definition. A weaker condition [13.] that includes bandlimited processes (see Section 2.9) is

$$\lim_{T\to\infty} \frac{1}{2T} \int_{-T}^{T} r_U(\alpha) d\alpha < \infty$$

Equation 2.86, therefore, gives two forms for $R_U(j\omega)$ of an ergodic process provided the stated condition on $r_U(\cdot)$ is satisfied. Inspection of Property 7 shows why $R_U(j\omega)$ is called a <u>power</u> spectral density; it has the dimension power per Hz. The analogy to $P_{cs}(\omega)$ in Section 2.5 is clear.

A similar sequence of steps for an ergodic sequence $\{V(n)\}$ yields a result analogous to Equation 2.86:

$$\lim_{N\to\infty} E\left[\frac{|V_N(\gamma)|^2}{2N+1} \right] = \sum_{k=-\infty}^{\infty} r_V(k) \gamma^{-k} = R_V(\gamma) . \quad (2.87)$$

Note that the dimension of $R_V(\gamma)$ is simply power, and the analogy to $P_{ds}(\theta)$ can be seen.

Example 2.20

a) For Example 2.17a

$$r_V(k) = v_o(k) \; .$$

Therefore,

$$R_V(z) = \mathfrak{Z}_2 r_V(k) = 1 \; .$$

b) From Example 2.17b,

$$r_V(k) = (\tfrac{1}{2})^{|k|} \; .$$

Therefore,

$$R_V(z) = \mathfrak{Z}_2 r_V(k) = -\frac{3}{2}\,\frac{z}{(z - \frac{1}{2})(z - 2)}$$

by using Table A.5 (Appendix A).

c) From Example 2.17c

$$r_U(\tau) = \sigma^2 \cos \tau$$

Therefore,

$$R_U(j\omega) = \mathfrak{F} r_U(\tau) = \pi\sigma^2[u_o(\omega-1) + u_o(\omega+1)]$$

by using Table A.2 (App. A). Also

$$R_U(s) = \pi\sigma^2[u_o(s^2 + 1)]$$

is a convenient form in some cases.

2.7.7 Special Processes

Several wide-sense stationary stochastic processes or sequences appear frequently enough in various applications to make it worthwhile to give their properties in summary form. The scalar forms are given here, and vector forms are given later in Chapter 4.

1. White Process or Sequence

A <u>white process</u> $\{U(t)\}$ is a process with a spectral density function that is constant. Thus

$$R_U(s) = K$$

and

$$r_U(\tau) = \mathcal{L}_2^{-1} R_U(s) = K u_o(\tau) .$$

This is a useful concept that can be approximated over a finite range of frequencies, and a number of physical examples appear in the references. See, for example, [4.].

A <u>white sequence</u> $\{V(n)\}$ is a sequence with a spectral density function that is constant. Thus

$$R_V(z) = K$$

and

$$r_V(k) = \mathcal{Z}_2^{-1} R_V(z) = K v_o(k)$$

Example 2.12 (and 2.20a) is a white sequence; it also is called a Bernoulli sequence. K is the <u>intensity</u> in both of these cases.

White stochastic signals are particularly useful as test inputs to systems. Approximate white processes $\{U(t)\}$ are often generated by using various physical phenomena such as electrical noise in a gas tube, and white sequences $\{V(n)\}$ are frequently generated by digital programs usually called random number generators.

2. Gaussian (Normal) Process or Sequence

A Gaussian process {U(t)} is a process such that for any n, the random variables $U(t_1)$, $U(t_2)$, ..., $U(t_n)$ have a joint normal density function. Furthermore the density function can be written in terms of mean values and covariance functions. Thus

$$p_{U_1,\ldots,U_n}(u_1,\ldots,u_n) = \frac{1}{(2\pi)^{n/2}\sqrt{|\underline{C}_{\underline{U}}|}} \cdot \exp(1/2(\underline{u}-\underline{m}_{\underline{U}})^T \underline{C}_{\underline{U}}^{-1}(\underline{u}-\underline{m}_{\underline{U}}))$$

where $\underline{U}^T = (U_1,\ldots,U_n)$, $\underline{u}^T = (u_1,\ldots,u_n)$, $\underline{m}_{\underline{U}}$ is the mean of $\underline{U}$, and $|C_{\underline{U}}|$ is the determinant of the covariance matrix of $\underline{U}$ (see Appendix C). Many stochastic phenomena can be represented by this Gaussian process, and, in addition, it has very useful mathematical properties, e.g., the process is completely described by first and second moments in $\underline{m}_{\underline{U}}$ and $\underline{C}_{\underline{U}}$. Other special properties are: wide-sense and strict stationarity are equivalent, a diagonal $C_{\underline{U}}$ implies $U_1,\ldots,U_n$ are uncorrelated, and uncorrelated $U_1,\ldots,U_n$ implies that $U_1,\ldots,U_n$ are independent.

A Gaussian sequence {V(n)} is a process such that for any m, the random variables $V(n_1),\ldots,V(n_m)$ have a joint normal density function. Remarks similar to those for Gaussian {U(t)} follow.

3. White Gaussian Process or Sequence

A white Gaussian process {U(t)} is both white and Gaussian. Therefore, the random variables $U(t_1)$, $U(t_2),\ldots,U(t_n)$, for any n, are normal and independent. A common example of a white Gaussian process is the thermal noise of a resistor [12.].

A white Gaussian sequence {V(n)} is both white and

Gaussian. Therefore, the random variables $V(n_1),\ldots,V(n_m)$, for any m, are normal and independent. The sequence of numbers generated by the "random function" statement of the FORTRAN digital programming language is an example.

4. Markov Process or Sequence
A Markov process $\{U(t)\}$ is a process such that for any n, and $t_n > t_{n-1} >,\ldots,> t_1$ the conditional distribution function

$$P_{U_n}(u(t_n)\,|\,U(t_{n-1}) = u(t_{n-1}),\ldots,U(t_1) = u(t_1))$$

equals

$$P_{U_n}(u(t_n)\,|\,U(t_{n-1}) = u(t_{n-1}))$$

or in simpler notation,

$$P_{U_n}(u_n\,|\,u_{n-1},\ldots,u_1) = P_{U_n}(u_n\,|\,u_{n-1})$$

where U is a continuous random variable. A similar equality holds where U is a discrete random variable. This means that in some sense the past history of U(t) for instants $t_{n-1},\ldots,t_1$ is contained in $U(t_{n-1})$. Also, the joint distribution function of $U_n,\ldots,U_1$ can be expressed as a product of n-1 conditional distribution functions and the distribution function of U_1. Thus, for a continuous random variable, we obtain

$$P_{\underline{U}}(u_1,\ldots,u_n) = \left[\prod_{i=2}^{n} P_{U_i}(u_i\,|\,u_{i-1})\right] P_{U_1}(u_1) \;.$$

Markov processes play an important role in describing the behavior of physical phenomena which evolve in a probabilistic manner [13.].

A Markov sequence $\{V(n)\}$ is a sequence such that for any m and $n_m > n_{m-1} >, \ldots, > n_1$,

$$p_{V_m}(v_m | v_{m-1}, \ldots, v_1) = p_{V_m}(v_m | v_{m-1})$$

where the notation is abbreviated as above, and where V is a continuous random variable. A similar equality holds when V is a discrete random variable and the sequence $\{V(n)\}$ is then called a Markov chain. Example 2.13 is a simple two-state Markov chain.

5. Gauss Markov Process or Sequence

A Gauss Markov process $\{U(t)\}$ is a stochastic process that is both Gaussian and Markov. This process occurs frequently in linear systems analysis as will be seen in later chapters; in fact, one useful way of generating a stationary Gauss Markov process is to drive a certain linear system with a white Gaussian process. The output is a stationary Gauss Markov process. It can be shown [12.] that a stationary Gauss Markov process has an autocorrelation function

$$r_U(\tau) = r(0)\exp(-a|\tau|), \quad a > 0$$

A Gauss Markov sequence $\{V(n)\}$ is a stochastic sequence that is both Gaussian and Markov. The remarks made above also are true for $\{V(n)\}$, and the corresponding autocorrelation function is

$$r_V(k) = r_V(0)(b)^{|k|}, \quad 0 < b < 1.$$

In addition to these stochastic processes and sequences, there are many other well-known examples in [4.][12.][13.]. Such processes as the Wiener process, Poisson processes, telegraph processes, and counting processes or sequences appear in a wide variety of engineering systems analyses.

2.7.8 Cross Correlation

It frequently is of interest to consider two or more stochastic processes or sequences that are not necessarily independent. The previous concepts of correlation and covariance functions and of spectral density functions are, therefore, extended to include this possibility. Starting with two processes $U_i(t)$ and $U_j(t)$, the cross-correlation functions are

$$r_{U_iU_j}(t_1,t_2) = E[U_1(t_1)U_j(t_2)] = E[U_{1i}U_{2j}] \tag{2.88a}$$

and

$$r_{U_jU_i}(t_1,t_2) = E[U_j(t_1)U_i(t_2)] = E[U_{1j}U_{2i}] \tag{2.88b}$$

where as a matter of convenience, $t_2 \geq t_1$. Similarly, for $\{V_i(n)\}$ and $\{V_j(n)\}$,

$$r_{V_iV_j}(n_1,n_2) = E[V_i(n_1)V_j(n_2)] = E[V_{1i}V_{2j}] \tag{2.89a}$$

and

$$r_{V_jV_i}(n_1,n_2) = E[V_j(n_1)V_i(n_2)] = E[V_{1j}V_{2i}] \tag{2.89b}$$

with $n_2 \geq n_1$. If the processes or sequences are wide-sense stationary, letting t_1=t and t_2=t+τ, and shortening the notation, we get

$$r_{ij}(\tau) = E[U_i(t)U_j(t+\tau)] \tag{2.90a}$$

$$r_{ji}(\tau) = E[U_j(t)U_i(t+\tau)] \tag{2.90b}$$

and the property $r_{ij}(\tau) = r_{ji}(-\tau)$ follows. Similarly, for n_1=n and n_2=n+k,

$$r_{ij}(k) = E[V_i(n)V_j(n+k)] \tag{2.91a}$$

$$r_{ji}(k) = E[V_j(n)V_i(n+k)] \tag{2.91b}$$

with $r_{ij}(k) = r_{ji}(-k)$. Furthermore, by taking transforms of 2.90 and 2.91, the cross-spectral density functions are defined. We obtain

$$\mathfrak{F} r_{ij}(\tau) = R_{ij}(j\omega)$$

$$\mathfrak{F} r_{ji}(\tau) = R_{ji}(j\omega)$$

with

$$R_{ij}(j\omega) = R_{ji}(-j\omega)$$

and

$$\mathfrak{F}_d r_{ij}(k) = R_{ij}(\gamma)$$

$$\mathfrak{F}_d r_{ji}(k) = R_{ji}(\gamma)$$

with

$$R_{ij}(\gamma) = R_{ji}(\gamma^{-1}),$$

and Laplace and Z-Transforms also may be used. In the same way, the cross-covariance functions are

$$c_{U_iU_j}(t_1,t_2) = E[(U_{1i}-m_{1i})(U_{2j}-m_{2j})]$$

$$c_{U_jU_i}(t_1,t_2) = E[(U_{1j}-m_{ij})(U_{2i}-m_{2i})]$$

and

$$c_{V_iV_j}(n_1,n_2) = E[(V_{1i}-m_{1i})(V_{2j}-m_{2j})]$$

$$c_{V_jV_i}(n_1,n_2) = E[(V_{1j}-m_{1j})(V_{2i}-m_{2i})]$$

and for wide-sense stationary processes and sequences, these become $c_{ij}(\tau)$, $c_{ji}(\tau)$, $c_{ij}(k)$, and $c_{ji}(k)$ using forms similar to 2.90 and 2.91. The cross-correlation and cross-covariance functions are a useful measure of the interdependence of two processes or sequences. In general, they are related by

$$c_{U_iU_j}(t_1,t_2) = r_{U_iU_j}(t_1,t_2) - m_{1i}m_{2j}$$

with similar forms for the other three pairs of functions. The processes $\{U_i(t)\}$ and $\{U_j(t)\}$ are <u>uncorrelated</u> if $c_{U_iU_j}(t_1,t_2) = 0$ for all t_1,t_2.

The counterparts to Eq. 2.86 and 2.87 for cross-spectral density functions are

$$R_{ij}(j\omega) = \lim_{T\to\infty} E\left[\frac{U_{iT}(-j\omega)U_{jT}(j\omega)}{2T}\right] \qquad (2.92)$$

and

$$R_{ij}(\gamma) = \lim_{N\to\infty} E\left[\frac{V_{iN}(\gamma^{-1})V_{jN}(\gamma)}{2N+1}\right] \qquad (2.93)$$

and Laplace and Z-Transforms also can be used.

2.7.9 Vector Processes

A vector stochastic process has components $\{U_i(t)\}$, $i=1,2,\ldots,m$, and using the notation introduced earlier in this chapter the process is $\{\underline{U}(t)\}$. Similarly, a vector stochastic sequence is $\{\underline{V}(n)\}$. Furthermore, the notation,

$$\underline{U}(t_1) = \underline{U}_1 \text{ with components } U_{1i} \text{ or } U_{1iq}$$

(according to whether the random variables are continuous or discrete), and

$$\underline{V}(n_1) = \underline{V}_1 \text{ with components } V_{1i} \text{ or } V_{1iq}$$

is used. The first subscript is a time index, the second is the component number, and the integer variable q indicates a discrete random variable as discussed in Appendix C.

The <u>correlation matrix</u> of the process $\{\underline{U}(t)\}$ is

$$R_{\underline{U}}(t_1,t_2) = E[\underline{U}(t_1)\underline{U}^T(t_2)] = E[\underline{U}_1\underline{U}_2{}^T] \qquad (2.94a)$$

and for the sequence $\{\underline{V}(n)\}$,

$$R_{\underline{V}}(n_1,n_2) = E[\underline{V}(n_1)\underline{V}^T(n_2)] = E[\underline{V}_1\underline{V}_2{}^T]. \qquad (2.94b)$$

These are mxm matrices whose elements on the main diagonal, $r_{ii}(\cdot)$, are correlation functions and off the main diagonal, $r_{ij}(\cdot)$ and $r_{ji}(\cdot)$, $j \neq i$, are cross-correlation functions. It can be shown that

$$R_{\underline{U}}(t_1,t_2) = R_{\underline{U}}^T(t_2,t_1) \text{ and } R_{\underline{V}}(n_1,n_2) = R_{\underline{V}}^T(n_2,n_1).$$

The <u>covariance kernel</u> of $\{\underline{U}(t)\}$ is

$$P_{\underline{U}}(t_1,t_2) = E[(\underline{U}_1-\underline{m}_{\underline{U}_1})(\underline{U}_2-\underline{m}_{\underline{U}_2})^T] \qquad (2.95a)$$

and for $\{\underline{V}(n)\}$ is

$$P_{\underline{V}}(n_1,n_2) = E[(\underline{V}_1-\underline{m}_{\underline{V}_1})(\underline{V}_2-\underline{m}_{\underline{V}_2})^T] \qquad (2.95b)$$

where both are mxm matrices whose elements are covariance functions and cross-covariance functions. If the component processes of $\{\underline{U}(t)\}$ and $\{\underline{V}(n)\}$ are wide-sense stationary, the correlation matrices and covariance kernels are functions of a single time variable. Thus for $\tau = t_2-t_1$ and $k = n_2-n_1$, the matrices are

$$R_{\underline{U}}(t_1,t_2) \rightarrow R_{\underline{U}}(\tau) \text{ or } R_{\underline{V}}(n_1,n_2) \rightarrow R_{\underline{V}}(k) \qquad (2.96a)$$

and

$$P_{\underline{U}}(t_1,t_2) \rightarrow P_{\underline{U}}(\tau) \text{ or } P_{\underline{V}}(n_1,n_2) \rightarrow P_{\underline{V}}(k). \qquad (2.96b)$$

Spectral density matrices follow from 2.96a and 2.96b by taking transforms. Thus we get

$$R_{\underline{U}}(s) = \mathcal{L}_2 R_{\underline{U}}(\tau) \tag{2.97a}$$

and

$$R_{\underline{V}}(z) = \mathcal{Z}_2 R_{\underline{V}}(k). \tag{2.97b}$$

and the elements of these matrices are spectral density functions and cross-spectral density functions. Also, it can be shown that $R_{\underline{U}}(s) = R_{\underline{U}}^T(-s)$ and $R_{\underline{V}}(z) = R_{\underline{V}}^T(z^{-1})$.

Cases of particular interest in Chapter 4 with regard to the correlation matrices and covariance kernels of non-stationary processes or sequences are those for $t_1=t_2=t$ and $n_1=n_2=n$. Equations 2.94a and 2.94b become

$$R_{\underline{U}}(t,t) = E[\underline{U}(t)\underline{U}^T(t)] = S_{\underline{U}}(t)$$

and

$$R_{\underline{V}}(n,n) = E[\underline{V}(n)\underline{V}^T(n)] = S_{\underline{V}}(n)$$

These are the second moment matrices (see Appendix C) of the random vectors $\underline{U}(t)$ and $\underline{V}(n)$. Similarly, Equations 2.95a and 2.95b become

$$P_{\underline{U}}(t,t) = E[(\underline{U}(t)-\underline{m}_{\underline{U}})(\underline{U}(t)-\underline{m}_{\underline{U}})^T] = C_{\underline{U}}(t) \tag{2.98a}$$

and

$$P_{\underline{U}}(n,n) = E[(\underline{V}(n)-\underline{m}_{\underline{V}})(\underline{V}(n)-\underline{m}_{\underline{V}})^T] = C_{\underline{V}}(n) \tag{2.98b}$$

These are the time varying covariance matrices (see Appendix C) of the random vectors $\underline{U}(t)$ and $\underline{V}(n)$. In the event that $\{\underline{U}(t)\}$ and $\{\underline{V}(n)\}$ are wide-sense stationary, we use $\tau=t_2-t_1=0$

and $k=n_2-n_1=0$. Therefore, the forms 2.96a and 2.96b become the constant matrices

$$R_{\underline{U}}(0) = S_{\underline{U}}, \; R_{\underline{V}}(0) = S_{\underline{V}} \tag{2.99a}$$

and

$$P_{\underline{U}}(0) = C_{\underline{U}}, \; P_{\underline{V}}(0) = C_{\underline{V}} \;. \tag{2.99b}$$

The matrices $C_{\underline{U}}(t)$ and $C_{\underline{V}}(n)$ are widely used as a measure of system response where $\{\underline{U}(t)\}$ and $\{\underline{V}(n)\}$ are the stochastic outputs (nonstationary) of a linear system driven by stochastic inputs. For stationary $\{\underline{U}(t)\}$ and $\{\underline{V}(n)\}$, $C_{\underline{U}}$ and $C_{\underline{V}}$ are used similarly. These ideas will be developed in detail in Chapter 4. Finally, it is noted again that the second moment and covariance matrices are symmetric, e.g., $C_{\underline{U}}(t) = C_{\underline{U}}^T(t)$. Also, we can show that

$$C_{\underline{U}} = S_{\underline{U}} - \underline{m}_{\underline{U}}\underline{m}_{\underline{U}}^T$$

and

$$C_{\underline{V}} = S_{\underline{V}} - \underline{m}_{\underline{V}}\underline{m}_{\underline{V}}^T \;,$$

so that the second moment and covariance matrices are equal when the mean values are zero.

Example 2.21

a) To show the notation, we consider a stationary white Gaussian process $\{\underline{U}(t)\}$ of dimension = 3. It has a covariance kernel

$$P_{\underline{U}}(\tau) = \begin{bmatrix} 2 & 1 & 0 \\ 1 & 3 & 0 \\ 0 & 0 & 4 \end{bmatrix} u_o(\tau) .$$

and a spectral density matrix

$$R_{\underline{U}}(s) = \begin{bmatrix} 2 & 1 & 0 \\ 1 & 3 & 0 \\ 0 & 0 & 4 \end{bmatrix},$$

The coefficient matrix of $P_{\underline{U}}$ is an <u>intensity</u> <u>matrix</u>.

b) A stationary white Gaussian sequence $\{V(n)\}$ of dimension = 2 has a covariance kernel

$$P_{\underline{V}}(k) = \begin{bmatrix} 1 & 0 \\ 0 & 2 \end{bmatrix} v_o(k)$$

and spectral density matrix

$$R_{\underline{V}}(z) = \begin{bmatrix} 1 & 0 \\ 0 & 2 \end{bmatrix}.$$

Furthermore,

$$P_{\underline{V}}(0) = C_{\underline{V}} = \begin{bmatrix} 1 & 0 \\ 0 & 2 \end{bmatrix}$$

and the variances of the first and second components of $\underline{V}$ are 1 and 2, and the two components are uncorrelated since the cross-covariances are zero.

The several matrices associated with one vector stochastic process or sequence are summarized in Table 2.5.

For two vector process $\{\underline{U}'(t)\}$ and $\{\underline{U}''(t)\}$ or two vector sequences $\{\underline{V}'(n)\}$ or $\{\underline{V}''(n)\}$, it is possible to define cross-correlation matrices and cross-covariance kernels, and the forms 2.95, 2.96, 2.97, and 2.98 carry over with

TABLE 2.5

		Process $\{\underline{U}(t)\}$	Sequence $\{\underline{V}(n)\}$	
Correlation Matrix (General)	$t_1 \neq t_2, \tau = t_2 - t_1$	$R_{\underline{U}}(t_1,t_2)$	$R_{\underline{V}}(n_1,n_2)$	$n_1 \neq n_2, k = n_2 - n_1$
Correlation Matrix (w-s stationary)		$R_{\underline{U}}(\tau)$	$R_{\underline{V}}(k)$	
Covariance Kernel (General)		$P_{\underline{U}}(t_1,t_2)$	$P_{\underline{V}}(n_1,n_2)$	
Covariance Kernel (w-s stationary)		$P_{\underline{U}}(\tau)$	$P_{\underline{V}}(k)$	
Second Moment Matrix (General)	$t_1 = t_2 = t$	$S_{\underline{U}}(t)$	$S_{\underline{V}}(n)$	$n_1 = n_2 = n$
Second Moment Matrix (w-s stationary)		$S_{\underline{U}}$	$S_{\underline{V}}$	
Covariance Matrix (General)		$C_{\underline{U}}(t)$	$C_{\underline{V}}(n)$	
Covariance Matrix (w-s stationary)		$C_{\underline{U}}$	$C_{\underline{V}}$	

obvious changes in notation. For example, for wide-sense stationary processes,

$$P_{\underline{U}'\underline{U}''}(\tau) = E[(\underline{U}'(t)-\underline{m}_{U'})(\underline{U}''(t+\tau)-\underline{m}_{U''})^T]$$

is a cross-covariance kernel.

The matrix counterparts of 2.86 and 2.87 provide alternative forms for 2.97a and 2.97b. The spectral density matrices are

$$R_{\underline{U}}(j\omega) = \lim_{T\to\infty} E\left[\frac{\underline{U}_T(-j\omega)\underline{U}_T^{\,T}(+j\omega)}{2T}\right] = \mathcal{F}\, R_{\underline{U}}(\tau) \qquad (2.100a)$$

and

$$R_V(\gamma) = \lim_{N\to\infty} E\left[\frac{\underline{V}_N(\gamma^{-1})\underline{V}_N^{\,T}(\gamma)}{2N+1}\right] = \mathcal{F}_d R_{\underline{V}}(k) \qquad (2.100b)$$

Similarly, a cross-spectral density matrix is

$$R_{\underline{U}'\underline{U}''}(j\omega) = \lim_{T\to\infty} E\left[\frac{(\underline{U}_T'(-j\omega))(\underline{U}_T''^T(+j\omega))}{2T}\right]$$

and $R_{\underline{U}'\underline{U}''}(j\omega) = R^T_{\underline{U}''\underline{U}'}(-j\omega)$.

In previous sections of this chapter, it was noted that one problem in signal analysis was that of approximating an actual signal (experimental data) with a model. For example, the method of least squares was mentioned with regard to 2.16 and 2.36, 2.38 and 2.40, and 2.42 and 2.43, 2.44; these are deterministic signal models. Also, finding the real spectral coefficients in an orthonormal function expansion was illustrated in Example 2.8, and frequency domain approximations using a finite number of poles and zeros was described. Finally, the amplitude spectra, e.g., $|U(j\omega)|$ or $|V(\gamma)|$, are often measured. When the signals are modelled as stochastic signals, however, the problem is more complicated. Even when ergodicity is assumed, only a finite amount of data is available on a sample function. Measurement of correlation functions in this case is discussed in [4.], and an interesting example is given in [14.]. Methods for estimating power density spectra are given in [15.] in considerable detail. For stochastic sequences there is a large amount of material available -- usually called time series analysis, e.g., [16.], [43.]. Examples of some of these methods are given in Chapter 6. One rather obvious method which makes use of previous results in this chapter is to compute an approximation to 2.80 or 2.81 to give a numerical approximation to $r_U(\tau)$ or $r_V(k)$. Models like 2.39 or 2.41 can then be used to provide an analytic form of correlation function. In general, it is desirable that the estimate make the best use of the available data in some sense, i.e., an optimal estimate of the parameters of the signal model is sought.

2.8 SUMMARY OF SIGNAL MODELS

The first seven sections of this chapter have covered a variety of signal models. Sections 2.2 through 2.6 were concerned with deterministic signals and Section 2.7 with stochastic signals. After a number of elementary functions were defined, linear combinations of these were used to develop general signal models involving real spectral functions. These ideas were then extended to include the frequency domain and other spectral functions. Table 2.4 summarized these results on spectral models. The auxiliary functions called translation functions were then defined and these also had frequency domain representations. In developing stochastic signal models, however, a parallel was not followed. Instead, the auxiliary functions called correlation functions were defined first and their frequency representations were the spectral density functions. It was pointed out earlier that these can then be used to define certain properties and, in some cases, spectral forms for stochastic signals that are the counterparts of those in Table 2.4 can then be defined. Only a few of these are mentioned here as examples, and references to more advanced books are provided for additional information.

Tables 2.6a and 2.6b are arranged to allow easy comparison of deterministic and stochastic signal models discussed in this book. The forms given are only for scalar signals, and wide-sense stationarity is assumed for the stochastic signals. Each entry has an equation number and the functions associated with that equation. In some cases further reference is made to commonly used forms as typical examples. In Table 2.6b, the entries A, B, C, and D have not been discussed previously; these are described briefly in this summary.

Entry A requires the expansion of a stochastic continuous-time (or discrete-time) signal in terms of a real spectral function of continuous argument. An example [12.] where the spectral function of U(t) is its Hilbert transform $\hat{U}(t)$ is of some general interest in finding the in-phase and quadrature components of $R_U(j\omega)$ with respect to a constant ω_o. Entry B requires the expansion of U(t) (or V(n)) in terms of a spectral function with a complex argument. The $\mathcal{F}$ and $\mathcal{F}_d$ transforms are in this class. An example is given in the derivations of Eq. 2.86 and 2.87, and these transforms exist in general for energy signals. Entry C requires the expansion of U(t) (or V(n)) in terms of a real spectral function with a discrete argument. Expansions in terms of sets of orthonormal functions are in this class. An example is the Karhunen-Loeve expansion of $U_T(t)$ where

$$U_T(t) = U(t) \qquad |t| < T/2$$

and $\{U(t)\}$ is a wide-sense stationary, zero mean process. For a set $\psi_i(t)$ (orthogonal on [-T/2,T/2]) satisfying

$$\lambda_i \psi_i(t) = \int_{-T/2}^{T/2} r_U(t-\tau)\psi_i(\tau)d\tau,$$

we have

$$U_T(t) = \sum_{i=1}^{\infty} A_i \psi_i(t). \qquad (2.101)$$

The random variables A_i are

$$A_i = \int_{-T/2}^{T/2} U(t)\psi_i(t)dt$$

with $E[A_i] = 0$ and $E[A_iA_j] = \lambda_i$, i=j, or $E[A_iA_j] = 0$, $i \neq j$. Thus the A_i are zero mean and uncorrelated with variance λ_i. A parallel for V(n) also can be shown. Entry D in Table 2.6b requires the expansion of U(t) (or V(n)) in terms of complex spectral functions with discrete argument. An example is the expansion of U(t) in a Fourier Series [12.]. U(t) is <u>periodic</u> in the mean square sense if

$$r_U(\tau+T_o) = r_U(\tau) \text{ for all } \tau.$$

TABLE 2.6a

DETERMINISTIC SIGNALS			
Signal	Spectral Forms		Auxiliary Function
$u(t)$	Eq. 2.52 $f(\lambda)$, $q_1(t,\lambda)$ Hilbert Transform	Eq. A.4 $U(j\omega)$, $\exp(jt\omega)$ $\mathfrak{F}$ (and $\mathfrak{L}$ Transform)	Eq. 2.69 $w_u(\tau)$
	Eq. 2.50 $\ell(k)$, $q_2(t,k)$ Orthonormal Function Exp.	Eq. 2.45 U_k, $\exp(jk\omega_0 t)$ $\mathfrak{F}$ Series	Eq. 2.71 $W_u(s)$
$v(n)$	Eq. 2.53 $g(\lambda)$, $q_3(n,\lambda)$	Eq. A.14 $V(\gamma)$, γ^{n-1} $\mathfrak{F}_d$ (and Z) Transform	Eq. 2.70 $w_v(i)$
	Eq. 2.51 $m(k)$, $q_4(n,k)$ Orthonormal Function Exp.	Eq. 2.46, 2.47 V_k, $\exp(jk\lambda_0 n)$ $\mathfrak{F}_d$ Series	Eq. 2.72 $W_v(z)$

TABLE 2.6b

STOCHASTIC SIGNALS			
Signal	Spectral Forms		Auxiliary Function
$u^i(t)$ $\{U(t)\}$	A Hilbert Transform	B $\mathfrak{F}$ Transform	Eq. 2.79a $r_U(\tau)$
	C Eq. 2.101 Karhunen-Loeve Exp.	D Eq. 2.102 $\mathfrak{F}$ Series	Eq. 2.82 $R_U(s)$
$v^i(n)$ $\{V(n)\}$	A	B $\mathfrak{F}_d$ Transform	Eq. 2.79b $r_V(k)$
	C	D $\mathfrak{F}_d$ Series	Eq. 2.83 $R_V(z)$

Then, for U(t) periodic in the mean square sense,

$$U(t) = \sum_{n=-\infty}^{\infty} A_n \exp(jn\omega_o t) \qquad (2.102)$$

where

$$\omega_o = \frac{2\pi}{T_o}$$

and

$$A_n = \frac{1}{T_o}\int_0^{T_o} U(\lambda)\exp(-jn\omega_o\lambda)\,d\lambda$$

with

$$E[A_n] = E[U(t)] = m_U, \quad n = 0$$
$$= 0 \quad , \quad n \neq 0$$

and

$$E[A_n A_j] = \alpha_n, \quad n = j$$
$$= 0 \ , \quad n \neq j$$

and

$$\alpha_n = \frac{1}{T_o}\int_0^{T_o} r_U(\tau)\exp(-jn\omega_o\tau)\,d\tau \ .$$

In comparing Tables 2.6a and 2.6b, it is noted that the entries under Auxiliary Function are all deterministic since an expection (or time average) has been taken in the case of stochastic signals. However, the entries under Spectral Forms are deterministic in the case of deterministic signals but contain random variables in the case of stochastic signals.

A further comparison concerns the energy and power spectra of deterministic and stochastic signals and their relation to the auxiliary functions. The more common analogy between the deterministic and stochastic models is on a power basis, and Table 2.7 shows the various forms defined in this chapter. It is important to observe that the autocorrelation functions dimensionally are signal power. The second possibility is an analogy on an energy basis, and Table 2.8 shows the forms. In this case, the autotranslation functions dimensionally are signal energy. Although both power or energy spectra have been found useful for certain deterministic signals, power spectra are widely used for stochastic signals since wide-sense stationarity (or even ergodicity) usually is assumed. Energy spectra for nonstationary stochastic signals with finite energy can be defined as shown, but time domain methods have proved more useful in these cases as shown in Chapter 4.

2.9 SIGNAL CONVERSIONS

The signal conversions introduced in Section 2.1 were sampling (S), quantizing (Q), reconstruction (R), and dequantizing (F). It is interesting to consider some of the effects of these conversions on various signal models. The definitions of the four conversions provide one way of describing an output of a converter if an input function u(t) or v(n) is given. For example, for uniform sampling at a rate $1/T_s$ an input u(t) to S yields an output $u(nT_s)$. Alternatively, if the input model is a spectral form, it would be useful to have the spectral function of the output. Also, if the input is an auxiliary function model, a corresponding auxiliary function of the output is of interest. Although there are many possibilities, only a few of the simpler cases are considered here: the inputs are deterministic energy signals (except as noted) or ergodic stochastic signals, the sampling is uniform, the quantization is uniform, and the reconstruction and dequantization are linear operations. Primary emphasis is on the effects on the Fourier spectra of deterministic signals or on the spectral density functions of stochastic signals. For each

TABLE 2.7
ANALOGY BETWEEN DETERMINISTIC AND STOCHASTIC SIGNALS

Power Spectra

	Deterministic	Stochastic (Ergodic)
cont.-time	$P_c = \lim_{T\to\infty} \frac{1}{2T}\int_{-T}^{T} u^2(t)\,dt = \frac{1}{2\pi}\int_{-\infty}^{\infty} P_{cs}(\omega)\,d\omega$ $P_{cs}(\omega) = \lim_{T\to\infty} \left[\frac{\lvert U_T(j\omega)\rvert^2}{2T}\right]$ Eq. 2.67	$r_U(0) = \lim_{T\to\infty} \frac{1}{2T}\int_{-T}^{T} (u^i(t))^2\,dt = E[U^2] = \frac{1}{2\pi}\int_{-\infty}^{\infty} R_U(j\omega)\,d\omega$ $R_U(j\omega) = \mathfrak{F}\, r_U(\tau) = \lim_{T\to\infty} E\left[\frac{\lvert U_T(j\omega)\rvert^2}{2T}\right]$ Eq. 2.86
discrete-time	$P_d = \lim_{N\to\infty} \frac{1}{2N+1}\sum_{-N}^{N} v^2(n) = \frac{1}{2\pi}\int_{-\pi}^{\pi} P_{ds}(\theta)\,d\theta$ $P_{ds}(\theta) = \lim_{N\to\infty} \left[\frac{\lvert V_N(\varepsilon^{j\theta})\rvert^2}{2N+1}\right]$ Eq. 2.68	$r_V(0) = \lim_{N\to\infty} \frac{1}{2N+1}\sum_{-N}^{N} (v^i(j))^2 = E[V^2] = \frac{1}{2\pi}\int_{-\pi}^{\pi} R_V(\varepsilon^{j\theta})\,d\theta$ $R_V(\varepsilon^{j\theta}) = \mathfrak{F}_d r_v(k) = \lim_{N\to\infty} E\left[\frac{\lvert V_N(\varepsilon^{j\theta})\rvert^2}{2N+1}\right]$ Eq. 2.87

TABLE 2.8
ANALOGY BETWEEN DETERMINISTIC AND STOCHASTIC SIGNALS

Energy Spectra

	Deterministic	Stochastic (Nonstationary)
cont.-time	$E_c = w_u(0) = \int_{-\infty}^{\infty} u^2(t)\,dt = \frac{1}{2\pi}\int_{-\infty}^{\infty} W_u(\omega)\,d\omega$ $W_u(\omega) = \lvert U(j\omega)\rvert^2$ Sect. 2.5.3	$E\left[\int_{-\infty}^{\infty} U(t)^2 dt\right] = \frac{1}{2\pi}\int_{-\infty}^{\infty} M_c(j\omega)\,d\omega$ $M_c(j\omega) = E[\lvert U(j\omega)\rvert^2]$
discrete-time	$E_d = w_v(0) = \sum_{-\infty}^{\infty} v(n)^2 = \frac{1}{2\pi}\int_{-\pi}^{\pi} W_v(\varepsilon^{j\theta})\,d\theta$ $W_v(\varepsilon^{j\theta}) = \lvert V(\varepsilon^{j\theta})\rvert^2$ Section 2.5.3	$E\sum_{-\infty}^{\infty} V(n)^2 = \frac{1}{2\pi}\int_{-\pi}^{\pi} M_d(\varepsilon^{j\theta})\,d\theta$ $M_d(\varepsilon^{j\theta}) = E[\lvert V(\varepsilon^{j\theta})\rvert^2]$

of the four conversions, then, it is necessary to define the conversion, describe its effect on the Fourier spectrum of deterministic signals, and describe its effect on the spectral density function of ergodic signals. The four cases are summarized in Figure 2.11.

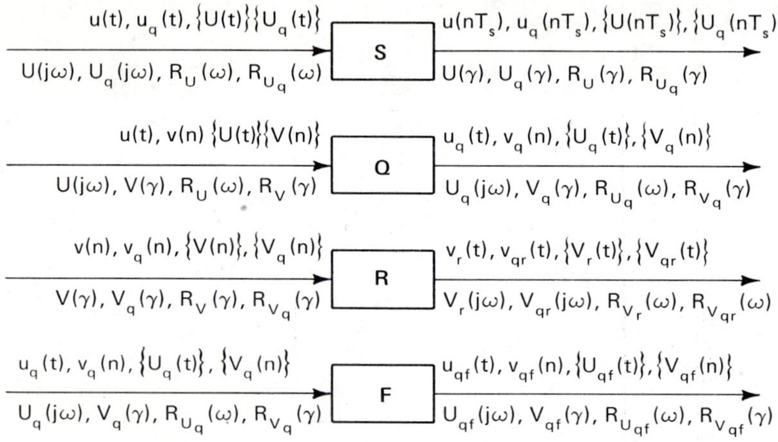

Figure 2.11 Summary of conversion input/output pairs.
(Time domain above the line; frequency domain below the line)

In each case, the time domain input signal is above the line, its spectrum or spectral density function is below the line and the corresponding output functions are in the equivalent positions on the output side. The notational convention of using the same symbol U or V on input and output is for convenience; the arguments and subscripts should make clear which is intended. They are not the same functions. Although there are many results available on sampling and reconstruction of both deterministic and stochastic signals, and some results on quantization, general results for all the cases suggested by Figure 2.11 are not available. This introduction is limited to some of the more common situations; references are cited which provide many interesting details.

2.9.1 Sampling

In general, sampling is defined as an operation on a continuous-time signal such that for each sample time t_n (n integer) of a set $S_n = \{t_n\}$ there is a unique sample. For example, for input u(t), at $t=t_n$ the sample is $u(t_n)$. The set S_n may be found by a deterministic procedure [19.] or the time between samples may be a random variable [20.]. In the former case, uniform sampling is common, and a definition is as follows:

S: For a continuous-time signal (e.g., u(t), $u_q(t)$, U(t), $U_q(t)$), $t \varepsilon T_t$, and a subset S_n of T_t, where

$$S_n = \{t_n | t_n = nT_s, \ t_{n+1} - t_n = T_s, \ n \text{ integer}\}$$

a unique sample $u(t_n)$ is defined for each t_n. T_s is the sampling period and $1/T_s$ the sampling rate, and $u(t_n)$ is a discrete-time signal or function.

Many uniform sampler models can be used, and the usual ones start with a signal called a sampling train or sequence of strobe "pulses". For example [19.] [20.] a sequence of unit amplitude flat-topped pulses of duration τ and uniform spacing T_s seconds is called a sampling function $p_{T_s}(t,\tau)$ and the continuous-time output of the sampler is

$$u^*(t) = u(t)p_{T_s}(t,\tau).$$

For $\tau << T_s$ and u(t) such that $|U(j\omega)|$ is small for $|\omega| >> \omega_s$, $\omega_s = 2\pi/T_s$,

$$\begin{aligned} u^*(t) &= u(nT_s), \quad & nT_s \leq t < nT_s + \tau \\ &= 0 \quad , & nT_s + \tau \leq t < (n+1)T_s \end{aligned}$$

so that the output samples $u(nT_s)$ are the amplitudes of $u^*(t)$, $nT_s \leq t < nT_s + \tau$. Other pulse shapes (e.g., a unit exponential) can be used in the sampling function. Some spectra of $u^*(t)$ are given in [19.]. A convenient and widely-used ideal sampling function is a sequence of unit impulses as introduced in Eq. 2.57. In this case, let

$$u_{T_s}(t) = \sum_{n=-\infty}^{\infty} u_o(t-nT_s)$$

then

$$u^*(t) = u(t)u_{T_s}(t) = \sum_{n=-\infty}^{\infty} u(nT_s)u_o(t-nT_s) \qquad (2.103)$$

and the samples are given by

$$u(nT_s) = \int_{-\infty}^{\infty} u(\lambda)u_o(\lambda-nT_s)d\lambda \ . \qquad (2.57)$$

Other integral forms have been used (e.g., see [22.] on Poisson transform sampling) but the uniform impulse sampler is a relatively simple mathematical idealization. Figure 2.12 illustrates this model.

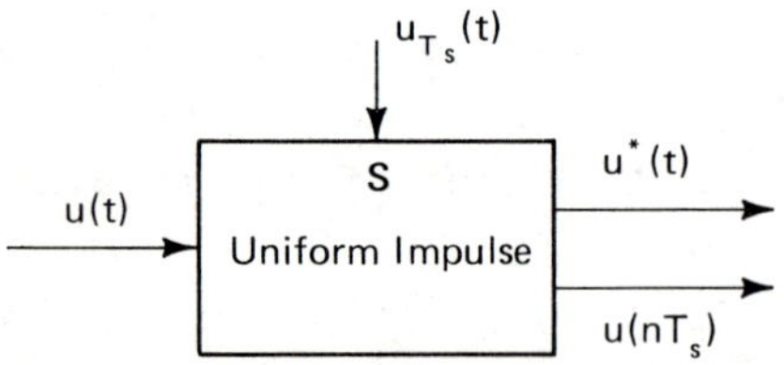

Figure 2.12

In view of Eq. 2.8, $u(\cdot)$ must be a continuous function for this sampler model to be valid. However, it is common to define the sample as the right hand limit if t_n occurs at a finite discontinuity of $u(\cdot)$. Assuming $u(t)$ has a Laplace Transform,

$$U(s) = \mathcal{L}\, u(t),$$

and

$$U^*(s) = \mathcal{L}\, u^*(t)$$
$$= \int_{-\infty}^{\infty} u(t) u_{T_s}(t) \varepsilon^{-st} dt$$
$$= \sum_{n=-\infty}^{\infty} u(nT_s) \exp(-nT_s s).$$

If the substitution $z = \varepsilon^{T_s s}$ introduced in Appendix A is used,

$$U^*(s) \Big|_{z=\varepsilon^{sT_s}} = \sum_{n=-\infty}^{\infty} u(nT_s) z^{-n} \equiv U(z) .$$

To obtain the amplitude spectrum of $u^*(t)$, we use a Fourier Series expansion of $u_{T_s}(t)$, thus

$$u_{T_s}(t) = \frac{1}{T_s} \sum_{k=-\infty}^{\infty} \exp(jk\omega_s t), \quad \omega_s = 2\pi/T_s$$

and therefore,

$$u^*(t) = \frac{1}{T_s} \sum_{k=-\infty}^{\infty} u(t) \exp(jk\omega_s t)$$

and

$$U^*(S) = \mathcal{L}\, u^*(t) = \frac{1}{T_s} \sum_{k=-\infty}^{\infty} U(s+jk\omega_s). \qquad (2.104)$$

The input amplitude spectrum is then $|U(j\omega)|$ and the output amplitude spectrum is $|U^*(j\omega)|$. These are illustrated in Figure 2.13 where it is clear that the uniform impulse sampler has two effects: attenuation by $1/T_s$ and the generation of an infinite number of sidebands. $U^*(j\omega)$ is

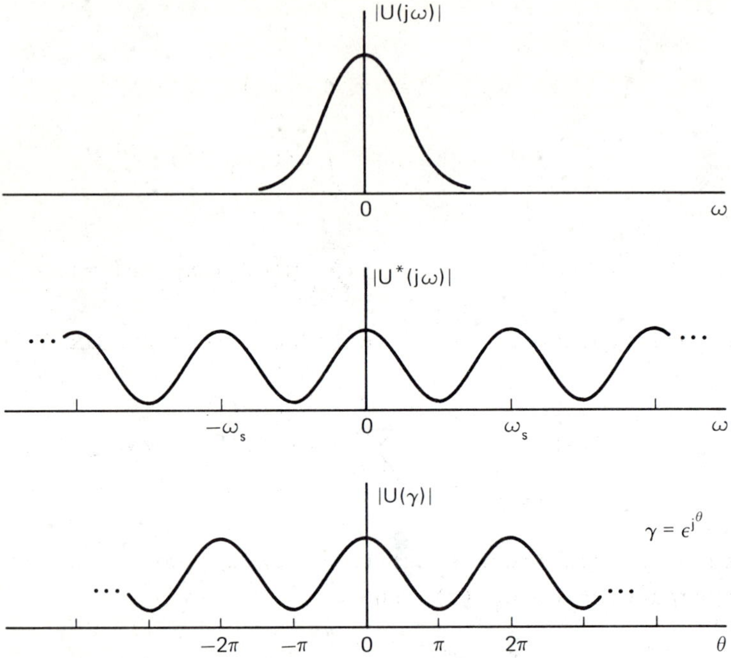

Figure 2.13 Input and output spectra of uniform impulse sampler

periodic with a period ω_s. The spectrum of $u(nT_s)$ is obtained by taking $\mathcal{F}_d\, u(nT_s) = U(\gamma)$ and if there are no poles of $U(z)$ on the unit circle,

$$U(\gamma) = U(z)\Big|_{z=\gamma=\varepsilon^{j\theta}}, \qquad \theta = \omega T_s.$$

Thus, one would plot $|U(\gamma)|$, $-\pi \leq \theta \leq \pi$ and the shape of this spectrum would be similar to $|U(j\omega)|$, $-\omega_s/2 \leq \omega \leq \omega_s/2$. For example, we may compare the spectra of Example 2.10i and Example 2.10j with $T_s=1$ and $b=\varepsilon^{-a}$.

In the special case where $u(t)$ is band-limited, i.e.,

$$U(j\omega) = 0, \qquad |\omega| > \omega_o,$$

a sampling theorem can be proved [19.]:

<u>Sampling Theorem (Deterministic Signals)</u>

> $u(t)$ such that $U(j\omega) = \mathcal{F}u(t)$ exists and
> $U(j\omega) = 0, \ |\omega| > \omega_o$, with $U(j\omega_o)$ finite
> $\Rightarrow$
> $u(t)$ is completely determined for all t by the samples $u(nT_s)$ if $\omega_s \geq 2\omega_o$ and sampler S is used.

Inspection of Figure 2.13 shows that the requirement $\omega_s \geq 2\omega_o$ means that there will be no overlap of the sidebands, and in this case, the output spectrum is simply a repetition of the fundamental (k=0) term with centers of the sidebands at $\omega = \pm k\omega_s$. The form of the exact representation of $u(t)$ from the samples $u(nT_s)$ implied by this theorem is given later in the discussion on reconstruction in Section 2.9.3.

In general, the effect of S is to produce many sidebands in the spectrum of the output signal. The exact effect depends on the particular sampling function, and in the special case of a band-limited signal, uniformly impulse-sampled at $\omega_s \geq 2\omega_o$, the sidebands are duplicates (except for a fixed $1/T_s$ attenuation) of the input spectrum. In other cases, the shapes of the sidebands may be different and a frequency-dependent attenuation may occur [19.]. It also should be noted that $u^*(t)$ is not an energy signal since the impulse sampler supplies "infinite energy".

The foregoing remarks, where $u(t)$ is the input to S, carry over to the case where $u_q(t)$ is the input except that $u_q(t)$ cannot be continuous. Thus the difficulty associated with impulse samples at discontinuities of $u_q(t)$ mentioned earlier can occur. The question is primarily one of mathematical interest since in the physical sampler, impulses cannot be used.

When the input to the sampler is an ergodic process, the outputs in Figure 2.12 become the process $\{U^*(t)\}$ and the sequence $\{U(nT_s)\}$ if uniform impulse sampling is used. In this situation, the sampling train is

$$u_{T_s}(t;\beta) = \sum_{n=-\infty}^{\infty} u_o(t-nT-\beta)$$

where β is a random variable uniformly distributed on the interval $[0,T_s]$. Then it can be shown that $\{U^*(t)\}$ and $\{U(nT_s)\}$ are wide-sense stationary. For $\{U(nT_s)\}$,

$$r_{U(nT_s)}(kT_s) = E[U(nT_s)U(nT_s+kT_s)]. \qquad (2.105)$$

Now consider the input with

$$r_U(\tau) = E[U(t)U(t+\tau)]$$

and for $t=nT_s$

$$r_U(\tau)\Big|_{t=nT_s} = E[U(nT_s)U(nT_s+\tau)].$$

For $\tau = kT_s$,

$$r_U(kT_s) = E[U(nT_s)U(nT_s+kT_s)] . \qquad (2.106)$$

Eq. 2.105 was obtained by sampling $U(t)$ and then taking an expectation, but Eq. 2.106 resulted from taking an expectation to find $r_U(\tau)$ and then "sampling" $r_U(\tau)$ to find $r_U(kT_s)$. In view of the right hand sides,

$$r_{U(nT_s)}(kT_s) = r_U(kT_s)$$

and an interchangeability of expectation and sampling for this situation has been shown. This could have been anticipated since sampling and expectation are linear operations. The spectral density functions of $\{U(t)\}$ and $\{U(nT_s)\}$ are

$$R_U(\tau) = \mathcal{F}\, r_U(\tau)$$

and

$$R_{U(nT_s)}(\gamma) = \mathfrak{F}_d r_{U(nT_s)}(kT_s) = \mathfrak{F}_d r_U(kT_s) .$$

Also, similar to Equation 2.104,

$$R_{U^*}(\tau) = \mathfrak{F}\, r_{U^*}(\tau) = \frac{1}{T_s^2} \sum_{k=-\infty}^{\infty} R_U(\omega + k\omega_s) \qquad (2.107)$$

A set of power spectra similar to the amplitude spectra of Figure 2.13 can be plotted from these. For input $\{U_q(t)\}$, the same remarks on discontinuities made previously for $u_q(t)$ are noted. In the special case where $\{U(t)\}$ is <u>band-limited</u>, i.e.,

$$R_U(\omega) = 0, \qquad |\omega| > \omega_o ,$$

A sampling theorem can be proved [12.]:

<u>Sampling Theorem (Stochastic Signals)</u>

> $\{U(t)\}$ such that it is wide-sense stationary and $R_U(\omega) = 0,\ |\omega| > \omega_o$
> $\Rightarrow$
> $U(t)$ is completely determined for all t by the samples $U(nT_s)$ if $\omega_s \geq 2\omega_o$ and sampler S is used.

It also may be seen that if the process $\{U(t)\}$ is band-limited, the sampling theorem for deterministic signals implies that $r_U(\tau)$ is completely determined by $r_{U(nT_s)}(kT_s) = r_U(kT_s)$ for $\omega_s \geq 2\omega_o$. This is shown in the discussion on reconstruction in Section 2.9.3.

A complete study of the great variety of sampler models and sampling theorems is not the intent here. However, it should be clear from the simpler cases discussed that the conversion S generates high-frequency signal components which were not in the original input and these must be removed if an accurate reconstruction is required.

2.9.2 Quantization

Quantization is defined as an operation on continuous-time or discrete-time signals ($u(t)$ or $v(n)$) such that for a set $L = \{L_1, \ldots, L_j, \ldots, L_p | L_{j+1} > L_j, L_j \text{ real}\}$ and set $A = \{A_1, \ldots, A_j, \ldots, A_{p+1} | A_{j+1} > A_j, A_j \text{ real}\}$ we have unique $u_q(t)$ or $v_q(n)$ for all $t \varepsilon T_c$ or $n \varepsilon T_d$. Thus, for example, $u_q(t)$ is obtained from $u(t)$ as

$$
\begin{array}{rl}
u(t) \le L_1 , & u_q(t) = A_1 \\
L_1 < u(t) \le L_2 , & u_q(t) = A_2 \\
\vdots & \vdots \\
L_{j-1} < u(t) \le L_j , & u_q(t) = A_j \\
\vdots & \vdots \\
L_{p-1} < u(t) \le L_p , & u_q(t) = A_p \\
L_p < u(t) \qquad , & u_q(t) = A_{p+1}
\end{array}
$$

with a similar specification for $v_q(n)$. Usually the sets L and A are chosen so that the quantizer is symmetric for p even with the middle level of A zero. Many quantizer characteristics have been used, and the choice depends on input signal models, errors, and cost. A <u>uniform quantizer</u> is defined as follows:

> Q: For the set L with $L_j = -L_{p-j+1}$, p even, and $L_{j+1} - L_j = K_1$ for all j and the set A with $A_j = -A_{p-j+2}$ for $j \neq p/2+1$, $A_j = 0$ for $j = p/2+1$, and $A_{j+1} - A_j = K_2$ for all j, the outputs u_q and v_q are given for inputs u and v as above.

This quantizer is shown graphically in Figure 2.14. This is a p+1 level quantizer, and it has a quantum size of K_2. Frequently $K_1 = K_2$ for a "unity gain" uniform quantizer.

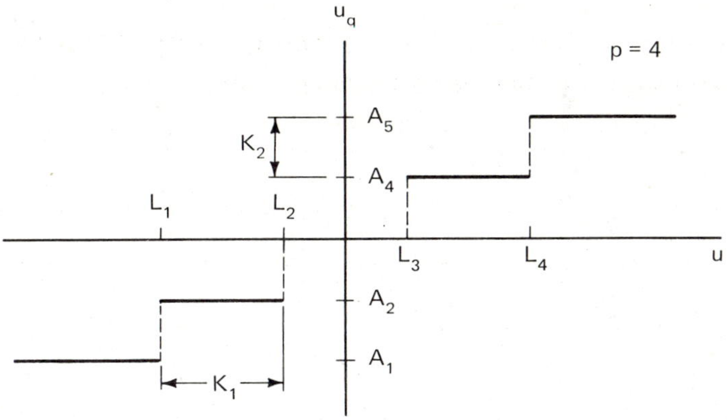

Figure 2.14 A uniform quantizer

A symmetric uniform quantizer for p odd also could be defined. Figure 2.20 illustrates the model that we shall use.

Figure 2.20

Since Q is a nonlinear operation, it is difficult to make general statements concerning input and output spectra. When the input is deterministic, it is possible to find the output spectrum for a particular input signal and quantizer, but the numerical results are not general; the spectrum of the output depends on the amplitude of the input. However, since one general effect of the quantizer is to produce a discontinuous $u_q(t)$ or a $v_q(n)$, it is expected that new spectral components may occur and that relative amplitudes may change with input amplitude. This is illustrated in the following simple examples.

Example 2.22

a) For $u(t) = K \sin \omega_o t$, $K >> 1$, the process of "infinite clipping" produces an approximate output of

$$u_q(t) = 1 \,, \quad u > 0$$
$$= -1 \,, \quad u \leq 0$$

where

$$L = \{0\}$$

$$A = \{-1,1\}, \qquad p = 1.$$

Thus $u_q(t)$ is a square wave with unit amplitude. The Fourier spectra from Eq. 2.45 are

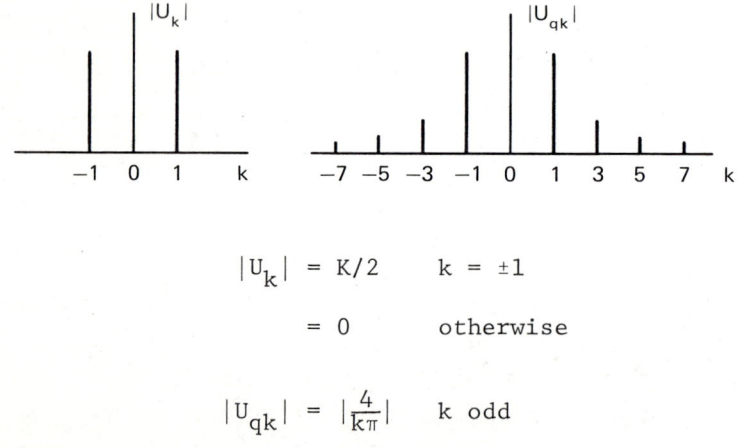

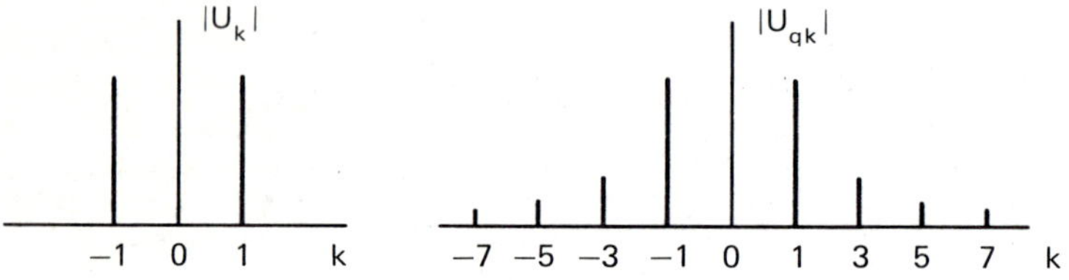

$$|U_k| = K/2 \qquad k = \pm 1$$
$$= 0 \qquad \text{otherwise}$$

$$|U_{qk}| = \left|\frac{4}{k\pi}\right| \qquad k \text{ odd}$$
$$= 0 \qquad k \text{ even} .$$

Thus, all the components for $k = \pm 3, \pm 5, \ldots,$ are generated by the quantization in this example. The Fourier Series expansion of a square wave is well known; this is simply a matter of looking at infinite clipping as a coarse quantization.

b) For input $v(n) = K \sin \frac{n\pi}{4}$, $N_o = 8$ and $\lambda_o = \frac{\pi}{4}$, and for a quantizer with $p = 2$, $K_1 = K_2 = 1$,

$$L = \{-1/2, \quad 1/2\}$$

and

$$A = \{-1 \quad 0, \quad 1\} .$$

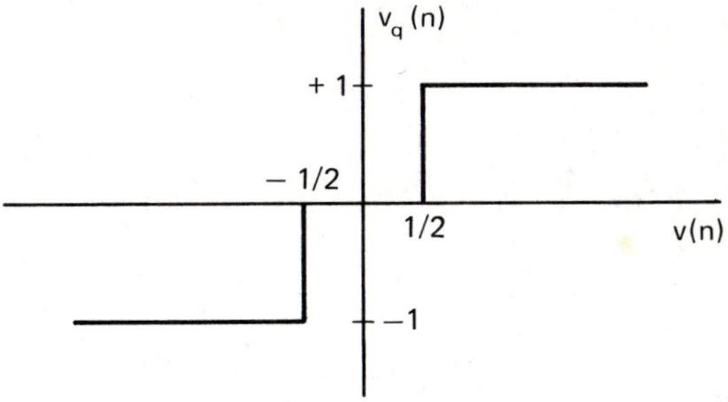

Quantizer Q

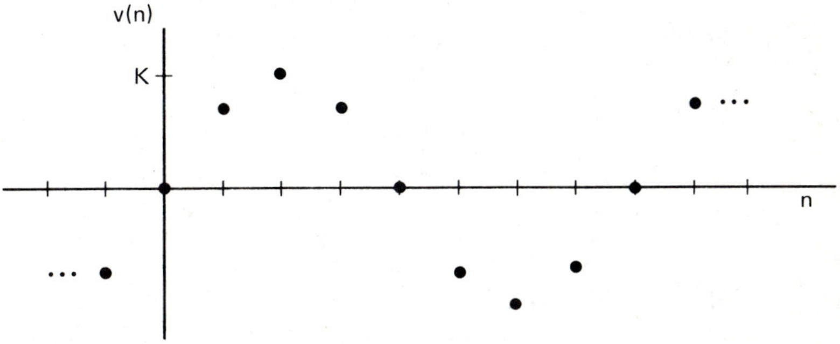

Input

For K = 2 the output is

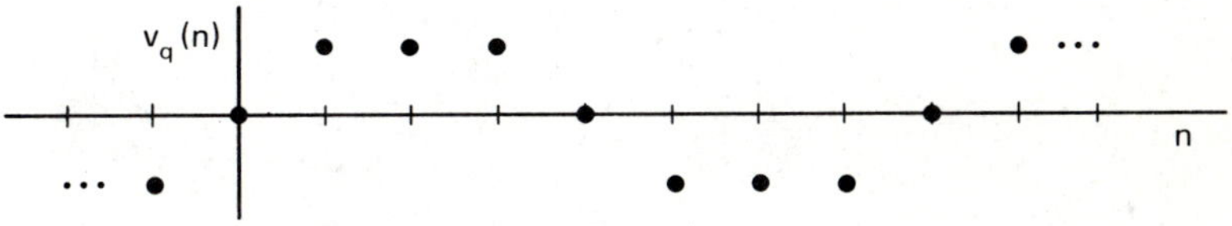

and for K = 0.6 the output is

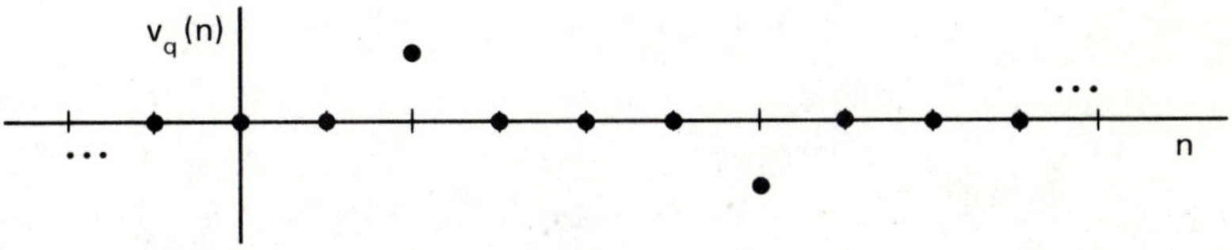

The output spectral components from Eq. 2.47 are for K = 2,

$$|V'_{qk}| = 0 \qquad k = 0,\ \pm 2,\ 4$$

$$|V'_{q,-1}| = |V'_{q1}| = 0.60$$

$$|V'_{q,-3}| = |V'_{q3}| = 0.10$$

and for K = 0.6,

$$|V'_{qk}| = 0 \qquad k = 0,\ \pm 2,\ 4$$

$$|V'_{q,-1}| = |V'_{q1}| = 0.25$$

$$|V'_{q,-3}| = |V'_{q3}| = 0.25 \ ,$$

and for the input,

$$|V'_1| = |V'_{-1}| = K/2$$

$$|V'_k| = 0, \qquad k = 0,\ \pm 2,\ \pm 3,\ 4\ .$$

We see that the output spectrum amplitudes depend on K, and that the quantization produces spectral lines at k = ±3 as well as k = ±1.

c) For input $u(t) = \varepsilon^{-t} u_1(t)$, and the quantizer of Part b),

$$\begin{aligned} u_q(t) &= 0, \qquad t < 0 \\ &= 1, \qquad 0 \le t < 0.69 \\ &= 0, \qquad 0.69 \le t \end{aligned}$$

and

$$|U(j\omega)| = \frac{1}{\sqrt{1+\omega^2}}$$

and

$$|U_q(j\omega)| = \frac{1}{\omega}\sqrt{(1-\cos 0.69\omega)^2+(\sin 0.69\omega)^2}$$

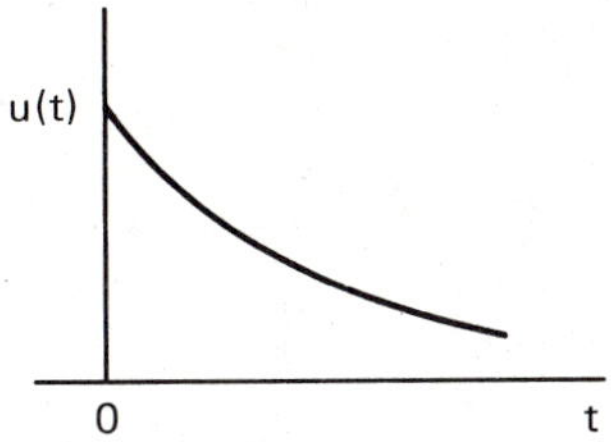

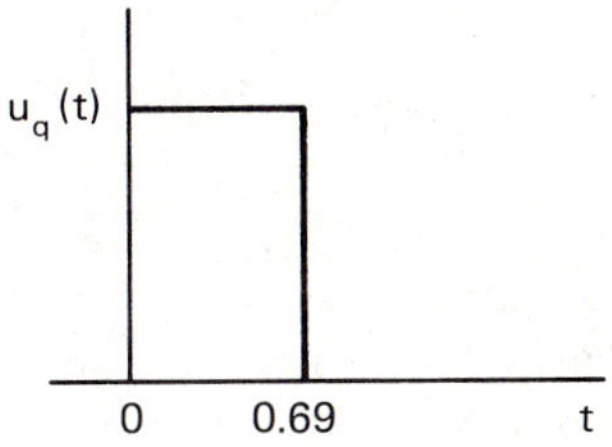

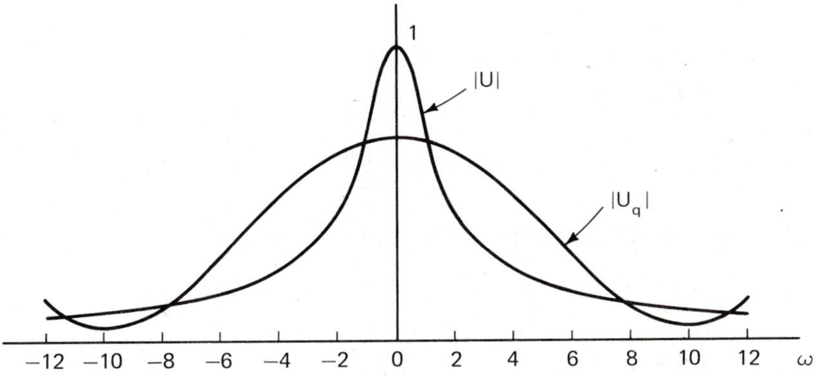

Here again we observe that the spectrum of u_q is "broader" than that of u.

When the input of the quantizer is stochastic, more general results can be obtained since performance in terms of moments, density functions or autocorrelation and spectral density functions can be found. The input stochastic signals are assumed wide-sense stationary in all cases. To illustrate the effect of coarse quantization on the signal bandwidth, the following simple example, which is analogous to Example 2.22a, is used.

Example 2.23

For {U(t)} a stationary Gaussian process with zero mean and the quantizer of Example 2.22a ("infinite clipper"), the output autocorrelation function

$$r_{U_q}(\tau) = E[U_q(t)U_q(t+\tau)]$$

$$= 1\cdot Pr(U_q(t)U_q(t+\tau)>0)$$

$$- 1\cdot Pr(U_q(t)U_q(t+\tau)<0)$$

and from this [12.][23.],

$$r_{U_q}(\tau) = \frac{2}{\pi} \arcsin \frac{r_U(\tau)}{r_U(0)} .$$

For $r_U(\tau) = K(\exp -a|\tau|)$, $K >> 1$, the output autocorrelation function is

$$r_{U_q}(\tau) = \frac{2}{\pi} \arcsin (\exp -a|\tau|).$$

A numerical illustration for a=2 is shown.

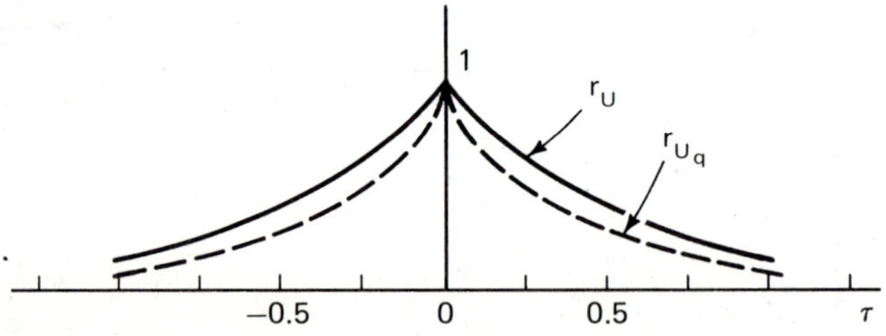

Since r_{U_q} is narrower than r_U, the corresponding spectral density $R_{U_q}(\omega)$ will be broader than $R_U(\omega)$.

Some early results on the effect of a uniform quantizer on a band-limited white Gaussian input were obtained in [24.] for the spectral density function of the error $E = U-U_q$. Data for 2^4 to 2^8 quantization levels show that the error or noise band width is much greater than that of the signal. A useful model of the effect of quantization was developed by Widrow [25.][26.] and later refined [27.][28.][29.] to show conditions under which the quantizer Q above can be approximated by a model where the quantization noise is an additive, uniformly-distributed independent (of the input) stochastic signal.

We start with the first density function of the input U(t), i.e., $p_U(\lambda)$, and we wish to develop an analogy to the uniform impulse sampling of the previous section. Furthermore, we will find it convenient to use a "frequency domain" rather than a "time domain", and we, therefore, use the characteristic functions of Appendix C, Section C.5. In analogy to $u_{T_s}(t)$, let

$$u_\Delta(\lambda) = \sum_{k=-\infty}^{\infty} u_o(\lambda - k\Delta)$$

where $\Delta = K_1 = K_2$ is the quantizer increment. We could then try to impulse sample $p_U(\lambda)$ to obtain an output density function. However, a sample of $p_U(\lambda)$, i.e., a number $p_U(\lambda_o)$, does not have the meaning we wish. (Recall the discussion of density functions for continuous random variables in Appendix C.) The form we want is, in analogy to 2.103:

$$p^*_{U_q}(\lambda) = \sum_{k=-\infty}^{\infty} p_{U_q}(k\Delta)u_o(\lambda - k\Delta) \qquad (2.108)$$

where U_q is the discrete random variable output and where

$$p_{U_q}(k\Delta) = \Pr(U_q = k\Delta)$$

is the first density function of U_q. However, rather than using the density functions of U and U_q directly, Widrow introduced the concept of "area sampling". We define

$$g_U(\lambda) = \Pr(\lambda - \tfrac{\Delta}{2} \leq U < \lambda + \tfrac{\Delta}{2})$$

which can be found from

$$g_U(\lambda) = P_U(\lambda + \tfrac{\Delta}{2}) - P_U(\lambda - \tfrac{\Delta}{2})$$

or

$$g_U(\lambda) = \int_{-\infty}^{\lambda} p_U(\sigma + \tfrac{\Delta}{2})\,d\sigma - \int_{-\infty}^{\lambda} p_U(\sigma - \tfrac{\Delta}{2})\,d\sigma \ . \qquad (2.109)$$

Then we impulse sample $g_U(\lambda)$ to obtain

$$g_U^*(\lambda) = g_U(\lambda)u_\Delta(\lambda) \ . \qquad (2.110)$$

Now using 2.109,

$$\begin{aligned} g_U(k\Delta) &= \Pr(k\Delta - \tfrac{\Delta}{2} \leq U < k\Delta + \tfrac{\Delta}{2}) \\ &= \int_{-\infty}^{k\Delta} p_U(\sigma + \tfrac{\Delta}{2})\,d\sigma - \int_{-\infty}^{k\Delta} p_U(\sigma - \tfrac{\Delta}{2})\,d\sigma \end{aligned}$$

and since

$$\Pr(U_q = k\Delta) = \Pr(k\Delta - \tfrac{\Delta}{2} \leq U < k\Delta + \tfrac{\Delta}{2}) \ ,$$

we have

$$g_U(k\Delta) = p_{U_q}(k\Delta)$$

so that 2.108 becomes

$$p_{U_q}^*(\lambda) = \sum_{k=-\infty}^{\infty} g_U(k\Delta)u_o(\lambda - k\Delta) \ .$$

We also can write 2.110 as

$$\begin{aligned} g_U^*(\lambda) &= \sum_{k=-\infty}^{\infty} g_U(k\Delta)u_o(\lambda - k\Delta) \\ &= p_{U_q}^*(\lambda) \ . \end{aligned} \qquad (2.111)$$

The analogy of 2.111 to 2.103 is now obvious; $g_U(\lambda)$, $g_U^*(\lambda)$, $g_U(k\Delta)$, and $u_\Delta(\lambda)$ have their counterparts in Figure 2.12. The $g_U(k\Delta)$ are the "area samples" of $p_U(\lambda)$ or the samples of $g_U(\lambda)$.

The characteristic function of the input U is

$$\psi_U(\nu) = E[\varepsilon^{j\nu U}]$$
$$= \int_{-\infty}^{\infty} p_U(\lambda)\varepsilon^{j\nu\lambda} d\lambda \ .$$

To find ψ_{U_q}, the characteristic function of U_q, we could use $\mathfrak{F}\, p_{U_q}^*(\lambda)$ from the result in 2.111. However, we can develop a more useful form for obtaining a model by starting with 2.109; we therefore, take

$$\mathfrak{F}\, g_U(\lambda) = \frac{(\exp\ j\nu\Delta/2)}{j\nu}\ \mathfrak{F}\, p_U(\lambda) - \frac{(\exp\ -j\nu\Delta/2)}{j\nu}\ \mathfrak{F}\, p_U(\lambda)$$
$$= \mathfrak{F}\, p_U(\lambda)\cdot\Delta\left(\frac{\sin\frac{\Delta}{2}\nu}{\frac{\Delta}{2}\nu}\right)$$

or

$$= \psi_U(-\nu)\cdot\Delta\left(\frac{\sin\ (-\frac{\Delta}{2}\nu)}{(-\frac{\Delta}{2}\nu)}\right) \ .$$

Now regarding the quantizer as a uniform "area sampler", the left side is

$$\left[\mathfrak{F}\, g_U(\lambda)\right]^* = \left[\mathfrak{F}\, g_U^*(\lambda)\right] = \psi_{U_q}(-\nu) \ .$$

We then have

$$\psi_{U_q}(\nu) = \left[\psi_U(\nu)\cdot\Delta\left(\frac{\sin\frac{\Delta}{2}\nu)}{\frac{\Delta}{2}\nu}\right)\right]^*$$

and using Equation 2.104 with $T_s \to \Delta$, $s \to j\nu$, and $\nu_s = 2\pi/\Delta$,

$$\psi_{U_q}(\nu) = \sum_{k=-\infty}^{\infty} \psi_U(\nu + k\nu_s) \frac{\sin \frac{\Delta}{2}(\nu + k\nu_s)}{\frac{\Delta}{2}(\nu + k\nu_s)} . \qquad (2.112)$$

Thus the characteristic function of the uniform quantizer output is obtained from the characteristic function of the input by summing terms of the form

$$\psi_U(\nu + k\nu_s) \ \mathrm{Sa}\left(\tfrac{\Delta}{2}(\nu + k\nu_s)\right) .$$

The form of Eq. 2.112 when compared to Eq. 2.107 suggests a quantizing theorem for cases where the input U has a "band-limited" characteristic function, i.e.,

$$\psi_U(\nu) = 0, \qquad |\nu| > \nu_o .$$

This is a fictitious condition since it implies that $p_U(\cdot)$ cannot be nonnegative as is required of a density function (see Section C.3). However, many practical cases of interest approximate this condition.

We have then:

<u>Quantizing Theorem</u> (<u>Stochastic Signals</u>)

> $\{U(t)\}$ such that $\psi_U(\nu) = 0$, $|\nu| > \nu_o$
>
> $\Rightarrow$
>
> the sections of ψ_{U_q} for each k do not overlap for $\nu_s \geq 2\nu_o$.

This quantizing theorem means that $\psi_U(\nu)$ and $p_U(\lambda)$ can be recovered from $\psi_{U_q}(\nu)$ (or $p^*_{U_q}(\lambda)$) when the theorem is satisfied. Thus, the first-order density function of the input is recoverable from that of the quantizer output in a manner similar to the recovery of u(t) from $u(nT_s)$ when u(t) is band-limited and sampled by S with $\omega_s \geq 2\omega_o$. Figure 2.15 illustrates this result.

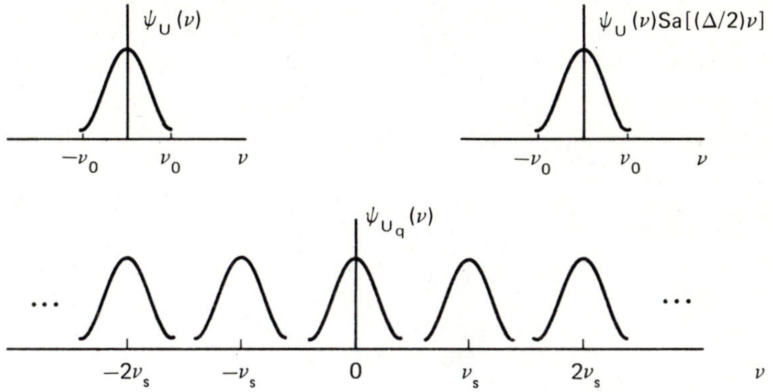

Figure 2.15 Effect of area sampling with $\nu_s > 2\nu_0$

Equation 2.112 relates the uniform quantizer input and output first-order statistics, and a similar sequence of steps leads to a relationship between the second-order statistics [27.]; here the concept of "volume sampling" is used:

$$\psi_{U_{1q}U_{2q}}(\nu_1,\nu_2) = \sum_{i=-\infty}^{\infty}\sum_{k=-\infty}^{\infty}\psi_{U_1U_2}(\nu_1+i\nu_s,\ \nu_2+k\nu_s)\cdot$$

$$\cdot\left(\frac{\sin\frac{\Delta}{2}(\nu_1+i\nu_s)}{\frac{\Delta}{2}(\nu_1+i\nu_s)}\right)\left(\frac{\sin\frac{\Delta}{2}(\nu_2+k\nu_s)}{\frac{\Delta}{2}(\nu_2+k\nu_s)}\right) \qquad (2.113)$$

A quantizing theorem for the second-order density functions also can be stated, and it also has been shown [27.][28.][29.] that if only the moments of the first and second-order density functions must be recovered, the condition $\nu_s \geq \nu_0$ is sufficient.

The quantization error or noise is $N = U-U_q$ and its characteristic function is [27.]

$$\psi_N(\nu) = \sum_{k=-\infty}^{\infty}\psi_U(k\nu_s)\left\{\frac{\sin\frac{\Delta}{2}(\nu-k\nu_s)}{\frac{\Delta}{2}(\nu-k\nu_s)}\right\} \qquad (2.114)$$

If $\psi_U(\nu) = 0,\ |\nu| \geq \nu_0 = \nu_s$, there is only one nonzero term in this sum, and

$$\psi_N(\nu) = \psi_U(0) \frac{\sin \frac{\Delta}{2} \nu}{\frac{\Delta}{2} \nu} = \frac{\sin \frac{\Delta}{2} \nu}{\frac{\Delta}{2} \nu} = Sa(\tfrac{\Delta}{2} \nu). \qquad (2.115)$$

Using the inverse Fourier transform of this,

$$p_N(\lambda) = \frac{1}{\Delta}, \qquad |\lambda| < \frac{\Delta}{2} \qquad (2.116)$$
$$= 0, \qquad |\lambda| > \frac{\Delta}{2}$$

Thus, when the quantizing theorem is satisfied (i.e., $\nu_s \geq 2\nu_o$) or even "half-satisfied" (i.e., $\nu_s \geq \nu_o$), the quantization noise N is a uniformly distributed zero mean random variable. Inspection of Eq. 2.112 shows that the term for k=0 is

$$\psi_U(\nu) \left(\frac{\sin \frac{\Delta}{2} \nu}{\frac{\Delta}{2} \nu} \right) = \psi_U(\nu)\psi_N(\nu) .$$

When the quantizing theorem is satisfied, the moments of U_q can be found from the derivatives of this product at $\nu = 0$ (see Appendix C). In addition, a simple model of the quantizer can be used where for

$$\psi_{U_q}(\nu)\Big|_{k=0} = \psi_U(\nu)\psi_N(\nu) = \psi_{U_q'}(\nu),$$

where U_q' is the sum of U and a statistically independent uniformly-distributed noise N. This model, which is equivalent as far as moments of the output are concerned, is as shown in Figure 2.16. (The circle, as we shall see in Chapter 3, represents a summer.)

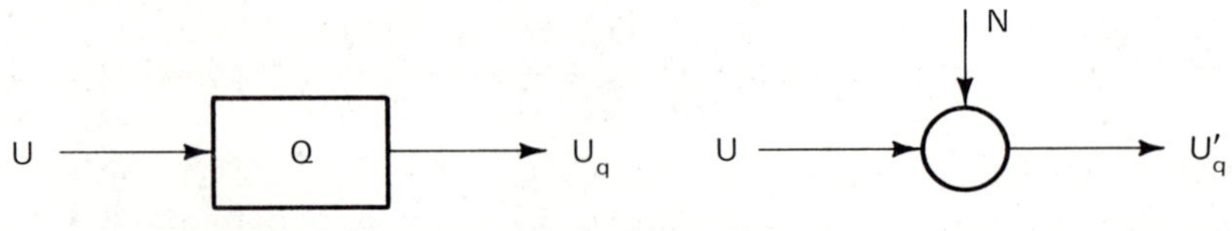

Figure 2.16 Quantizer model for moment calculation

The result on statistical independence follows from the product form of $\psi_{U_q'}(\nu)$ as shown in Appendix C, Theorem C.1. The mean square U_q' is then

$$S_{U_q'}^{\ 2} = S_U^{\ 2} + S_N^{\ 2}$$

and using $p_N(\lambda)$ given by Eq. 2.116,

$$S_N^{\ 2} = \frac{1}{12}\,\Delta^2 \ .$$

Thus when $\nu_s \geq 2\nu_o$,

$$S_{U_q}^{\ 2} = S_U^{\ 2} + \frac{1}{12}\,\Delta^2 \ . \qquad (2.117)$$

When $\psi_U(\nu)$ is not "band-limited", Equation 2.117 is an approximation and for U a zero-mean, Gaussian random variable, the size of the error between $S_{U_q}^{\ 2}$ and $S_{U_q'}^{\ 2}$ has been computed [27.].

Example 2.24

For $\{U(t)\}$ a zero-mean, Gaussian process,

$$\psi_U(\nu) = (\exp - \frac{\sigma^2\nu^2}{2}).$$

Since this is not "band-limited", an arbitrary choice of ν_o is required. If $\nu_o = \frac{\pi}{2\sigma}$ is chosen,

$$\psi_U(\nu_o) \cong (\exp -1.235) = 0.176$$

and $\psi_U(\nu_o)/\psi_U(0) = 0.176$. For

$$\nu_s = \frac{2\pi}{\Delta} = 2\nu_o \ ,$$

we get

$$\Delta = \frac{\pi}{\nu_o} = 2\sigma \ .$$

Using this,

$$S_N^2 = \frac{\Delta^2}{12} = \frac{\sigma^2}{3}$$

and

$$S_{U'_q}^2 = \sigma^2 + \frac{\sigma^2}{3} = \frac{4}{3}\sigma^2 .$$

Using [27.], the error in this mean square is $3.1 \times 10^{-2}\sigma^2$. This is, of course, a very coarse quantization. A suitable practical choice of p=2 gives the quantization characteristic shown. The probability that U is within the 3σ limits

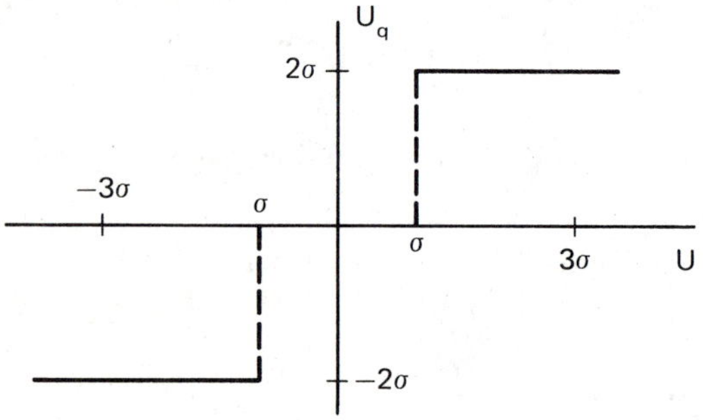

is 0.9973 and the error estimate on the mean square output is then slightly in error for this choice of finite p. A useful rule suggested in [30.] is that the range of U be at least 8Δ so that the error in the mean square for a Gaussian process is less than $10^{-6}\sigma^2$.

Although this summary of the "area sampling" model of Q has shown how to make input/output mean square calculations, the correlation functions and spectral density functions have not been introduced. Equation 2.113 gives the output second-order characteristic function, and the output second density function is then

$$p_{U_{1q}U_{2q}}(\lambda_1,\lambda_2) = \frac{1}{4\pi^2}\int\int_{-\infty}^{\infty} \psi_{U_{1q}U_{2q}}(\nu_1,\nu_2)\cdot$$

$$\cdot(\exp -j\nu_1\nu_1)(\exp -j\lambda_2\nu_2)d\nu_1 d\nu_2$$

and using this, the output autocorrelation function is

$$r_{U_q}(t_1,t_2) = E[U_q(t_1)U_q(t_2)]$$
$$= \int_{-\infty}^{\infty}\int u_{1q}u_{2q}p_{U_{1q}U_{2q}}(u_{1q},u_{2q})du_{1q}du_{2q}. \quad (2.118)$$

Since $\{U(t)\}$ was assumed wide-sense stationary, $\{U_q(t)\}$ is also, and the output spectral density function is

$$R_{U_q}(\omega) = \mathfrak{F}\, r_{U_q}(\tau), \quad \tau = t_2-t_1. \quad (2.119)$$

The actual steps for carrying this out are not done here although a general result is given in [29.].

In terms of Figure 2.11, cases where the input to Q are u(t), v(n), and $\{U(t)\}$ have been considered. When the input is $\{V(n)\}$, a parallel to the preceding development could be carried out. Basically, the change is one of replacing U by V in all expressions. The discrete-time argument appears only in the autocorrelation function.

Comparison of the results for S and Q suggests a set of analogous terms when "area sampling" is used to model Q. The following table summarizes some of these.

S - Time Sampling	Q - Area Sampling
T_s	Δ
$\omega_s = 2\pi/T_s$	$\nu_s = 2\pi/\Delta$
u(t), U(t)	$g_U(\lambda)$ (from $p_U(\cdot)$)
$U(j\omega)$, $R_U(\omega)$	$\psi_U(\nu)$
$\omega_s \geq 2\omega_o$	$\nu_s \geq 2\nu_o$

Since S and Q have both been defined at this point, the question of their interchangeability mentioned in Section 2.1 can now be considered. Because S and Q as they are defined here are "zero memory" operations, they may be interchanged. However, when a uniform impulse sampler is used, there are mathematical difficulties when u(t) (or

U(t)) is discontinuous. Attempts to use right or left limits when a sample time t_n occurs at a discontinuity invalidate the zero memory condition. The lack of interchangeability occurs then only in the very unlikely event that t_n is such that $u(t_n)$ equals a level L_j. For all practical purposes, S and Q may be interchanged. Interesting results on the combined effect of sampling and quantizing are given in [24.], [31.], and [32.]. The design of optimal uniform quantizers is discussed in [33.].

2.9.3 Reconstruction

The third signal conversion, reconstruction or R, is limited here to linear operations. Basically, for any discrete-time inputs v(n), $v_q(n)$, or V(n), or $V_q(n)$ it is necessary to construct corresponding signals for all t. This is, in a sense, an inverse operation to S, and recalling that one effect of S was to generate high-frequency components in the spectrum of the sampled signal, the operation R should remove these components if good reconstruction is to be achieved. R generally will have the effect of a low-pass filter.

R: For equally-spaced discrete-time data, $n \varepsilon T_d$, (e.g., v(n) or V(n)), a function (e.g., $v_r(t)$ or $V_r(t)$) is defined for $nT_s \le t < (n+1)T_s$.

It was pointed out in the sampling theorem for deterministic signals that a signal u(t), band-limited to ω_o, was completely determined by its samples $u(nT_s)$. This, of course, would constitute ideal reconstruction. We consider this first. Starting with a band-limited u(t) and using the inverse Fourier Transform, we have

$$u(t) = \frac{1}{2\pi}\int_{-\infty}^{\infty} U(j\omega)(\exp j t\omega)d\omega$$

$$= \frac{1}{2\pi}\int_{-\omega_o}^{\omega_o} U(j\omega)(\exp j t\omega)d\omega \ .$$

Therefore,

$$u(nT_s) = \frac{1}{2\pi}\int_{-\omega_o}^{\omega_o} U(j\omega)(\exp\ jnT_s\omega)d\omega \ .$$

For $u^*(t) = u(t)u_{T_s}(t)$,

$$U^*(j\omega) = \sum_{n=-\infty}^{\infty} u(nT_s)(\exp\ -jnT_s\omega)$$

as shown in the discussion on sampling in Section 2.9.1. Since $U^*(j\omega)$ is periodic, it has a Fourier Series expansion

$$U^*(j\omega) = \sum_{n=-\infty}^{\infty} U_n(\exp\ -jn\omega T_s)$$

with

$$U_n = \frac{T_s}{2\pi}\int_{-\frac{\omega_s}{2}}^{\frac{\omega_s}{2}} U^*(j\omega)(\exp\ jn\ T_s)d\omega, \quad \omega_s = \frac{2\pi}{T_s} \ .$$

From Eq. 2.104,

$$U^*(j\omega) = \frac{1}{T_s}\ U(j\omega), \quad \frac{-\omega_s}{2} \le \omega \le \frac{\omega_s}{2} \ ,$$

since there is no overlap of the sidebands, or since $\omega_s = 2\omega_o$, we have

$$U_n = \frac{1}{2\pi}\int_{-\frac{\omega_s}{2}}^{\frac{\omega_s}{2}} U(j\omega)(\exp\ jn\omega T_s)d\omega = u(nT_s) \ .$$

Then for $-\frac{\omega_s}{2} \le \omega \le \frac{\omega_s}{2}$, we get

$$U(j\omega) = T_s U^*(j\omega) = T_s \sum_{n=-\infty}^{\infty} u(nT_s)(\exp\ -jn\omega T_s)$$

and

$$U(j\omega) = 0, \qquad |\omega| > \omega_o \ .$$

Then using the inverse Fourier Transform,

$$u(t) = \mathfrak{F}^{-1} U(j\omega)$$

$$= \frac{T_s}{2\pi} \int_{-\frac{\omega_s}{2}}^{\frac{\omega_s}{2}} \left[\sum_{n=-\infty}^{\infty} u(nT_s)(\exp -jnT_s) \right] (\exp jt\omega) d\omega$$

which can be written

$$u(t) = \sum_{n=-\infty}^{\infty} u(nT_s) \frac{\sin \frac{\omega_s}{2}(t-nT_s)}{\frac{\omega_s}{2}(t-nT_s)}$$

$$= \sum_{n=-\infty}^{\infty} u(nT_s)\ Sa(\frac{\omega_s}{2}(t-nT_s)) \ . \qquad (2.120)$$

Thus, we see that u(t) can be reconstructed exactly from its samples if u(t) is band-limited, $\omega_s \geq 2\omega_0$, and <u>all</u> samples $u(nT_s)$ are used. Comparison with Eq. 2.58 shows that the <u>ideal</u> hold is

$$u_h(t,n) = Sa(\frac{\omega_s}{2}(t-nT_s)) \ .$$

Furthermore, if $u(t) = Sa(\frac{\omega_s t}{2})$, we get

$$u(nT_s) = 1 \qquad n = 0$$

$$= 0 \qquad n \neq 0$$

or

$$u(nT_s) = v_o(n).$$

From this, it is obvious in view of Eq. 2.120 that $Sa(\frac{\omega_s t}{2})$ is the unit pulse response of an ideal reconstructor and its transfer function (see Chapter 3) is (Table A.1 in App. A),

$$H_i(j\omega) = \frac{2\pi}{\omega_s}[u_1(\omega + \frac{\omega_s}{2}) - u_1(\omega - \frac{\omega_s}{2})] = \mathfrak{F}^{-1} Sa(\frac{\omega_s t}{2})$$

as illustrated in Figure 2.19. This is an ideal low pass

filter. An approximate reconstruction could be done if a finite number of terms in Eq. 2.120 were used. For example, if only one term is used, a simple choice is

$$u_r(t) = u(nT_s)Sa(\frac{\omega_s}{2}(t-nT_s), \quad nT_s \le t < (n+1)T_s$$

but this is very crude. More terms could be used, but the Sa function is still not a convenient choice to implement. Therefore, other practical reconstructors which use a finite number of samples are considered.

One class of reconstructors is the polynomial type. This requires that samples of the original u(t) and samples of its derivatives be known. Thus we let

$$u_r(t) = u(nT_s) + u'(nT_s)(t-nT_s) + \ldots + \frac{u^{(k)}(nT_s)}{k!}(t-nT_s)^k \qquad (2.121)$$

$$nT_s \le t < (n+1)T_s \quad .$$

Theoretically the required $u(nT_s),\ldots,u^{(k)}(nT_s)$ could be obtained using sampling trains of $u_o(\cdot),\ldots,u_{-k}(\cdot)$ as suggested by Equation 2.11. However, a more practical reconstructor would use only the samples $u(nT_s)$ and approximate the derivatives with higher-order differences. The simplest reconstructor of this form follows from Eq. 2.121 for k=0. Thus, we have

$$u_r(t) = u(nT_s), \quad nT_s \le t < (n+1)T_s. \qquad (2.122)$$

This is the well-known zero-order hold, and an example is shown in Figure 2.5b. A real spectral form for this is given by Eq. 2.58. For k=1, if $u'(nT_s)$ is replaced by

$$\frac{1}{T_s}(\Delta_b\, u(nT_s)) = \frac{1}{T_s}[u(nT_s)-u((n-1)T_s],$$

we obtain

$$u_r(t) = u(nT_s) + \frac{1}{T_s}[u(nT_s)-u((n-1)T_s)](t-nT_s) \qquad (2.123)$$

$$nT_s \le t < (n+1)T_s$$

and this is a <u>first-order</u> <u>hold</u>. An example of this reconstruction is shown in Figure 2.17.

This proceduee could be continued for k = 2,3,... by using Δ_b^2, Δ_b^3,..., but only the first two have found very wide application. One important feature of this class of

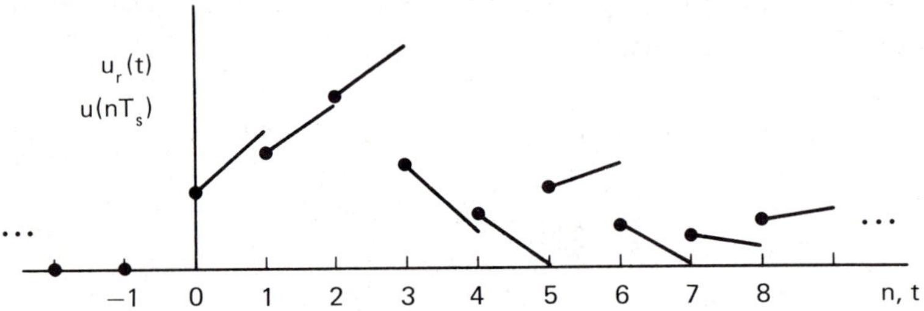

Figure 2.17 First-order hold reconstruction

reconstructors is that only $u(nT_s)$, $u((n-1)T_s)$, $u((n-2)T_s)$, ..., i.e., present and past inputs, are used to extrapolate over one sample period. This is a desirable property for real-time applications. Since the zero-order hold tends to produce a $u_r(t)$ which does not follow changes in the data fast enough, and the first-order hold tends to produce a $u_r(t)$ which changes too quickly, a <u>fractional-order</u> <u>hold</u> is sometimes used [34.]:

$$u_r(t) = u(nT_s) + \frac{\alpha}{T_s}[u(nT_s)-u((n-1)T_s)](t-nT_s) \qquad (2.124)$$

$$nT_s \le t < (n+1)T_s$$

$$0 \le \alpha \le 1$$

and α = 0.3 to 0.5 has been found to be a good compromise.

Another type of reconstructor is a polygonal reconstructor where the adjacent samples are connected by straight lines. Thus

$$u_r(t) = \frac{1}{T_s} [u((n+1)T_s)(t-nT_s)+u(nT_s)((n+1)T_s-t)] \quad (2.125)$$

$$nT_s \leq t < (n+1)T_s$$

This is illustrated in Figure 2.18. This type of reconstruction usually is more accurate, but it requires $u((n+1)T_s)$ to find $u_r(t)$, $nT_s \leq t < (n+1)T_s$ and this means a delay of T_s seconds in a real-time application. Other

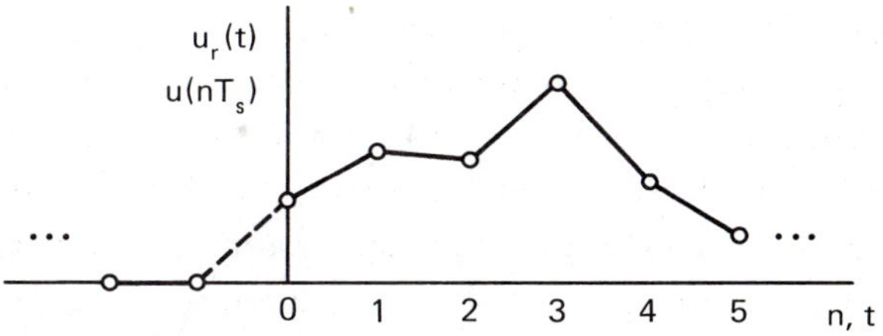

Figure 2.18 Polygonal reconstruction

reconstructors such as exponential holds and optimal linear (minimum mean square error) reconstructors [35.] also have been used.

Figure 2.11 indicates reconstruction for both deterministic and stochastic signals. Since the above reconstructors are linear operations, the methods of Chapter 4 for calculating responses for both deterministic and stochastic inputs can be used. These ideas are not developed in detail here although Example 2.25 shows one case. A point of special interest regarding the reconstruction of stochastic signals concerns the stationarity of the output when the input is a wide-sense stationary sequence. If for the ensemble $\{U(t)\}$, all realizations $u^i(t)$ are sampled relative to the same set of sample points, the process $\{U_r(t)\}$ is not stationary but has a special periodicity [6.]. However, if the phase of the samplers is a random variable, the process

$\{U_r(t)\}$ is wide-sense stationary. For example, for a zero-order hold, if $u_h(t,n)$ of Eq. 2.58 is changed to

$$u_h(t,n;\beta) = u_1(t-nT_s-\beta)-u_1(t-n+1)T_s-\beta)$$

where β is a random variable uniformly distributed for $0 \leq \beta < T_s$, then

$$U_r(t) = \sum_{n=-\infty}^{\infty} U(nT_s+\beta)u_h(t,n,\beta). \tag{2.126}$$

This is used in Example 2.25.

In choosing a reconstructor for a specified input signal model, one generally is interested in making the reconstruction error, e.g., $u(t)-u_r(t)$, small. For deterministic signals, there may be an exact reconstruction. For example, a first-order hold could provide an exact reconstruction of the samples of the unit ramp $u_2(t)$. However, it is more likely that reconstruction for several kinds of signals is needed, and in this case a compromise is made between accuracy and complexity of implementation. For stochastic signals, comparisons of reconstructors for wide-sense stationary Markov processes and band-limited white noise inputs have been calculated using mean square error as a measure of quality [36.]. For the Markov process case, the zero-order hold is very much better than the first-order hold but inferior to the polygonal hold.

In addition to comparing reconstructors on the basis of the reconstructed wave forms as in Figures 2.5b, 2.17, and 2.18 or on the basis of reconstruction errors, the response to a unit pulse $v_o(n)$ can be used. The output in this case will be called $h(t)$. Thus for the zero-order hold,

$$\begin{aligned} h_o(t) &= 1 \qquad 0 \leq t < T_s \\ &= 0 \qquad \text{otherwise} \quad . \end{aligned}$$

Taking the Laplace transform,

$$\mathcal{L}\, h_o(t) = H_o(s) = \frac{1-(\exp -sT_s)}{s} \,. \qquad (2.127)$$

Similarly, for the first-order hold.

$$\begin{aligned} h_1(t) &= 1 + t/T_s & 0 \le t < T_s \\ &= -\frac{1}{T_s}(t-T_s) & T_s \le t < 2T_s \\ &= 0 & \text{otherwise} \end{aligned}$$

and

$$H_1(s) = H_o(s)^2 \left(\frac{1+T_s s}{T_s}\right) \,. \qquad (2.128)$$

Similar results can be obtained for the other reconstructors. Comparisons can then be made using the magnitudes and angles of $H_o(j\omega)$, $H_1(j\omega)$, etc. For example, we have

$$|H_o(j\omega)| = \frac{2\pi}{\omega_s} \left| \frac{\sin \frac{\pi\omega}{\omega_s}}{\frac{\pi\omega}{\omega_s}} \right|$$

and

$$|H_1(j\omega)| = \frac{2\pi}{\omega_s} \left(1 + \frac{4\pi^2\omega^2}{\omega_s^2}\right) \left(\frac{\sin \frac{\pi\omega}{\omega_s}}{\frac{\pi\omega}{\omega_s}}\right)^2 \,. \qquad (2.129)$$

These are plotted in Figure 2.19, and the low-pass character of R suggested earlier is apparent. The "high frequency" emphasis of H_1 also can be seen. The ideal is for the case when the sampling theorem is satisfied for a band-limited input to a uniform impulse sampler.

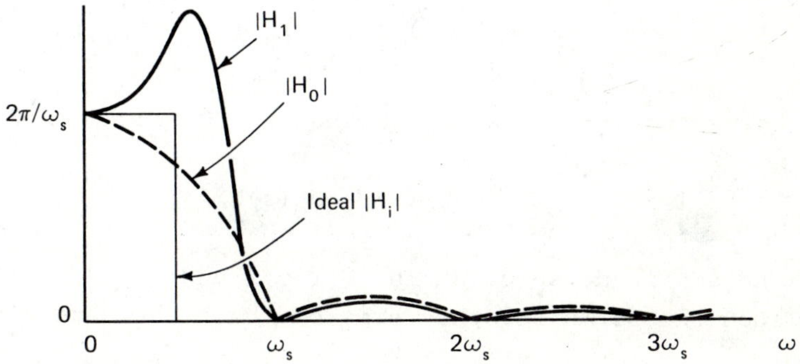

Figure 2.19

Example 2.25

For a zero-order hold, find the output amplitude spectrum when the input is $u(nT_s)$, $n \geq 0$, and find the output spectral density function when the input is a wide-sense stationary sequence $\{U(nT_s)\}$.

a) Starting with Eq. 2.58,

$$
\begin{aligned}
u_r(t) &= \sum_{n=0}^{\infty} u(nT_s)\, u_h(t,n) \\
&= u(0) \qquad 0 \leq t < T_s \\
&= u(T_s) \qquad T_s \leq t < 2T_s \\
&\ \ \vdots \\
&= u(nT_s) \quad nT_s \leq t < (n+1)T_s \\
&\ \ \vdots
\end{aligned}
$$

and

$$
\begin{aligned}
U_r(s) &= \mathcal{L}\, u_r(t) = u(0) \int_0^{T_s} \varepsilon^{-st} dt + \ldots \\
&\quad + u(nT_s) \int_{nT_s}^{(n+1)T_s} \varepsilon^{-st} dt + \ldots \\
&= u(0)\left(\frac{1-(\exp -sT_s)}{s}\right) + \ldots \\
&\quad + u(nT_s)\,(\exp -nT_s s)\left(\frac{1-(\exp -sT_s)}{s}\right) + \ldots \\
&= H_o(s) \sum_{n=0}^{\infty} u(nT_s)(\exp -nT_s s) .
\end{aligned}
$$

Using the star notation introduced in Section 2.9.1, we have

$$U_r(s) = H_o(s) U^*(s) .$$

The amplitude spectrum is $|U_r(j\omega)|$ and Figures 2.13 and 2.19 show the forms for the terms on the right hand side.

b) Starting with Eq. 2.126,

$$U_r(t) = \sum_{n=-\infty}^{\infty} U(nT_s + \beta)\, u_h(t,n;\beta)$$

$$= \sum_{n=-\infty}^{\infty} U(nT_s + \beta)[u_1(t-nT_s-\beta)-u_1(t-(n+1)T_s-\beta)].$$

Now to find the output mean we take

$$m_{U_r} = E[U_r],$$

and since

$$p_\beta(\alpha) = \frac{1}{T_s}, \qquad 0 \le \alpha < T_s$$

$$= 0 \qquad \text{otherwise},$$

and β is independent of U and $\{U(nT_s)\}$ is a wide-sense stationary sequence, we obtain

$$m_{U_r} = m_U \sum_{n=-\infty}^{\infty} \frac{1}{T_s} \int_0^{T_s} [u_1(t-nT_s-\alpha)-u_1(t-(n+1)T_s-\alpha)]\,d\alpha$$

$$= m_U \ .$$

Also,

$$r_{U_r}(t_1,t_2) = E[U_r(t_1)U_r(t_2)] =$$

$$= E \sum_{n=-\infty}^{\infty} U(nT_s+\beta)[u_1(t_1-nT_s-\beta)$$

$$- u_1(t_1-(n+1)T_s-\beta)]$$

$$\times \sum_{m=-\infty}^{\infty} U(mT_s+\beta)[u_1(t_2-mT_s-\beta)$$

$$- u_1(t_2-(m+1)T_s-\beta)]$$

or

$$r_{U_r}(t_1,t_2) = \sum_{n=-\infty}^{\infty}\sum_{m=-\infty}^{\infty} E[U(nT_s+\beta)U(mT_s+\beta)]F_1(n,m)$$

where

$$F_1(n,m) = \frac{1}{T_s}\int_0^{T_s}[u_1(t_1-nT_s-\alpha)-u_1(t_1-(n+1)T_s-\alpha]\cdot$$

$$\cdot[u_1(t_2-mT_s-\alpha)-u_1(t_2-(m+1)T_s-\alpha)]d\alpha \ .$$

Now let $m = n+k$ and $t_2 = t_1+\tau$, and we obtain

$$r_{U_r}(t_1,t_1+\tau) = \frac{1}{T_s}\sum_{k=-\infty}^{\infty} r_U(kT_s)F_2(n,k,\tau)$$

where

$$F_2(n,k,\tau) = \sum_{n=-\infty}^{\infty}\int_0^{T_s}[u_1(t_1-nT_s-\alpha)-u_1(t_1-(n+1)T_s-\alpha)]\cdot$$

$$\cdot[u_1(t_1+\tau-kT_s-\alpha)-u_1(t_1+\tau-(k+1)T_s-\alpha)]d\alpha \ .$$

Inspection of this shows that the summation over n can be changed to give

$$r_{U_r}(t_1,t_1+\tau) = \frac{1}{T_s}\sum_{k=-\infty}^{\infty} r_U(kT_s)\int_{-\infty}^{\infty}[u_1(t_1-\alpha)-u_1(t_1-T_s-\alpha)]\cdot$$

$$\cdot[u_1(t_1+\tau-kT_s-\alpha)-u_1(t_1+\tau-(k+1)T_s-\alpha]d\alpha.$$

Now we let

$$q(\tau) = \int_{-\infty}^{\infty}[u_1(\lambda)-u_1(\lambda-T_s)][u_1(\lambda+\tau)-u_1(\lambda+\tau-T_s)]d\lambda \ .$$

The right hand side is a function of τ only, and

$$r_{U_r}(\tau) = \frac{1}{T_s}\sum_{k=-\infty}^{\infty} r_U(kT_s)q(\tau-kT_s)$$

and $\{U_r\}$ is wide-sense stationary.

Finally,

$$R_{U_r}(s) = \mathcal{L}\, r_{U_r}(\tau)$$

$$= \frac{1}{T_s} Q(s) \sum_{k=-\infty}^{\infty} r_U(kT_s)(\exp -ksT_s)$$

where

$$Q(s) = \mathcal{L}\, q(\tau)$$

$$= \frac{1}{s^2} [\varepsilon^{sT_s} + \varepsilon^{-sT_s} - 2].$$

$$= H_o(-s)H_o(s)$$

and

$$R_{U_r}(s) = \frac{1}{T_s} H_o(-s)H_o(s)R_U^*(s) .$$

This is a special case of a general form we shall derive in Chapter 4. As in part a), Figure 2.17 and the stochastic counterpart to Figure 2.13 show the forms that are factors for $|R_{U_r}(\omega)|$.

2.9.4 Dequantizing

Dequantizing, the fourth signal conversion to be considered, is difficult to express as a general operation. The removal of the effects of quantization is a general goal, and it is expected then that any rule that "smooths the discontinuities" in $u_q(t)$ or $U_q(t)$ or "reduces the rate of change" of $v_q(n)$ or $V_q(n)$ would be useful. One could use ad hoc rules to obtain $u_{qf}(t)$ from $u_q(t)$ (or $v_{qf}(n)$ from $v_q(n)$) in deterministic cases and $u^i_{qf}(t)$ from $u^i_q(t)$ (or $v^i_{qf}(n)$ from $v^i_q(n)$) in stochastic cases. For example, the midpoints of the segments of zero slope could be connected by straight lines. However, such rules are not useful in general, and a more restricted view of dequantizing as a linear filtering operation is introduced. The definition of a transfer function is given in Chapter 3.

F: For input signals $u_q(t)$ or $U_q(t)$, the dequantized signals $u_{qf}(t)$ or $U_{qf}(t)$ are the outputs of a linear filter with transfer function $H_{fc}(s)$. For input signals $v_q(n)$ or $V_q(n)$, the dequantized signals $v_{qf}(n)$ or $V_{qf}(n)$ are the outputs of a linear filter with transfer function $H_{fd}(z)$.

It is clear from the discussion of Q (where it was shown that quantization noise or error is high frequency) that F generally is a low-pass filter. The implementation of $H_{fc}(s)$ is an analog filter and of $H_{fd}(z)$, a digital filter. These are discussed in Chapter 6. The results of Chapter 4 can be used to find the outputs of F. We have

$$U_{qf}(s) = \mathcal{L}\, u_{qf}(t) = H_{fc}(s)U_q(s) \tag{2.130}$$

and

$$V_{qf}(z) = \mathcal{Z}\, v_{qf}(n) = H_{fd}(z)V_q(z) \tag{2.131}$$

for the deterministic cases. Also, for stochastic signals the output spectral density functions are

$$R_{U_{qf}}(s) = H_{fc}(-s)R_{U_q}(s)H_{fc}(s) \tag{2.132}$$

and

$$R_{V_{qf}}(z) = H_{fd}(z^{-1})R_{V_q}(z)H_{fd}(z) . \tag{2.133}$$

Some interesting general results are given in [37.][35.] [39.] for band-limited signals where Q is a zero-memory nonlinearity. Optimal nonlinear filtering is discussed in [40.].

Returning to Figure 2.11, it can be seen that partial results have been presented for all of the sixteen cases introduced there. More general results are beyond the intent of this book, but it is obvious that there is a need for further research on signal conversions.

REFERENCES

1. Erdelyi, A., "Modern Mathematics for the Engineer," (E. F. Beckenbach (ed.)). New York: McGraw-Hill Book Co., 1961.
2. Hamming, R. W., "Numerical Methods for Scientists and Engineers." New York: McGraw-Hill Book Co., 1962.
3. Pennington, R. H., "Introductory Computer Methods and Numerical Analysis." New York: Macmillan Co., 1965.
4. Laning, J. and Battin, R., "Random Processes in Automatic Control." New York: McGraw-Hill Book Co., 1956.
5. McDonough, R. N. and Huggins, W. H., "Best Least-Squares Representation of Signals by Exponentials," IEEE Transactions on Automatic Control, v. AC-13, n. 4, August, 1968, pp. 408-412.
6. Franks, L. E., "Signal Theory." Englewood Cliffs, N.J.: Prentice-Hall, Inc., 1969.
7. Harmuth, H. F., "Applications of Walsh Functions in Communications," IEEE Spectrum, v. 16, n. 11, Nov., 1969, pp. 82-91.
8. Truxal, J. G., "Control System Synthesis." New York: McGraw-Hill Book Co., 1955.
9. Burington, R. and Torrance, C., "Higher Mathematics." New York: McGraw-Hill Book Co., 1939.
10. Newton, G., Gould, L., and Kaiser, J., "Analytical Design of Linear Feedback Controls." New York: John Wiley & Sons, 1957.
11. Cramer, H. and Leadbetter, M. R., "Stationary and Related Stochastic Processes," New York: John Wiley & Sons, 1967.

12. Papoulis, A., "Probability, Random Variables, and Stochastic Processes," New York: McGraw-Hill Book Co., 1965.

13. Parzen, E., "Stochastic Processes," San Francisco: Holden-Day, Inc., 1962.

14. James, H. M., Nichols, N. B., and Phillips, R. S., "Theory of Servomechanisms," New York: McGraw-Hill Book Co., 1947.

15. Blackman, R. B. and Tukey, J. W., "The Measurement of Power Spectra," New York: Dover, 1958.

16. Hannan, E. J., "Time Series Analysis," New York: John Wiley & Sons, 1960.

17. Aigrain, P. and Williams., "Synthesis of n-Reactive Networks for Desired Transient Response," Journ. Appl. Physics, v. 20, June, 1949, pp. 597-600.

18. Miller, G., "An Iterative Solution of the Equations of Aigrain and Williams," IEEE Trans. Circuit Theory, v. CT-17, February, 1970, pp. 155-158.

19. Tou, J. T., "Digital and Sampled-Data Control Systems," New York: McGraw-Hill Book Co., 1959.

20. Beutler, F. J. and Leneman, O., "The Theory of Stationary Point Processes," Information and Control, v. 9, 1966, pp. 325-346.

21. Kuo, B. C., "Discrete-Data Control Systems," Englewood Cliffs, N.J.: Prentice-Hall, Inc., 1970.

22. Bolgiano, L. and Provoso, M., "Poisson Transform Signal Analysis," IEEE Trans. Inf. Theory, v. IT-14, July, 1968, pp. 600-601.

23. Price, R., "A Useful Theorem for Nonlinear Devices Having Gaussian Inputs," IRE Trans. of Inf. Theory, v. IT-4, June, 1957, pp. 69-72.

24. Bennett, W. R., "Spectra of Quantized Signals," Bell Sys. Tech. Jour., v. 27, 1948, pp. 446-472.

25. Widrow, B., "A Study of Rough Amplitude Quantization by Means of Nyquist Sampling Theory," IRE Trans., PGCT, v. CT-3, n. 4, December, 1956, pp. 266-276.

26. Widrow, B., "Propagation of Statistics in Systems," IRE WESCON Conv. Record, Pt. 2, 1957, pp. 114-121.
27. Widrow, B., "Statistical Analysis of Amplitude-Quantized Sampled-Data Systems," AIEE Trans., Pt. II, January, 1961, pp. 555-567.
28. Kosyakin, A., "The Statistical Theory of Amplitude Quantization," Automatica i Telemechanica, v. 22, n. 6, June, 1961, pp. 722-729.
29. Watts, D. G., "A General Theory of Amplitude Quantization with Applications to Correlation Determination," IEEE Proc., v. 109, Pt. C, 1962, pp. 209-218.
30. Susskind, A. (Ed.), "Notes on Analog-Digital Conversion Techniques," New York: John Wiley & Sons, 1957.
31. Katzenelson, J., "On Errors Introduced by Combined Sampling and Quantization," IRE Trans., PGAC, v. AC-7, n. 3, April, 1962, pp. 58-68.
32. Goodman, L., "Optimum Sampling and Quantizing Rates," Proc. IEEE, January, 1966, pp. 90-92.
33. Max, J., "Quantizing for Minimum Distortion," IRE Trans., PGIT, v. IT-6, n. 1, March, 1960, pp. 7-12.
34. Ragazzini, J. and Franklin, G., "Sampled-Data Control Systems," New York: McGraw-Hill Book Co., 1958.
35. Leneman, O. and Lewis, J., "Random Sampling of Random Processes: Mean-Square Comparison of Various Interpolators," IEEE Trans., PGAC, v. AC-11, n. 3, July, 1966, pp. 396-403.
36. Leneman, O. and Lewis, J., "On Mean-Square Reconstruction Error," IEEE Trans., v. AC-11, n. 2, April, 1966, pp. 324-325.
37. Sandberg, I., "On the Properties of Some Systems that Distort Signals - I", Bell Sys. Tech. Journ., v. XLII, n. 5, September, 1963, pp. 2033-2046.
38. Masry, E., "The Recovery of Distorted Band-Limited Stochastic Processes," IEEE Trans., PGIT, v. IT-19, n. 4, July, 1973, pp. 398-403.
39. Brown, J. L. Jr., "Some Results on Nonlinear Transfor-

mation of Bandlimited Stochastic Signals," Proc. 8th Princeton Conf. on Info. Sci. and Sys., March, 1974.

40. Tung, F. and Schwarz, R., "Optimum Nonlinear Filters for Quantized Inputs," IRE Trans., PGIT, v. IT-7, n. 4, October, 1961, pp. 257-265.

41. Carlson, A. B., "Communication Systems," New York: McGraw-Hill Book Co., 1968.

42. Oppenheim, A. and Schafer, R., "Digital Signal Processing," Englewood Cliffs, N.J.: Prentice-Hall, 1975.

43. Schwarz, M. and Shaw, L., "Signal Processing," New York: McGraw-Hill Book Co., 1975.

PROBLEMS

1. A signal $u(t) = \varepsilon^{-2t}$, $t \in T_{c1}$, is sampled at a rate of five per second.
   (a) Sketch $u(nT_s)$.
   (b) If $u(nT_s)$ is reconstructed as in Figure 2.50, sketch $u_r(t)$.
   (c) If $u(t)$ is quantized so that amplitude values are given to the nearest 0.1, sketch $u_q(t)$.
2. (a) For $u(t)$ of Problem 1., find E_c and P_c.
   (b) For $u(nT_s) = v(n)$ of Problem 1., find E_d and P_d.
3. Find

$$\int_{-\infty}^{\infty} u_o(t) \, \frac{\sin 3t}{t} \, dt \, .$$

4. Demonstrate the validity of Eq. 2.11 by integrating the left side by parts for $u(t) = \cos \omega_o t$ for k=1 and k=2.
5. A signal $u(t)$ is to be modeled using a sum of singularity functions u_j, $j = 1,2,\ldots$. The measured data at $t_1=1$, $t_2=2$, and $t_3=3$ are:

$$u_x(1) = 1.2$$

$$u_x(2) = 2.2$$

$$u_x(3) = 3.1$$

(a) If N+1=2, what is the "exact fit" for the data at t_1 and t_2?

(b) If N+1=2, what is the least squares fit to the data at t_1, t_2, and t_3?

6. A signal

$$u(t) = \begin{cases} 0, & t < 0 \\ t, & 0 \le t \le 1 \\ 2, & 1 < t < 3 \\ -t+4, & 3 \le t \le 4 \\ 0, & t > 4 \end{cases}$$

Find a model for u(t) in the form of Eq. 2.19.

7. A least squares fit to equally-spaced data $v_x(n)$, $n=n_1, n_2, \ldots, n_M$ can be obtained using a model of the form Eq. 2.36 for M>N+1. The coefficients are

$$\underline{a}^T = (a_0 \; a_1 \; \ldots \; N! \; a_N)$$

and

$$\underline{a} = G_D^{-1} \; \underline{v}_x(n) .$$

Find expressions for the elements g_{ij} of G_D and v_{xj} of $\underline{v}_x$, i, j, = 1,2,...,N+1.

8. Starting with $v_3(n)$, find

$$\Delta_b v_3(n),$$

$$\Delta_b(\Delta_b v_3(n)), \text{ and}$$

$$\Delta_b[\Delta_b(\Delta_b v_3(n))] .$$

Compare these results to Eq. 2.25, 2.26, and 2.29.

9. In Example 2.3, give two other possible models for v(n).

10. A signal u(t) is to be modeled as $u_{(2)}(t)$ in Eq. 2.39. If

$$u(t) = 2\varepsilon^{-3t} + 3\varepsilon^{-t} + \varepsilon^{-5t}, \quad t \ge 0,$$

choose a value for c and find B_1 and B_2. Find $u_{(2)}(t)$.

11. In Eq. 2.41, derive the expression

$$E_i = \sum_{n=0}^{\infty} v(n)\phi_i(n) .$$

12. A periodic signal u(t) is

$$u(t) = \frac{4}{T}t + 2 \qquad -\frac{T}{2} \leq t \leq 0$$
$$= -\frac{4}{T}t + 2 \qquad 0 \leq t \leq \frac{T}{2}$$

Find a model in the form of Eq. 2.42.

13. A periodic signal v(n) is

$$\ldots = v(-5) = v(0) = v(5) = v(10) = \ldots = 0$$
$$\ldots = v(-4) = v(1) = v(6) = \ldots = 1$$
$$\ldots = v(-3) = v(2) = v(7) = \ldots = 2$$
$$\ldots = v(-2) = v(3) = v(8) = \ldots = 2$$
$$\ldots = v(-1) = v(4) = v(9) = \ldots = 1$$

Obtain a Fourier Series expansion for v(n).

14. Find frequency domain models for the signals in Problems 5. and 6.

15. Express u(t) of Problem 6. resolved in terms of u_0. Repeat for u_1.

16. A signal $v(n) = 0, \quad n < 0$
$$= n, \quad 0 \leq n \leq 5$$
$$= 0, \quad n \geq 6$$

(a) Express v(n) as a sum of $v_0(\cdot)$.
(b) Express v(n) as a sum of $v_1(\cdot)$.
(c) Verify answer to (b) by evaluating the sum.

17. A signal $u(t) = \varepsilon^{3t} \quad t \leq 0$
$$= \varepsilon^{-2t} \quad t \geq 0$$

(a) Find the amplitude spectrum $|U(j\omega)|$.
(b) Find the amplitude spectrum of $\dot{u}(t)$.

18. A signal $v(n) = 0 \quad n \leq 0$
$$v(1) = 1$$
$$v(2) = 4$$
$$v(n) = 0 \quad n \geq 3$$

What are the coefficients a, b, c, and d when the model is

$$v(n) = av_3(n) + bv_3(n-3) + cv_2(n-3) + dv_1(n-3) \ ?$$

19. A discrete-time signal has a Z-Tranform

$$V(z) = \frac{z^2}{(z-1)(z+2)}, \qquad |z| > 2$$

Find v(n) for all n.

20. (a) What is the total energy of a two-sided signal $u(t) = 2(\exp -|t|)$?

(b) What is the energy between $\omega=0$ and $\omega=2$?

21. Two signals $f_1(t)$ and $f_2(t)$ have the same

$$F(s) = \frac{s}{(s+1)(s-2)},$$

but the regions of convergence are $\sigma>2$ and $-1<\sigma<2$. Find f_1 and f_2 for all t.

22. A signal $u(t) = 2, \quad |t| \leq 3$
$\quad = 0, \quad |t| > 3$

What is U(s)?

23. A signal

$$u(t) \; (\varepsilon^{-2t} - \varepsilon^{-3t})u_1(t)$$

is uniformly sampled at a rate of 2/sec. What is the Z-Transform of the sampled signal?

24. A signal v(n) has a transform

$$V(z) = \frac{z^4-1}{z^3(z-1)}, \qquad |z| > 1$$

What is v(n)? Consider several possible methods of finding the inverse transform.

25. A signal $v(n) = (0.5)^n \quad n \geq 0$
$\quad = (3)^n \quad n < 0$

(a) Find V(z) and the signal energy.

(b) Find $V(\gamma)$ and sketch $|V(\gamma)|$, $-\pi\leq\theta\leq\pi$, $\gamma=\varepsilon^{j\theta}$.

26. Find U(s) for

(a) $u(t) = (t \sin\omega_o t) \, u_1(t)$

(b) $u(t) = (t^n \, \varepsilon^{-at}) \, u_1(t)$

(c) $u(t) = (\varepsilon^{-at} \sin\omega_o t) \, u_1(t)$.

27. A signal $u(t)$ has a Fourier Transform

$$U(j\omega) = 1 \qquad |\omega| \le 2$$
$$= 0 \qquad |\omega| > 2$$

Find the Fourier Transform of $u^2(t)$.

28. A wide-sense stationary sequence has a spectral density function

$$R_V(z) = \frac{-Kz}{(z-\frac{1}{3})(z-3)}, \qquad \frac{1}{3} < |z| < 3$$

What is the autocorrelation sequence and the mean square value of V?

29. A process $\{U(t)\}$ is uniformly quantized so that the amplitude quantum is q. Assume the random quantization error E is uniformly distributed on $[-\frac{q}{2}, \frac{q}{2}]$. What is the mean square quantization error?

30. A process $\{U(t)\}$ has

$$U(t) = At+b, \qquad t \ge 0,$$

where A is a random variable with zero mean and variance σ^2, and b is a constant.

(a) Find $m_U(t_1)$.

(b) Find $r_U(t_1,t_2)$.

Is $\{U\}$ a wide-sense stationary process?

31. A random sequence V(n) is such that V(n) is independent of V(j), $n \ne j$ and V(n) equals 2, 0, or -1 with

$$Pr(V=2) = \frac{1}{3}$$
$$Pr(V=0) = \frac{1}{3}$$
$$Pr(V=-1) = \frac{1}{3}$$

Find the mean and the autocorrelation sequence. Is the sequence wide-sense stationary?

32. Prove Properties 1., 2., 3., and 4. of Eq. 2.84.

33. If an elementary function $\varepsilon^{-t}u_1(t) = u_e(t)$ is used, a real spectral model for u(t) is

$$u(t) = \int_{-\infty}^{\infty} f(\lambda)u_e(t-\lambda)d\lambda \ .$$

Show that $f(\lambda) = u(\lambda)-\dot{u}(\lambda)$.

34. Obtain a discrete-time result analogous to that of Problem 33.

35. If $U(t) = \sin(t+\Theta)$ where Θ is a random variable uniformly distributed on $[0,2\pi]$:
(a) Is $\{U(t)\}$ wide-sense stationary?
(b) Is the ergodic assumption plausible?

36. If $U(t) = A \sin(\omega_o t+\Theta)$ where A and Θ are independent random variables with A uniformly distributed on $[0,1]$ and Θ uniformly distributed on $[0,\pi]$, find m_U. Is $\{U(t)\}$ wide-sense stationary?

37. A signal u(t) is

$$\begin{aligned} u(t) &= \tfrac{1}{2}t+2 \qquad && 0 \le t \le 4 \\ &= 0 && t > 4 \end{aligned}$$

and is an even function. Express u(t) as a linear combination of $u_j(t-\tau_j)$.

38. A signal u(t) is

$$\begin{aligned} u(t) &= 0, \qquad && 0 \le t < 1 \\ &= 1, && 1 \le t \le 2 \\ &= 0, && t > 2 \end{aligned}$$

and is an even function. Find the amplitude spectrum $|U(j\omega)|$.

39. A stochastic sequence $\{V(n)\}$ has

$$\begin{aligned} V(n) &= A\ (0.5)^n + B(0.3)^n \ , \qquad && n \ge 0 \\ &= 0 && n < 0 \end{aligned}$$

with A and B independent random variables with zero means and variances σ_A^2 and σ_B^2. Find $r_V(n_1,n_2)$. Is the sequence wide-sense stationary?

40. Signal u(t) is

$$u(t) = 2+3t+4t^3 \qquad t \geq 0$$
$$= 0 \qquad t < 0$$

(a) If u(t) is modeled using a set of singularity functions $u_j(t)$, what is the real-spectral function?

(b) If u(t) is modeled by resolving in terms of $u_1(t)$, what is the real spectral function?

41. A signal $u(t) = (3\varepsilon^{-2t}+2\varepsilon^{-5t})u_1(t)$. Find the auto-translation function of u(t) and the total signal energy.

42. A signal v(n) is to be approximated by

$$v_{(2)}(n) = a_o v_1(n)+a_1 v_2(n), \qquad n \geq 0$$

Three experimental readings are $v_x(1) = 6$, $v_x(3) = 10$, $v_x(5) = 18$. Find a_o and a_1 such $v_{(2)}(n)$ is the best approximation in the sense of least square error.

43. A signal u(t) has a Laplace Transform

$$U(s) = -\frac{s^2+4s+1}{(s+1)(s+2)(s-3)} \qquad -1 < \sigma < 3$$

Find u(t) for all t.

Chapter 3

LINEAR SYSTEM MODELS

3.1 DEFINITIONS

Systems, like the signals in the previous chapter, are of great interest in engineering studies. As before, a distinction is made between a system model and an actual physical system. The system model is a mathematical representation, and there are certain general conditions on the nature of the related physical system which are assumed here. These restrictions mean that many common "systems" are not considered. However, the kinds of system models developed in this chapter are used to study an amazing variety of problems with good results.

3.1.1 System

In general, a system model or simply system as used in

this book represents a collection of interacting physical objects. We assume that there is a boundary which includes all the elements of the system and excludes objects not in the system. The system is oriented and there are designated input signals and output signals at the boundary; their roles cannot be interchanged. In addition, there are internal signals or variables and there are parameters, properties of the physical objects in the system. In some cases the parameters may be abstractions; for example, constants in a mathematical algorithm.

The mathematical forms of the system models discussed in this chapter usually are differential or difference equations or their solutions. In a few cases, algebraic equations occur. These mathematical forms subsequently will be called systems in the same way the idealizations in the previous chapter were called signals. The problem of relating actual systems to the models is an important one which we shall not consider in detail. A few examples are given in Chapter 4 as was done previously with signals in Chapter 2 to emphasize the need for methods of relating experimental data to a model. For example, the determination of the system parameters by direct measurement (if possible) or by the use of input and output signals -- an identification problem -- is vital to the use of the system models in analyzing actual systems.

There are several properties of actual systems and their models which commonly are defined. In using these it is important to distinguish between verification of these properties on an actual system by experimenting, and verification of the properties from a system model. In the latter case, since a mathematical representation is available, it is relatively easy to apply mathematical conditions to test a hypothesis. In the former case, however, it rarely is possible to be absolutely sure that the experiments performed have covered all possible situations such as "for all time," "for all amplitudes," or "for all initial conditions," Put another way, it is the difference between being able to examine the actual system and to perform tests on it or being given a mathematical description for verification of the properties in question. The difficult, but essential, task of finding a suitable model requires both experience with the actual system and a knowledge of various system models, several of which are developed in this chapter.

3.1.2 Causality

The concept of causality is an old one in classical physics. Basically, a system is causal for any input-output pair (A,B) when "if A happens, then (and only then) B is always produced by it". This broad definition is difficult to apply, and in practice, experience with a particular system guides the user. However, a failure in the system would certainly produce misleading results, and an understanding of the nature of proper or nonfailed operation is important. To be more specific, it has been suggested [1] that a system is causal if and only if

1. there is a boundary
2. changes in the system variables are produced externally
3. the system plus its environment are isolated
4. it is oriented
5. there is a one-one relationship between input-output pairs.

Of these, the first and fourth have been assumed earlier. The second, which frequently is taken as equivalent to causality, expresses the idea of a nonanticipatory system. A system is nonanticipatory if the output at a given time does not depend on the input applied after that time. For real input and output signals, a nonanticipative system is also called realizable. The fifth condition implies a unique relationship between input and output, and systems meeting this condition are deterministic. By contrast, there are systems of practical interest where the input-output relation is known probabilistically, e.g., where a parameter is a random variable. The third, along with other properties, is illustrated in Figure 3.1.

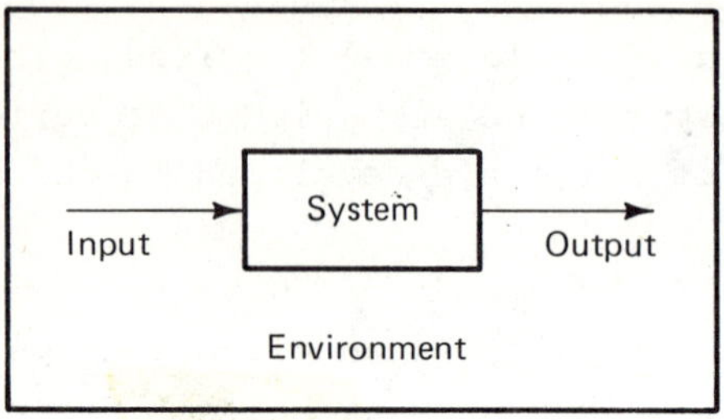

Figure 3.1 A system and its environment

3.1.3 Realizability

By characterizing a system as an operator N_c, the above and other definitions can be stated more explicitly. Thus for input u(t) and output y(t), we have

$$y(t) = N_c u(t) \tag{3.1}$$

and the system is deterministic if for every u(t) there is one, and only one, y(t). The system is realizable if for real $u_a(t)$, $u_b(t)$, $y_a(t)$, $y_b(t)$ and any time t_o,

$$u_a(t) = u_b(t), \qquad t < t_o$$

implies

$$y_a(t) = N_c u_a(t) = N_c u_b(t) = y_b(t), \qquad t < t_o. \tag{3.2}$$

This definition includes active as well as passive systems [2.]. For a passive system, that is a system with no internal energy sources, the system is realizable if for real u(t) and y(t),

$$u(t) = 0, \qquad t < t_o$$

implies

$$N_c u(t) = y(t) = 0, \qquad t < t_o . \tag{3.3}$$

For $t \geq t_o$, the output is nonzero for at least some finite interval.

3.1.4 Stationarity

A system is <u>stationary</u> if its input-output relationship is invariant under a translation in time. Thus, using 3.1, the system is stationary if for an input $u(t-\tau)$,

$$N_c u(t-\tau) = y(t-\tau) \tag{3.4}$$

for all t and τ. Systems that are not stationary are <u>time-varying</u>.

3.1.5 Linearity

A system is <u>linear</u> if it satisfies the additivity and homogeneity requirements of a linear operation (see Appendix B). Thus, changing from N_c to L_c, a linear system is such that

$$L_c(u_a(t) + u_b(t)) = L_c u_a(t) + L_c u_b(t) \tag{3.5a}$$

and

$$L_c \alpha u(t) = \alpha L_c u(t) \tag{3.5b}$$

for all $u_a(t)$, $u_b(t)$, $u(t)$, and α. A much more complete discussion of system linearity is given later in this chapter. In applying these definitions, it is assumed that the system is <u>relaxed</u>, that is it contains no stored energy, before the application of the input $u(t)$ or inputs $u_a(t)$ and $u_b(t)$. In other words, the response $y(t)$ or responses $y_a(t)$ and $y_b(t)$ are due only to the associated inputs. In general, it is desirable to include arbitrary initial conditions, and this is discussed later also.

3.1.6 Stability

A definition of a stable system is needed. This subject is covered in much greater detail in Chapter 5 where several definitions are given, but for this and the following chapter, a convenient and common definition is as follows:

A system is b.i.b.o. stable if for each bounded input, the output is bounded, i.e.,

$$|u(t)| \leq M_1 < \infty$$

implies

$$|y(t)| = |N_c u(t)| \leq M_2 < \infty \tag{3.6}$$

for all $t \geq t_o$.

The definitions 3.1 through 3.6 apply to continuous-time systems since the inputs and outputs are continuous-time signals. A similar set of definitions when the input is v(n) and the output is z(n) can be written to describe the corresponding properties of discrete-time systems. The general operator is N_d and the linear operator is L_d. In some cases a system will have both continuous-time and discrete-time signals at its boundary and this will be called a hybrid system.

3.1.7 Linearization

The linearity property is a very useful idealization of actual physical systems, and much of what follows in this and later chapters depends on it. There are many ad hoc methods for finding linear approximations to actual systems, and a short introduction to these ideas is given in Section 3.7 after the forms of linear system models have been developed. The method of small perturbations is probably most common and is familiar in such applications as transistor equivalent circuits, small amplitude vibration analysis, and aircraft modeling (see Chapter 6). In addition, some methods of obtaining a best (in the least squares sense) linear models are discussed in Chapter 4 in the section on system identification, Section 4.4.

3.2 LINEAR SYSTEMS AND RESPONSE FUNCTIONS

3.2.1 Spectral Forms of System Output

Equations 3.5a and 3.5b give the conditions for the linearity of a relaxed system, and linear systems have properties which make it possible to obtain many useful general results. One class of linear system models can be obtained by using the spectral forms of Equations 2.50, 2.51, 2.52, and 2.53 for the input signal. We start by using 2.50,

$$u(t) = \sum_k \ell(k) q_2(t,k) \ ,$$

and let $Q_2(t,k)$ be the response of a continuous-time linear system to the elementary function $q_2(t,k)$. Using the linear operator L_c on both sides of this equation, where L_c operates on functions of t,

$$L_c u(t) = y(t) = L_c \sum_k \ell(k) q_2(t,k)$$

and using additivity,

$$y(t) = \sum_k L_c \ell(k) q_2(t,k)$$

and using homogeneity,

$$y(t) = \sum_k \ell(k) L_c q_2(t,k)$$

or

$$y(t) = \sum_k \ell(k) Q_2(t,k) \qquad (3.7)$$

from the definition of $Q_2(t,k)$. Thus the response of any input signal u(t) is given in terms of the spectral function $\ell(k)$ of u(t) (with respect to $q_2(t,k)$) and the response of the system to $q_2(t,k)$. This is called a superposition sum, and the linear system model is $Q_2(t,k)$. Similarly, using

2.51, we let $Q_4(n,k)$ be the response of a discrete-time linear system to the elementary function $q_4(n,k)$. Then using L_d, where L_d operates on functions of n,

$$L_d v(n) = z(n) = L_d \sum_k m(k) q_4(n,k)$$

which leads to

$$z(n) = \sum_k m(k) Q_4(n,k) \ . \tag{3.8}$$

Similarly, using 2.52 and 2.53, two superposition integrals are

$$y(t) = \int f(\lambda) Q_1(t,\lambda) d\lambda \tag{3.9}$$

where $Q_1(t,\lambda)$ is the continuous-time linear system response to $q_1(t,\lambda)$, and

$$z(n) = \int g(\lambda) Q_3(n,\lambda) d\lambda \tag{3.10}$$

where $Q_3(n,\lambda)$ is the discrete-time linear system response to $q_3(n,\lambda)$. In all four forms the summation or integration is over the spectral variable of v(n) or u(t) in the signal model.

An examination of Equations 3.7, 3.8, 3.9, and 3.10 shows that there are four general system models that can be developed for real spectral arguments. Recalling that in Table 2.4 both real and complex arguments are used, it is clear that eight general forms can be found. The choice of the particular elementary functions and their corresponding response functions is a matter of convenience to the systems analyst and some examples are given below. It also should be noted in 3.7, 3.8, 3.9, and 3.10 that the outputs y(t) and z(n) are in a spectral form. However, in many cases it is more useful to express the outputs in the forms 2.50, 2.51, 2.52, and 2.53 where the same elementary function that was used for the input is again used for the output. Thus, for example, if

$$u(t) = \sum_k \ell_u(k) q_2(t,k)$$

it would be useful to have

$$y(t) = \sum_k \ell_y(k) q_2(t,k), \tag{3.11}$$

and be able to find $\ell_y(k)$ from $\ell_u(k)$ and $Q_2(t,k)$.

Some examples of particular choices of elementary functions and the corresponding response functions that will be developed in the following sections are given here.

1. Using 2.52, with $q_1(t,\lambda) \to u_o(t-\lambda)$, and from 2.55, $f(\lambda) = u(\lambda)$. Equation 3.9 becomes

$$y(t) = \int_{-\infty}^{\infty} u(\lambda) h(t,\lambda) d\lambda \tag{3.12}$$

   where $h(t,\lambda)$ is the response of the relaxed linear system to $u_o(t-\lambda)$, i.e., $h(t,\lambda)$ is the <u>unit impulse response</u>.
2. Using 2.52, with $q_1(t,\lambda) \to u_1(t-\lambda)$, and from 2.55a, $f(\lambda) = \dot{u}(\lambda)$. Equation 3.9 becomes

$$y(t) = \int_{-\infty}^{\infty} \dot{u}(\lambda) a(t,\lambda) d\lambda \tag{3.13}$$

   where $a(t,\lambda)$ is the response of the relaxed linear system to $u_1(t-\lambda)$, i.e., $a(t,\lambda)$ is the <u>unit step function response</u>.
3. Using 2.51, with $q_4(n,k) \to v_o(n-k)$, and from 2.56, $m(k)=v(k)$. Equation 3.8 becomes

$$z(n) = \sum_{k=-\infty}^{\infty} v(k) h(n,k) \tag{3.14}$$

where h(n,k) is the response of the relaxed linear system to $v_o(n-k)$, i.e., h(n,k) is the unit pulse response.

4. Using 2.51, with $q_4(n,k) \to v_1(n-k)$, and from 2.56a, $m(k)=\Delta_b v(k)$. Equation 3.8 becomes

$$z(n) = \sum_{k=-\infty}^{\infty} \Delta_b v(k) a(n,k) \tag{3.15}$$

where a(n,k) is the response of the relaxed linear system to $v_1(n-k)$, i.e., a(n,k) is the discrete unit step function response.

Note that in all four of these cases the lower limits on the sums and integrals are $-\infty$ which is consistent with the requirement of considering only relaxed systems where the inputs start in the infinite past. Forms for finite initial times t_o and n_o are developed later. Two more forms of response functions are obtained by using complex spectral arguments.

5. Using Equation A.4, the inverse Fourier transform, where the elementary function is $\varepsilon^{jt\omega}$, the input is

$$u(t) = \frac{1}{2\pi}\int_{-\infty}^{\infty} U(j\omega)\varepsilon^{jt\omega}d\omega$$

and the output is

$$L_c u(t) = y(t) = \frac{1}{2\pi}\int_{-\infty}^{\infty} U(j\omega)L_c\varepsilon^{jt\omega}d\omega$$

$$= \frac{1}{2\pi}\int_{-\infty}^{\infty} U(j\omega)R(t,j\omega)\,d\omega$$

where $R(t,j\omega)$ is a response of the linear system to $\varepsilon^{jt\omega}$. A similar set of equations using the Laplace transform can be written.

6. Using Equation A.14, the inverse $\mathfrak{F}_d$ transform, where the elementary function is γ^n, the input is

$$v(n) = \frac{1}{2\pi j}\int V(\gamma)\gamma^{n-1}d\gamma$$

and the output is

$$L_d v(n) = z(n) = \frac{1}{2\pi j}\int V(\gamma)L_d\gamma^{n-1}d\gamma$$

$$= \frac{1}{2\pi j}\int V(\gamma)R(n,\gamma)\gamma^{-1}d\gamma$$

where $R(n,\gamma)$ is a response of the linear system to γ^n. A similar set of equations using the Z-transform can be written.

We also would like to have an analogy to Equation 3.11 for the case of complex arguments in the spectral function. It would be useful, for example, where

$$u(t) = \frac{1}{2\pi}\int_{-\infty}^{\infty} U(j\omega)\,\varepsilon^{jt\omega}d\omega$$

to have

$$y(t) = \frac{1}{2\pi}\int_{-\infty}^{\infty} Y(j\omega)\,\varepsilon^{jt\omega}\,d\omega \qquad (3.16)$$

and to be able to find $Y(j\omega)$ from $U(j\omega)$ and $R(t,j\omega)$. The fact that it is quite easy to do this when the Fourier, $\mathfrak{F}_d$, Laplace, and Z-transforms are used in the analysis of stationary linear systems is a reason for the wide use of these transforms. We shall see this in Chapter 4.

3.2.2 Six Common Forms

Our results can be summarized as shown in Table 3.1. The response functions listed are the first ones to be considered in the following sections. Observe that the notation for the response functions is chosen to emphasize the similarity

TABLE 3.1

Elementary Function	Response Function
$u_o(t-\lambda)$	$h(t,\lambda)$
$v_o(n-k)$	$h(n,k)$
$u_1(t-\lambda)$	$a(t,\lambda)$
$v_1(n-k)$	$a(n,k)$
ε^{st}	$R(t,s)$
z^n	$R(n,z)$

of forms. For example, the functions $h(t,\lambda)$ and $h(n,k)$ are not the same, and this applies also to the functions $a(\cdot,\cdot)$ and $R(\cdot,\cdot)$. The first argument indicates whether the response function is for a continuous-time or discrete-time system.

It was pointed out in Chapter 2, Section 2.3 that the resolution of signals in terms of real exponentials also can be done. Thus for

$$q_1(t,\lambda) = (\exp -(t-\lambda))u_1(t-\lambda)$$

in Eq. 2.52 and

$$q_4(n,k) = a^{(n-k)}v_1(n-k), \qquad 0 < a < 1,$$

in Eq. 2.51 with spectral functions $f(\lambda)$ and $m(k)$, respectively, the corresponding response functions are $Q_1(t,\lambda) = e(t,\lambda)$ and $Q_4(n,k) = e(n,k)$, the responses to these elementary functions. The investigation of these is suggested in an exercise at the end of this chapter (Problem 7).

3.3 LINEAR DIFFERENTIAL AND DIFFERENCE EQUATIONS

3.3.1 General Forms

In the previous section the linear continuous-time and discrete-time systems were first characterized by linear operators. We had

$$L_c u(t) = y(t) \tag{3.17a}$$

and

$$L_d v(n) = z(n) \ . \tag{3.17b}$$

Let the linear system of 3.17a be described for all t by an ordinary, linear differential equation

$$(D^r + \alpha_{r-1} D^{r-1} + \ldots \alpha_o) y(t) = (\beta_m D^m + \ldots + \beta_o) u(t) \tag{3.18}$$

where D is the time derivative operator and $m<r$ for a <u>strictly</u> <u>proper</u> system. The coefficients $\alpha_{r-1}, \ldots, \alpha_o$ and $\beta_m, \ldots, \beta_o$ may be functions of t or constants. Using M and N to represent the polynomials in D in 3.18,

$$M(D,t) y(t) = N(D,t) u(t) \tag{3.19a}$$

for a time varying linear system, or

$$M(D) y(t) = N(D) u(t) \tag{3.19b}$$

for a stationary linear system. The solutions of these equations may be written symbolically as

$$y(t) = M(D,t)^{-1} N(D,t) u(t)$$

or

$$y(t) = M(D)^{-1} N(D) u(t)$$

which may be compared with 3.17a to give one form of L_c.

Similarly, we can start with an ordinary, linear difference equation for all n(with $\alpha_o \neq 0$)

$$(E^r+\alpha_{r-1}E^{r-1}+\ldots+\alpha_o)z(n) = (\beta_m E^m+\ldots+\beta_o)v(n) \quad (3.20)$$

where E is a unit advance operator ($Ev(n) = v(n+1)$) and $m \leq r$. Again we use M and N to represent the polynomials in E in 3.20,

$$M(E,n)z(n) = N(E,n)v(n) \quad (3.21a)$$

or

$$M(E)z(n) = N(E)v(n) \quad (3.21b)$$

with symbolic solutions

$$z(n) = M(E,n)^{-1}N(E,n)v(n)$$

and

$$z(n) = M(E)^{-1}N(E)v(n)$$

which gives one form of L_d in 3.17b. It is left as an exercise to show the following: first, that forms 3.19 and 3.21 describe linear systems as defined by 3.5 (or a similar form for discrete-time systems), and second, that 3.19b and 3.21b describe stationary systems.

3.3.2 Some System Examples

Equations 3.18 and 3.20 are the first specific forms of linear system models to be introduced, and these will be referred to as rth-order forms of differential or difference equations. The linear operators L_c and L_d and the response functions, Q_1, Q_2, Q_3 and Q_4 are only symbolic at this point. From these specific forms the particular response functions in Table 3.1 can be derived. However, before doing this, it is important that we keep in mind that in an actual analysis of a system, these models must first be found in some way.

Certainly, the commonest way of obtaining these is by writing a set of differential or difference equations from a knowledge of the components of the system and the manner of their interconnection. Obviously this step is crucial in any system analysis, and there are many excellent references [3.][4.] on electrical, mechanical, and other physical systems. The original set of equations is then combined to give the rth-order forms which describe the single input-single output linear systems. If the parameters are not known, they must be measured, and if the basic form of the model is not known and cannot be observed directly, an experimental identification procedure is required. In practice, it usually is necessary to combine several approaches to obtain the model, and this can be a difficult procedure requiring a thorough knowledge of the physical properties of the system and the mathematical forms of signal and system models. Some simple examples illustrate these ideas.

Example 3.1

a) A permanent-magnet meter movement is shown. Find an ordinary differential equation relating the input voltage e(t) and an output angular displacement θ(t). Specify M(D) and N(D). Moment of interia of moving coil assembly = J,

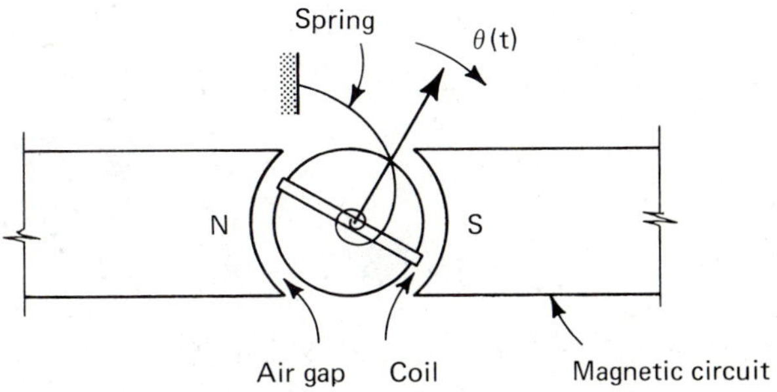

uniform radial magnetic flux density in gap = B, linear spring constant=K, and rectangular coil as shown:

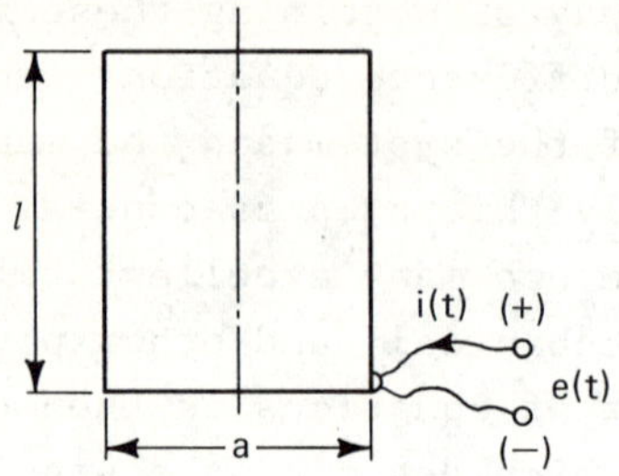

The force on one conductor in the coil side is

$$f = Bi\ell$$

and the corresponding torque is

$$\tau = \frac{1}{2}\, aBi\ell .$$

For 2N conductors, the total torque produced is

$$T = 2N\tau = NaBi\ell .$$

Neglecting bearing friction, the counter torque is

$$J\frac{d^2\theta}{dt^2} + K(\theta - \theta_o)$$

where θ_o is a reference position. Then for equilibrium,

$$J\frac{d^2\theta}{dt^2} + K\theta = NaBi\ell + K\theta_o .$$

For the coil,

$$e = Ri + L\frac{di}{dt} + e_b$$

where

$$e_b = (2N)B\ell\left(\frac{a}{2}\right)\frac{d\theta}{dt} = NaB\ell\frac{d\theta}{dt}$$

and for $NaB\ell = C$, there are two equations

$$J\frac{d^2\theta}{dt^2} + K\theta = Ci + K\theta_o$$

$$e = Ri + L\frac{di}{dt} + C\frac{d\theta}{dt} .$$

Using the D-operator,

$$(JD^2 + K)\theta = Ci + K\theta_o$$

$$e = (R + LD)i + CD\theta \ .$$

From the second equation, the internal variable

$$i = \frac{1}{R+LD}(e - CD\theta).$$

Therefore, from the first equation,

$$(JD^2 + K)\theta = \frac{C}{R+LD}(e - CD\theta) + K\theta,$$

or

$$(R+LD)(JD^2+K)\theta = Ce - C^2D\theta + (R+LD)K\theta_o \ .$$

Then

$$[LJD^3 + RJD^2 + (LK + C^2)D + RK]\theta = Ce + RK\theta_o$$

or for convenience, let $\theta_1 = \theta - \theta_o$, then

$$[LJD^3 + RJD^2 + (LK + C^2)D + RK]\theta_1(t) = Ce(t)$$

and dividing by LJ, we have

$$M(D) = D^3 + (\frac{R}{L})D^2 + (\frac{LK+C^2}{LJ})D + (\frac{RK}{LJ})$$

$$N(D) = \frac{C}{LJ} \ .$$

The rth-order forms 3.18 and 3.19b (for r=3) have been found.

b) For the electric circuit shown, find an ordinary differential equation relating the input voltage e(t) and the output voltage $v_o(t)$.

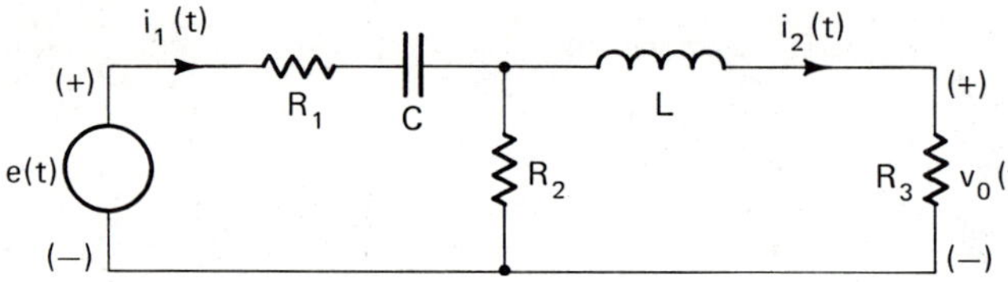

Using the Kirchhoff voltage law,

$$e = R_1 i_1 + \frac{1}{C}\int i_1 dt + R_2(i_1 - i_2)$$

$$o = -R_2(i_1 - i_2) + L\frac{di_2}{dt} + v_o$$

with $v_o = R_3 i_2$. Using the D-operator,

$$e = (R_1 + R_2 + \frac{1}{CD})i_1 - \frac{R_2}{R_3}v_o$$

$$o = -R_2 i_1 + (\frac{R_2}{R_3} + \frac{L}{R_3}D + 1)v_o.$$

Using the second equation,

$$i_1 = \frac{1}{R_2}(\frac{L}{R_3}D + \frac{R_2}{R_3} + 1)v_o$$

and substituting in the first equation,

$$e = \frac{[(R_1+R_2)CD + 1][LD + (R_2R_3)]}{R_2R_3CD}v_o - \frac{R_2}{R_3}v_o$$

or

$$\{(R_1+R_2)LCD^2+[(R_1+R_2)(R_2+R_3)C+L-R_2^2C]D+(R_2+R_3)\}v_o = R_2R_3CDe$$

Dividing by the coefficient of D^2,

$$M(D) = D^2 + \frac{[(R_1+R_2)(R_2+R_3)C+L-R_2^2C]}{(R_1+R_2)LC} D + \frac{(R_2+R_3)}{(R_1+R_2)LC}$$

$$N(D) = \frac{R_2R_3}{(R_1+R_2)L} D \quad .$$

Note that i_1 and i_2 are internal variables in this example.

c) For the mass-spring-dashpot system shown, find an ordinary differential equation relating the input force f(t) and the output displacement x(t).

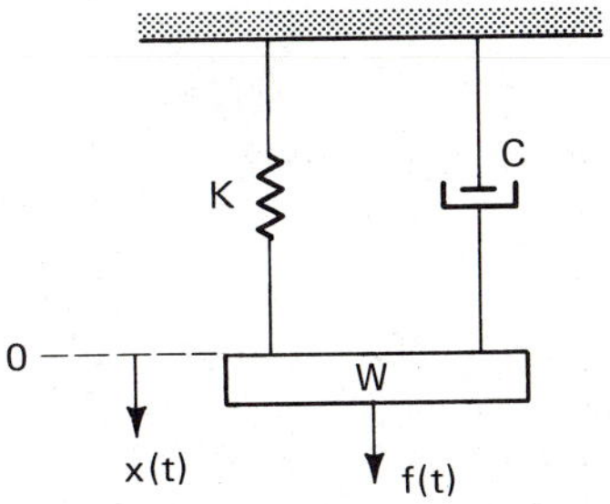

K lb/ft
C lb-sec/ft
W lb
f lb

The reference x=0 corresponds to an initial equilibrium position with f=0. Then

$$\frac{W}{g}\frac{d^2x}{dt^2} + C\frac{dx}{dt} + Kx = f$$

or

$$(\frac{W}{g} D^2 + CD + K)x = f$$

and

$$M(D) = D^2 + \frac{Cg}{W} D + \frac{Kg}{W}$$

$$N(D) = \frac{g}{W} .$$

d) A simple bellows-type pressure gage is shown. Find an ordinary differential equation relating input pressure p(t) and output displacement x(t). Assume ideal flow conditions

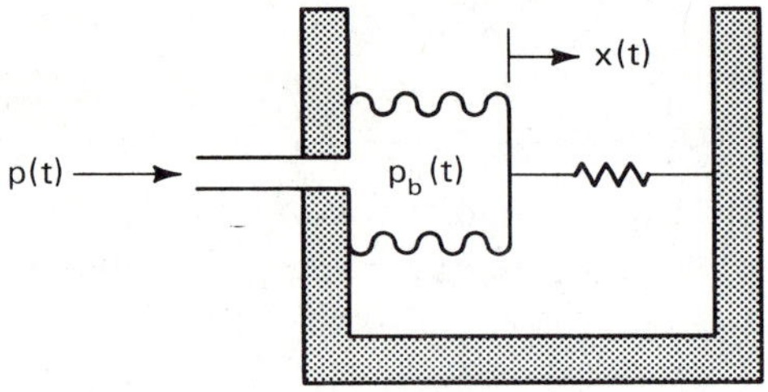

Inlet diameter = d

Bellows diameter = d_b

Spring constant = K

Bellows pressure = p_b

For q the input flow rate,

$$qdt = \frac{\pi}{4} d_b{}^2 dx .$$

Also,

$$p_b \frac{\pi}{4} d_b^2 = K(x-x_o)$$

and

$$p - p_b = K_r q.$$

These equations follow from a material balance, a force balance, and a transport rate equation. K_r depends on d, on the inlet tube length, and on the viscosity of the fluid. Then,

$$K(x-x_o) = \frac{\pi}{4} d_b^2 (p - K_r q)$$

$$= \frac{\pi}{4} d_b^2 (p - K_r \frac{\pi}{4} d_b^2 \frac{dx}{dt})$$

or

$$[(\frac{\pi}{4}d_b^2)^2 K_r] \frac{dx}{dt} + Kx = Kx_o + (\frac{\pi}{4} d_b^2)p$$

or

$$A \frac{dx}{dt} + Kx = Kx_o + Bp .$$

Using the D-operator and letting $x_1 = x-x_o$,

$$(AD + K)x_1 = Bp .$$

Then

$$M(D) = D + \frac{K}{A}$$

$$N(D) = \frac{B}{A} .$$

In all of these examples, the polynomials M(D) and N(D) were obtained by applying various physical laws to a particular description of the system. There are many excellent references available on these procedures, see for example, [3.] and [4.].

e) Simpson's Rule provides a simple way of computing a numerical approximation to a definite integral

$$q(t) = \int_{t_o}^{t} p(\tau)d\tau .$$

If the equally-spaced samples of p(t) are v(0), v(1),...,

v(n), n even, and the corresponding numerical approximation to the integral q(t) is z(n), Simpson's Rule is

$$z(n) = \frac{1}{3}T[(v(0)+v(n))+4(v(1)+v(3)+...+v(n-1)) + 2(v(2)+v(4)+...+v(n-2))]$$

where T is the time between adjacent samples of p(t), i.e., $nT = t-t_o$. To develop a difference equation of the form 3.21b, add two samples v(n+1) and v(n+2). Then we have

$$z(n+2) - z(n) = \frac{1}{3}T\ [v(n+2)+4v(n+1)+v(n)].$$

This recursive form can be used to find z(n), $n \geq 2$, by starting with z(0)=0. Using the E-operator we have

$$M(E) = E^2 - 1$$

$$N(E) = \frac{1}{3}T\ (E^2 + 4E + 1).$$

f) A digital filter is described by the diagram given. Find M(E) and N(E).

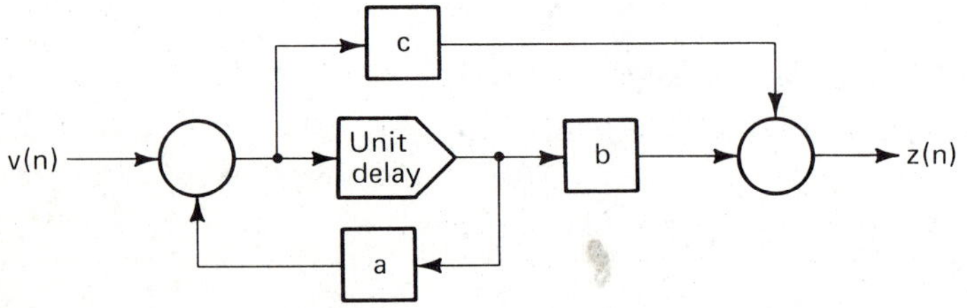

Let q(n) be the input signal to the unit delay, then

$$q(n) = v(n) + aq(n-1)$$

$$z(n) = cq(n) + bq(n-1)$$

or

$$q(n+1) = v(n+1) + aq(n)$$

$$z(n+1) = cq(n+1) + bq(n).$$

Using the E-operator,

$$Eq(n) = Ev(n) + aq(n)$$

$$Ez(n) = cEq(n) + bq(n).$$

From the first equation,

$$q(n) = \frac{E}{E-a} v(n)$$

and therefore,

$$z(n) = \frac{cE+b}{E-a} v(n)$$

or

$$(E-a)\ z(n) = (cE+b)\ v(n)\ .$$

$$M(E) = E-a$$

$$N(E) = cE+b\ .$$

g) A simple problem in the design of an optimal control system is as follows. For a linear plant described by

$$\frac{dy}{dt} = ay + bu, \qquad a < 0, \quad b > 0\ ,$$

where u is the control input and y is the plant output, find the control u such that the functional

$$J = \int_0^T u(t)^2 dt$$

is minimal and $y(T) = 0$. The solution is $u(t)=K\varepsilon^{-at}y(0)$, $0 \le t \le T$, or in terms of $y(t)$

$$u(t) = \frac{K\varepsilon^{-2at}}{[1 + \frac{bK}{2a}(1-\varepsilon^{-2at})]} y(t), \quad 0 \le t \le T ,$$

where

$$K = \frac{-2a}{b(1-\varepsilon^{-2aT})} .$$

This is a feedback control law which uses a time-varying gain in the form

$$u(t) = G(t)y(t), \quad 0 \le t \le T.$$

Substituting in the plant equation,

$$\frac{dy}{dt} = ay + G(t)y$$

or

$$[D - (a+G(t)]y = 0 .$$

is the differential equation describing the plant operating with an optimal feedback control. In the notation of Eq. 3.19a,

$$M(D,t) = D - (a+G(t))$$

$$N(D,t) = 0.$$

Examples 3.1a through 3.1f are all stationary systems. Example 3.1g is a time-varying system.

3.3.3 Solutions for Response Functions

In order to use the response forms Eq. 3.7, 3.8, 3.9, and

3.10, the system models in rth-order form must be solved to find system models that are response functions. Specifically, this important step in this development of system models is the use of the polynomials M and N of Eq. 3.19 and 3.21 to find the response functions Q_1, Q_2, Q_3, and Q_4. Figure 3.2 summarizes the cases of particular interest where the inner circle represents the forms already developed and the outer ring contains forms to be derived from them. Since these response models are directly related to the spectral models of signals summarized in Tables 2.6a

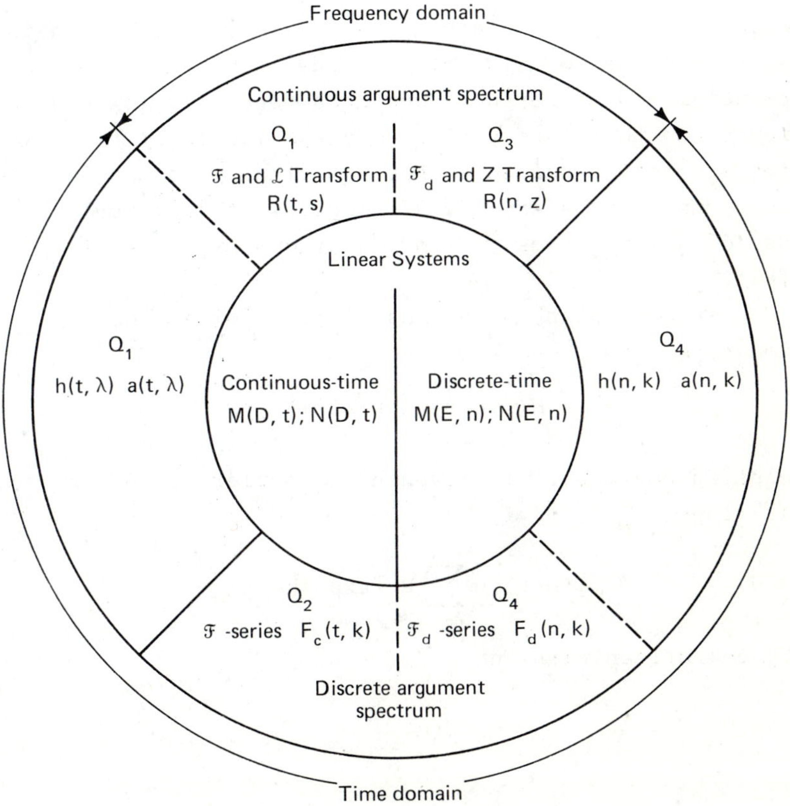

Figure 3.2 Classification of system response functions

and 2.6b, continuous and discrete spectral arguments as well as real (time domain) and complex (frequency domain) arguments will occur. Each sector of Figure 3.2 indicates the general response function of interest and the specific ones of Table 3.1 which apply. The number of possible results which could be obtained is very large; only a few of the most widely-used forms will be discussed. For example, the response functions $e(t,\lambda)$ and $e(n,k)$ mentioned earlier could be included in the "time domain" sectors of Figure 3.2, and response functions for various sets of orthonormal functions (in addition to sine, cosine sets) mentioned in Chapter 2 for Eq. 2.59 and 2.60 could be included in the two lower sectors.

A major simplification occurs by considering only stationary systems. The solution of rth-order time-varying linear differential and difference equations generally is very difficult and the state variable system models of the next section are better suited to the numerical methods that are necessary in most cases. Good summaries of classical methods of solving time-varying equations are given in [5.] and [6.].

As a starting point, a response $Q_2(t,k)$ of a linear, stationary system

$$M(D)y(t) = N(D)u(t)$$

is required where $u(t)$ is given by a Fourier Series (Eq. 2.45). Thus

$$u(t) = \sum_{k=-\infty}^{\infty} U_k(\exp jk\omega_0 t) ,$$

and by comparison with Eq. 3.7,

$$\ell(k) \to U_k$$

and for

$$q_2(t,k) \to (\exp jk\omega_0 t) ,$$

$Q_2(t,k) \rightarrow F_c(t,k)$ is the required response to $(\exp jk\omega_0 t)$. Substituting F_c for y and $(\exp jk\omega_0 t)$ for u,

$$M(D)F_c(t,k) = N(D)(\exp jk\omega_0 t)$$

and $F_c(t,k)$ in this instance is the forced or <u>steady-state response</u> only. We assume an

$$F_c(t,k) = K(\exp jk\omega_0 t)$$

and substituting,

$$M(D)K(\exp jk\omega_0 t) = N(D)(\exp jk\omega_0 t)$$

and carrying out the D-operations,

$$M(jk\omega_0)\ K(\exp jk\omega_0 t) = N(jk\omega_0)(\exp jk\omega_0 t).$$

Therefore,

$$K = \frac{N(jk\omega_0)}{M(jk\omega_0)}$$

and

$$F_c(t,k) = \frac{N(jk\omega_0)}{M(jk\omega_0)}\ (\exp jk\omega_0 t), \qquad (3.22)$$

which is the required response function. For any periodic input u(t) expanded in a Fourier Series (Eq. 2.45), the linear system steady-state response is then

$$\begin{aligned} y(t) &= \sum_{k=-\infty}^{\infty} U_k F_c(t,k) \\ &= \sum_{k=-\infty}^{\infty} U_k \frac{N(jk\omega_0)}{M(jk\omega_0)}\ (\exp jk\omega_0 t) \end{aligned}$$

and for

$$Y_k = U_k \frac{N(jk\omega_o)}{M(jk\omega_o)} ,$$

$$y(t) = \sum_{k=-\infty}^{\infty} Y_k \,(\exp jk\omega_o t) \ . \qquad (3.23)$$

Thus the spectral function of the output y(t), i.e., Y_k, is found directly from U_k, the input spectral function, and M and N, the polynomials of the rth-order system model. The property suggested by Eq. 3.16 occurs in this well-known Fourier form. In fact, for k=1, the common method of phasor analysis of linear electric circuits follows immediately where the ratio N/M is an immittance function when u(t) and y(t) are a voltage-current or current-voltage pair.

A completely parallel development for the case of a discrete-time system also can be worked out. For the system

$$M(E)z(n) = N(E)v(n)$$

with input

$$v(n) = \sum_{k=-K}^{K} V_k(\exp jk\lambda_o n)$$

as given in Eq. 2.46 for $N_o = 2\pi/\lambda_o$ odd, and using the form of Eq. 3.8, with

$$m(k) \rightarrow V_k$$

$$q_4(n,k) \rightarrow (\exp jk\lambda_o n)$$

and

$$Q_4(n,k) \rightarrow F_d(n,k) ,$$

we obtain

$$F_d(n,k) = \frac{N(\exp jk\lambda_o)}{M(\exp jk\lambda_o)} (\exp jk\lambda_o n), \qquad (3.24)$$

as the required response. Thus for any periodic v(n) (N_o odd),

$$z(n) = \sum_{k=-K}^{K} V_k F_d(n,k)$$

$$= \sum_{k=-K}^{K} Z_k (\exp jk\lambda_o n) \qquad (3.25)$$

where

$$Z_k = \frac{N(\exp jk\lambda_o)}{M(\exp jk\lambda_o)} V_k$$

is the steady-state response. A similar development for N_o even follows from Eq. 2.47. In analogy to the case of k=1 in Eq. 3.23 where U_1 and Y_1 are phasors, Z_1 and V_1 in Eq. 3.25 are "discrete-time phasors" and the ratio $N(\varepsilon^{j\lambda_o})/M(\varepsilon^{j\lambda_o})$ can be called a "discrete-time immittance function".

These results for $F_c(t,k)$ and $F_d(n,k)$ are examples of the lower two sectors of Figure 3.2. Note that in these two cases, the responses (steady-state) are for all t or n and initial conditions are not considered.

The next response functions to be developed are the others on the time-domain sectors of Figure 3.2. These are the unit impulse response $h(t,\lambda)$, the unit pulse response h(n,k), and the unit step function responses $a(t,\lambda)$ and a(n,k). In all four cases the definition of the response function includes the assumption that the system is relaxed before either u_o, u_1, v_o, or v_1 is applied. Furthermore, the inputs u(t) and v(n) are one-sided as in Eq. 2.55, 2.55a, 2.56, and 2.56a and u(t) is assumed to be piecewise continuous with finite discontinuities.

Starting with the stationary linear system

$$M(D)\,y(t) = N(D)\,u(t)$$

with

$$u(t) = \int_{t_o}^{\infty} u(\lambda)u_o(t-\lambda)d\lambda$$

and by comparison with Eq. 3.9 and 2.52,

$$f(\lambda) \to u(\lambda)$$

$$q_1(t,\lambda) \to u_o(t-\lambda)$$

and

$$Q_1(t,\lambda) \to h(t,\lambda) \ .$$

Therefore, we have

$$M(D)h(t,\lambda) = N(D)u_o(t-\lambda) \ , \qquad (3.26)$$

or since the system is stationary, $h(t,\lambda) = h(t-\lambda,0)$, which is defined as $h(t-\lambda)$ [5.], and

$$M(D)h(t-\lambda) = N(D)u_o(t-\lambda).$$

For $t-\lambda=\tau$ and for $D = \frac{d}{d\tau}$,

$$M(D)h(\tau) = N(D)u_o(\tau). \qquad (3.27)$$

A classical solution for $h(\tau)$ is obtained by solving

$$M(D)h(\tau) = 0, \qquad \tau > 0 \qquad (3.27a)$$

using initial conditions

$$h^{(r-1)}(0+),\ h^{(r-2)}(0+),\ldots,h(0+).$$

The general solution is

$$h(\tau) = k_1(\exp \lambda_1\tau)+\ldots+k_r(\exp \lambda_r\tau), \qquad \tau \geq 0$$

where the $\lambda_1, \lambda_2,\ldots, \lambda_r$ are simple zeros of $M(\lambda)$. If there are repeated zeros of $M(\lambda)$, a different form of solution is required. For example, if only λ_1 is of multiplicity j,

$$h(\tau) = k_1(\exp \lambda_1\tau) + k_2\tau(\exp \lambda_1\tau)+\ldots+ k_j\tau^{j-1}(\exp \lambda_1\tau)$$

$$+ k_{j+1}(\exp \lambda_{j+1}\tau)+\ldots+ k_r(\exp \lambda_r\tau), \quad \tau \geq 0.$$

Also, if the system of Eq. 3.27 is realizable,

$$h(\tau) = 0, \qquad \tau < 0.$$

It is necessary, of course, to obtain the r initial conditions in order to find the r constants in the general solutions. These initial conditions may be found by using a power series expansion of $h(\tau)$ about $\tau=0$, but a simpler method is derived below. The following two examples illustrate classical solutions when the initial conditions are known.

Example 3.2

A numerical form of Example 3.1c is

$$(D^2 + 3D + 2)y = u.$$

Find $h(t-\lambda)$ for this stationary linear system where $M(D) = D^2 + 3D + 2$, $N(D) = 1$, and $r = 2$. Solve first for $h(\tau)$ with (see Example 3.4)

$$Dh(0+) = 1$$

$$h(0+) = 0,$$

and

$$M(\lambda) = \lambda^2 + 3\lambda + 2.$$

Then for

$$M(\lambda) = 0, \ \lambda_1 = -1, \ \lambda_2 = -2,$$

and

$$h(\tau) = k_1\varepsilon^{-\tau} + k_2\varepsilon^{-2\tau}, \qquad \tau \geq 0$$

Using the initial conditions,

$$h(0+) = k_1 + k_2 = 0$$

$$Dh(0+) = -k_1 - 2k_2 = 1,$$

and from these $k_1 = 1$ and $k_2 = -1$. Then,

$$h(\tau) = \varepsilon^{-\tau} - \varepsilon^{-2\tau}, \qquad \tau \geq 0$$

$$= 0 \qquad , \qquad \tau < 0$$

or

$$h(t-\lambda) = [(\exp -(t-\lambda)) - (\exp -2(t-\lambda)]u_1(t-\lambda).$$

<u>Example 3.3</u>

Find $h(t-\lambda)$ for the system represented by

$$(D^2 + 3D + 2)y = (2D + 3)u$$

with initial conditions (see Example 3.4)

$$h(0+) = 2$$

$$Dh(0+) = -3.$$

Then,

$$h(\tau) = k_1\varepsilon^{-\tau} + k_2\varepsilon^{-2\tau}, \qquad \tau \geq 0 .$$

Using the initial conditions,

$$2 = k_1 + k_2$$

$$-3 = -k_1 - 2k_2$$

and $k_1 = 1$, $k_2 = 1$. Then,

$$\begin{aligned} h(\tau) &= \varepsilon^{-\tau} + \varepsilon^{-2\tau} \qquad & \tau \geq 0 \\ &= 0 & \tau < 0 \end{aligned}$$

or

$$h(t-\lambda) = (((\exp -(t-\lambda)) + (\exp -2(t-\lambda)))u_1(t-\lambda).$$

Equation 3.27 also can be solved using the Laplace Transform, and following the notation of Appendix A, the equation can be solved using $\mathcal{L}_{o+}$ or $\mathcal{L}_{o-}$. In general

$$\mathcal{L}_{o+}D^k h(\tau) = s^k H(s) - \sum_{i=0}^{k-1} s^i h^{(k-i-1)}(o+), \qquad k \geq 1$$

and

$$\mathcal{L}_{o-}D^k h(\tau) = s^k H(s)$$

where for strictly proper systems, $\mathcal{L}_{o+}h(\tau) = \mathcal{L}_{o-}h(\tau) = H(s)$. Then for Eq. 3.27, (sum from 0 to -1 defined as 0)

$$M(s)H(s) = \sum_{k=0}^{r} \alpha_k \sum_{i=0}^{k-1} s^i h^{(k-i-1)}(0+) \qquad (3.28a)$$

using $\mathcal{L}_{0+}$, and

$$M(s)H(s) = \sum_{j=0}^{m} \beta_j s^j = N(s) \qquad (3.28b)$$

using $\mathcal{L}_{0-}$. Note that $\alpha_r = 1$ and the notation

$$h^{(k-i-1)}(0+) \equiv D^{k-i-1}h(0+) \ .$$

Also defining $\beta_{m+1}, \ldots, \beta_{r-1} = 0$, the equality of the two forms means

$$\sum_{k=0}^{r} \alpha_k \sum_{i=0}^{k-1} s^i h^{(k-i-1)}(0+) = \sum_{j=0}^{r-1} \beta_j s^j \ ,$$

or

$$\sum_{k=1}^{r} \sum_{i=0}^{k-1} \alpha_k s^i h^{(k-i-1)}(0+) = \sum_{i=0}^{r-1} \beta_i s^i \ ,$$

where again a sum from 0 to -1 is defined as zero. Then, interchanging the order of summation,

$$\sum_{i=0}^{r-1} \sum_{k=i+1}^{r} \alpha_k s^i h^{(k-i-1)}(0+) = \sum_{i=0}^{r-1} \beta_i s^i$$

Equating coefficients of s^i,

$$\sum_{k=i+1}^{r} \alpha_k h^{(k-i-1)}(0+) = \beta_i, \qquad i = 0, \ldots, r-1 \ ,$$

and changing the variable of summation, we get

$$\sum_{j=0}^{r-i-1} \alpha_{j+i+1} h^{(j)}(0+) = \beta_i, \qquad i = 0, \ldots, r-1 \ .$$

This set of r linear algebraic equations can be solved for

$h(0+)$, $h^{(1)}(0+), \ldots, h^{(r-1)}(0+)$. In matrix notation,

$$A_\alpha \, \underline{h}(0+) = \underline{\beta}$$

where $\underline{h}^T(0+) = [h(0+) \; \ldots \; h^{(r-1)}(0+)]$

$$\underline{\beta}^T = [\beta_o \; \cdots \; \beta_{r-1}]$$

$$A_\alpha = \begin{bmatrix} \alpha_1 & \alpha_2 & \alpha_3, & \ldots, & \alpha_{r-1} & 1 \\ \alpha_2 & \alpha_3 & , & \ldots, & 1 & 0 \\ \alpha_3 & & & & & \\ \vdots & & & & & \\ \alpha_{r-1} & 1 & 0, & \ldots, & & 0 \\ 1 & 0 & , & \ldots, & 0 & 0 \end{bmatrix}$$

Therefore,

$$\underline{h}(0+) = A_\alpha^{-1} \, \underline{\beta} \, . \tag{3.29}$$

<u>Example 3.4</u>

The results of Examples 3.2 and 3.3 can be checked using Eq. 3.29. Thus for $\alpha_1 = 3$, $\beta_o = 1$, $\beta_1 = 0$.

$$\underline{h}(0+) = \begin{bmatrix} 3 & 1 \\ 1 & 0 \end{bmatrix}^{-1} \begin{bmatrix} 1 \\ 0 \end{bmatrix}$$

$$= \begin{bmatrix} 0 & 1 \\ 1 & -3 \end{bmatrix} \begin{bmatrix} 1 \\ 0 \end{bmatrix} = \begin{bmatrix} 0 \\ 1 \end{bmatrix}$$

and for $\alpha_1 = 3$, $\beta_0 = 3$, $\beta_1 = 2$, we have

$$\underline{h}(0+) = \begin{bmatrix} 0 & 1 \\ 1 & -3 \end{bmatrix} \begin{bmatrix} 3 \\ 2 \end{bmatrix} = \begin{bmatrix} 2 \\ -3 \end{bmatrix}$$

Summarizing, the impulse response $h(\tau)$ can be found from Eq. 3.27 by (1) solving the homogeneous equation $M(D)h(\tau)=0$ using the classical method with initial conditions found by using Eq. 3.29, or (2) by using $\mathcal{L}_{o+}$ with the initial conditions found in the same way, or (3) by using $\mathcal{L}_{o-}$ as in Eq. 3.28b where

$$h(\tau) = [\mathcal{L}^{-1} N(s)/M(s)]u_1(\tau). \qquad (3.30)$$

Eq. 3.30 obviously is the simplest to use since the initial conditions at $\tau=0+$ are not needed.

Example 3.5

From Example 3.3,

$$M(s) = s^2 + 3s+2 = (s+1)(s+2)$$

$$N(s) = 2s + 3.$$

Therefore,

$$h(\tau) = [\mathcal{L}^{-1} \frac{2s+3}{(s+1)(s+2)}] u_1(\tau)$$

$$= (\varepsilon^{-\tau} + \varepsilon^{-2\tau})u_1(\tau).$$

The development of solutions for $a(t,\lambda)$ can be done quite easily by using the results for $h(t,\lambda)$. Thus for

$$M(D)y(t) = N(D)u(t)$$

and

$$u(t) = \int_{t_o}^{\infty} \dot{u}(\lambda) u_1(t-\lambda) d\lambda$$

by comparison with Eq. 3.9 and 2.52,

$$f(\lambda) \to \dot{u}(\lambda)$$

$$q_1(t,\lambda) \to u_1(t-\lambda)$$

and

$$Q_1(t,\lambda) \to a(t,\lambda) \ .$$

Thus we have

$$M(D)a(t,\lambda) = N(D)u_1(t-\lambda)$$

and as before, for a stationary system, $a(t,\lambda) = a(t-\lambda,0) \equiv a(t-\lambda)$ and

$$M(D)a(\tau) = N(D)u_1(\tau) \ , \tag{3.31}$$

with $\tau=t-\lambda$. For $\tau \geq 0$,

$$M(D)a(\tau) = N(D)$$

and a classical solution is obtained by using initial conditions

$$a^{(r-1)}(0+), \ldots, a(0+).$$

A general solution is (for a realizable system)

$$\begin{aligned} a(\tau) &= \beta_o/\alpha_o + k_1(\exp \lambda_1\tau) + \ldots + k_r(\exp \lambda_r\tau), && \tau \geq 0 \\ &= 0 && \tau < 0 \end{aligned}$$

when the zeros of $M(\lambda)$ are simple. For zeros of multiplicity greater than one, other forms as shown above are required.

Example 3.7

For the system of Example 3.2, find $a(t-\lambda)$ with

$$Da(0+) = 0$$

$$a(0+) = 0$$

and

$$a(\tau) = 1/2 + k_1\varepsilon^{-\tau} + k_2\varepsilon^{-2\tau}, \qquad \tau \geq 0 .$$

Using the initial conditions, we have

$$0 = 1/2 + k_1 + k_2$$

$$0 = -k_1 - 2k_2$$

and from these, $k_1 = -1$, $k_2 = 1/2$. Then

$$a(\tau) = (1/2 - \varepsilon^{-\tau} + 1/2\varepsilon^{-2\tau})u_1(\tau)$$

and

$$a(t-\lambda) = (1/2 - \varepsilon^{-(t-\lambda)} + 1/2\varepsilon^{-2(t-\lambda)})u_1(t-\lambda).$$

Example 3.8

For the system of Example 3.3, find $a(t-\lambda)$. Using

$$a(0+) = 0$$

$$Da(0+) = 2,$$

$$a(\tau) = (3/2 + k_1\varepsilon^{-\tau} + k_2\varepsilon^{-2\tau})u_1(\tau)$$

and using the initial conditions, we have

$$k_1 = -1 \text{ and } k_2 = -1/2$$

and

$$a(\tau) = (\tfrac{3}{2} - \varepsilon^{-\tau} - \tfrac{1}{2}\,\varepsilon^{-2\tau})u_1(\tau)$$

or

$$a(t-\lambda) = (\tfrac{3}{2} - (\exp -(t-\lambda)) - \tfrac{1}{2}(\exp -2(t-\lambda))u_1(t-\lambda).$$

Equation 3.31 also can be solved using the Laplace Transform. Thus for

$$\mathcal{L}_{o+}D^k a(\tau) = s^k A(s) - \sum_{i=0}^{k-1} s^i a^{(k-i-1)}(0+), \qquad k \geq 1$$

and

$$\mathcal{L}_{o-}D^k a(\tau) = s^k A(s) ,$$

and the transformed differential equations are

$$M(s)A(s) = \beta_o s^{-1} + \sum_{k-1}^{r} \alpha_k \sum_{i=0}^{k=0} s^i a^{(k-i-1)}(0+) \qquad (3.32a)$$

for $\mathcal{L}_{o+}$, or

$$M(s)A(s) = N(s)s^{-1} = \sum_{j=0}^{m} \beta_j s^{j-1} \qquad (3.32b)$$

for $\mathcal{L}_{o-}$, and $\mathcal{L}_{o+}a(\tau) = \mathcal{L}_{o-}a(\tau) = A(s)$. Then using the same conventions as in the previous development for h, we have

$$\sum_{k=0}^{r} \alpha_k \sum_{i=0}^{k-1} s^i a^{(k-i-1)}(0+) = \sum_{j=0}^{r-1} \beta_j s^{j-1} - \beta_o s^{-1}$$

$$= \sum_{j=1}^{r-1} \beta_j s^{j-1}$$

$$= \sum_{i=0}^{r-2} \beta_{i+1} s^i .$$

Proceeding as before,

$$\sum_{j=0}^{r-i-1} \alpha_{j+i+1} a^{(j)}(0+) = \beta_{i+1}, \qquad i = 0, 1, \ldots, r-2$$
$$= 0 \quad , \qquad i = r-1$$

and in matrix notation,

$$A_\alpha \underline{a}(0+) = \underline{\beta}_o$$

where

$$\underline{a}^T(0+) = [a(0+), \ldots, a^{(r-1)}(0+)]$$

$$\underline{\beta}_o^T = [\beta_1 \beta_2, \ldots, \beta_{r-1}\ 0]$$

and A_α is as before in Eq. 3.29. Therefore,

$$\underline{a}(0+) = A_\alpha^{-1} \underline{\beta}_o \ . \qquad (3.33)$$

<u>Example 3.9</u>

For the system of Example 3.3 and using A_α^{-1} of Example 3.4,

$$\underline{a}(0+) = \begin{bmatrix} 0 & 1 \\ 1 & -3 \end{bmatrix} \begin{bmatrix} 2 \\ 0 \end{bmatrix} = \begin{bmatrix} 0 \\ 2 \end{bmatrix}$$

which checks Example 3.8.

Summarizing, the response $a(\tau)$ can be found from Equation 3.31 by (1) solving it using the classical method with initial conditions found by using Equation 3.33, or (2) by using $\mathcal{L}_{o+}$ with initial conditions, or (3) by using $\mathcal{L}_{o-}$ as in Equation 3.32b to get

$$a(\tau) = [\mathcal{L}^{-1} N(s)/sM(s)] u_1(\tau). \qquad (3.34)$$

Again that is obviously the simplest way to find $a(\tau)$.

Example 3.10

From Example 3.3,

$$M(s) = (s+1)(s+2)$$

$$N(s) = 2s+3.$$

Therefore,

$$a(\tau) = [\mathcal{L}^{-1} \frac{2s+3}{s(s+1)(s+2)}]\, u_1(\tau)$$

$$= (\frac{3}{2} - \varepsilon^{-\tau} - \frac{1}{2}\varepsilon^{-2\tau}) u_1(\tau).$$

Since u_0 and u_1 are related as in Equation 2.14 and 2.15, the response functions $h(t-\lambda)$ and $a(t-\tau)$ of the stationary linear system

$$M(D)y = N(D)u$$

should also be related. Starting with

$$u_1(t-\lambda) = \int_{-\infty}^{t} u_0(\alpha-\lambda)\,d\alpha$$

and using the linear operator L_c as $M(D)^{-1}N(D)$, we have

$$L_c u_1(t-\lambda) = L_c \int_{-\infty}^{t} u_0(\alpha-\lambda)\,d\alpha \ .$$

Using the definition of $a(t-\lambda)$ on the left and letting $\alpha=\sigma+t$ on the right, we obtain

$$a(t-\lambda) = L_c \int_{-\infty}^{0} u_0(\sigma+t-\lambda)\,d\sigma \ .$$

Using the additivity of L_c and the definition of h

$$a(t-\lambda) = \int_{-\infty}^{0} h(t-\lambda+\sigma)d\sigma$$

or for $\tau=t-\lambda$,

$$a(\tau) = \int_{-\infty}^{0} h(\tau+\sigma)d\sigma.$$

For $\alpha=\tau+\sigma$ and for a realizable system, for $t_o=0$,

$$a(\tau) = \int_{0}^{\tau} h(\alpha)d\alpha \tag{3.35}$$

and differentiating with respect to τ,

$$\frac{da(\tau)}{d\tau} = h(\tau). \tag{3.36}$$

Example 3.11

a) From Example 3.3,

$$h(\tau) = (\varepsilon^{-\tau} + \varepsilon^{-2\tau})u_1(\tau).$$

Therefore, using Equation 3.35,

$$a(\tau) = \int_{0}^{\tau} (\varepsilon^{-\alpha} + \varepsilon^{-2\alpha})d\alpha$$

$$= (\tfrac{3}{2} - \varepsilon^{-\tau} - \tfrac{1}{2}\varepsilon^{-2\tau})u_1(\tau)$$

with checks Example 3.8.

b) For

$$a(\tau) = (\tfrac{3}{2} - \varepsilon^{-\tau} - \tfrac{1}{2}\varepsilon^{-2\tau})u_1(\tau)$$

using Equation 3.36,

$$h(\tau) = (\varepsilon^{-\tau} + \varepsilon^{-2\tau})u_1(\tau) + (\tfrac{3}{2} - \varepsilon^{-\tau} - \tfrac{1}{2}\varepsilon^{-2\tau})u_o(\tau).$$

The second terms results from the use of u_1 in writing $a(\tau)$. Since the coefficient of u_o is zero for $\tau=0$, the second term is understood to be zero. A term like this might be expected since $a(\tau)$ is not differentiable at $\tau=0$. $h(0)$ also can be verified by using Equation 3.36 for $\tau>0$ and defining

$h(0) = \lim_{\tau \to 0+} h(\tau)$. Thus we get

$$h(\tau) = (\varepsilon^{-\tau} + \varepsilon^{-2\tau})u_1(\tau).$$

The response functions $h(n,k)$ and $a(n,k)$ can be found for the stationary linear system

$$M(E)z(n) = N(E)v(n)$$

by carrying out a parallel development to the one just completed for a continuous-time system. For

$$v(n) = \sum_{k=n_o}^{\infty} v(k)v_o(n-k)$$

and by comparison with Equation 3.8 and 2.51,

$$m(k) \to v(k)$$

$$q_4(n,k) \to v_o(n-k)$$

and

$$Q_4(n,k) \to h(n,k).$$

Substituting in the system difference equation we obtain

$$M(E)\, h(n,k) = N(E)\, v_o(n-k).$$

For a stationary system, using the same notational convention as for continuous-time systems with $j = n-k$ and with $Ef(j) = f(j+1)$, the problem is one of solving

$$M(E)\ h(j) = N(E)\ v_o(j) \tag{3.37}$$

or

$$M(E)\ h(j) = 0, \qquad j \geq 1 \tag{3.37a}$$

with M(E) and N(E) given in Equation 3.20. The realizability condition implies $h(j) = 0$, $j<0$, and it is necessary to find $h(0), h(1),\ldots,h(r)$. These are used in evaluating constants in the general solution. Starting with $j = -r$ in Equation 3.37,

$$h(0) + \alpha_{r-1}h(-1) +\ldots+ \alpha_o h(-r) = \beta_m v_o(m-r) +\ldots + \beta_o v_o(-r).$$

Then

$$h(0) = \beta_m, \qquad m = r$$
$$= 0 \qquad m < r.$$

Next let $j = -r+1$ in Equation 3.37,

$$h(1) + \alpha_{r-1}h(0) + \alpha_{r-2}h(-1) +\ldots+ \alpha_o h(-r+1) =$$

$$\beta_m v_o(m-r+1) +\ldots+ \beta_q v_o(q-r+1) +\ldots+ \beta_o v_o(-r+1).$$

Then

$$h(1) = -\alpha_{r-1}h(0) + \beta_{r-1}.$$

This procedure is continued for $h(2),\ldots,h(r)$ with $j = -r+2$,

-r+3,..., 0. The general solution for h(j) is then

$$h(j) = k_1 \rho_1^j + \dots + k_r \rho_r^j, \qquad j \geq 1$$
$$= h(0) \qquad\qquad , \qquad j = 0$$

where $\rho_1, \dots, \rho_r$ are the zeros, assumed to be simple, of $M(\rho)$. The constants $k_1, \dots, k_r$ are found by using $h(1), \dots, h(r)$. The method is illustrated in Example 3.12. A general form for the constants is suggested in Section 3.4, Example 3.26. If one zero, e.g., ρ_1, is of multiplicity i, the solution is

$$h(j) = k_1 \rho_1{}^j + k_2 j \rho_1{}^j + \dots + k_i j^{i-1} \rho_1{}^j +$$
$$+ k_{i+1} \rho_{i+1}{}^j + \dots + k_r \rho_r{}^j$$

and similar forms are used if more than one zero is of multiplicity greater than one.

Example 3.12

Find h(n-k) for the system described by

$$(E^2 - 0.75E + 0.125)z(n) = (E-0.4)v(n).$$

The equation to solve is

$$(E^2 - 0.75E + 0.125)h(j) = (E-0.4)v_o(j).$$

For j = -r = -2,

$$h(0) - 0.75h(-1) + 0.125h(-2) = v_o(-1)-0.4v_o(-2)$$

and

$$h(0) = 0$$

since all other terms are zero. For j = -1

$$h(1) - 0.75h(0) + 0.125h(-1) = v_o(0)-0.4v_o(-1)$$

and

$$h(1) = 1.$$

For $j = 0$

$$h(2) - 0.75h(1) + 0.125h(0) = v_o(1) - 0.4v_o(0)$$

and

$$h(2) = 0.75 - 0.4 = 0.35.$$

For

$$M(\rho) = \rho^2 - 0.75\rho + 0.125 = 0 \ ,$$

$$\rho_1 = 0.5, \quad \rho_2 = 0.25$$

and

$$\begin{aligned} h(j) &= k_1(0.5)^j + k_2(0.25)^j & j \geq 1 \\ &= 0 & j \leq 0. \end{aligned}$$

Using h(1) and h(2),

$$1 = 0.5k_1 + 0.25k_2$$

$$0.35 = 0.25k_1 + 0.0625k_2$$

and from these

$$k_1 = 0.8, \quad k_2 = 2.4.$$

Then

$$\begin{aligned} h(j) &= 0.8\ (0.5)^j + 2.4(0.25)^j & j \geq 1 \\ &= 0 & j \leq 0 \end{aligned}$$

and

$$h(n-k) = 0.8\ (0.5)^{n-k} + 2.4(0.25)^{n-k} \qquad n \geq k+1$$

$$= 0 \qquad n \leq k.$$

Equation 3.37 also can be solved using the Z-transform. Doing this we have

$$z^r H(z) - z^r \sum_{i=0}^{r-1} h(i) z^{-i} +$$

$$+ \alpha_{r-1}(z^{r-1} H(z) - z^{r-1} \sum_{i=0}^{r-2} h(i) z^{-i}) +$$

$$+ \ldots + \alpha_0 H(z) = \beta_0.$$

Therefore,

$$M(z)H(z) = \beta_0 + \sum_{k=1}^{r} \alpha_k \sum_{i=0}^{k-1} h(i) z^{k-i}$$

or

$$M(z)H(z) = \beta_0 + \sum_{k=1}^{r} \alpha_k \sum_{i=1}^{k} h(k-i) z^{i}. \qquad (3.38a)$$

This form is analogous to Eq. 3.28a. Expanding the right hand side

$$M(z)H(z) = \beta_0 + [\alpha_1 h(0) + \alpha_2 h(1) + \ldots + h(r-1)]z$$

$$+ [\alpha_2 h(0) + \alpha_3 h(1) + \ldots + h(r-2)]z^2$$

$$+ \ldots + [h(0)]z^r .$$

Now the equations for h(0), h(1),...,h(r) for m=r are

$$h(0) = \beta_r$$

$$h(1) = -\alpha_{r-1}h(0) + \beta_{r-1}$$

$$\vdots$$

$$h(j) = -\alpha_{r-1}h(j-1) - \ldots - \alpha_{r-j}h(0) + \beta_{r-j}$$

$$\vdots$$

$$h(r) = -\alpha_{r-1}h(r-1) - \ldots - \alpha_o h(0) + \beta_o$$

or rearranging and writing in matrix form

$$\underline{\beta}_d = B_\alpha \underline{h}_d$$

where

$$\underline{\beta}_d^T = [\beta_r \ldots \beta_o]$$

$$\underline{h}_d^T = [h(0) \ldots h(r)] \quad ,$$

and

$$B_\alpha = \begin{bmatrix} 1 & 0 & 0 & \ldots & 0 \\ \alpha_{r-1} & 1 & 0 & \ldots & 0 \\ \alpha_{r-2} & \alpha_{r-1} & 1 & \ldots & 0 \\ \vdots & & & \ddots & \vdots \\ \alpha_1 & & & & 0 \\ \alpha_o & \alpha_1 & \ldots & \alpha_{r-1} & 1 \end{bmatrix}$$

From these equations we find that the expanded right hand side is

$$\beta_o + \beta_1 z + \ldots + \beta_r z^r = N(z) \quad ,$$

and therefore,

$$M(z)H(z) = N(z). \tag{3.38b}$$

This is analogous to Eq. 3.28b, and

$$\underline{h}_d = B_\alpha^{-1} \underline{\beta}_d \tag{3.39}$$

which is analogous to Eq. 3.29. For m<r, $\beta_{m+1}, \beta_{m+2}, \ldots, \beta_r$ are defined to be zero and the previous results are still true. Note that h(r) is included in Eq. 3.39 for convenience; it is not used in deriving Eq. 3.38b. Eq. 3.38b also can be obtained directly to provide an analogy to the direct derivation of Eq. 2.28b which used $\mathcal{L}_{o-}$. To do this we operate on both sides of Eq. 3.37 with E^{-r} and then use the first form of the "time shift" property of the Z-transform. (See Appendix A, Section A.3.3). Doing this we obtain

$$(1+\alpha_{r-1}E^{-1} + \ldots + \alpha_o E^{-r})h(j) = (\beta_m E^{m-r} + \ldots + \beta_o E^{-r})v_o(j),$$

and taking the Z-transform gives

$$(1+\alpha_{r-1}z^{-1} + \ldots + \alpha_o z^{-r})H(z) = (\beta_m z^{m-r} + \ldots + \beta_o z^{-r})$$

and finally multiplying both sides by z^r leads to

$$M(z)H(z) = N(z).$$

Example 3.13

The results of Example 3.12 can be checked using Eq. 3.39. Thus for

$$\alpha_o = 0.125, \ \alpha_1 = -0.75, \ \alpha_2 = \alpha_r = 1$$

$$\beta_o = -0.4, \ \beta_1 = 1, \ \beta_2 = 0 \ ,$$

we have

$$\underline{h}_d = \begin{bmatrix} 1 & 0 & 0 \\ -0.75 & 1 & 0 \\ 0.125 & -0.75 & 1 \end{bmatrix}^{-1} \begin{bmatrix} 0 \\ 1 \\ -0.4 \end{bmatrix}$$

$$= \begin{bmatrix} 1 & 0 & 0 \\ 0.75 & 1 & 0 \\ 0.44 & 0.75 & 1 \end{bmatrix} \begin{bmatrix} 0 \\ 1 \\ -0.4 \end{bmatrix} = \begin{bmatrix} 0 \\ 1 \\ 0.35 \end{bmatrix}$$

Summarizing, the unit pulse response h(j) can be found from Eq. 3.37 by (1) using the classical solution and evaluating the constants $k_1,\ldots,k_r$ with $h(1),\ldots,h(r)$ found with Eq. 3.39, or (2) by using the Z-transform leading to Eq. 3.38a or (3) by using the Z-transform leading to Eq. 3.38b to get

$$h(j) = [\mathcal{Z}^{-1}N(z)/M(z)]. \tag{3.40}$$

Eq. 3.40 is obviously the simplest form to use.

Example 3.14

From Example 3.12,

$$M(z) = z^2 - 0.75z + 0.125$$

$$= (z - 0.5)(z - 0.25)$$

$$N(z) = (z - 0.4) .$$

Therefore,

$$h(j) = [\mathcal{Z}^{-1} \frac{(z-0.4)}{(z-0.5)(z-0.25)}]$$

$$= 0.8(0.5)^j + 2.4(0.25)^j, \quad j \geq 1$$

$$= 0 \quad , \quad j \leq 0$$

Solutions for a(n,k) can be found similarly. For

$$M(E)z(n) = N(E)v(n)$$

with

$$v(n) = v(n_o)v_1(n-n_o) + \sum_{k=n_o+1}^{\infty} v_1(n-k)\Delta_b v(k)$$

by comparison with Eq. 3.8 and 2.51

$$m(k) \rightarrow \Delta_b v(k)$$

$$q_4(n,k) \rightarrow v_1(n-k)$$

and

$$Q_4(n,k) \rightarrow a(n,k) \ .$$

Substituting,

$$M(E)a(n,k) = N(E)v_1(n-k)$$

and for a stationary system using j=n-k as before, the problem is one of solving

$$M(E)a(j) = N(E)v_1(j). \qquad (3.41)$$

Realizability implies a(j)=0, j<0 and it is necessary to find a(0), a(1),...,a(r-1). Starting with j=-r in Eq. 3.41 and assuming m=r,

$$a(0) + \alpha_{r-1}a(-1) +\ldots+ \alpha_o h(-r) = \beta_r v_1(r-r) +\ldots+ \beta_o v_1(-r)$$

Then $a(0) = \beta_r$. Next let $j=-r+1$,

$$a(1) + \alpha_{r-1}a(0) +\ldots+ \alpha_o a(-r+1) =$$

$$\beta_r v_1(r-r+1) +\ldots+ \beta_o v_1(-r+1).$$

Then $a(1) = -\alpha_{r-1}a(0) + \beta_r + \beta_{r-1}$. This procedure is continued for $a(2),\ldots,a(r-1)$ with $j=-r+2,\ldots,-1$. For $m<r$, the convention introduced previously is used, i.e., $\beta_{m+1}=\beta_{m+2}=\ldots=\beta_r=0$. The general solution for $a(j)$ is then

$$a(j) = k_o + k_1\rho_1^j +\ldots+ k_r\rho_r^j \qquad j \geq 0$$

for $\rho_1,\ldots,\rho_r$ simple zeros of $M(\rho)$ and with similar forms when there are multiple zeros as shown in the solutions for $h(j)$. Also,

$$k_o = \sum_{i=0}^{m} \beta_i \Big/ \sum_{i=0}^{r} \alpha_i .$$

Example 3.15

Find $a(n-k)$ for the system of Example 3.12. For $j=-2$,

$$a(0) = \beta_2 = 0$$

and for $j=-1$,

$$a(1) = -\alpha_1 a(0) + \beta_2 + \beta_1$$

$$= 1 .$$

The general solution is

$$a(j) = 1.6 + k_1(0.5)^j + k_2(0.25)^j \qquad j \geq 0$$

and using $a(0)$ and $a(1)$,

$$0 = 1.6 + k_1 + k_2$$

$$1 = 1.6 + 0.5k_1 + 0.25k_2$$

and from these, $k_1 = k_2 = -0.8$. Therefore,

$$a(j) = 1.6 - 0.8(0.5)^j - 0.8(0.25)^j \qquad j \geq 0$$

$$= 0 \qquad j < 0$$

and

$$a(n-k) = [1.6 - 0.8(0.5)^{n-k} - 0.8(0.25)^{n-k}]v_1(n-k) \ .$$

Equation 3.41 also can be solved using the Z-transform. We have

$$z^r A(z) - z^r \sum_{i=0}^{r-1} a(i)z^{-i} +$$

$$+ \alpha_{r-1}(z^{r-1}A(z) - z^{r-1} \sum_{i=0}^{r-2} a(i)z^{-i}) +$$

$$+\ldots+ \alpha_o A(z) = (\beta_r +\ldots+ \beta_o) \frac{z}{z-1}$$

and therefore,

$$M(z)A(z) = (\beta_r+\ldots+\beta_o)\frac{z}{z-1} + \sum_{k=1}^{r} \alpha_k \sum_{i=0}^{k-1} a(i)z^{k-i}$$

$$= (\sum_{i=0}^{r} \beta_i)\frac{z}{z-1} + \sum_{k=1}^{r} \alpha_k \sum_{i=1}^{k} a(k-i)z^i \ . \qquad (3.42a)$$

Expanding the right hand side

$$M(z)A(z) = (\sum_{i=0}^{r} \beta_i)\frac{z}{z-1} + [\alpha_1 a(0)+\ldots+a(r-1)]z +$$

$$+ [\alpha_2 a(0)+\ldots+a(r-2)]z^2 +\ldots+[a(0)z^r].$$

Now the equations for a(0), a(1),...,a(r-1) for m=r are

$$a(0) = \beta_r$$

$$a(1) = -\alpha_{r-1}a(0) + \beta_r + \beta_{r-1}$$

$$\vdots$$

$$a(j) = -\alpha_{r-1}a(j-1) - \ldots - \alpha_{r-j}a(0) + \beta_r + \ldots + \beta_{r-j}$$

$$\vdots$$

$$a(r-1) = -\alpha_{r-1}a(r-2) - \ldots - \alpha_1 a(0) + \beta_r + \ldots + \beta_1$$

or rearranging and writing in matrix form

$$\underline{\beta}_{d1} = B_{\alpha 1}\underline{a}_d$$

where

$$\underline{\beta}_{d1} = \begin{bmatrix} \beta_r \\ \beta_r + \beta_{r-1} \\ \vdots \\ \sum_{i=1}^{r} \beta_i \end{bmatrix}$$

$$\underline{a}_d^T = [a(0) \quad \ldots \quad a(r-1)]$$

and $B_{\alpha 1}$ is obtained from B_α by deleting the last row and column. From these equations, the expanded right hand side is

$$\frac{z}{z-1}(\beta_0 + \beta_1 z + \ldots + \beta_r z^r) = \frac{z}{z-1} N(z)$$

and we get

$$M(z)A(z) = N(z)\frac{z}{z-1} \tag{3.42b}$$

This equation also can be obtained directly by use of a "time shift" as was pointed out after Eq. 3.38b.

Finally then

$$\underline{a}_d = B_{\alpha 1}^{-1} \underline{\beta}_{d1} . \qquad (3.43)$$

Example 3.16

The results of Example 3.15 for a(0) and a(1) can be checked using Eq. 3.43. Thus we have

$$\underline{a}_d = \begin{bmatrix} 1 & 0 \\ -0.75 & 1 \end{bmatrix}^{-1} \begin{bmatrix} \beta_2 \\ \beta_2+\beta_1 \end{bmatrix}$$

$$= \begin{bmatrix} 1 & 0 \\ 0.75 & 1 \end{bmatrix} \begin{bmatrix} 0 \\ 1 \end{bmatrix} = \begin{bmatrix} 0 \\ 1 \end{bmatrix} .$$

Summarizing, the unit discrete-time step function response a(j) can be found from Eq. 3.41 by (1) using the classical solution and evaluating the constants $k_1, \ldots, k_r$ with a(0), ..., a(r-1) found with Eq. 3.43, or (2) by using the Z-transform leading to Eq. 3.42a or (3) by using the Z-transform leading to Eq. 3.42b to get

$$a(j) = [\mathfrak{z}^{-1} \frac{N(z)}{M(z)} \frac{z}{z-1}] \qquad (3.44)$$

Eq. 3.44 is obviously the simplest form to use.

Example 3.17

From Example 3.12

$$M(z) = (z - 0.5)(z - 0.25)$$

$$N(z) = (z - 0.4).$$

Therefore,

$$a(j) = [\mathcal{Z}^{-1} \frac{z(z - 0.4)}{(z-1)(z - 0.5)(z - 0.25)}]$$

or

$$a(j) = [1.6 - 0.8(0.5)^j - 0.8(0.25)^j]v_1(j) \ .$$

Since v_o and v_1 are related as in Eq. 2.33 and 2.34, the response functions h(n-k) and a(n-k) of the stationary linear system

$$M(E)z = N(E)v$$

should also be related. Starting with

$$v_1(n-k) = \sum_{i=-\infty}^{n} v_o(i-k)$$

and using the linear operator L_d as $M(E)^{-1}N(E)$,

$$L_d v_1(n-k) = L_d \sum_{i=-\infty}^{n} v_o(i-k).$$

Using the definition of a(n-k) on the left and letting i=q+n on the right,

$$a(n-k) = L_d \sum_{q=-\infty}^{o} v_o(n-k+q) \ .$$

Using the additivity of L_d and the definition of h,

$$a(n-k) = \sum_{q=-\infty}^{o} h(n-k+q)$$

or

$$a(j) = \sum_{q=-\infty}^{o} h(j+q).$$

For i=j+q and for a realizable system,

$$a(j) = \sum_{i=0}^{j} h(i) \tag{3.45}$$

and taking Δ_b with respect to j,

$$\Delta_b a(j) = h(j). \tag{3.46}$$

Example 3.18

a) From Example 3.12

$$h(j) = 0.8(0.5)^j + 2.4(0.25)^j \qquad j \geq 1$$
$$= 0 \qquad j \leq 0.$$

Therefore, using Eq. 3.45,

$$a(0) = 0$$

$$a(j) = \sum_{i=1}^{j} (0.8(0.5)^i + 2.4(0.25)^i) \qquad j \geq 1$$

$$= 0.8(0.5+0.5^2+...+0.5^j) + 2.4(0.25+0.25^2+...+0.25^j)$$

$$= 0.4[(1+0.5+0.5^2+...)-(0.5^j+0.5^{j+1}+...)] + 0.6[(1+0.25+0.25^2+...)-(.25^j+0.25^{j+1}+...)]$$

$$= 0.4[\frac{1}{1-0.5} - \frac{1}{1-0.5}(0.5)^j] + 0.6[\frac{1}{1-0.25} - \frac{1}{1-0.25}(0.25)^j]$$

or

$$a(j) = 1.6-0.8(0.5)^j - 0.8(0.25)^j \qquad j \geq 1 .$$

For j=0, this gives a(0) = 0, therefore,

$$a(j) = (1.6-0.8(0.5)^j - 0.8(0.25)^j)v_1(j)$$

which checks Example 3.15.

b) Using Eq. 3.46,

$$\Delta_b a(0) = a(0)-a(-1) = 0 = h(0)$$

and for $j \geq 1$

$$\begin{aligned}\Delta_b a(j) &= [1.6-0.8(0.5)^j-0.8(0.25)^j] \\ &\quad - [1.6-0.8(0.5)^{j-1}-0.8(0.25)^{j-1}] \\ &= 0.8(.5)^j+2.4(0.25)^j .\end{aligned}$$

Therefore,

$$\begin{aligned}h(j) &= 0.8(0.5)^j+2.4(0.25)^j \qquad & j \geq 1 \\ &= 0 & j \leq 0\end{aligned}$$

which checks Example 3.12.

The final pair of response functions in Figure 3.2 which we wish to develop are R(t,s) and R(n,z). Starting with

$$M(D)y(t) = N(D)u(t)$$

where u(t) is given by an inverse Laplace Transform:

$$u(t) = \frac{1}{2\pi j}\int_{c-j\infty}^{c+j\infty} U(s)\varepsilon^{ts}ds, \qquad t \geq 0 .$$

By comparison with Eq. 3.9 and 2.52

$$f(\lambda) \to U(s)$$

$$q_1(t,\lambda) \to \varepsilon^{ts}$$

and

$$Q_1(t,\lambda) \to R(t,s)$$

is the required response to ε^{ts}. Thus, on substitution we have

$$M(D)R(t,s) = N(D)\varepsilon^{ts}$$

and R(t,s) in this instance is the <u>forced response</u> or particular integral. Assuming

$$R(t,s) = H(s)\varepsilon^{ts}$$

and substituting,

$$M(D)H(s)\varepsilon^{ts} = N(D)\varepsilon^{ts}$$

and the required H(s) = N(s)/M(s). The notational choice of H(s) here has anticipated this result and this choice agrees with Eq. 3.28b. The response function is, therefore,

$$R(t,s) = (N(s)/M(s))\varepsilon^{ts} \quad . \tag{3.47}$$

It is now apparent that H(s) has at least two interpretations: it is the Laplace Transform of $h(\tau)$ as in Eq. 3.30 and it is the coefficient of ε^{ts} in R(t,s), the forced response to ε^{ts}, as in Eq. 3.47. A third interpretation occurs if for

$$M(D)y(t) = N(D)u(t)$$

with any piecewise continuous input u(t), $\mathcal{L}_{0-}u(t) = U(s)$, and a corresponding output y(t), $\mathcal{L}_{0-}y(t) = Y(s)$, where the system starts from rest, i.e.,

$$D^{r-1}y(0-) = \ldots = y(0-) = 0,$$

and

$$D^{r-2}u(0-) = \ldots = u(0-) = 0,$$

we have

$$M(s)Y(s) = N(s)U(s)$$

and, therefore,

$$Y(s) = \frac{N(s)}{M(s)} U(s) = H(s)U(s). \qquad (3.48)$$

H(s) is called the <u>system transfer function</u>, and these results apply to stationary linear systems. For time-varying systems the impulse response is $h(t,\lambda)$ and it is common to define an

$$H(t,s) = \int_0^\infty h(t,\lambda)(\exp -s(t-\lambda))d\lambda = \mathcal{L}_\lambda h(t,t-\lambda)$$

so that

$$R(t,s) = H(t,s)\varepsilon^{ts} .$$

The notation $\mathcal{L}_\lambda$ means that the real variable in the transform integral is λ. Of course in this form, it still is necessary to find $h(t,\lambda)$ from $M(D,t)$ and $N(D,t)$, and this usually is difficult.

A similar development for $R(n,z)$ also can be done. Starting with

$$M(E)z(n) = N(E)v(n)$$

where $v(n)$ is given by an inverse Z-Transform:

$$v(n) = \frac{1}{2\pi j} \oint_C V(z)z^{n-1}dz, \qquad n \geq 0.$$

By comparison with Eq. 3.10 and 2.53

$$q(\lambda) \rightarrow V(z)$$

$$q_3(n,\lambda) \rightarrow z^n$$

and

$$Q_3(n,\lambda) \to R(n,z)$$

is the required response to z^n. Thus we have

$$M(E)R(n,z) = N(E)z^n$$

and $R(n,z)$ in this instance is the forced response or particular sum. Assuming

$$R(n,z) = H(z)z^n$$

and substituting,

$$M(E)H(z)z^n = N(E)z^n$$

and carrying out the operations with E,

$$M(z)H(z)z^n = N(z)z^n$$

and the required $H(z) = N(z)/M(z)$. Again the notational choice of $H(z)$ here has anticipated this result and agrees with Eq. 3.38b. The response function is, therefore,

$$R(n,z) = (N(z)/M(z))z^n \quad . \tag{3.49}$$

As before it is apparent that $H(z)$ has three interpretations: it is the Z-Transform of $h(j)$ as in Eq. 3.40, it is the coefficient of z^n in $R(n,z)$, the forced response of z^n, as in Eq. 3.49, and again for an input $v(n)$, $\mathcal{Z}\, v(n) = V(z)$, and output $z(n)$, $\mathcal{Z}\, z(n) = Z(z)$, where the system starts from rest, i.e.,

$$z(-r) = z(-r+1) = \ldots = z(-1) = 0,$$

and

$$v(-r) = v(-r+1) = \ldots = v(-1) = 0,$$

we have

$$M(z)Z(z) = N(z)V(z)$$

and

$$Z(z) = \frac{N(z)}{M(z)} V(z) = H(z)V(z) \quad . \tag{3.50}$$

H(z) is again a transfer function. These results apply to stationary systems, and for time-varying systems, given h(n,k),

$$H(n,z) = \sum_{k=0}^{\infty} h(n,k)z^{-(n-k)} = \mathfrak{z}_k h(n,n-k)$$

so that

$$R(n,z) = H(n,z)z^n.$$

Example 3.19

a) Find the transfer function for the system of Example 3.1a. Since

$$M(s) = s^3 + (\tfrac{R}{L})s^2 + (\frac{LK+C^2}{LJ})s + (\tfrac{RK}{LJ})$$

and

$$N(s) = \frac{C}{LJ} ,$$

$$H(s) = \frac{N(s)}{M(s)} = \frac{\frac{C}{LJ}}{s^3 + (\frac{R}{L})s^2 + (\frac{LK+C^2}{LJ})s + (\frac{RK}{LJ})} \quad .$$

b) Find the transfer function for the numerical integrator of Example 3.1e.

$$M(z) = z^2-1$$

$$N(z) = \tfrac{1}{3}T(z^2+4z+1)$$

and

$$H(z) = \frac{\frac{1}{3}T(z^2+4z+1)}{(z^2-1)} .$$

As was pointed out earlier in discussing Table 3.1, developments for response functions $R(t,j\omega)$ and $R(n,\gamma)$ can be done by starting with Eq. A.4, the inverse Fourier Transform, and with Eq. A.14, the inverse $\mathfrak{F}_d$-Transform. In some system analyses, Fourier representation of nonperiodic (continuous-time or discrete-time) signals is useful. The forms for $H(j\omega)$ and $H(\gamma)$ are then

$$H(j\omega) = N(j\omega)/M(j\omega) \text{ and } H(\gamma) = N(\gamma)/M(\gamma)$$

for stationary systems.

This completes the development of response forms in Figure 3.2. These are all obtained from system models of the form given in Eq. 3.19b or Eq. 3.21b. The systems all have one input and one output.

3.3.4 Multiple Input/Output Systems

The extension of the forms of the previous section to multiple input and multiple output system models leads to simultaneous differential or difference equations. Only stationary systems are discussed, and the models are introduced primarily to show their relationship to the scalar forms already discussed, to introduce the definitions of impulse response matrix and transfer matrix, and to relate the forms to the well-known equations used in electrical network analysis. The state variable methods introduced in the next section are an alternative, and usually more convenient, way to model multiple input/multiple output systems.

For a continuous-time system with q inputs and p outputs, the input vector is $\underline{u}$ and the output vector is $\underline{y}$. To write a set of consistent simultaneous differential equations, a set of m internal variables w_i, $i = 1,2,\ldots,m$, is used.

Thus the internal variable vector is $\underline{w}$, and the equations for the system are

$$M(D)\underline{w} = N(D)\underline{u} = \underline{f} \qquad (3.51)$$

and

$$\underline{y} = P(D)\underline{w}$$

where the dimensions are

$$M(D)_{mxm},\ N(D)_{mxq},\ P(D)_{pxm}\ .$$

A symbolic solution is

$$\underline{w} = M(D)^{-1}N(D)\underline{u}$$

$$\underline{y} = P(D)M(D)^{-1}N(D)\underline{u}\ .$$

Using the Laplace Transform with all initial conditions zero,

$$\underline{Y}(s) = P(s)M(s)^{-1}N(s)\underline{U}(s)$$

and letting $H(s) = P(s)M(s)^{-1}N(s)$, the pxq <u>transfer matrix</u>, we have

$$\underline{Y}(s) = H(s)\underline{U}(s), \qquad (3.52)$$

the vector-matrix counterpart of Eq. 2.48. Taking the inverse Laplace Transform and making use of the real convolution theorem,

$$\underline{y}(t) = \int_0^t H(t-\lambda)\underline{u}(\lambda)\,d\lambda$$

where $H(t-\lambda)$ is the <u>impulse response matrix</u>, and analogous to Eq. 3.30,

$$H(\tau) = [\mathcal{L}^{-1}H(s)]u_1(\tau)\ . \qquad (3.53)$$

The elements $h_{ij}(\tau)$ are responses at the ith output to $u_o(\tau)$ at the jth input with all other inputs inactive.

Two cases of particular interest in network analysis (see Chapter 6, Section 6.3) are those where the internal variables are loop currents or node to reference voltages. For ℓ loop currents, and ℓ equivalent voltage sources,

$$\underline{w} \to \underline{i}$$

$$N(D)\underline{u} \to \underline{e}_s$$

$$M(D) \to Z(D),*$$

and

$$Z(D)\underline{i} = \underline{e}_s.$$

Usually the outputs are the loop currents and P(D) is an identity matrix. Similarly, for n node voltages, and n equivalent current sources,

$$\underline{w} \to \underline{e}$$

$$N(D)\underline{u} \to \underline{i}_s$$

$$M(D) \to Y(D),*$$

and

$$Y(D)\underline{e} = \underline{i}_s.$$

A similar sequence of steps for discrete-time systems starts with

$$M(E)\underline{w} = N(E)\underline{v} = \underline{f} \tag{3.54}$$

and

$$\underline{z} = P(E)\underline{w},$$

---

*If the voltages and currents are exponential forms $k\varepsilon^{st}$, these become Z(s) and Y(s), the usual impedance and admittance matrices.

to obtain

$$\underline{z} = P(E)M(E)^{-1}N(E)\underline{v} .$$

Using the Z-Transform with all initial conditions zero,

$$\underline{Z}(z) = P(z)M(z)^{-1}N(z)\underline{V}(z)$$

and letting

$$H(z) = P(z)M(z)^{-1}N(z),$$

$$\underline{Z}(z) = H(z)\underline{V}(z), \tag{3.55}$$

the counterpart of Eq. 3.50. Taking the inverse Z-Transform,

$$\underline{z}(n) = \sum_{i=0}^{n} H(n-i)\underline{v}(i)$$

and

$$H(j) = [\mathfrak{z}^{-1} H(z)], \tag{3.56}$$

the matrix counterpart of Eq. 3.40. H(z) is a <u>transfer matrix</u> and H(j) a unit <u>pulse response matrix</u>.

3.4 VECTOR-MATRIX EQUATIONS AND STATE VARIABLES

The system models developed in Section 3.3 have been used for many years, and satisfactory methods of finding system responses using these will be given in Chapter 4. However, there are several reasons for developing models of greater generality. A fundamental limitation of some of the previous models (e.g., transfer functions, impulse responses, and step-function responses) is that they only model the controllable-observable part of the linear system. These terms will be defined later in this section; the difficulty is associated with the well-known problem of pole-zero cancellations.

Second, the system models developed so far are primarily concerned with single input-single output systems. Although extensions using simultaneous differential or difference equations have been introduced, the vector-matrix state variable forms have turned out to be more convenient -- particularly when the order of the system is large and computer solutions are needed. Third, the introduction of arbitrary initial conditions is usually easier in the vector-matrix form since the initial conditions appear explicitly as the initial state vector. Of course, arbitrary initial conditions can be included in the previous models by use of appropriate classical or transform methods for solving differential or difference equations. Fourth, because these vector-matrix equations are time domain models, both stationary and time-varying problems can be handled with relative ease by computational techniques. This was suggested in the previous section where the response functions were found only for stationary, linear systems. Fifth, a more abstract advantage in some cases, is in the greater emphasis placed on the internal structure of a linear system or its model. The large reservoir of mathematical concepts associated with linear vector spaces and linear operators can be used to discuss many interesting properties of linear systems models, e.g., the ideas of controllability and observability mentioned above.

3.4.1 Definitions and State Equations

The form of the vector-matrix or <u>state equations</u> describing a continuous-time linear system is:

$$\dot{\underline{x}}(t) = A\,\underline{x}(t) + B\underline{u}(t), \qquad \dot{\underline{x}}(t) = \frac{d}{dt}\,\underline{x}(t) \tag{3.57a}$$

$$\underline{y}(t) = C\,\underline{x}(t) + D\underline{u}(t) \tag{3.57b}$$

where $\underline{x}$ is an r-dimension vector, the state vector, $\underline{u}$ is a q-dimension input vector, and $\underline{y}$ is a p-dimension output vector. Thus the dimensions of the four matrices are

$$A_{rxr},\ B_{rxq},\ C_{pxr},\ D_{pxq}$$

and A is a system coefficient matrix, B is an input matrix, and C an output matrix. Equation 3.57a is a system differential equation and Equation 3.57b is an output equation. The components of $\underline{x}$, i.e., x_i, i = 1,2,...,r, are state variables. In general, $\underline{x}$, $\underline{u}$, and $\underline{y}$ are functions of time t for t on a specified interval T_t, e.g., T_{c1} or T_{c2} of Chapter 2. The matrices A, B, C, and D may be constants, as in a stationary system, or known functions of time as in time-varying systems. The real finite-dimensional linear vector space R_r is the space of all vectors $\underline{x}(t)$, $t \varepsilon T_t$ (see Appendix B). R_r is the state space, and by similar definitions, R_q is an input space and R_p is an output space. It is clear from Equation 3.57b that C and D are linear operators that map $\underline{x}$ and $\underline{u}$ to the output space, but the nature of the mapping of $\underline{u}$ that produces $\underline{x} \varepsilon R_r$ must be examined in detail.

Although Equations 3.57 specify a form of linear system model, the conditions under which $\underline{x}(t)$, $t \varepsilon T_t$, qualifies as a state vector have not been stated; $\underline{u}(t)$ and $\underline{y}(t)$ can be described by the signal models of Chapter 2. In general, the concept of state implies that if the state of a system is known at some time t_o, the response of the system for $t>t_o$ can be found for a specified input, $t \geq t_o$. More explicity, the state of a linear system, as given by $\underline{x}(t_o)$, $t_o \varepsilon T_t$, is a set of numbers $x_1(t_o),\ldots,x_r(t_o)$ for minimal r such that $\underline{x}(t)$, $t>t_o$, can be found for a specified $\underline{u}(t)$, $t \geq t_o$. This definition implies that there is a function $\underline{f}_c$, called a state transition function, such that $\underline{x}(t)$, $t \geq t_o$, can be written

$$\underline{x}(t) = \underline{f}_c(t;\underline{x}(t_o),\ \underline{u}(t)). \qquad (3.58)$$

This is in the form of a solution of Equation 3.57a, and the solution for the special case of $\underline{u}(t)=\underline{0}$, $t \in T_t$, will result in the matrix response function $\Phi(t,t_o)$, the <u>state transition matrix</u>. This is discussed in detail in this section; the result will be

$$\underline{x}(t) = \Phi(t,t_o)\underline{x}(t_o) \tag{3.59}$$

and it will be shown in Chapter 4 that $\underline{x}(t)$, $t \geq t_o$, also can be written in terms of $\Phi(t,t_o)$ when $\underline{u}(t)$ is not zero for $t \geq t_o$.

A parallel set of definitions can be made for discrete-time linear systems. Thus the state equations are:

$$\underline{x}(n+1) = A\underline{x}(n) + B\underline{v}(n), \quad \underline{x}(n+1) = E\underline{x}(n) \tag{3.60a}$$

$$\underline{z}(n) = C\underline{x}(n) + D\underline{v}(n). \tag{3.60b}$$

Here $\underline{x}$, $\underline{v}$, and $\underline{z}$ are functions of n, $n \in T_n$, e.g., T_{d1} or T_{d2} of Chapter 2. The state space R_r is now the space of all $\underline{x}(n)$, $n \in T_n$. The state vector $\underline{x}(n_o)$ has components $x_1(n_o),\ldots,x_r(n_o)$, a set of numbers for r minimal such that $\underline{x}(n)$, $n>n_o$, can be found for a specified $\underline{v}(n)$, $n \geq n_o$. An associated state transition function is $\underline{f}_d$ so that we have

$$\underline{x}(n) = \underline{f}_d(n;\underline{x}(n_o), \underline{v}(n)) \tag{3.61}$$

and the state transition matrix will be $\Phi(n,n_o)$, and for $\underline{v}(n) = \underline{0}$, $n \in T_n$,

$$\underline{x}(n) = \Phi(n,n_o)\underline{x}(n_o). \tag{3.62}$$

It is again important to observe that the notation has been chosen to emphasize the similarity of forms in continuous and discrete time as was done earlier with response functions. The exact meaning may be inferred from the context and from the arguments specified. Thus $\underline{x}$ is always a state vector, Φ is always a transition matrix, and A, B, C, D are the matrices in both models.

The definitions of state given here imply that t_o and n_o are the left ends of the intervals T_t and T_n and thus the system state in the future (i.e., $t>t_o$ or $n>n_o$) is required. Actually, the concept is more general and transitions to the past (i.e., $t<t_o$ or $n<n_o$) also are included. Thus, t_o and n_o can be any point in their respective intervals T_t and T_n.

A state $\underline{x}(t_o)$ is called an <u>initial state</u>. A state $\underline{\theta}$ is a <u>zero state</u> if for $\underline{x}(t_o) = \underline{\theta}$ and $\underline{u}(t) = \underline{0}$, $t \geq t_o$.

$$\underline{y}(t) = C\underline{f}_c(t;\underline{\theta},\underline{0}) = \underline{0}, \qquad t \geq t_o.$$

A <u>ground state</u> $\underline{\gamma}$ is that <u>unique</u> state, if it exists, such that for $\underline{u}(t) = \underline{0}$, $t \geq t_o$, and any $\underline{x}(t_o)$,

$$\lim_{t\to\infty} \underline{x}(t) = \underline{\gamma} = \lim_{t\to\infty} \underline{f}_c(t;\underline{x}(t_o),\underline{0})$$

An <u>equilibrium state</u> $\underline{\lambda}$ is any state such that for $\underline{u}(t) = \underline{0}$, $t \geq t_o$,

$$\underline{x}(t) = \underline{\lambda} = \underline{f}_c(t;\underline{\lambda},\underline{0}), \qquad t > t_o.$$

A similar set of definitions can be stated for discrete-time systems.

Examinations of these definitions shows that a zero state is any initial state such that for zero input, the output is zero. A ground state is a unique state (which may not exist for some systems) to which the system always converges for large t starting from any initial state when the input is zero. An equilibrium state is any state, for zero input, such that the system remains in that state if it starts in that state. A ground state is an equilibrium state, but the converse is not always true.

Example 3.20

Consider the simple circuit shown

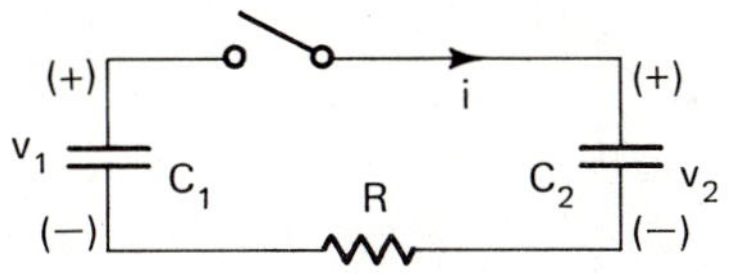

There is no input. Let the state $\underline{x}^T = (v_1\ v_2)$ and let i = y be the output. Then we have

Zero States: Any state such that $v_1(t_o)=v_2(t_o)$
Ground State: None
Equilibrium States: $v_1=v_2$

If C_2 is replaced by a short circuit and $x = v_1$:

Zero State: $v_1(t_o)=0$
Ground State: $v_1=0$
Equilibrium State: $v_1=0$

It is now possible to give a more general definition of system linearity than that given in Section 3.1. In fact, the earlier definition can now be seen to be one of zero-state linearity. A system is zero-state linear if for the definition

$$f_c^o(t;\underline{x}(t_o),\underline{u}(t)) = C\underline{f}_c(t;\underline{x}(t_o),\underline{u}(t)) + D\underline{u}(t),$$

we obtain

$$\alpha\underline{y}(t) = \underline{f}_c^o(t;\underline{\theta},\alpha\underline{u}(t)), \quad \alpha \text{ any constant},$$

where

$$\underline{y}(t) = \underline{f}_c^o(t;\underline{\theta},\underline{u}(t)), \quad t \geq t_o$$

and we obtain

$$\underline{y}_1(t) + \underline{y}_2(t) = \underline{f}_c^o(t;\underline{\theta},\underline{u}_1(t) + \underline{u}_2(t)) ,$$

where

$$\underline{y}_1(t) = \underline{f}_c^o(t;\underline{\theta},\underline{u}_1(t))$$

and

$$\underline{y}_2(t) = \underline{f}_c^o(t;\underline{\theta},\underline{u}_2(t)).$$

Thus a system is zero-state linear if it satisfies the homogeneity and additivity conditions for $\underline{x}(t_o) = \underline{\theta}$. A system is <u>zero input linear</u> if we obtain

$$\alpha\underline{y}(t) = C\underline{f}_c(t;\alpha\underline{x}(t_o), \underline{0})$$

where

$$\underline{y}(t) = C\underline{f}_c(t;\underline{x}(t_o), \underline{0}), \qquad t \geq t_o$$

and we obtain

$$\underline{y}_1(t) + \underline{y}_2(t) = C\underline{f}_c(t;\underline{x}_1(t_o) + \underline{x}_2(t_o), \underline{0})$$

where

$$\underline{y}_1(t) = C\underline{f}_c(t;\underline{x}_1(t_o), \underline{0})$$

and

$$\underline{y}_2(t) = C\underline{f}_c(t;\underline{x}_2(t_o), \underline{0}).$$

Finally, a system has the <u>decomposition property</u> if for any $\underline{x}(t_o)$ and $\underline{u}(t)$, $t>t_o$, with

$$\underline{y}_1(t) = C\underline{f}_c(t;\underline{x}(t_o), \underline{0})$$

and

$$\underline{y}_2(t) = \underline{f}_c^o(t;\underline{\theta},u(t)) ,$$

we obtain

$$\underline{y}_1(t) + \underline{y}_2(t) = \underline{f}_c^o(t;\underline{x}(t_o),\ \underline{u}(t)) \ .$$

A system is then defined as linear if it is zero-state linear, zero-input linear, and has the decomposition property. Analogous definitions apply to discrete-time systems. For a more advanced discussion of the concept of system state and associated definitions, see [7.].

Some examples of writing state equations are now discussed before proceeding to the solutions for the new response functions -- the state transition matrices $\Phi(t,t_o)$ and $\Phi(n,n_o)$. As in the previous section, the equations may originate with a description of a physical system or with a mathematical form such as the rth-order forms of differential or difference equations. When the state equations are written from a description of physical system, the dimension of $\underline{x}$, i.e., r, depends on the number of independent energy storage elements in the system. Thus, for example, in an electric circuit, r is usually the sum of the number of capacitors and the number of inductors. Special cases occur, however, when there are tie sets made up entirely of capacitors and voltage sources or cut sets which consist only of inductors and current sources (see Chapter 6, Section 6.3). Similarly, in a mechanical system, the number of independent mass elements and spring elements determines r; and similar rules apply to other lumped physical systems or combinations of electrical, mechanical, and other types of elements. More complete rules for some of these cases are given in Chapter 6 where several detailed applications are discussed. When the state equations are written from a single differential or difference equation, the dimension r equals the degree of M. It already has been shown in the previous section that r initial conditions are required to find a specific solution for $t>t_o$ or $n>n_o$ when inputs u(t) or v(n) are given for $t \geq t_o$ or $n \geq n_o$. This, of course, is in agreement with the concepts of state introduced in this section. In general, the choice of state variables is not unique, but the dimension of any $\underline{x}$ for a given system is always the same.

Example 3.21

For the circuit shown, choose a set of state variables and find the corresponding state equations.

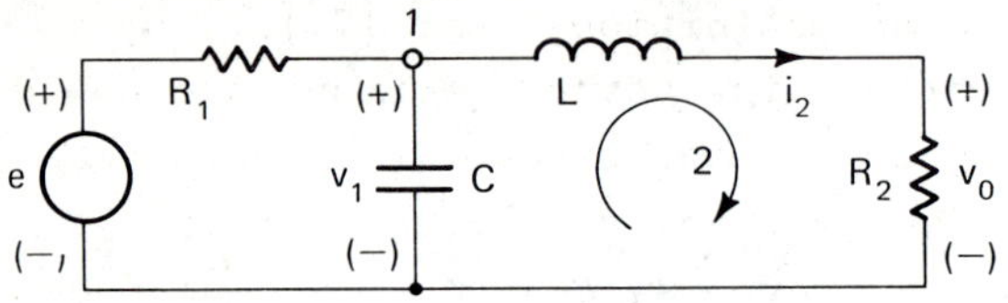

The input is $e=u$ and the output is $v_o=y$. Choose $v_1=x_1$ and $i_2=x_2$. At node 1,

$$\frac{1}{R_1}(e-v_1) = C\frac{dv_1}{dt} + i_2$$

and for loop 2,

$$v_1 = L\frac{di_2}{dt} + R_2 i_2$$

and solving for the derivatives,

$$\frac{dv_1}{dt} = -\frac{1}{R_1C}v_1 - \frac{1}{C}i_2 + \frac{1}{R_1C}e$$

$$\frac{di_2}{dt} = \frac{1}{L}v_1 - \frac{R_2}{L}i_2.$$

Thus we have

$$\dot{x}_1 = -\frac{1}{R_1C}x_1 - \frac{1}{C}x_2 + \frac{1}{R_1C}u$$

$$\dot{x}_2 = \frac{1}{L}x_1 - \frac{R_2}{L}x_2$$

and

$$y = R_2 x_2 .$$

Putting these in the form of Eq. 3.57,

$$\dot{\underline{x}} = A\underline{x} + \underline{b}u$$

$$y = \underline{c}^T\underline{x}$$

where

$$A = \begin{bmatrix} -\frac{1}{R_1 C} & -\frac{1}{C} \\ \frac{1}{L} & -\frac{R_2}{L} \end{bmatrix} \qquad B=\underline{b} = \begin{bmatrix} \frac{1}{R_1 C} \\ 0 \end{bmatrix}$$

$$C = \underline{c}^T = [0 \quad R_2] \qquad D = d = 0.$$

Here

$$r = 2,\ q = 1,\ p = 1,\ \underline{x}^T = [x_1 \quad x_2] .$$

<u>Example 3.22</u>

For the equation

$$\ddot{y} + \alpha_1\dot{y} + \alpha_o y = \beta_1\dot{u} + \beta_o u$$

find a state variable model. Here $r = 2$ and the two state variables must be chosen so that no derivatives of u appear if a form such as Eq. 3.57 is to be obtained. One procedure is to let

$$x_1 = y + au$$

$$\dot{x}_1 = x_2 + bu$$

$$\dot{x}_2 = -\alpha_o x_1 - \alpha_1 x_2 + cu$$

where a, b, and c must be found. Using the first equation

$$y = x_1 - au$$

$$\dot{y} = \dot{x}_1 - a\dot{u}$$

$$= x_2 + bu - a\dot{u}$$

from the second equation

$$\ddot{y} = \dot{x}_2 + b\dot{u} - a\ddot{u}$$

$$= -\alpha_o x_1 - \alpha_1 x_2 + cu + b\dot{u} - a\ddot{u}$$

from the third equation. Substituting y, $\dot{y}$, and $\ddot{y}$ in the system equation

$$[-\alpha_o x_1 - \alpha_1 x_2 + cu + b\dot{u} - a\ddot{u}]$$

$$+ \alpha_1 [x_2 + bu - a\dot{u}]$$

$$+ \alpha_o [x_1 - au] = \beta_1 \dot{u} + \beta_o u.$$

Equating coefficients of u, $\dot{u}$, and $\ddot{u}$,

$$a = 0$$

$$b = \beta_1$$

$$c = \beta_0 - \alpha_1\beta_1 .$$

Therefore,

$$\dot{x}_1 = x_2 + \beta_1 u$$

$$\dot{x}_2 = -\alpha_0 x_1 - \alpha_1 x_2 + (\beta_0 - \alpha_1\beta_1)u$$

$$y = x_1$$

or

$$\dot{\underline{x}} = A\underline{x} + \underline{b}u$$

$$y = \underline{c}^T\underline{x}$$

$$A = \begin{bmatrix} 0 & 1 \\ -\alpha_0 & -\alpha_1 \end{bmatrix}, \qquad \underline{b} = \begin{bmatrix} \beta_1 \\ \beta_0 - \alpha_1\beta_1 \end{bmatrix}$$

$$\underline{c}^T = [1 \quad 0], \qquad d = 0.$$

A general version of this method is developed later in this chapter.

Example 3.23

From Example 3.1e,

$$z(n+2) - z(n) = \frac{1}{3}T[v(n+2) + 4v(n+1) + v(n)].$$

Here r=2 and x_1 and x_2 must be chosen so that the terms v(n+2) and v(n+1) do not appear if a form such as Eq. 3.60 is to be obtained. Similar to Example 3.22, let

$$x_1(n) = z(n) + av(n)$$

$$x_1(n+1) = x_2(n) + bv(n)$$

$$x_2(n+1) = x_1(n) + cv(n).$$

The $x_2(n)$ term is missing in the third equation since there is no z(n+1) term in the original difference equation. Then

$$z(n) = x_1(n) - av(n)$$

$$\begin{aligned} z(n+2) &= x_1(n+2) - av(n+2) \\ &= x_2(n+1) + bv(n+1) - av(n+2) \\ &= x_1(n) + cv(n) + bv(n+1) - av(n+2). \end{aligned}$$

Equating coefficients of v(n), v(n+1), v(n+2) and solving the three equations,

$$a = -\frac{1}{3}T$$

$$b = \frac{4}{3}T$$

$$c = \frac{2}{3}T\ .$$

Therefore,

$$x_1(n+1) = x_2(n) + \tfrac{4}{3}T\, v(n)$$

$$x_2(n+1) = x_1(n) + \tfrac{2}{3}T\, v(n)$$

$$z(n) = x_1(n) + \tfrac{1}{3}T\, v(n)$$

or

$$\underline{x}(n+1) = A\underline{x}(n) + \underline{b}v(n)$$

$$z(n) = \underline{c}^T\underline{x}(n) + dv(n)$$

$$A = \begin{bmatrix} 0 & 1 \\ 1 & 0 \end{bmatrix}, \qquad \underline{b} = \begin{bmatrix} \tfrac{4}{3}T \\ \tfrac{2}{3}T \end{bmatrix}$$

$$\underline{c}^T = [1 \quad 0], \qquad d = \tfrac{1}{3}T .$$

As a simple illustration of numerical integration using this state-variable model of Simpson's Rule of Ex. 3.1e, let

$$p(t) = t^2,$$

$$t_o = 0 ,$$

then

$$q(t) = \int_o^t \tau^2 d\tau = \tfrac{1}{3}t^3 .$$

$z(n)$ is the numerical approximation to $q(t)$ and for $T=1$, $v(n) = p(nT) = n^2$. The table shows the first four steps in the calculation. We may make several observations. First, we have inferred the values of z(n), n odd, although the particular algorithm we started with in Ex. 3.1e was only for z(n), n even. The original Simpson's Rule approximates the integrand by constructing non-overlapping second degree curves. The model here still approximates the integrand by using second degree curves, but they overlap, i.e., we find

n	$x_1(n)$	$x_2(n)$	v(n)	z(n)
0	0	0	0	0
1	0	0	1	1/3
2	4/3	2/3	4	8/3
3	6	4	9	9
4	16	12	16	64/3

z(n) for all n. Second, we have "exact" results in this example. This occurs in part because p(t) is of second degree and Simpson's Rule is a second degree method. Also, the use of an initial state $\underline{x}(0) = \underline{0}$ provides a starting condition such that z(0) = 0 and subsequent z(n) are correct. Other inputs and other initial states, even for integrands of degree two or less, do not necessarily give exact results. We shall consider these cases in Ex. 4.5 and 4.9 in Chapter 4. (See also Problem 17, Chapter 4.)

<u>Example 3.24</u>

a) To show that the choice of state variables is not unique, consider the system of Ex. 3.22 again:

$$\ddot{y} + \alpha_1\dot{y} + \alpha_o y = \beta_1\dot{u} + \beta_o u$$

or

$$(D^2 + \alpha_1 D + \alpha_o)y = (\beta_1 D + \beta_o)u$$

or

$$D^2y + D(\alpha_1 y - \beta_1 u) + (\alpha_o y - \beta_o u) = 0$$

or operating with D^{-2},

$$y + D^{-1}(\alpha_1 y - \beta_1 u) + D^{-2}(\alpha_o y - \beta_o u) = 0$$

or

$$y = D^{-1}(\beta_1 u - \alpha_1 y) + D^{-2}(\beta_o u - \alpha_o y) .$$

Now let

$$\dot{x}_1' = -\alpha_1 y + x_2' + \beta_1 u$$

$$\dot{x}_2' = -\alpha_o y + \beta_o u,$$

then

$$y = x_1' .$$

Therefore,

$$\dot{x}_1' = -\alpha_1 x_1' + x_2' + \beta_1 u$$

$$\dot{x}_2' = -\alpha_o x_1' + \beta_o u$$

or

$$\dot{\underline{x}}' = A' \, \underline{x}' + \underline{b}' u$$

$$y = (\underline{c}')^T \underline{x}'$$

where

$$A = \begin{bmatrix} -\alpha_1 & 1 \\ -\alpha_0 & 0 \end{bmatrix} \qquad \underline{b} = \begin{bmatrix} \beta_1 \\ \beta_0 \end{bmatrix}$$

$$(\underline{c}')^T = [1 \quad 0] \qquad d' = 0 .$$

b) Another choice for state variables is

$$\dot{x}_1'' = x_2''$$

$$\dot{x}_2'' = -\alpha_0 x_1'' - \alpha_1 x_2'' + u$$

and

$$y = \beta_0 x_1'' + \beta_1 x_2'' ,$$

thus for

$$\dot{\underline{x}}'' = A''\underline{x}'' + \underline{b}''u$$

$$y = (\underline{c}'')^T \underline{x}''$$

$$A'' = \begin{bmatrix} 0 & 1 \\ -\alpha_0 & -\alpha_1 \end{bmatrix} \qquad \underline{b}'' = \begin{bmatrix} 0 \\ 1 \end{bmatrix}$$

$$(\underline{c}'')^T = [\beta_0 \quad \beta_1] \qquad d'' = 0$$

c) Another choice for state variables is

$$\dot{x}_1''' = -\alpha_1 x_1''' + x_2'''$$

$$x_2''' = -\alpha_0 x_1''' + u$$

and

$$y = (\beta_0 - \alpha_1\beta_1)x_1''' + \beta_1 x_2'''$$

Thus for

$$\dot{\underline{x}}''' = A'''\underline{x}''' + \underline{b}'''u$$

$$y = (\underline{c}''')^T \underline{x}'''$$

$$A''' = \begin{bmatrix} -\alpha_1 & 1 \\ -\alpha_0 & 0 \end{bmatrix} \qquad \underline{b}''' = \begin{bmatrix} 0 \\ 1 \end{bmatrix}$$

$$(\underline{c}''')^T = [\beta_0 - \alpha_1\beta_1 \quad \beta_1].$$

Comparison of Ex. 3.22 and Ex. 3.24a, Ex. 3.24b, and Ex. 3.24c shows four standard forms for the same system. The forms given in Ex. 3.24a and Ex. 3.24b generalize easily and provide a simple way of writing two state variable models directly from the rth-order model of a single input-single output linear system. The state vectors in these four cases are all related by linear transformations, for example

$$\underline{x} = \begin{bmatrix} 1 & 0 \\ -\alpha_1 & 1 \end{bmatrix} \underline{x}'$$

Canonical or standard forms such as these are discussed in the next section.

Example 3.25

A model of a linear discrete-time two-input, two-output system is given by the equations below. A state-variable model is required. (v_1 and v_2 are input components -- not "singularity functions".)

$$z_1(n+2)+\alpha_{11}z_1(n+1)+\alpha_{o1}z_1(n)+\gamma_{11}z_2(n+1)+\gamma_{o1}z_2(n) =$$

$$= \beta_{11}v_1(n+1)+\beta_{o1}v_1(n)+\delta_{o1}v_2(n)$$

$$z_2(n+1)+\alpha_{o2}z_2(n)+\gamma_{12}z_1(n+1)+\gamma_{o2}z_1(n) =$$

$$= \delta_{o2}v_1(n)+\beta_{o2}v_2(n) .$$

Since $z_1(n+2)$ appears in the first equation and $z_2(n+1)$ appears in the second equation, r=3; also q=p=2. Finally, a term $v_1(n+1)$ appears on the right of the first equation. Let

$$x_1(n) = z_1(n) + av_1(n)$$

$$x_1(n+1) = x_2(n) + bv_1(n)$$

$$x_2(n+1) = -\alpha_{o1}x_1(n)-\alpha_{11}x_2(n)+cv_1(n)+f+\delta_{o1}v_2(n) ,$$

where

$$f = -\gamma_{11}z_2(n+1) -\gamma_{o1}z_2(n).$$

Also let

$$x_3(n) = z_2(n)$$

$$x_3(n+1) = -\alpha_{o2}x_3(n)+g+\delta_{o2}v_1(n)+\beta_{o2}v_2(n)$$

where

$$g = -\gamma_{12}z_1(n+1) - \gamma_{02}z_1(n).$$

Proceeding as in Example 3.23,

$$a = 0, \; b = \beta_{11}, \; c = \beta_{01} - \alpha_{11}\beta_{11} \; .$$

Also,

$$f = -\gamma_{11}[-\alpha_{02}x_3(n) + g + \delta_{02}v_1(n) + \beta_{02}v_2(n)]$$

$$- \gamma_{01}x_3(n)$$

$$g = -\gamma_{12}[x_2(n) + \beta_{11}v_1(n)] - \gamma_{02}x_1(n).$$

Combining these results,

$$\underline{x}(n+1) = \begin{bmatrix} 0 & 1 & 0 \\ \gamma_{11}\gamma_{02}-\alpha_{01} & \gamma_{11}\gamma_{12}-\alpha_{11} & \gamma_{11}\alpha_{02}-\gamma_{01} \\ -\gamma_{02} & -\gamma_{12} & -\alpha_{02} \end{bmatrix} \underline{x}(n)$$

$$+ \begin{bmatrix} \beta_{11} & 0 \\ \gamma_{12}\gamma_{11}\beta_{11}-\gamma_{11}\delta_{02}-\alpha_{11}\beta_{11}+\beta_{01} & \delta_{01}-\gamma_{11}\beta_{02} \\ \delta_{02}-\gamma_{12}\beta_{11} & \beta_{02} \end{bmatrix} \underline{v}(n)$$

$$\underline{z}(n) = \begin{bmatrix} 1 & 0 & 0 \\ 0 & 0 & 1 \end{bmatrix} \underline{x}(n) \; .$$

3.4.2 The State Transition Matrix

To find the transition matrix $\Phi(t,t_o)$, a zero-input response function, Equation 3.57a with $\underline{u} = \underline{0}$ is the starting point, i.e.,

$$\dot{\underline{x}}(t) = A(t)\underline{x}(t), \qquad t \in T_t, \tag{3.63}$$

and the form of the solution is given by Equation 3.59. Some general results from the theory of differential equations are given first, then a basis set of vectors for the space R_r is given, and $\Phi(t,t_o)$ is finally specified in terms of a fundamental matrix formed from these basis vectors. The results are merely summarized here; for proofs the reader may consult other references [7.], [8.], [9.].

I: Equation 3.63 has a unique solution $\underline{x}=\underline{\phi}$ for $t \in T_t$ and for given $\underline{x}(t_o)=\underline{x}_o$, $t_o \in T_t$, if the elements of A are continuous functions of t, $t \in T_t$.

II: If the solution $\underline{\phi}(t_1)=\underline{0}$ for some $t_1 \in T_t$, the solution $\underline{\phi}=\underline{0}$ for all $t \in T_t$. Conversely, if $\underline{\phi}(t_1)\neq\underline{0}$ for some $t_1 \in T_t$, $\underline{\phi}\neq\underline{0}$ for all $t \in T_t$. Thus if $\underline{x}_o\neq\underline{0}$, the corresponding solution $\underline{\phi}\neq\underline{0}$.

III: If $\underline{\phi}_1, \underline{\phi}_2, \ldots, \underline{\phi}_j$ are solutions of Equation 3.63,

$$\underline{\phi} = c_1\underline{\phi}_1 + c_2\underline{\phi}_2 + \ldots + c_j\underline{\phi}_j$$

is a solution.

IV: A set of solutions $\underline{\phi}_1, \underline{\phi}_2, \ldots, \underline{\phi}_j$ is linearly independent at t_1 if the vectors $\underline{\phi}_1(t_1)$, $\underline{\phi}_2(t_1), \ldots, \underline{\phi}_j(t_1)$ are linearly independent. (See Appendix B for a definition of linear independence.)

It already has been stated that the set of all solutions of the state equations, i.e., $\underline{x}=\underline{\phi}$, forms a linear vector space

R_r. To form a basis we consider a set of solutions $\underline{\phi}_1, \underline{\phi}_2, \ldots, \underline{\phi}_i, \ldots, \underline{\phi}_r$, and let $\underline{\phi}_i(t_o) = \underline{e}_i$, $i=1,2,\ldots,r$. The vectors $\underline{e}_i$ are the unit vectors that form a basis of an r-dimension linear vector space (See Appendix B), in this case for $t=t_o$. Then the following results are true:

V: The solutions $\underline{\phi}_1, \underline{\phi}_2, \ldots, \underline{\phi}_r$ are linearly independent for all $t \varepsilon T_t$.

VI: Every solution $\underline{\phi}$ can be written uniquely as

$$\underline{\phi} = c_1\underline{\phi}_1 + c_2\underline{\phi}_2 + \ldots + c_r\underline{\phi}_r .$$

From these results if follows that the set of $\underline{\phi}_i$, $i=1,2,\ldots,r$, forms a basis in R_r, $t \varepsilon T_t$. This linearly independent set of solutions $\underline{\phi}_i$ is called a <u>fundamental set</u>, and using such a set, a <u>fundamental matrix</u> X is formed as:

$$X = [\underline{\phi}_1 \ \underline{\phi}_2 \ \ldots \ \underline{\phi}_r].$$

X is a fundamental matrix of Equation 3.63 and we have

$$\dot{X}(t) = A(t)X(t), \qquad t \ \varepsilon \ T_t .$$

Note that X is not unique and it can be shown that det $X(t) \neq 0$ for all $t \varepsilon T_t$. The final step is to define $\Phi(t,t_o)$ such that

$$\frac{\partial}{\partial t} \Phi(t,t_o) = A(t)\Phi(t,t_o) \qquad t,t_o \ \varepsilon \ T_t$$

and

$$\Phi(t_o,t_o) = I .$$

Then the unique solution of Equation 3.63 is

$$\underline{x}(t) = \Phi(t,t_o)\underline{x}(t_o), \qquad t,t_o \ \varepsilon \ T_t . \tag{3.63a}$$

Furthermore, $\Phi(t,t_o)$ can be written in terms of a fundamental matrix X. Thus for

$$\dot{X}(t) = A(t)X(t)$$

with specified $X(t_o)$, we have

$$X(t) = \Phi(t,t_o)X(t_o)$$

and

$$\Phi(t,t_o) = X(t)X(t_o)^{-1} .$$

Since the fundamental set of solutions was chosen such that $\underline{\phi}_i(t_o) = \underline{e}_i$, we obtain $X(t_o)=I$. Initial bases other than the $\underline{e}_i$ could be used although this particular choice is convenient.

This development has shown that the transition matrix $\Phi(t,t_o)$ is an rxr matrix whose columns are the linearly independent solutions $\underline{\phi}_i$, i=1,...,r, of the zero input state equation with $\underline{\phi}_i(t_o)=\underline{e}_i$. However, this is not an efficient way of constructing $\Phi(t,t_o)$; alternative methods are discussed below. Three properties of the transition matrix are:

1. $\Phi(t_3,t_1) = \Phi(t_3,t_2)\Phi(t_2,t_1)$ group
2. $[\Phi(t_2,t_1)]^{-1} = \Phi(t_1,t_2)$ inversion
3. $\Phi(t_1,t_2) = X(t_1)[X(t_2)]^{-1}$ separation

A parallel development can be carried out for Equation 3.60a with $\underline{v}=\underline{0}$. For

$$\underline{x}(n+1) = A(n)\underline{x}(n), \quad n \in T_n, \tag{3.64}$$

a solution of the form of Equation 3.62 and a definition of $\Phi(n,n_o)$ are required. The first result concerning the uniqueness of the solution of the differential equation 3.63

(item I. after Equation 3.63) must be changed for the case of the difference Equation 3.64. First, of course, the previous requirement of continuity of the elements of A(t) has no counterpart for A(n). Instead we have

I: Equation 3.64 has a unique solution $\underline{x}=\underline{\phi}$ for $n \varepsilon T_n$ and for given $\underline{x}(n_o)=\underline{x}_o$, $n_o \varepsilon T_n$, if A(n) is nonsingular for all $n \varepsilon T_n$.

The analogous results II, III, IV, V, and VI then follow with the fundamental set of solutions such that $\underline{\phi}_i(n_o)=\underline{e}_i$, $i=1,2,\ldots,r$. The $\underline{\phi}_i$, $i=1,2,\ldots,r$, form a basis in the space R_r of all solutions $\underline{x}=\underline{\phi}$, $n \varepsilon T_n$. A fundamental matrix X of Equation 3.64 is formed as:

$$X = [\underline{\phi}_1 \; \underline{\phi}_2 \; \cdots \; \underline{\phi}_r]$$

and

$$X(n+1) = A(n)X(n), \qquad n \; \varepsilon \; T_n.$$

Again, X(n) is not unique and det $X(n) \neq 0$ for all $n \varepsilon T_n$. In analogy to the previous case, the final step is to define $\Phi(n,n_o)$ such that

$$\Phi(n+1,n_o) = A(n)\Phi(n,n_o), \qquad n,n_o \; \varepsilon \; T_n,$$

and

$$\Phi(n_o,n_o) = I \; .$$

The unique solution of Equation 3.64 is

$$\underline{x}(n) = \Phi(n,n_o)\underline{x}(n_o), \qquad n,n_o \; \varepsilon \; T_n. \tag{3.64a}$$

Furthermore, $\Phi(n,n_o)$ can be written in terms of a fundamental matrix X. Thus for

$$X(n+1) = A(n)X(n)$$

with specified $X(n_o)$, we have

$$X(n) = \Phi(n,n_o)X(n_o)$$

and

$$\Phi(n,n_o) = X(n)X(n_o)^{-1}$$

and the use of the initial basis $\underline{e}_i$, $i=1,...,r$ at n_o means, as before, that $X(n_o)^{-1}=I$. Three properties of the transition matrix are:

1. $\Phi(n_3,n_1) = \Phi(n_3,n_2)\Phi(n_2,n_1)$ group
2. $[\Phi(n_2,n_1)]^{-1} = \Phi(n_1,n_2)$ inversion
3. $\Phi(n_1,n_2) = X(n_1)[X(n_2)]^{-1}$ separation

The requirement that A(n) be nonsingular is always met in one important class of discrete-time systems. It is the case where Equation 3.60 follows from a sampled-data model of Equation 3.57 and this is discussed later in this chapter. In other cases of interest, solutions for $n \geq n_o$ can still be found when A(n) is singular.

The definitions of $\Phi(t,t_o)$ and $\Phi(n,n_o)$ are general in the sense that the matrices A(t) and A(n) are time variable. However, finding the transition matrices in closed form in these instances usually is difficult. The method suggested at this point is to form a fundamental matrix from a fundamental set of solutions. This method has a counterpart in the solution of the rth-order forms 3.19 and 3.21. Results for stationary systems can be seen in the solutions of 3.27a and 3.37a and some time-varying cases are discussed in [5.]

and [6.]. To illustrate the use of fundamental sets of solutions before discussing other methods of finding $\Phi(t,t_o)$ and $\Phi(n,n_o)$, two simple examples for a stationary second-order continuous-time system and a stationary second-order discrete-time system are worked out. Both rth-order models and state variable models are used. It should be apparent that this method of constructing solutions for the state equations is not convenient. Other methods for obtaining $\Phi(t,t_o)$ and $\Phi(n,n_o)$ can be categorized according to their suitability for digital computation, their applicability to time-varying systems or not, the ease of using them when computers are not available, and whether or not a closed form is obtained.

<u>Example 3.26</u>

a) Classical solution

For the system of Example 3.22, the homogeneous equation is:

$$\ddot{y} + \alpha_1\dot{y} + \alpha_o y = 0, \qquad t,t_o \;\varepsilon\; T_t.$$

Now assume

$$y = k\varepsilon^{\lambda t}, \qquad t \geq t_o$$

then

$$\dot{y} = k\varepsilon^{\lambda t}$$

and

$$\ddot{y} = \lambda^2 k\varepsilon^{\lambda t} .$$

Substituting,

$$(\lambda^2 + \alpha_1 \quad + \alpha_o)k\varepsilon^{\lambda t} = 0$$

and for nontrivial solutions (i.e., k≠0), we require

$$(\lambda^2 + \alpha_1\lambda + \alpha_o) = 0$$

or

$$(\lambda-\lambda_1)(\lambda-\lambda_2) = 0$$

For $\lambda_1 \neq \lambda_2$, solutions are

$$y_1 = \varepsilon^{\lambda_1 t} \quad \text{and} \quad y_2 = \varepsilon^{\lambda_2 t}, \qquad t \geq t_o$$

and

$$y = k_1 y_1 + k_2 y_2.$$

y_1 and y_2 are linearly independent for $t \geq t_o$. Every solution y for given $y(t_o)$ and $\dot{y}(t_o)$ can be written uniquely in this form. Now

$$y(t_o) = k_1(\exp \lambda_1 t_o) + k_2(\exp \lambda_2 t_o)$$

and

$$\dot{y}(t_o) = \lambda_1 k_1(\exp \lambda_1 t_o) + \lambda_2 k_2(\exp \lambda_2 t_o)$$

or putting these together,

$$\begin{bmatrix} (\exp \lambda_1 t_o) & (\exp \lambda_2 t_o) \\ \lambda_1(\exp \lambda_1 t_o) & \lambda_2(\exp \lambda_2 t_o) \end{bmatrix} \begin{bmatrix} k_1 \\ k_2 \end{bmatrix} = \begin{bmatrix} y(t_o) \\ \dot{y}(t_o) \end{bmatrix}$$

or

$$\begin{bmatrix} y_1(t_o) & y_2(t_o) \\ \dot{y}_1(t_o) & \dot{y}_2(t_o) \end{bmatrix} \begin{bmatrix} k_1 \\ k_2 \end{bmatrix} = \begin{bmatrix} y(t_o) \\ \dot{y}(t_o) \end{bmatrix}$$

or defining the matrix we have

$$W_c(t_o) \begin{bmatrix} k_1 \\ \\ k_2 \end{bmatrix} = \begin{bmatrix} y(t_o) \\ \\ \dot{y}(t_o) \end{bmatrix} .$$

Now det $W_c(t_o) = w_c(t_o)$ is the Wronskian at t_o; furthermore, $w_c(t) \neq 0$ if, and only if, $y_1(t)$ and $y_2(t)$ are linearly independent.

As noted above, y_1 and y_2 are linearly independent for $\lambda_1 \neq \lambda_2$, thus

$$\begin{bmatrix} k_1 \\ \\ k_2 \end{bmatrix} = W_c(t_o)^{-1} \begin{bmatrix} y(t_o) \\ \\ \dot{y}(t_o) \end{bmatrix}$$

and finally,

$$y(t) = [y_1 \quad y_2] \begin{bmatrix} k_1 \\ k_2 \end{bmatrix}$$

$$= [y_1 \quad y_2] \; W_c(t_o)^{-1} \begin{bmatrix} y(t_o) \\ \dot{y}(t_o) \end{bmatrix} .$$

For an rth-order, stationary system with distinct λ_i, $i=1,2,\ldots,r$, there is an obvious general form that follows from this. One special case of interest is the solution of Equation 3.27a for h(t) with $t_o=0$ and distinct λ_i. Using Equation 3.29 for $\underline{h}(0+)$,

$$h(t) = [(\exp \lambda_1 t)\ldots(\exp \lambda_r t)] \; V_c^{-1} A_\alpha^{-1} \underline{\beta}, \qquad t > 0 .$$

where

$$W_c(0) = V_c = \begin{bmatrix} 1 & \cdots & 1 \\ \lambda_1 & \cdots & \lambda_r \\ \vdots & & \vdots \\ \lambda_1^{r-1} & & \lambda_r^{r-1} \end{bmatrix},$$

a Van der Monde matrix.

b) State equation solution

A state-variable model for the same example is:

$$\dot{\underline{x}} = \begin{bmatrix} 0 & 1 \\ -\alpha_o & -\alpha_1 \end{bmatrix} \underline{x} + \begin{bmatrix} 0 \\ 1 \end{bmatrix} u$$

$$y = [\beta_o \quad \beta_1]\underline{x}$$

and we let u=0 for this calculation of $\Phi(t,t_o)$ and require $\underline{x}(t_o)=\underline{x}_o\neq\underline{0}$. Now assume

$$\underline{x}(t) = \underline{a}\varepsilon^{\lambda t}, \quad \underline{a} \text{ constant,}$$

then

$$\dot{\underline{x}}(t) = \lambda\underline{a}\varepsilon^{\lambda t} .$$

Thus we require

$$\lambda\underline{a}\varepsilon^{\lambda t} = A\underline{a}\varepsilon^{\lambda t}$$

or

$$(\lambda I-A)\underline{a}\varepsilon^{\lambda t} = 0 .$$

For $\varepsilon^{\lambda t} \neq 0$, we, therefore, require

$$(\lambda I - A)\underline{a} = \underline{0}$$

or

$$A\underline{a} = \lambda \underline{a} \ .$$

This is satisfied if λ is a characteristic value of A, and $\underline{a}$ is a corresponding characteristic vector (See Appendix B). For a nontrivial solution of $(\lambda I - A)\underline{a} = \underline{0}$,

$$\det(\lambda I - A) = 0 = p(\lambda)$$

and since

$$\lambda I - A = \begin{bmatrix} \lambda & -1 \\ \alpha_o & \lambda + \alpha_1 \end{bmatrix}$$

we have

$$p(\lambda) = \lambda(\lambda + \lambda_1) + \alpha_o = \lambda^2 + \alpha_1 \lambda + \alpha_o$$

$$= (\lambda - \lambda_1)(\lambda - \lambda_2) \ .$$

Assuming $\lambda_1 \neq \lambda_2$, a general solution of the state equation is

$$\underline{x}(t) = \underline{a}_1(\exp \lambda_1 t) + \underline{a}_2(\exp \lambda_2 t) \ .$$

The two terms on the right are <u>modes</u> of the system response. At $t = t_o$,

$$\underline{x}_o = \underline{a}_1(\exp \lambda_1 t_o) + \underline{a}_2(\exp \lambda_2 t_o)$$

and

$$\dot{\underline{x}}(t_o) = A\underline{x}_o = \lambda_1 \underline{a}_1(\exp \lambda_1 t_o) + \lambda_2 \underline{a}_2(\exp \lambda_2 t_o) .$$

From these two equations we get

$$\underline{a}_1 = \frac{(\exp -\lambda_1 t_o)}{\lambda_1-\lambda_2} (A-\lambda_2 I)\underline{x}_o$$

$$\underline{a}_2 = \frac{(\exp -\lambda_2 t_o)}{\lambda_2-\lambda_1} (A-\lambda_1 I)\underline{x}_o \quad .$$

These can be shown to be characteristic vectors of A. Therefore,

$$\underline{x}(t) = \frac{1}{\lambda_1-\lambda_2} \Big\{ (\exp \lambda_1(t-t_o))(A-\lambda_2 I) - $$

$$- (\exp \lambda_2(t-t_o))(A-\lambda_1 I) \Big\} \; \underline{x}_o$$

$$= \frac{1}{\lambda_1-\lambda_2} \Big\{ (\exp \lambda_1(t-t_o)) \begin{bmatrix} -\lambda_2 & 1 \\ -\alpha_o & -\lambda_2-\alpha_1 \end{bmatrix}$$

$$- \; (\exp \lambda_2(t-t_o) \begin{bmatrix} -\lambda_1 & 1 \\ -\alpha_o & -\lambda_1-\alpha_1 \end{bmatrix} \Big\} \; \underline{x}_o$$

$$= \frac{1}{\lambda_1-\lambda_2} \begin{bmatrix} -\lambda_2 \exp(\lambda_1(t-t_o)) & \exp(\lambda_1(t-t_o) \\ \quad +\lambda_1 \exp(\lambda_2(t-t_o)) & \quad -\exp(\lambda_2(t-t_o)) \\ & \\ -\alpha_o \exp(\lambda_1(t-t_o)) & -(\lambda_2+\alpha_1)\exp(\lambda_1(t-t_o)) \\ \quad +\alpha_o \exp(\lambda_2(t-t_o)) & \quad +(\lambda_1+\alpha_1)\exp(\lambda_2(t-t_o)) \end{bmatrix}$$

or from the form of Eq. 3.63a,

$$\underline{x}(t) = \Phi(t,t_o)\underline{x}_o$$

and

$$\underline{\phi}_1(t-t_o) = \frac{1}{\lambda_1-\lambda_2}\begin{bmatrix} -\lambda_2(\exp\lambda_1(t-t_o))+\lambda_1(\exp\lambda_2(t-t_o)) \\ \\ -\alpha_o((\exp\lambda_1(t-t_o))-(\exp\lambda_2(t-t_o))) \end{bmatrix}$$

$$\underline{\phi}_2(t-t_o) = \frac{1}{\lambda_1-\lambda_2}\begin{bmatrix} (\exp\lambda_1(t-t_o))-(\exp\lambda_2(t-t_o)) \\ \\ -(\lambda_2+\alpha_1)(\exp\lambda_1(t-t_o)) \\ \quad + (\lambda_1+\alpha_1)(\exp\lambda_2(t-t_o)) \end{bmatrix}.$$

Note too that $\Phi(t_o,t_o)=I$. It is left as an exercise to show that $y(t)=[\beta_o \quad \beta_1]\underline{x}(t)$ checks part a).

<u>Example 3.27</u>

An example (see Example 3.12) similar to the previous one can be worked out for a discrete-time system.

a) Classical solution

For

$$z(n+2) + \alpha_1 z(n+1) + \alpha_o z(n) = 0, \qquad n,n_o \in T_n.$$

assume

$$z(n) = k\rho^n, \qquad n \geq n_o+1 ,$$

then we have

$$z(n+1) = k\rho^{n+1}$$

$$z(n+2) = k\rho^{n+2}$$

and therefore,

$$(\rho^2 + \alpha_1\rho + \alpha_o)k\rho^n = 0.$$

For $k \neq 0$, require

$$\rho^2 + \alpha_1\rho + \alpha_o = (\rho-\rho_1)(\rho-\rho_2) = 0 .$$

For $\rho_1 \neq \rho_2$, $n \geq n_o+1$,

$$z(n) = k_1\rho_1^n + k_2\rho_2^n = k_1z_1(n) + k_2z_2(n) .$$

Proceeding as before, we define W_d and get

$$\begin{bmatrix} k_1 \\ \\ k_2 \end{bmatrix} = W_d^{-1}(n_o) \begin{bmatrix} z(n_o+1) \\ \\ z(n_o+2) \end{bmatrix}$$

and

$$z(n) = [z_1(n) \;\; z_2(n)] \; W_d^{-1}(n_o) \begin{bmatrix} z(n_o+1) \\ \\ z(n_o+2) \end{bmatrix} , \quad n \geq n_o+1$$

where

$$W_d(n_o) = \begin{bmatrix} z_1(n_o+1) & z_2(n_o+1) \\ \\ z_1(n_o+2) & z_2(n_o+2) \end{bmatrix} .$$

Again, as a special case, the solution of Equation 3.37a for h(n), with $n_o=0$, and for distinct ρ_i is:

$$h(n) = [\rho_1^n \ \dots \ \rho_r^n] \ V_d^{-1} \ QB_\alpha^{-1} \underline{\beta}_d \qquad n \geq 1$$

and

$$h(0) = \beta_r$$

where B_α and $\underline{\beta}_d$ are given with Equation 3.39 and

$$W_d(0) = V_d = \begin{bmatrix} \rho_1 & \dots & \rho_r \\ \rho_1^2 & \dots & \rho_r^2 \\ \vdots & & \vdots \\ \rho_1^r & \dots & \rho_r^r \end{bmatrix}$$

and

$Q = [\underline{0} \ \vdots \ I_r]$, with I_r an rxr unit matrix.

b) State equation solution

A state variable model for this example is:

$$\underline{x}(n+1) = \begin{bmatrix} 0 & 1 \\ \\ -\alpha_o & -\alpha_1 \end{bmatrix} \underline{x}(n) + \begin{bmatrix} 0 \\ \\ 1 \end{bmatrix} v(n)$$

$$z(n) = [\beta_o \quad \beta_1] \ \underline{x}(n)$$

and we let v=0 for this calculation of $\Phi(n,n_o)$ and $\underline{x}(n_o)=\underline{x}_o\neq\underline{0}$. Now assume

$$\underline{x}(n) = \underline{a}\rho^n \ , \quad \underline{a} \text{ constant}$$

then

$$\underline{x}(n+1) = \underline{a}\rho^{n+1}.$$

Thus require

$$\underline{a}\rho^{n+1} = A\underline{a}\rho^n$$

or

$$(\rho I-A)\underline{a}\rho^n = \underline{0}.$$

For $\rho^n\neq 0$, require

$$(\rho I-A)\underline{a} = \underline{0}$$

or

$$A\underline{a} = \rho\underline{a}.$$

As in the previous example, require

$$\det\ (\rho I-A) = 0 = p(\rho)$$

and since

$$(\rho I-A) = \begin{bmatrix} \rho & -1 \\ \alpha_o & \rho+\alpha_1 \end{bmatrix}$$

$$p(\rho) = \rho^2 + \alpha_1\rho + \alpha_o = (\rho-\rho_1)(\rho-\rho_2) \ .$$

Assuming $\rho_1 \neq \rho_2$,

$$\underline{x}(n) = \underline{a}_1 \rho_1^n + \underline{a}_2 \rho_2^n .$$

At $n = n_o$,

$$\underline{x}_o = \underline{a}_1 \rho_1^{n_o} + \underline{a}_2 \rho_2^{n_o}$$

and

$$\underline{x}(n_o+1) = A\underline{x}_o = \rho_1 \underline{a}_1 \rho_1^{n_o} + \rho_2 \underline{a}_2 \rho_2^{n_o} .$$

From these, two characteristic vectors of A are

$$\underline{a}_1 = \frac{\rho_1^{-n_o}}{\rho_1 - \rho_2} (A - \rho_2 I)\underline{x}_o$$

$$\underline{a}_2 = \frac{\rho_2^{-n_o}}{\rho_2 - \rho_1} (A - \rho_1 I)\underline{x}_o$$

and therefore,

$$\underline{x}(n) = \frac{1}{\rho_1 - \rho_2} \left\{ \rho_1^{(n-n_o)} (A - \rho_2 I) - \rho_2^{(n-n_o)} (A - \rho_1 I) \right\} \underline{x}_o$$

$$= \frac{1}{\rho_1 - \rho_2} \left\{ \rho_1^{(n-n_o)} \begin{bmatrix} -\rho_2 & 1 \\ -\alpha_o & -\rho_2 - \alpha_1 \end{bmatrix} - \rho_2^{(n-n_o)} \begin{bmatrix} -\rho_1 & 1 \\ -\alpha_o & -\rho_1 - \alpha_1 \end{bmatrix} \right\} \underline{x}_o$$

$$= \frac{1}{\rho_1-\rho_2}\begin{bmatrix} -\rho_2\rho_1^{(n-n_0)}+\rho_1\rho_2^{(n-n_0)} & \rho_1^{(n-n_0)}-\rho_2^{(n-n_0)} \\ -\alpha_0(\rho_1^{(n-n_0)}-\rho_2^{(n-n_0)}) & -(\rho_2+\alpha_1)\rho_1^{(n-n_0)} \\ & +(\rho_1+\alpha_1)\rho_2^{(n-n_0)} \end{bmatrix}\underline{x}_0$$

or using Eq. 3.64, $\underline{x}(t)=\Phi(n,n_0)\underline{x}_0$.

As in the previous example, $\underline{\phi}_1(n-n_0)$ and $\underline{\phi}_2(n-n_0)$ are the columns of Φ, and $\Phi(n_0,n_0)=I$.

3.4.3 Computation of Transition Matrix

There are many methods of finding $\Phi(t,t_0)$ and $\Phi(n,n_0)$, but most of them apply only to the case where A is a constant matrix. The methods introduced are listed below, and each method is examined for both continuous-time and discrete-time systems.

A: variable	i)	Iterative
	ii)	Perturbation
A: constant	i)	Series (c.-t.) or Product (d.-t.)
	ii)	Sylvester Theorem
	iii)	Cayley-Hamilton Theorem
	iv)	Resolvent Matrix
	v)	Jordan Matrix

When A is not a constant, the only practical general methods are iterative and except in simple examples, numerical routines are used. For continuous-time systems, an

<u>iterative</u> <u>method</u> [5.] is used to find a series representation of the matrizant which is a solution of

$$\frac{d}{dt} G(A) = A(t)G(A)$$

and

$$G(A) = I+Q(A)+Q(AQ(A))+Q(AQ(AQ(A)))+...$$

where

$$Q(\cdot) = \int_{t_o}^{t} (\cdot)d\lambda \ .$$

Since for $t=t_o$, $G(A)$ (which is a function of t and t_o) equals I, it is the required transition matrix:

$$\Phi(t,t_o) = G(A), \qquad t,t_o \ \varepsilon \ T_t \ .$$

The analogous iterative procedure for discrete-time systems uses the definition

$$\Phi(n+1,n_o) = A(n)\Phi(n,n_o)$$

with $\Phi(n_o,n_o)=I$, $A(n)$ nonsingular, and $n,n_o \varepsilon T_d$. A <u>perturbation</u> <u>method</u> also can be used as an alternative where

$$A(t) = A_o(t) + A_1(t)$$

and $A_1(t)$ is a perturbation of $A_o(t)$, and $A_o(t)$ satisfies the condition

$$A_o(t_1) \ A_o(t_2) = A_o(t_2) \ A_o(t_1).$$

For discrete-time systems

$$A(n) = A_o + A_1(n)$$

where $A_1(n)$ is a perturbation and A_0 is constant. In both cases, series like that for G(A) can be developed, and convergence is relatively fast if suitable choices can be found to make $A_1(t)$ and $A_1(n)$ "small". [See Reference 5, page 365 and page 433].

When A is constant, $\Phi(t,t_0)$ becomes $\Phi(t-t_0)$ and $\Phi(n,n_0)$ becomes $\Phi(n-n_0)$ as shown below. A solution of

$$\dot{\underline{x}} = A\underline{x}, \quad \underline{x}(t_0) = \underline{x}_0 ,$$

can be found using the matrix exponential series (See Appendix B)

$$\varepsilon^{At} = I + At + \frac{A^2t^2}{2!} + \ldots .$$

Thus if a solution $\underline{x}(t) = \varepsilon^{At}\underline{\alpha}_0$ is assumed,

$$\dot{\underline{x}} = (A + A^2t + \frac{A^3t^2}{2!} + \ldots)\underline{\alpha}_0$$

$$= A(I + At + \frac{A^2t^2}{2!} + \ldots)\underline{\alpha}_0$$

$$= A\varepsilon^{At}\underline{\alpha}_0 ,$$

and substitution in the differential equation confirms the assumption. Then at $t=t_0$,

$$\underline{x}(t_0) = \underline{x}_0 = \varepsilon^{At_0}\underline{\alpha}_0$$

or

$$\underline{\alpha}_0 = [\varepsilon^{At_0}]^{-1} \underline{x}_0 .$$

Since $[\varepsilon^{At_0}]^{-1} = \varepsilon^{-At_0}$, (see exercise at end of this chapter),

$$\underline{x}(t) = \varepsilon^{At}(\exp -At_0)\underline{x}_0 = (\exp A(t-t_0))\underline{x}_0 .$$

Comparing this to Eq. 3.63a, i.e.,

$$\underline{x}(t) = \Phi(t,t_o)\underline{x}(t_o),$$

it is clear that

$$\Phi(t-t_o) = (\exp A(t-t_o)), \qquad t,t_o \in T_t, \qquad (3.65)$$

which can be computed from the series for a matrix exponential. It is also noted that

$$\frac{\partial}{\partial t}\Phi(t-t_o) = A\Phi(t-t_o)$$

and $\Phi(t_o-t_o)=I$, thus $(\exp A(t-t_o))$ is the required transition matrix. The usefulness of a direct application of this result depends on being able to recognize closed forms for the scalar series representations of the $\phi_{ij}(t-t_o)$, and this becomes difficult for higher order systems. However, many computational methods based on the matrix exponential have been published. Some of these develop a sum of a finite number of terms of the series and include a suitable stopping condition [10.] (See Appendix D). Other methods are based on theorems associated with functions of a matrix, and some of these are discussed below. For A constant, and

$$\underline{x}(n+1) = A\underline{x}(n), \qquad \underline{x}(n_o) = \underline{x}_o,$$

a solution $\underline{x}(n) = A^n\underline{\alpha}_o$ is assumed, and substitution confirms the assumption. Then at $n=n_o$,

$$\underline{x}(n_o) = \underline{x}_o = A^{n_o}\underline{\alpha}_o$$

or

$$\underline{\alpha}_o = A^{-n_o}\underline{x}_o, \qquad \text{(A nonsingular)}$$

and

$$\underline{x}(n) = A^n A^{-n_o} \underline{x}_o = A^{(n-n_o)} \underline{x}_o .$$

Comparing this to Eq. 3.64a; i.e.,

$$\underline{x}(n) = \Phi(n-n_o)\, \underline{x}(n_o) ,$$

it is clear that

$$\Phi(n-n_o) = A^{(n-n_o)}, \qquad n, n_o \in T_d. \tag{3.66}$$

Also,

$$\Phi(n+1-n_o) = A\, \Phi(n-n_o)$$

and

$$\Phi(n_o-n_o) = A^o \equiv I .$$

Here again, the usefulness of the form A^n depends on being able to recognize closed forms for the $\phi_{ij}(n-n_o)$. However, the computation is quite simple to implement as an iterative program.

Having shown that when A is constant, $\Phi(t-t_o)=(\exp A(t-t_o))$ for continuous-time systems and $\Phi(n-n_o)=A^{(n-n_o)}$ for discrete-time systems, we see that alternatives to the use of a series (for c.-t.) or product (for d.-t.) for the computation of the transition matrices may exist. In fact, theorems associated with functions of a matrix can be used as shown in the following.

The second method for finding the transition matrix when A is constant is based on the Sylvester Theorems -- Theorems B.2 and B.3 of Appendix B. For a continuous-time system, let

$$F(A) = \varepsilon^{At}$$

and then

$$\Phi(t-t_o) = (\exp A(t-t_o)) .$$

For a discrete-time system, let

$$F(A) = A^n$$

and then

$$\Phi(n-n_o) = A^{(n-n_o)} .$$

The method requires that the characteristic values λ_j of A and constituent idempotents A_j of A be found.

The third method [5.] uses the Cayley-Hamilton Theorem -- Theorem B.1 of Appendix B. For $p(\lambda)$ the characteristic polynomial of the rxr matrix A, let $f(\lambda)$ be the scalar function associated with the required matrix function F(A), and let the quotient

$$\frac{f(\lambda)}{p(\lambda)} = q(\lambda) + \frac{r(\lambda)}{p(\lambda)}$$

where $r(\lambda)$ is the remainder polynomial of degree at most r-1:

$$r(\lambda) = \gamma_o + \gamma_1\lambda + \ldots + \gamma_{r-1}\lambda^{r-1} .$$

Then,

$$f(\lambda) = p(\lambda)q(\lambda) + r(\lambda) ,$$

and for any characteristic value λ_j of A, $p(\lambda_j) = 0$. Then

$$f(\lambda_j) = r(\lambda_j).$$

For r distinct characteristic values, j=1,2,...,r, r equations are obtained. These equations are solved for the r values $\gamma_0, \gamma_1, \ldots, \gamma_{r-1}$. Then since $P(A) = \theta$ by the Cayley-Hamilton Theorem, the corresponding matrix equality is

$$F(A) = R(A) = \gamma_0 I + \gamma_1 A + \ldots + \gamma_{r-1} A^{r-1} .$$

$F(A)$ is again ε^{At} or A^n as above. If any λ_j is of multiplicity $m_j > 1$, the derivatives

$$\left.\frac{d^k f(\lambda)}{d\lambda^k}\right|_{\lambda=\lambda_j} = \left.\frac{d^k r(\lambda)}{d\lambda^k}\right|_{\lambda=\lambda_j}, \qquad k = 1,2,\ldots,m_j-1$$

must be used to complete the required set of r equations. This method requires the computation of the characteristic values of A as well as the solution of r linear algebraic equations for $\gamma_0, \gamma_1, \ldots, \gamma_{r-1}$. See [11.] for a computational method based on the Cayley-Hamilton Theorem.

The fourth method of finding the transition matrix when A is constant uses the Laplace Transform or the Z-Transform. For

$$\dot{\underline{x}} = A\underline{x}, \qquad \underline{x}(0) = \underline{x}_0,$$

taking the Laplace Transform gives

$$s\underline{X}(s) = A\underline{X}(s) + \underline{x}_o$$

or

$$[sI-A]\ \underline{X}(s) = \underline{x}_0$$

and

$$\underline{X}(s) = [sI-A]^{-1}\ \underline{x}_0 .$$

Taking the inverse transform

$$\underline{x}(t) = \mathcal{L}^{-1}[sI-A]^{-1}\ \underline{x}_0, \qquad t \geq 0,$$

and comparing with the general solution of the homogeneous equation for $t_o=0$, i.e.,

$$\underline{x}(t) = \Phi(t)\underline{x}_o ,$$

it is clear that

$$\Phi(t) = \mathcal{L}^{-1}[sI-A]^{-1} = \varepsilon^{At}$$

and

$$\Phi(s) = [sI-A]^{-1} , \quad \text{a \underline{resolvent matrix},}$$

From the uniqueness of the solution, the $\Phi(t)$ found as the inverse transform of $\Phi(s)$ can be extended to $\Phi(t-t_o)$, $t_o, t\varepsilon T_t$. The principal computational difficulty is the required inverse. Leverrier's Algorithm, shown below, is useful for computing a resolvent matrix when r is greater than three or four. For a discrete-time system,

$$\underline{x}(n+1) = A\underline{x}(n), \qquad \underline{x}(0) = \underline{x}_o,$$

taking the Z-Transform gives

$$z\underline{X}(z) = A\underline{X}(z) + z\underline{x}_o$$

or

$$[zI-A]\underline{X}(z) = z\underline{x}_o$$

and

$$\underline{X}(z) = z[zI-A]^{-1} \underline{x}_o .$$

Taking the inverse transform

$$\underline{x}(n) = \mathcal{Z}^{-1} z[zI-A]^{-1} \underline{x}_o, \qquad n \geq 0,$$

and comparing with the general solution

$$\underline{x}(n) = \Phi(n)\underline{x}_o, \qquad \text{where } n_o=0,$$

the result is

$$\Phi(n) = \mathfrak{z}^{-1} z[zI-A]^{-1} = A^n$$

and

$$\Phi(z) = z[zI-A]^{-1}.$$

For the general form $[\beta I-A]^{-1}$, Leverrier's Algorithm gives (for A rxr)

$$[\beta I-A]^{-1} = \frac{\beta^{r-1}F_1 + \beta^{r-2}F_2 + \ldots + F_{r-1} + F_r}{\beta^r + \theta_1\beta^{r-1} + \ldots + \theta_{r-1}\beta + \theta_r}$$

where the rxr matrices $F_1,\ldots,F_r$ and the scalars $\theta_1,\ldots,\theta_r$ are

$$\begin{aligned} F_1 &= I & \theta_1 &= -\text{tr } AF_1/1 \\ F_2 &= AF_1 + \theta_1 I & \theta_2 &= -\text{tr } AF_2/2 \\ &\vdots & &\vdots \\ F_r &= AF_{r-1} + \theta_{r-1}I & \theta_r &= -\text{tr } AF_r/r \end{aligned}$$

with a check $AF_r + \theta_r I = \Theta$. (See Appendix B for the definition of tr M = trace of the matrix M.)

Finally, a method of finding the transition matrix for constant A can be developed from the Jordan form for A. For a linear transformation T such that (see Appendix B),

$$J = T^{-1}AT,$$

we get

$$\Phi(t) = \varepsilon^{At} = T\varepsilon^{Jt}T^{-1}$$

for a continuous-time system and

$$\Phi(n) = A^n = TJ^nT^{-1}$$

for a discrete-time system. When A is a matrix of simple structure (i.e., J is diagonal -- including the case of distinct characteristic values of A),

$$\varepsilon^{Jt} = \text{diag}\ [\varepsilon^{\lambda_1 t} \ldots \varepsilon^{\lambda_r t}]$$

and

$$J^n = \text{diag}\ [\lambda_1{}^n \ldots \lambda_r{}^n] \quad .$$

When J is not diagonal, the forms for ε^{Jt} and J^n are more complicated [12.] and unless r is large, other methods of finding the transition matrices are preferred. The computation of T requires the characteristic values of A and a set of characteristic (or generalized) vectors of A.

There are, of course, other methods for finding the elements of the transition matrices when A is a constant. For example, if a simulation diagram (see next section) is prepared, the elements of Φ are

$\phi_{ij}(t)$ = output of integrator i due to input $u_o(t)$ at integrator j.

for continuous-time systems, and

$\phi_{ij}(n)$ = output of unit delay element i due to input $v_o(n+1)$ at unit delay element j.

for discrete-time systems. In these cases, all inputs other than j are inactive, and Laplace Transforms and Z-Transforms are a useful way of finding the $\phi_{ij}(\cdot)$ where the required transfer functions are obtained by inspection of the simulation diagram, e.g., by using flow graph techniques. A method based on the Heaviside Expansion form of a matrix partial fraction expansion has been found to be suited to digital computation [13.].

Example 3.28

To illustrate the use of the five methods of finding the transition matrix when A is constant, two simple second-order equations are used. For the continuous-time case, we use Example 3.22 with $\alpha_1=a>0$, $\alpha_o=\beta_1=0$, $\beta_o=1$:

$$\ddot{y}(t) + a\dot{y}(t) = u(t),$$

and

$$A = \begin{bmatrix} 0 & 1 \\ 0 & -a \end{bmatrix}$$

For the discrete-time case, we use Example 3.23:

$$z(n+2) - z(n) = \frac{1}{3}T\ [v(n+2) + 4v(n+1) + v(n)]$$

and

$$A = \begin{bmatrix} 0 & 1 \\ 1 & 0 \end{bmatrix}$$

a) For

$$A = \begin{bmatrix} 0 & 1 \\ 0 & -a \end{bmatrix},$$

$$\Phi(t) = \varepsilon^{At} = I + \begin{bmatrix} 0 & 1 \\ 0 & -a \end{bmatrix} t + \begin{bmatrix} 0 & -a \\ 0 & a^2 \end{bmatrix} \frac{t^2}{2!}$$

$$+ \begin{bmatrix} 0 & a^2 \\ 0 & -a^3 \end{bmatrix} \frac{t^3}{3!} + \ldots$$

Thus

$$\phi_{11}(t) = 1 + 0 + 0 + 0 + \ldots = 1$$

$$\phi_{12}(t) = 0 + t - a\frac{t^2}{2!} + \frac{a^2t^3}{3!} = -\frac{1}{a}\varepsilon^{-at} + \frac{1}{a}$$

$$\phi_{21}(t) = 0 + 0 + 0 + \ldots = 0$$

$$\phi_{22}(t) = 1 - at + \frac{a^2t^2}{2!} - \frac{a^3t^3}{3!} + \ldots = \varepsilon^{-at}$$

and

$$\Phi(t-t_o) = \begin{bmatrix} 1 & -\frac{1}{a}(\exp -a(t-t_o)) + \frac{1}{a} \\ 0 & (\exp -a(t-t_o)) \end{bmatrix}$$

For

$$A = \begin{bmatrix} 0 & 1 \\ 1 & 0 \end{bmatrix},$$

$\Phi(n) = A^n$ and

$$A^n = \begin{bmatrix} 0 & 1 \\ 1 & 0 \end{bmatrix}, \quad n \text{ odd}$$

$$= \begin{bmatrix} 1 & 0 \\ 0 & 1 \end{bmatrix}, \quad n \text{ even}.$$

By inspection

$$\phi_{11}(n) = \phi_{22}(n) = \tfrac{1}{2}(1 + (-1)^n)$$

$$\phi_{12}(n) = \phi_{21}(n) = \tfrac{1}{2}(1 - (-1)^n)$$

and

$$\Phi(n-n_o) = \begin{bmatrix} \frac{1}{2}(1 + (-1)^{n-n_o}) & \frac{1}{2}(1 - (-1)^{n-n_o}) \\ \frac{1}{2}(1 - (-1)^{n-n_o}) & \frac{1}{2}(1 + (-1)^{n-n_o}) \end{bmatrix}$$

b) Using the Sylvester Theorem with

$$F(A) = \varepsilon^{At} \text{ and } A = \begin{bmatrix} 0 & 1 \\ 0 & -a \end{bmatrix}$$

with λ_1, $\lambda_2 = 0, -a$ we have

$$\varepsilon^{At} = \varepsilon^{\lambda_1 t} A_1 + \varepsilon^{\lambda_2 t} A_2 = A_1 + \varepsilon^{-at} A_2$$

and

$$A_1 = \frac{A-\lambda_2 I}{\lambda_1-\lambda_2}, \qquad A_2 = \frac{A-\lambda_1 I}{\lambda_2-\lambda_1} .$$

Then

$$A_1 = \frac{1}{a}\begin{bmatrix} a & 1 \\ 0 & 0 \end{bmatrix}, \qquad A_2 = -\frac{1}{a}\begin{bmatrix} 0 & 1 \\ 0 & -a \end{bmatrix},$$

and

$$\Phi(t) = \varepsilon^{At} = \begin{bmatrix} 1 & \frac{1}{a}(1 - \varepsilon^{-at}) \\ 0 & \varepsilon^{-at} \end{bmatrix}$$

Using the Sylvester Theorem with

$$F(A) = A^n \text{ and } A = \begin{bmatrix} 0 & 1 \\ 1 & 0 \end{bmatrix}$$

with $\lambda_1, \lambda_2 = 1, -1$ we have

$$A^n = (1)^n A_1 + (-1)^n A_2 \ .$$

Then

$$A_1 = \frac{1}{2}\begin{bmatrix} 1 & 1 \\ 1 & 1 \end{bmatrix}, \qquad A_2 = -\frac{1}{2}\begin{bmatrix} -1 & 1 \\ 1 & -1 \end{bmatrix}$$

and

$$\Phi(n) = A^n = \begin{bmatrix} \frac{1}{2}(1 + (-1)^n) & \frac{1}{2}(1 - (-1)^n) \\ \frac{1}{2}(1 - (-1)^n) & \frac{1}{2}(1 + (-1)^n) \end{bmatrix} .$$

c) Using the Cayley-Hamilton Theorem with

$$F(A) = \varepsilon^{At} \text{ and } A = \begin{bmatrix} 0 & 1 \\ 0 & -a \end{bmatrix}$$

with $\lambda_1, \lambda_2 = 0, -a$ and $r(\lambda) = \gamma_0 + \gamma_1 \lambda$,

$$f(\lambda_1) = 1 = r(\lambda_1) = \gamma_0$$

$$f(\lambda_2) = \varepsilon^{-at} = r(\lambda_2) = \gamma_0 - a\gamma_1 = 1 - a\gamma_1$$

and

$$\gamma_1 = \frac{1}{a}(1 - \varepsilon^{-at}) \ .$$

Therefore,

$$\Phi(t) = F(A) = \varepsilon^{At} = \gamma_o I + \gamma_1 A$$

$$= \begin{bmatrix} 1 & 0 \\ 0 & 1 \end{bmatrix} + \frac{1}{a}(1 - \varepsilon^{-at}) \begin{bmatrix} 0 & 1 \\ 0 & -a \end{bmatrix}$$

$$= \begin{bmatrix} 1 & \frac{1}{a}(1 - \varepsilon^{-at}) \\ 0 & \varepsilon^{-at} \end{bmatrix}$$

Using the Cayley-Hamilton Theorem with

$$F(A) = A^n \text{ and } A = \begin{bmatrix} 0 & 1 \\ 1 & 0 \end{bmatrix}$$

with $\lambda_1, \lambda_2, = 1, -1$ and $r(\lambda) = \gamma_o + \gamma_1\lambda$,

$$f(\lambda_1) = (1)^n = r(\lambda_1) = \gamma_o + \gamma_1$$

$$f(\lambda_2) = (-1)^n = r(\lambda_2) = \gamma_o - \gamma_1$$

and

$$\gamma_o = \frac{1}{2}(1 + (-1)^n)$$

$$\gamma_1 = \frac{1}{2}(1 - (-1)^n) \text{ .}$$

Therefore,

$$\Phi(n) = F(A) = A^n = \gamma_o I + \gamma_1 A$$

with checks the above results.

d) For

$$A = \begin{bmatrix} 0 & 1 \\ 0 & -a \end{bmatrix}$$

we get

$$[sI-A] = \begin{bmatrix} s & -1 \\ 0 & s+a \end{bmatrix}$$

and

$$[sI-A]^{-1} = \phi(s) = \frac{1}{s(s+a)} \begin{bmatrix} s+a & 1 \\ 0 & s \end{bmatrix}.$$

Taking the inverse transform

$$\Phi(t) = \mathcal{L}^{-1}\Phi(s) = \begin{bmatrix} 1 & \frac{1}{a}(1 - \varepsilon^{-at}) \\ 0 & \varepsilon^{-at} \end{bmatrix}, \qquad t \geq 0.$$

If Leverrier's Algorithm were used,

$$[sI-A]^{-1} = \frac{sF_1 + F_2}{s^2 + \theta_1 s + \theta_2}$$

$$F_1 = \begin{bmatrix} 1 & 0 \\ 0 & 1 \end{bmatrix} \qquad \theta_1 = -\mathrm{tr} \begin{bmatrix} 0 & 1 \\ 0 & -a \end{bmatrix} = a$$

$$F_2 = \begin{bmatrix} a & 1 \\ 0 & 0 \end{bmatrix} \qquad \theta_2 = -\mathrm{tr} \begin{bmatrix} 0 & 0 \\ 0 & 0 \end{bmatrix} = 0$$

therefore,

$$[sI-A]^{-1} = \frac{1}{s(s+a)} \begin{bmatrix} s+a & 1 \\ 0 & s \end{bmatrix}.$$

For

$$A = \begin{bmatrix} 0 & 1 \\ 1 & 0 \end{bmatrix}$$

we get

$$[zI-A] = \begin{bmatrix} z & -1 \\ -1 & z \end{bmatrix}$$

and

$$[zI-A]^{-1} = \frac{1}{(z^2-1)} \begin{bmatrix} z & 1 \\ 1 & z \end{bmatrix}$$

and

$$\Phi(z) = \frac{z}{(z^2-1)} \begin{bmatrix} z & 1 \\ 1 & z \end{bmatrix}.$$

Taking the inverse transform

$$\Phi(n) = \mathfrak{Z}^{-1}\Phi(z) = \begin{bmatrix} \frac{1}{2}(1 + (-1)^n) & \frac{1}{2}(1 - (-1)^n) \\ \frac{1}{2}(1 - (-1)^n) & \frac{1}{2}(1 + (-1)^n) \end{bmatrix}, \quad n \geq 0.$$

e) For

$$A = \begin{bmatrix} 0 & 1 \\ 0 & -a \end{bmatrix} \quad \text{with } \lambda_1, \lambda_2 = 0, -a$$

a set of characteristic vectors of A are found from the columns of $C(\lambda_1)$ and $C(\lambda_2)$ (see Appendix B). Thus

$$C(\lambda) = \text{adj } (\lambda I - A) = \begin{bmatrix} \lambda+a & 1 \\ 0 & \lambda \end{bmatrix}$$

Therefore, let

$$\underline{t}_1 = \begin{bmatrix} a \\ 0 \end{bmatrix}, \quad \underline{t}_2 = \begin{bmatrix} 1 \\ -a \end{bmatrix}, \quad \text{and}$$

$$T = \begin{bmatrix} a & 1 \\ 0 & -a \end{bmatrix}.$$

Also,

$$T^{-1} = -\frac{1}{a^2} \begin{bmatrix} -a & -1 \\ 0 & a \end{bmatrix} = \begin{bmatrix} \frac{1}{a} & \frac{1}{a^2} \\ 0 & -\frac{1}{a} \end{bmatrix}$$

The Jordan Form for A is

$$J = T^{-1}AT = \begin{bmatrix} 0 & 0 \\ 0 & -a \end{bmatrix}$$

and

$$e^{Jt} = \begin{bmatrix} 1 & 0 \\ 0 & \varepsilon^{-at} \end{bmatrix} .$$

Therefore,

$$\Phi(t) = \varepsilon^{At} = T\,^{Jt}T^{-1} = \begin{bmatrix} a & 1 \\ 0 & -a \end{bmatrix} \begin{bmatrix} 1 & 0 \\ 0 & \varepsilon^{-at} \end{bmatrix} \begin{bmatrix} \frac{1}{a} & \frac{1}{a^2} \\ 0 & -\frac{1}{a} \end{bmatrix}$$

$$= \begin{bmatrix} 1 & \frac{1}{a}(1 - \varepsilon^{-at}) \\ 0 & \varepsilon^{-at} \end{bmatrix} .$$

For

$$A = \begin{bmatrix} 0 & 1 \\ 1 & 0 \end{bmatrix} \quad \text{with } \lambda_1, \lambda_2 = 1, -1$$

$$C(\lambda) = \text{adj } (\lambda I - A) = \begin{bmatrix} \lambda & 1 \\ 1 & \lambda \end{bmatrix} .$$

Therefore, let

$$\underline{t}_1 = \begin{bmatrix} 1 \\ 1 \end{bmatrix} , \quad \underline{t}_2 = \begin{bmatrix} 1 \\ -1 \end{bmatrix} \quad \text{and}$$

$$T = \begin{bmatrix} 1 & 1 \\ 1 & -1 \end{bmatrix} .$$

Also,

$$T^{-1} = -\frac{1}{2}\begin{bmatrix} -1 & -1 \\ -1 & 1 \end{bmatrix} = \begin{bmatrix} \frac{1}{2} & \frac{1}{2} \\ \frac{1}{2} & -\frac{1}{2} \end{bmatrix}.$$

The Jordan Form for A is

$$J = T^{-1}AT = \begin{bmatrix} 1 & 0 \\ 0 & -1 \end{bmatrix}$$

and

$$J^n = \begin{bmatrix} 1 & 0 \\ 0 & (-1)^n \end{bmatrix}.$$

Therefore,

$$\Phi(n) = A^n = TJ^nT^{-1} = \begin{bmatrix} 1 & 1 \\ 1 & -1 \end{bmatrix}\begin{bmatrix} 1 & 0 \\ 0 & (-1)^n \end{bmatrix}\begin{bmatrix} \frac{1}{2} & \frac{1}{2} \\ \frac{1}{2} & -\frac{1}{2} \end{bmatrix}$$

$$= \begin{bmatrix} \frac{1}{2}(1 + (-1)^n) & \frac{1}{2}(1 - (-1)^n) \\ \frac{1}{2}(1 - (-1)^n) & \frac{1}{2}(1 + (-1)^n) \end{bmatrix}$$

It is obvious from these simple examples that the five methods require different computations, and that as the dimension r increases, some are difficult to apply. Furthermore, they are not all suitable for machine computation. Method a) is probably the most popular for this purpose.

3.4.4 Standard Forms

The discussion of state-variable models up to this point has not been concerned with any particular form for the matrices A, B, C, and D in Eq. 3.57 or Eq. 3.60. It frequently is convenient to have the state model in some standard form as a starting point for further use of the model, e.g., in the design of a control system or a filter. Also, it is convenient to be able to write down a standard state-variable model by inspection of a given rth-order form, e.g., 3.18 or 3.20. Examples 3.22 and 3.24 illustrate four forms that are obtained from the same 2nd-order differential equation.

When there is one input ($u(t)$ or $v(n)$) and one output ($y(t)$ or $z(n)$) and the system is stationary, there are at least three forms of general interest: primal forms, a diagonal (or Jordan) form, and a cascade form. Starting with an rth-order differential equation

$$(D^r+\alpha_{r-1}D^{r-1} +\ldots+ \alpha_o)y(t) = (\beta_r D^r +\ldots+ \beta_o)u(t) \quad (3.67)$$

a set of state variables can be chosen as

$$\begin{aligned}
\dot{x}_1 &= x_2 \\
\dot{x}_2 &= x_3 \\
&\vdots \\
\dot{x}_{r-1} &= x_r \\
\dot{x}_r &= -\alpha_o x_1 - \ldots - \alpha_{r-1}x_r + u
\end{aligned}$$

and the output is

$$y = (\beta_o-\alpha_o\beta_r)x_1 +\ldots+ (\beta_{r-1}-\alpha_{r-1}\beta_r)x_r + \beta_r u \quad .$$

Then in vector-matrix form

$$\dot{\underline{x}} = \begin{bmatrix} 0 & 1 & 0 & & \cdots & 0 \\ 0 & 0 & 1 & 0 & \cdots & 0 \\ \vdots & & & & & \vdots \\ 0 & \cdots & & 0 & & 1 \\ -\alpha_o & -\alpha_1 & & \cdots & & -\alpha_{r-1} \end{bmatrix} \underline{x} + \begin{bmatrix} 0 \\ \vdots \\ 0 \\ 1 \end{bmatrix} u \quad (3.68)$$

$$y = [\beta_o - \alpha_o\beta_r \quad \cdots \quad \beta_{r-1} - \alpha_{r-1}\beta_r]\ \underline{x} + [\beta_r]u \quad .$$

This defines A, $\underline{b}$, $\underline{c}^T$, and d for this form. A is in phase variable (or companion) form [14.], and for $\underline{b}$, $\underline{c}^T$, and d as given here, this will be called the <u>first primal</u> form. For reasons discussed in the next section, this also is called a controllable form. Note that Eq. 3.57 is more general than Eq. 3.18 since the order of N(D) may be as high as the order of M(D). Example 3.24b shows this form. A second phase variable form is obtained from Eq. 3.67 by a choosing a set of state variables as

$$\begin{aligned} \dot{x}_1 &= -\alpha_{r-1}x_1 + x_2 + (\beta_{r-1} - \alpha_{r-1}\beta_r)u \\ \dot{x}_2 &= -\alpha_{r-2}x_1 + x_3 + (\beta_{r-2} - \alpha_{r-2}\beta_r)u \\ &\vdots \\ \dot{x}_r &= -\alpha_o x_1 \qquad\qquad + (\beta_o - \alpha_o\beta_r)u \end{aligned}$$

and the output is

$$y = x_1 + \beta_r u.$$

Then in vector-matrix form*

$$\dot{\underline{x}} = \begin{bmatrix} -\alpha_{r-1} & 1 & 0 & & \cdots & 0 \\ -\alpha_{r-2} & 0 & 1 & 0 & \cdots & 0 \\ \vdots & & & & & \\ -\alpha_1 & 0 & \cdots & & & 1 \\ -\alpha_0 & 0 & \cdots & & & 0 \end{bmatrix} \underline{x} + \begin{bmatrix} \beta_{r-1}-\alpha_{r-1}\beta_r \\ \\ \\ \\ \beta_0-\alpha_0\beta_r \end{bmatrix} u \tag{3.69}$$

$$y = [1 \quad 0 \quad \cdots \quad 0]\underline{x} + [\beta_r]u \quad .$$

This defines A, $\underline{b}$, $\underline{c}^T$, and d for this form, and it will be called the <u>second primal</u> form. This also is called an observable form. The verification of the first and second primal forms are suggested as problems at the end of this chapter, and Example 3.24a shows this form.

Before introducing other forms, it is convenient to construct <u>simulation diagrams</u> for 3.68 and 3.69; these diagrams will then suggest other possibilities. For linear, stationary systems, three ideal elements are introduced:

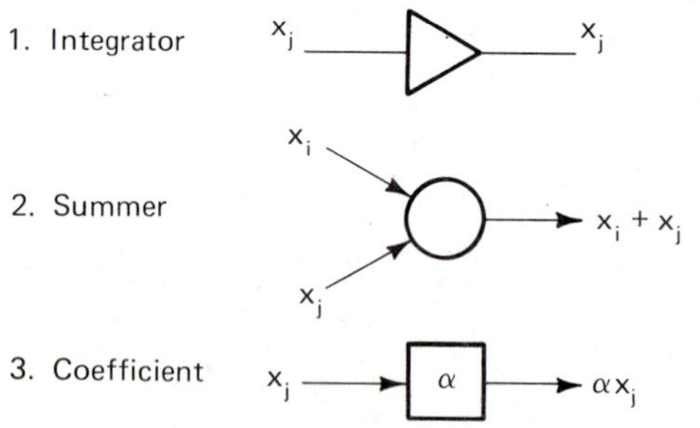

Figure 3.15

---

*The use of $\underline{x}$ for the state vectors in this discussion is for convenience. In general, the state variables are different for each model.

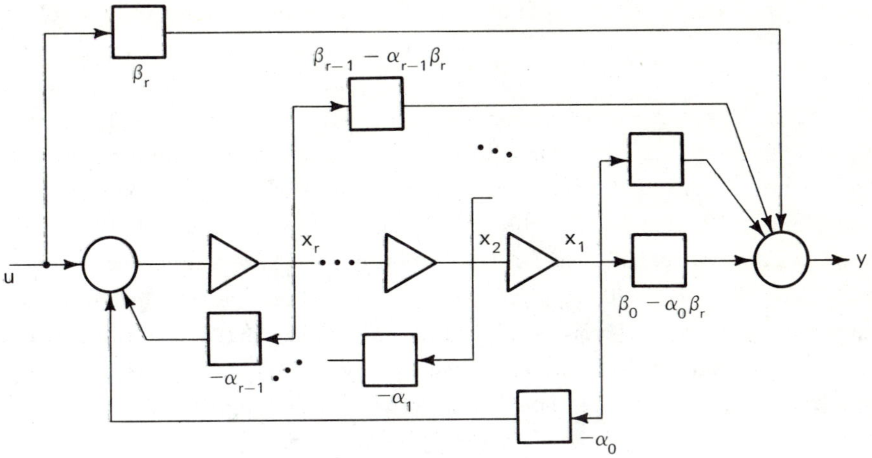

Figure 3.3 First primal form

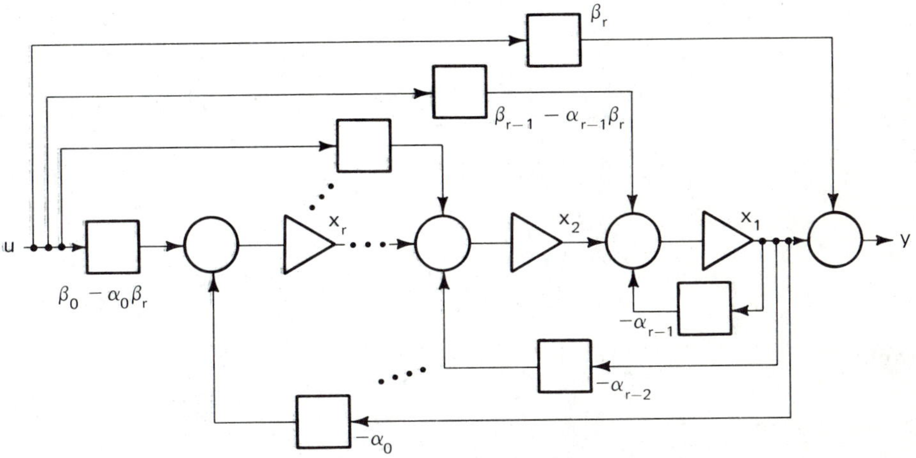

Figure 3.4 Second primal form

Feedback	Feedforward: To y	Feedforward: From u
From all x_j	First form	Third form
From x_1	Fourth form	Second form

Figure 3.5 Primal form classification

Note that these differ from the usual analog computer elements in that there is no sign inversion in the integrator or the summer, and α may be any real number. Using these, Eq. 3.68 has the simulation diagram shown in Figure 3.3 and Eq. 3.69 has the diagram shown in Figure 3.4. Inspection of these two figures shows that there are both feedback and feedforward paths. In the first form there is feedback from each state variable while in the second, the feedback links are all from x_1. Furthermore, in the first form there is a feedforward link from each state variable while in the second the feedforward links are from u. By classifying a form according to both its feedback and feedforward links, four forms are suggested, of which two have been found. Thus, from Figure 3.5, it is clear that there are possible third and fourth primal forms.

Again starting with Eq. 3.67, let

$$\begin{aligned} \dot{x}_1 &= x_2 + f_1 u \\ \dot{x}_2 &= x_3 + f_2 u \\ &\vdots \\ \dot{x}_r &= -\alpha_o x_1 - \ldots - \alpha_{r-1} x_r + f_r u \end{aligned}$$

and

$$y = x_1 + f_o u.$$

The derivatives of y, i.e., $D^i y$, $i=0,1,\ldots,r$, are found from these equations and a sequence of substitutions is carried out so that no derivatives of the state variables appear. These derivatives are then substituted in Eq. 3.67 and coefficients of the derivatives of u, i.e., $D^i u$, $i=0,1,\ldots,r$, are equated (See Example 3.22), and the resulting r+1 equations are written:

$$\begin{bmatrix} \alpha_o & \alpha_1 & \alpha_2 & \cdots & \alpha_{r-1} & 1 \\ \alpha_1 & \alpha_2 & & \cdots & 1 & 0 \\ \alpha_2 & & & & & \\ \vdots & & & & & \\ \alpha_{r-1} & 1 & 0 & \cdots & & 0 \\ 1 & 0 & & \cdots & & 0 \end{bmatrix} \begin{bmatrix} f_o \\ f_1 \\ \vdots \\ \\ f_r \end{bmatrix} = \begin{bmatrix} \beta_o \\ \beta_1 \\ \vdots \\ \\ \beta_r \end{bmatrix}$$

or

$$A_3 \underline{f} = \underline{\beta}$$

and

$$\underline{f} = A_3^{-1} \underline{\beta} \quad . \tag{3.70}$$

Then

$$\dot{\underline{x}} = \begin{bmatrix} 0 & 1 & 0 & \cdots & 0 \\ 0 & 0 & 1 & 0 & 0 \\ \vdots & & & & \\ -\alpha_o & -\alpha_1 & & \cdots & -\alpha_{r-1} \end{bmatrix} \underline{x} + \begin{bmatrix} f_1 \\ f_2 \\ \vdots \\ f_r \end{bmatrix} u \tag{3.71}$$

$$y = [1 \quad 0 \quad \cdots \quad 0]\underline{x} + [f_o]u \; .$$

This defines A, $\underline{b}$, $\underline{c}^T$, and d for this form, which will be called the <u>third primal</u> form. Figure 3.6 shows the simulation diagram.

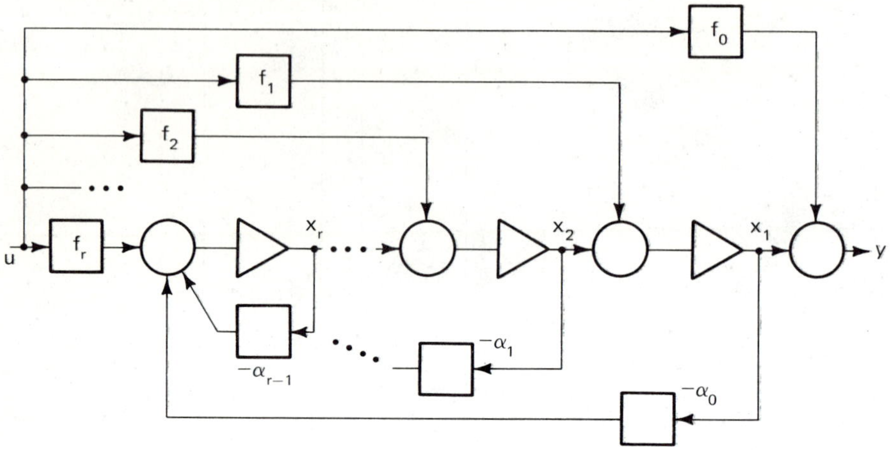

Figure 3.6 Third primal form

For the fourth primal form let

$$\dot{x}_1 = -\alpha_{r-1}x_1 + x_2$$
$$\dot{x}_2 = -\alpha_{r-2}x_1 + x_3$$
$$\vdots$$
$$\dot{x}_r = -\alpha_o x_1 + u$$

and

$$y = g_1x_1 + \ldots + g_rx_r + g_{r+1}u.$$

As before, the derivatives of y, i.e., D^iy, i=1,2,...,r are found from these equations and substitutions are made so that no derivatives of the state variables appear. These derivatives are then substituted in Eq. 3.67 and coefficients of the derivatives of u, i.e., D^iu, i=0,1,2,...,r are equated (see Example 3.24c) and the resulting r+1 equations are written:

$$\begin{bmatrix} 1 & \alpha_{r-1} & \cdots & \alpha_2 & \alpha_1 & \alpha_o \\ 0 & 1 & \alpha_{r-1} & \cdots & \alpha_2 & \alpha_1 \\ \cdot & & & & & \alpha_2 \\ \cdot & & & & & \cdot \\ \cdot & & & & & \cdot \\ & & & & & \cdot \\ & & & & & \alpha_{r-1} \\ 0 & \cdots & & & 0 & 1 \end{bmatrix} \begin{bmatrix} g_1 \\ g_2 \\ \cdot \\ \cdot \\ \cdot \\ \\ \\ g_{r+1} \end{bmatrix} = \begin{bmatrix} \beta_o \\ \beta_1 \\ \cdot \\ \cdot \\ \cdot \\ \\ \\ \beta_r \end{bmatrix}$$

or

$$A_4 \underline{g} = \underline{\beta}$$

and

$$\underline{g} = A_4^{-1} \underline{\beta} \ . \tag{3.72}$$

Then

$$\dot{\underline{x}} = \begin{bmatrix} -\alpha_{r-1} & 1 & 0 & \cdots & 0 \\ -\alpha_{r-2} & 0 & 1 & 0 & 0 \\ \cdot & & & & \\ \cdot & & & & \\ \cdot & & & & \\ -\alpha_1 & 0 & \cdots & & 1 \\ -\alpha_o & 0 & \cdots & & 0 \end{bmatrix} \underline{x} + \begin{bmatrix} 0 \\ \cdot \\ \cdot \\ \cdot \\ 0 \\ 1 \end{bmatrix} u \tag{3.73}$$

$$y = [g_1 \ \cdots \ g_r]\underline{x} + [g_{r+1}]u$$

This defines A, $\underline{b}$, $\underline{c}^T$, and d for the <u>fourth primal</u> form. Figure 3.7 shows the simulation diagram.
Note that A for the first and third forms is the same and A for the second and fourth forms is the same. Other equalities for the $\underline{b}$, $\underline{c}^T$, and d can be seen. Furthermore, A_3 of

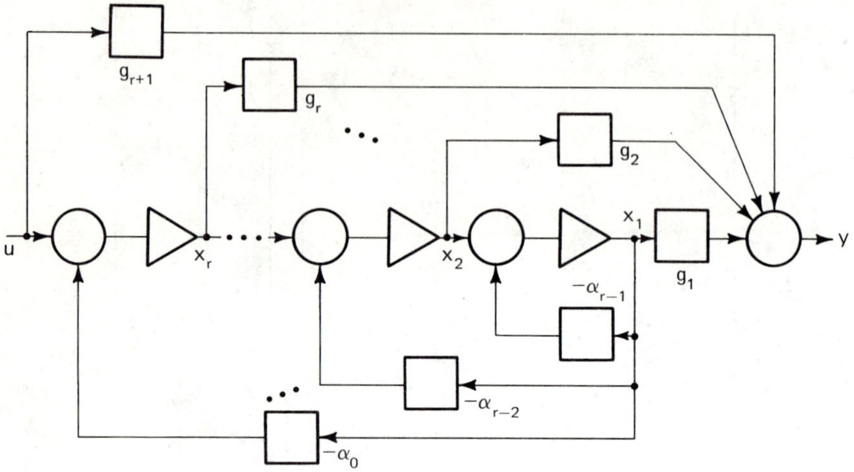

Figure 3.7 Fourth primal form

Eq. 3.70 and A_4 of Eq. 3.72 are related by column interchange, i.e., $A_4 = A_3T$ where

$$T = \begin{bmatrix} 0 & \cdots & 0 & 1 \\ 0 & \cdots & 1 & 0 \\ \vdots & & & \vdots \\ 1 & 0 & \cdots & 0 \end{bmatrix} . \qquad (3.74)$$

Another detail concerns the numbering of the state variables. In the four simulation diagrams of Fig. 3.3, 3.4, 3.6, and 3.7 the state variables are numbered right to left, a convention used throughout this book. Other references sometimes number left to right with a consequent change in the matrices of the models since the usual convention (as in Appendix B) is to number the components of $\underline{x}^T$ as $(x_1\ x_2\ \ldots\ x_r)$. The renumbered models are related by a similarity transformation with T as in Eq. 3.74. Thus for $\underline{x} = T\underline{x}'$,

$$\dot{\underline{x}}' = (T^{-1}AT)\underline{x}' + (T^{-1}\underline{b})u \qquad (3.75)$$

$$y = (\underline{c}^T T)\underline{x}' + du .$$

This transformation on the first and second primal forms produces models with simulation diagrams that are the same (with appropriate renumbering) as those for two dual system models discussed later in this section. Equation 3.75 is, in fact, a general result for any change such that $\underline{x}=T\underline{x}'$ for single input/output systems. The derivation of the general result 3.76 (for multiple input/output)

$$\dot{\underline{x}}' = (T^{-1}AT)\underline{x}' + (T^{-1}B)\underline{u} \tag{3.76}$$

$$y = (CT)\underline{x}' + D\underline{u}$$

is given as a problem at the end of this chapter. Usually, it is the transformation T that must be found if the new model is to be in some required standard form.

The second standard form of interest is a <u>diagonal</u> (or Jordan) form; this also is called a <u>modal</u> form [5.]. (See Appendix B for discussion of Jordan forms.) If Eq. 3.67 is the starting point, a diagonal form of state variable model can be found by using a partial fraction expansion of the system transfer function. Thus, from 3.67, using 3.48,

$$H(s) = \frac{\beta_r s^r + \beta_{r-1}s^{r-1} + \ldots + \beta_o}{s^r + \alpha_{r-1}s^{r-1} + \ldots + \alpha_o} = \frac{Y(s)}{U(s)}$$

and it is assumed that there are no cancellations of poles and zeros. Then by division,

$$H(s) = \beta_r + \frac{\gamma_{r-1}s^{r-1} + \ldots + \gamma_o}{s^r + \alpha_{r-1}s^{r-1} + \ldots + \alpha_o} .$$

The proper fraction is now expanded using the method of Section A.2.2, Appendix A. Thus if all poles of H(s) are simple, we have

$$H(s) = \beta_r + \sum_{j=1}^{r} \frac{A_j}{s-\lambda_j} \tag{3.77a}$$

or if some poles of H(s) are multiple,

$$H(s) = \beta_r + \sum_{j=1}^{m} \sum_{i=1}^{m_j} \frac{A_{ij}}{(s-\lambda_j)^i} \qquad (3.77b)$$

Using 3.77a, the simulation diagram will include one element, as shown in Figure 3.8a, for each pole λ_j of H(s), and each element has the common input u. The output y is the sum of the element outputs plus a term $\beta_r u$. The state equations have the form:

$$\dot{\underline{x}} = \text{diag } [\lambda_1 \ \ldots \ \lambda_r]\underline{x} + [1 \ \ldots \ 1]^T u \qquad (3.78)$$

$$y = [A_1 \ \ldots \ A_r]\underline{x} + \beta_r u.$$

This defines A, $\underline{b}$, $\underline{c}^T$, and d of the diagonal form for $m_j=1$ for all λ_j. If there are complex pairs λ_j, λ_j^*, some of the coefficients and state variables are complex. A modified diagonal form using elements as shown in Figure 3.8b can then be used to form a simulation diagram with real coefficients. The state equations change accordingly. For example if $\lambda_2=\lambda_1^*$ and all other λ_j, j=3,...,r, are real, the equations are:

$$\dot{\underline{x}}' = \begin{bmatrix} \sigma_1 & \omega_1 & 0 & \ldots & 0 \\ -\omega_1 & \sigma_1 & 0 & \ldots & 0 \\ 0 & \ldots & \lambda_3 & 0 & 0 \\ \vdots & & & & \vdots \\ 0 & & & & \lambda_r \end{bmatrix} \underline{x}' + \begin{bmatrix} 1 \\ \cdot \\ \cdot \\ \cdot \\ 1 \end{bmatrix} u \qquad (3.79)$$

$$y = [(a_1-b_1)(a_1+b_1) \ A_3 \ \ldots \ A_r] \ \underline{x}' + \beta_r u$$

When there are poles λ_j of multiplicity $m_j>1$, the expansion is given by Eq. 3.77b. For a particular multiple pole λ_j,

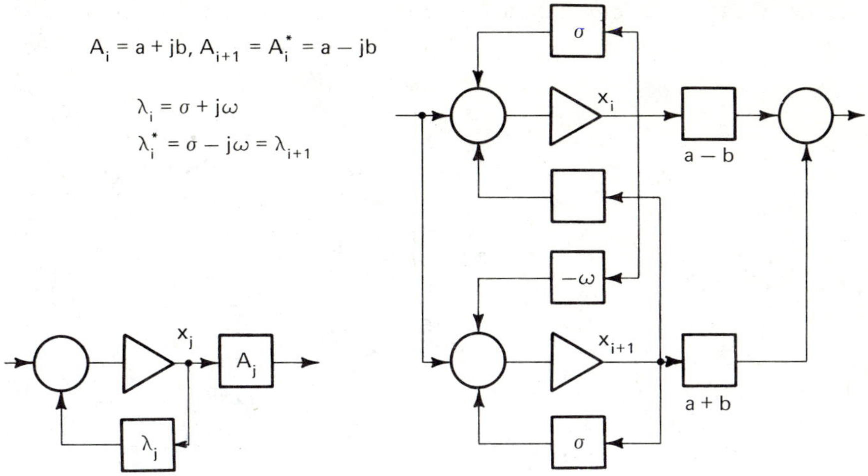

(a) Real poles (b) Complex pair poles

Figure 3.8 Simulation diagram elements for simple poles

Figure 3.9a shows the form of that part of the simulation diagram. The A-matrix is now in Jordan form (see Appendix B). For example, if $m_1>1$ and all other poles are of multiplicity $m_j=1$, the state equations are

$$\dot{\underline{x}} = \begin{bmatrix} \lambda_1 & 1 & 0 & 0 & \cdots & 0 \\ & \lambda_1 & 1 & 0 & \vdots & \\ & & \lambda_1 & 1 & 0 \cdots & 0 \\ & & & \lambda_1 & 0 \cdots & 0 \\ & 0 & & & \lambda_2 & \\ & & & & & \vdots \\ & & & & & \lambda_p \end{bmatrix} \underline{x} + \begin{bmatrix} 0 \\ \vdots \\ 0 \\ 1 \\ 1 \\ \vdots \\ 1 \end{bmatrix} u \qquad (3.80)$$

$$y = [A_{m_1 1} \ \cdots \ A_{11} \ A_2 \ \cdots \ A_p]\underline{x} + \beta_r u$$

with $p = r - m_1 + 1$. When there are complex pairs of poles λ_j, λ_j^* of multiplicity $m_j>1$, a modified diagonal form with all real coefficients can be developed. An example where $m_j=2$ is shown in Figure 3.9b.

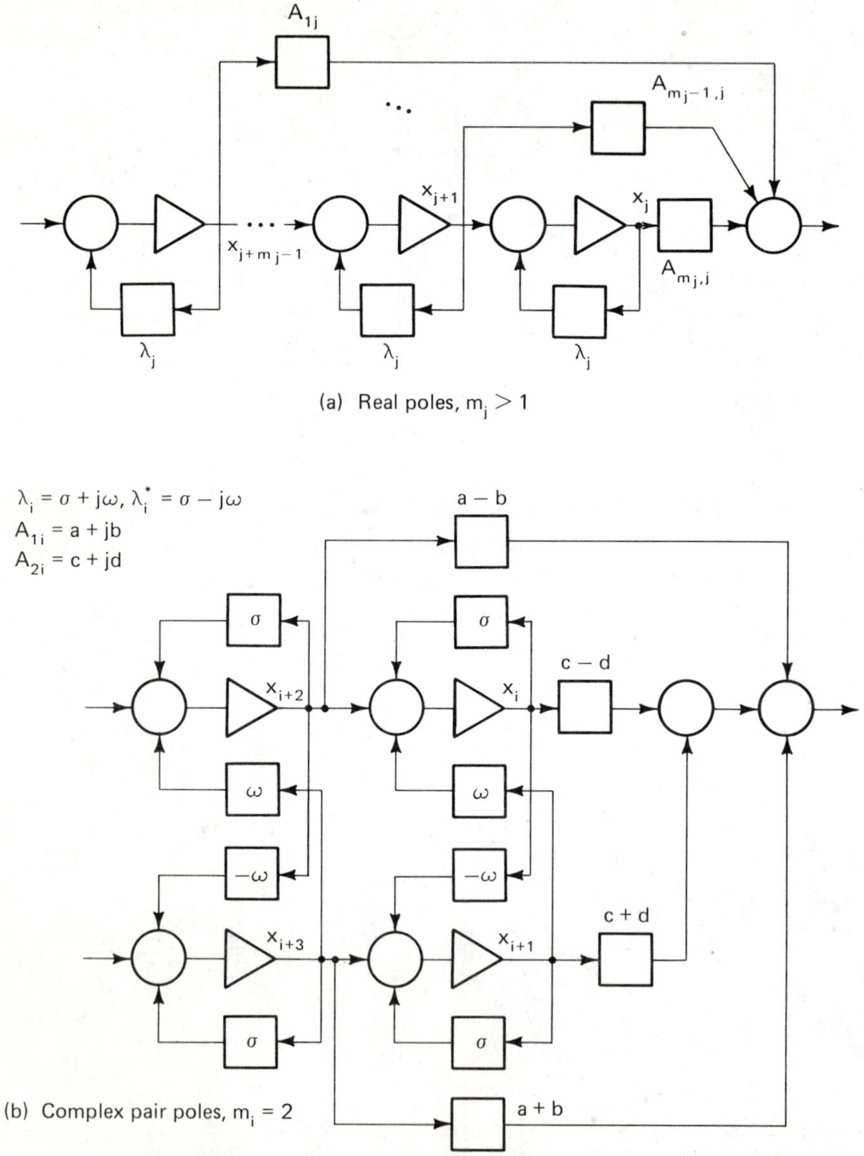

Figure 3.9 Simulation diagram elements for multiple poles

The third standard form of interest is the cascade form. If the numerator and denominator polynomials of H(s) are factored, we have

$$H(s) = \frac{\beta_q(s-\mu_1)\ \dots\ (s-\mu_q)}{(s-\lambda_1)\ \dots\ (s-\lambda_r)}, \quad q \le r$$

A simulation diagram for the case of q=r-1 is shown in Figure 3.10, and the corresponding state equations are given in Eq. 3.81.

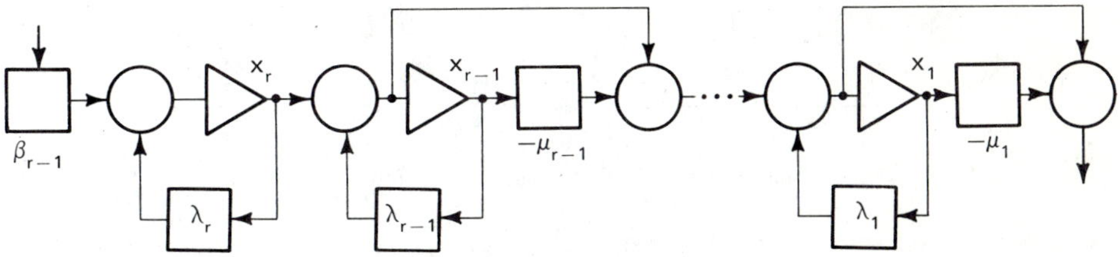

Figure 3.10 A cascade form for q = r − 1

$$\dot{\underline{x}} = \begin{bmatrix} \lambda_1 & \eta_2 & \eta_3 & \cdots & \eta_{r-1} & 1 \\ 0 & \lambda_2 & \eta_3 & \cdots & \eta_{r-1} & 1 \\ \vdots & & & & & \\ 0 & \cdots & & \lambda_{r-2} & \eta_{r-1} & 1 \\ 0 & \cdots & & 0 & \lambda_{r-1} & 1 \\ 0 & \cdots & & 0 & 0 & \lambda_r \end{bmatrix} \underline{x} + \begin{bmatrix} 0 \\ \vdots \\ \\ \\ 0 \\ \beta_{r-1} \end{bmatrix} u \qquad (3.81)$$

$$y = [\eta_1\ \eta_2\ \cdots\ \eta_{r-1}\ \ 1]\ \underline{x}\ .$$

In this, $\eta_j = \lambda_j - \mu_j$, $j=1,2,\dots,r-1$. If the element for the pole λ_r is placed on the right instead of the left, different, but similar, forms for A, $\underline{b}$, and $\underline{c}^T$ occur. It is possible that some of the λ_j and μ_j will occur in complex conjugate pairs λ_j, λ_j^* and μ_j, μ_j^*. Modifications of the cascade form also can found so that only real coefficients are used. For example, for one pair of zeros μ_i, μ_i^* = $a \pm jb$ and one pair of poles λ_i, $\lambda_i^* = \sigma \pm j\omega$, a second-order cascade element derived from the first primal form is shown in Figure 3.11.

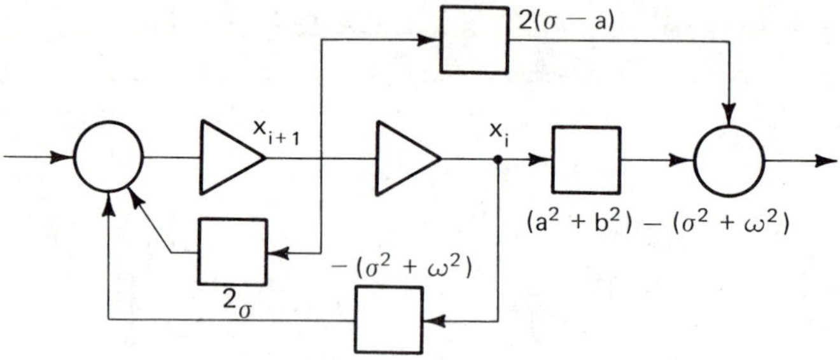

Figure 3.11 Modified cascade element

Thus a complete modified cascade form may contain some elements as shown in Figure 3.10 and some as shown in Figure 3.11.

A parallel development for discrete-time system models with one input v(n) and one output z(n) can be done. Starting with the rth-order difference equation

$$(E^r + \alpha_{r-1}E^{r-1} + \ldots + \alpha_o)z(n) = (\beta_r E^r + \ldots + \beta_o)v(n) \quad (3.82)$$

the four primal forms, the diagonal forms, and the cascade forms are obtained. For example, by choosing a set of state variables as

$$x_1(n+1) = x_2(n)$$
$$x_2(n+1) = x_3(n)$$
$$\vdots$$
$$x_{r-1}(n+1) = x_r(n)$$
$$x_r(n+1) = -\alpha_o x_1(n) - \ldots - \alpha_{r-1}x_r(n) + v(n)$$

and

$$z(n) = (\beta_o - \alpha_o \beta_r)x_1(n) + \ldots + (\beta_{r-1} - \alpha_{r-1}\beta_r)x_r(n) + \beta_r v(n)$$

the A, $\underline{b}$, $\underline{c}^T$, and d of the first primal form result. Similar results for the other three primal forms follow. The simulation diagrams for the linear discrete-time systems also contain three ideal elements.

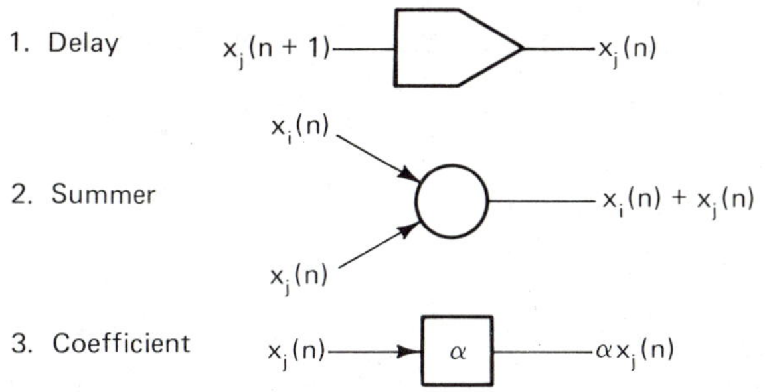

Figure 3.16

The complete simulation diagrams are quite similar to those for continuous-time systems except that the integrators are replaced by delay elements. Partial fraction expansions of H(z) found from Eq. 3.82 produces the diagonal (or Jordan) forms, and factoring the numerator and denominator polynomials of H(z) leads to the cascade forms as before. Again, it is assumed that there are no cancellations of poles and zeros. Table 3.2 summarizes these results for single-input, single-output linear stationary system models.

The extension of the single input/single output continuous-time and discrete-time forms to vector input and vector output forms is quite difficult. All the previous forms summarized in Table 3.2 were obtained by starting with the models of Eq. 3.67 or Eq. 3.82, and an analogous procedure would obtain the multivariable forms from the equations

$$\underline{y}(t) = P(D)M(D)^{-1}N(D)\underline{u}(t)$$

or

$$\underline{z}(n) = P(E)M(E)^{-1}N(E)\underline{v}(n)$$

discussed in Section 3.3.4. However, this is much more complicated. The methods are discussed briefly in Section 3.6. Multiple input/output standard forms have been developed which are obtained by similarity transformations on state models such as Eq. 3.57 and Eq. 3.60, and most of the standard forms are analogous to the first or third primal forms. For multiple inputs, the pair (A,B) is specified,

TABLE 3.2

SUMMARY OF STANDARD STATE VARIABLE FORMS SINGLE-INPUT/SINGLE-OUTPUT

Continuous-Time and Discrete-Time

		Equation†	Figure*
Primal Forms	First	3.68	3.3
	Second	3.69	3.4
	Third	3.71	3.6
	Fourth	3.73	3.7
Jordan Forms	$m_j = 1$	3.78	3.8
	$m_j > 1$	3.80	3.9
Cascade Forms	Original	3.81	3.10
	Modified	--	3.11

*with integrator replaced by delay for discrete-time.

†with $\dot{\underline{x}}$, $\underline{x}$ replaced by $\underline{x}(n+1)$, $\underline{x}(n)$ for discrete-time.

and for multiple outputs, the pair (A,C) is specified. Forms that are all related to the first or third primal forms are given here, and other forms, as well as the associated transformations, are introduced later in this chapter after the discussion of controllability and observability. The unspecified matrices in the five following standard forms are found as in Eq. 3.76.

Two multiple <u>input</u> forms that have been developed are described, and the first [15.][16.][22.] is a "block form" analogy to Eq. 3.68. One possibility is:

$$\dot{\underline{x}} = A_1\underline{x} + B_1\underline{u} \qquad (3.83)$$

$$\underline{y} = C_1\underline{x} + D_1\underline{u}$$

with

$$A_1 = \left[\begin{array}{c:c:c:c:c:c}
\multicolumn{2}{c:}{A_{11}} & & & & \\
\hdashline
\underline{\beta}_{21} & \Theta & A_{12} & & & 0 \\
\hdashline
\vdots & & & \ddots & & \\
\hdashline
\underline{\beta}_{q1} & \Theta & \cdots & \underline{\beta}_{q,q-1} & \Theta & A_{1q}
\end{array}\right]$$

and

$$B_1 = \left[\begin{array}{c:c:c}
 & & 0 \\
 & & \vdots \\
 0 & & 0 \\
 & & 1 \\
\hdashline
 & 0 & \\
 & \vdots & \\
 & 1 & \\
\hdashline
0 & & \\
\vdots & & 0 \\
0 & & \\
1 & &
\end{array}\right]$$

and each A_{1j}, $j=1,\ldots,\nu$, is phase variable as in Eq. 3.68. A_1 is rxr, the A_{1j} are $\alpha_j x \alpha_j$ such that $\Sigma\alpha_j = r$, and B_1 is rxq where q is the input dimension. The α_j are found in the construction of the transformation. Other transformations can be constructed which make A_1 an upper triangular form or ones which are neither upper or lower triangular but which yield other useful forms for B_1. The second input form [17.] [18.][19.] is a "matrix form" analogy to Eq. 3.68:

$$\dot{\underline{x}} = A_2\underline{x} + B_2\underline{u}' \qquad (3.84)$$

$$\underline{y} = C_2\underline{x} + D_2\underline{u}'$$

with

$$A_2 = \left[\begin{array}{c:c} \Theta & I_{r-q} \\ \hdashline \multicolumn{2}{c}{A_{20}} \end{array}\right]$$

and

$$B_2 = \left[\begin{array}{c} \Theta \\ \hdashline I_q \end{array}\right]$$

where I_{r-q} is an (r-q) unit matrix and A_{20} is qxr.

Two multiple output forms also are described here, and the first [20.] is a "block form" analogy to Eq. 3.71:

$$\dot{\underline{x}} = A_3\underline{x} + B_3\underline{u} \qquad (3.85)$$

$$\underline{y} = C_3\underline{x} + D_3\underline{u}$$

with

$$A_3 = \left[\begin{array}{c|c|c|c}
A_{31} & & & \\ \hline
\Theta & A_{32} & & 0 \\ \cline{1-1}
\underline{\gamma}_{21}^T & & & \\ \hline
\vdots & & \ddots & \\ \hline
\Theta & & \Theta & A_{3p} \\ \cline{1-1}\cline{3-3}
\underline{\gamma}_{p1}^T & & \underline{\gamma}_{p,p-1}^T &
\end{array}\right]$$

and

$$C_3 = \left[\begin{array}{cccc}
1 \quad 0 \quad \cdots \quad 0 & & & 0 \\ \cline{1-1}
 & \ddots & & \\ \cline{4-4}
0 & & & 1 \quad 0 \quad \cdots \quad 0
\end{array}\right]$$

and each A_{3j}, j=1,...,p is phase variable as in Eq. 3.71. A_3 is rxr, the A_{3j} are $\delta_j x \delta_j$ such that $\Sigma\delta_j = r$, and C_3 is pxr where p is the output dimension. The second output form [12.] is a "matrix form" analogy to Eq. 3.71:

$$\dot{\underline{x}} = A_4\underline{x} + B_4\underline{u} \tag{3.86}$$

$$\underline{y} = C_4\underline{x} + D_4\underline{u}$$

with

$$A_4 = \left[\begin{array}{c|c}
\Theta & I_{r-p} \\ \hline
\multicolumn{2}{c}{A_{40}}
\end{array}\right]$$

and

$$C_4 = \left[\, I_p \;\middle|\; \Theta \, \right]$$

where I_{r-p} is an (r-p) unit matrix and A_{40} is pxr. All four of these forms follow by construction of an appropriate linear transformation such that the state variable model of the original system is similar to one of these, and conditions on the controllability or observability of the original system must be met. Examples are given later in this chapter in Section 3.6.3.

In addition to the primal or phase variable multiple-input/multiple output forms, a diagonal (or Jordan) form can be constructed as in the single-input/single-output case. Using the method given in Appendix B:

$$\dot{\underline{x}} = A_5\underline{x} + B_5\underline{u} \qquad (3.87)$$

$$\underline{y} = C_5\underline{x} + D_5\underline{u}$$

and

$$A_5 = \begin{bmatrix} A_{51} & & & & 0 \\ & A_{52} & & & \\ & & \cdot & & \\ & 0 & & \cdot & \\ & & & \cdot & \\ & & & & A_{5m} \end{bmatrix}$$

where $A_{51}, A_{52}, \ldots, A_{5m}$ are Jordan blocks, $m \leq r$. The remarks made previously concerning the corresponding discrete-time system models still apply, and Table 3.3 summarizes these results for multiple-input/multiple-output standard forms.

Renumbering state variables as suggested previously will again provide additional relationships between these forms or other new forms.

3.4.5 Adjoint Systems

For systems represented by the homogeneous equations 3.63 and 3.64, it sometimes is useful to define the corresponding adjoint systems. Thus for

$$\frac{d\underline{x}(t)}{dt} = A(t)\underline{x}(t) \qquad t_o \leq t \leq t_1 ,$$

with transition matrix $\Phi(t,t_o)$, the adjoint is

$$\frac{d\underline{x}'(t)}{dt} = -A^T(t)\underline{x}'(t) \qquad t_o \leq t \leq t_1, \tag{3.88}$$

with transition matrix $\psi(t,t_o) = \Phi^T(t_o,t)$. Also, for

$$\underline{x}(n+1) = A(n)\underline{x}(n) \qquad n_o \leq n \leq n_1$$

with transition matrix $\Phi(n,n_o)$, the adjoint is

$$\underline{x}'(n+1) = [A^T(n)]^{-1}\underline{x}'(n), \qquad n_o \leq n \leq n_1 \tag{3.89}$$

with transition matrix $\psi(n,n_o) = \Phi^T(n_o,n)$. A(n) is required to be nonsingular as noted in the derivation of $\Phi(n,n_o)$. Adjoint systems are useful in writing the solutions of linear time-varying differential or difference equations [5.] and they also occur in certain optimization problems.

3.4.6 Dual Systems

Adjoint systems can be used to define dual systems. Starting with Equation 3.88 with $a \leq t_o \leq t \leq t_1 \leq b$ where a and b are constants and $b=t_o+t_1$, we let $\tau=b-t=t_o+t_1-t$. Then

$$\frac{d\underline{x}'(\tau)}{d\tau} = -A^T(\tau)\underline{x}'(\tau), \qquad t_o \leq \tau \leq t_1$$

TABLE 3.3

SUMMARY OF STANDARD STATE VARIABLE FORMS MULTIPLE-INPUT/MULTIPLE-OUTPUT

Continuous-Time and Discrete-Time

		Equation*
Primal Forms	First	3.83 3.84
	Third	3.85 3.86
Jordan Forms	--	3.87

*with $\dot{\underline{x}}$, $\underline{x}$ replaced by $\underline{x}(n+1)$, $\underline{x}(n)$ for discrete-time.

or

$$\frac{d\underline{x}'(b-t)}{d\tau} = -A^T(b-t)\underline{x}'(b-t)$$

or since $d\tau = -dt$,

$$\frac{d\underline{x}'(b-t)}{dt} = A^T(b-t)\underline{x}'(b-t), \qquad t_o \le t \le t_1. \tag{3.90}$$

For $\underline{x}'(b-t) = \underline{x}'(\tau) = \underline{x}(t)$, we have

$$\underline{x}'(b-t_1) = \underline{x}'(t_o) = \underline{x}(t_1)$$

and

$$\underline{x}'(b-t_o) = \underline{x}'(t_1) = \underline{x}(t_o).$$

If instead of $\underline{x}(t_o)$, $\underline{x}(t_1)$ is given and $\underline{x}(t)$, $t_o \le t \le t_1$, is required for the original system 3.63, it would be necessary to integrate "backward". An equivalent operation is to integrate 3.90 "forward" from $\underline{x}'(t_o)=\underline{x}(t_1)$. Furthermore, simulation diagrams of Equation 3.63 and Equation 3.90 suggest the interpretation of interchanging the inputs and outputs to go from an original system to its dual. For the system

$$\frac{d\underline{x}(t)}{dt} = A(t)\underline{x}(t) + B(t)\underline{u}(t) \tag{3.91}$$

$$\underline{y}(t) = C(t)\underline{x}(t)$$

a dual is

$$\frac{d\underline{x}'(b-t)}{dt} = A^T(b-t)\underline{x}'(b-t)+C^T(b-t)\underline{u}'(b-t) \tag{3.92}$$

$$\underline{y}'(b-t) = B^T(b-t)\underline{x}'(b-t).$$

For a stationary system

$$\frac{d\underline{x}}{dt} = A\underline{x} + B\underline{u} \tag{3.93}$$

$$\underline{y} = C\underline{x}$$

a dual is

$$\frac{d\underline{x}^*}{dt} = A^T\underline{x}^* + C^T\underline{u}^* \tag{3.94}$$

$$\underline{y}^* = B^T\underline{x}^*$$

with $\underline{x}^* = \underline{x}'(b-t)$ and similar meanings for $\underline{u}^*$ and $\underline{y}^*$.

Example 3.29

For

$$\dot{\underline{x}} = \begin{bmatrix} 0 & 1 \\ -\alpha_0 & -\alpha_1 \end{bmatrix} \underline{x} + \begin{bmatrix} 0 \\ 1 \end{bmatrix} u$$

$$y = [\beta_0 \quad \beta_1] \; \underline{x}$$

a dual is

$$\dot{\underline{x}}^* = \begin{bmatrix} 0 & -\alpha_0 \\ 1 & -\alpha_1 \end{bmatrix} \underline{x}^* + \begin{bmatrix} \beta_0 \\ \beta_1 \end{bmatrix} u^*$$

$$y^* = [0 \quad 1]\underline{x}^*$$

The simulation diagram for the original system is Figure 3.12

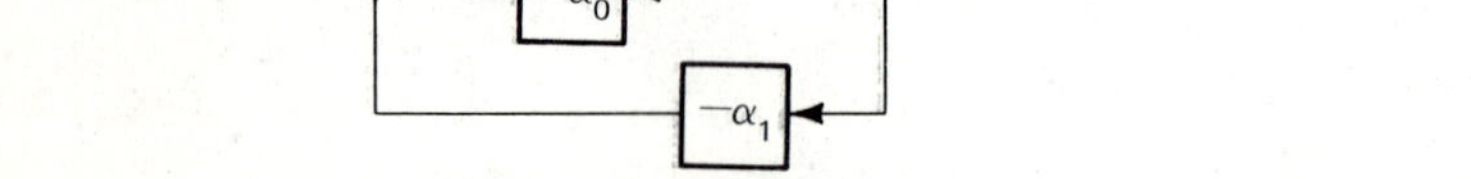

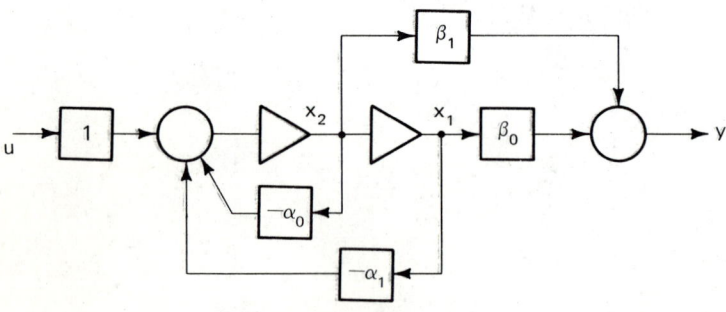

Figure 3.12

and by reversal, we get Figure 3.13

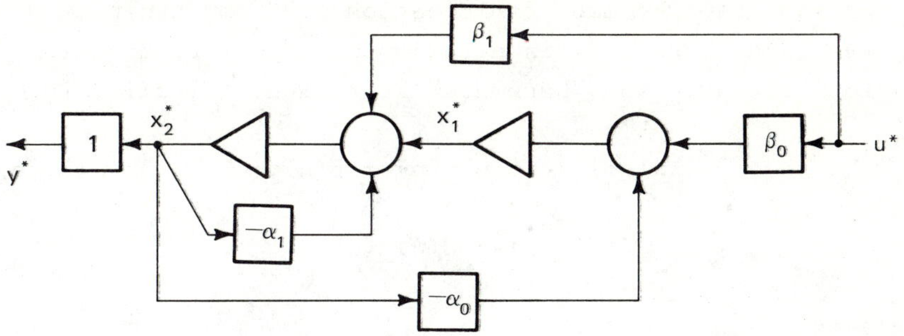

Figure 3.13

which is the simulation diagram of the dual.

The original system in Example 3.29 is in first primal (or phase variable) form and its dual is called dual first primal (or dual phase variable) form. These two forms have been particularly useful as the starting points of certain control and filter designs using feedback gains around single-input/single-output dynamical systems. In fact, all the standard forms previously developed have duals according to the definition above. For discrete-time systems, starting with Equation 3.89 with $a \le n_o \le n \le n_1 \le b$ where a and b are integer constants and $b = n_o + n_1$, let $k = b - n = n_o + n_1 - n$. Then

$$\underline{x}'(k+1) = [A^T(k)]^{-1}\underline{x}'(k), \qquad n_o \le k \le n_1$$

or

$$\underline{x}'(b-n+1) = [A^T(k)]^{-1}\underline{x}'(b-n)$$

or

$$\underline{x}'(b-n) = [A^T(k)]^{-1}\underline{x}'(b-(n-1)) \ . \qquad (3.95)$$

For $\underline{x}'(b-n) = \underline{x}'(k) = \underline{x}(n)$,

$$\underline{x}'(b-n_1) = \underline{x}'(n_o) = \underline{x}(n_1)'$$

$$\underline{x}'(b-n_o) = \underline{x}'(n_1) = \underline{x}(n_o)$$

and the same remarks made for Equation 3.90 concerning "forward" and "backward" integration now apply to Equation 3.95 for "forward" and "backward" iteration. Finally, for

$$\underline{x}(n+1) = A(n)\underline{x}(n) + B(n)\underline{v}(n) \tag{3.96}$$

$$\underline{z}(n) = C(n)\underline{x}(n)$$

a dual is

$$\underline{x}'(b-n) = A^T(b-n)\underline{x}'(b-(n-1))+C^T(b-n)\underline{v}\ (b-(n-1)) \tag{3.97}$$

$$\underline{z}'(b-n) = B^T(b-n)\underline{x}'(b-n).$$

For a stationary system

$$\underline{x}(n+1) = A\underline{x}(n) + B\underline{v}(n) \tag{3.98}$$

$$\underline{z}(n) = C\underline{x}(n)$$

a dual is

$$\underline{x}^*(n+1) = A^T\underline{x}^*(n) + C^T\underline{v}^*(n) \tag{3.99}$$

$$\underline{z}^*(n) = B^T\underline{x}^*(n)$$

with $\underline{x}^*(n) = \underline{x}'(b-n)$ and similar changes for $\underline{v}^*$ and $\underline{z}^*$. The duals of all four primal forms for single-input/single-output systems can be defined from these.

3.5 CONTROLLABILITY AND OBSERVABILITY

In the introduction to this chapter, several properties of system models were defined: realizability, stationarity, linearity, and stability. Generally, the system models that have been developed, i.e., rth-order forms, impulse and step responses, transfer functions, and state variable models are examined for these properties. Some models are better suited than others for testing for certain properties. For example, the general test for linearity is given in terms of zero-state and zero-input linearity plus decomposition. Realizability and stationarity were defined at the beginning of this chapter in terms of linear operators and these properties also may be verified with other model forms. Stability has been defined earlier as a bounded-input, bounded-output requirement. A more complete discussion of system stability is deferred until Chapter 5 so that the system response results in Chapter 4 can be used. In that way, both zero-input, zero-state, and total stability can be examined in a common framework. Two other system properties are controllability and observability. These must be defined and evaluated relative to state variable models.

The concept of controllability of a linear system is related to the way the input $\underline{u}(t)$ (or $\underline{v}(n)$) is coupled to the internal structure of the system. Thus it might be expected that the A and B matrices must be examined. More specifically, it is important to know whether the input to a system can excite all modes of response of the system. The formal definition below puts this in terms of being able to transfer any initial state $\underline{x}(t_o)\varepsilon R_r$ to another state $\underline{x}(t_1)\varepsilon R_r$, $t_o \le t_1 < \infty$. The concept of observability is related to the way the output $\underline{y}(t)$ (or $\underline{z}(n)$) is coupled to the internal structure of the system. Thus it is apparent that the A and C matrices must be considered. More specifically, it is important to know whether all response modes of the system can be detected by examination of the output. (Note that in general C is not square, and therefore, C may be singular.) The formal definition below puts this in terms of being able to determine an initial state $\underline{x}(t_o)\varepsilon R_r$ by examination of $\underline{y}(t)$, $t_o \le t \le t_1 < \infty$.

In the following, definitions for system controllability and observability are stated first; then a number of theorems are given (without proof) to test for these properties. Although there are many equivalent forms in the literature, only one set of theorems is given to include both time-varying and stationary systems for both continuous-time and discrete-time. For alternatives see, for example, [7.], [8.], [9.], [23.].

Controllability

System S_1: $\dot{\underline{x}}(t) = A(t)\underline{x}(t)+B(t)\underline{u}(t)$

S_1 is completely (state) controllable if for any t_o and $\underline{x}(t_o)\varepsilon R_r$ there exists a finite $t_1 \geq t_o$ and a piecewise-continuous $\underline{u}(t)$, $t_o \leq t \leq t_1$, such that $\underline{x}(t_1) = \underline{0}$.

The terminal state $\underline{x}(t_1) = \underline{0}$ is only for convenience; an appropriate translation could be made for any other $\underline{x}(t_1)\varepsilon R_r$. It also is common to characterize S_1 as the pair $A(t)$, $B(t)$.

Observability

System S_2: $\dot{\underline{x}}(t) = A(t)\underline{x}(t)$

$$\underline{y}(t) = C(t)\underline{x}(t)$$

S_2 is completely observable if for any t_o and $\underline{x}(t_o)\varepsilon R_r$ there exists a finite $t_1 \geq t_o$ such that $\underline{x}(t_o)$ can be determined from

$$\underline{y}(t), \quad t_o \leq t \leq t_1 .$$

This form of S_2 with $\underline{u}(t)=\underline{0}$ for all t is chosen for convenience since if $\underline{u}(t)$ is not zero, but known for $t_o \leq t \leq t_1$, the observability property is not changed. A related property called reconstructability [34.] also has come into use -- particularly where state estimators based on past outputs, i.e., $\underline{y}(t)$, $t \leq t_o$, are to be designed. Observability and reconstructability are equivalent for stationary systems. It also is common to characterize S_2 as the pair A(t), C(t).

Controllability

System S_3: $\underline{x}(n+1) = A(n)\underline{x}(n) + B(n)\underline{v}(n)$

S_3 is completely (state) controllable if for any n_0 and $\underline{x}(n_0) \varepsilon R_r$ there exists a finite $n_1 \geq n_0$ and $\underline{v}(n)$, $n_0 \leq n \leq n_1$, such that $\underline{x}(n_1) = \underline{0}$.

Observability

System S_4: $\underline{x}(n+1) = A(n)\underline{x}(n)$

$$\underline{z}(n) = C(n)\underline{x}(n)$$

S_4 is completely observable if for any n_0 and $\underline{x}(n_0) \varepsilon R_r$ there exists a finite $n_1 \geq n_0$ such that $\underline{x}(n_0)$ can be determined from $\underline{z}(n)$, $n_0 \leq n \leq n_1$.

Since much of this book is concerned with linear, stationary systems, two theorems are given for each of the above definitions. The first in each case is a general theorem for time-varying systems, and the second is a simpler form for stationary systems.

Theorem 3.1:

S_1 is completely controllable if and only if for all t_0 there exists a t_1, $t_0 < t_1 < \infty$, such that the matrix $M(t_0, t_1)$ is nonsingular where

$$M(t_0, t_1) = \int_{t_0}^{t_1} \Phi(t_1, \lambda) B(\lambda) B^T(\lambda) \Phi^T(t_1, \lambda) d\lambda \ .$$

A stronger form of controllability, uniform controllability, requires that there exist a δ such that for all t

$$M(t-\delta, t) \geq \alpha(\delta) I > \Theta$$

where $\alpha(\delta)$ is any arbitrary function which maps from positive real numbers to positive real numbers [9.]. (The notation $A \geq B$, for two matrices, means $(A-B)$ is positive semidefinite -- see Appendix B.)

<u>Theorem 3.2</u>:

$$S_1': \quad \underline{\dot{x}}(t) = A\underline{x}(t) + B\underline{u}(t)$$

S_1' is completely controllable if and only if the controllability matrix P_c has rank r where

$$P_c = [B \vdots AB \vdots \ \ldots \ \vdots A^{r-1}B] \ .$$

<u>Theorem 3.3</u>:

S_2 is completely observable if and only if for all t_o there exists a t_1, $t_o < t_1 < \infty$, such that the matrix $N(t_o, t_1)$ is nonsingular where

$$N(t_o, t_1) = \int_{t_o}^{t_1} \Phi^T(\lambda, t_o) C^T(\lambda) C(\lambda) \Phi(\lambda, t_o) d\lambda \quad .$$

A stronger form of observability, uniform observability, requires that there exist a δ such that for all t,

$$N(t-\delta, t) \geq \beta(\delta) I > \Theta$$

where $\beta(\delta)$ has the properties of $\alpha(\delta)$ in Theorem 3.1.

<u>Theorem 3.4</u>:

$$S_2': \quad \underline{\dot{x}}(t) = A\underline{x}(t)$$
$$\underline{y}(t) = C\underline{x}(t)$$

S_2' is completely observable if and only if the observability matrix P_o has rank r where

$$P_o = [C^T \vdots A^T C^T \vdots \ \ldots \ \vdots (A^T)^{r-1} C^T] \ .$$

Theorem 3.5:

S_3 is completely controllable if and only if for all n_o there exists an n_1, $n_o < n_1 < \infty$, such that the matrix $M(n_o,n_1)$ is nonsingular where

$$M(n_o,n_1) = \sum_{k=n_o}^{n_1-1} \Phi(n_1,k+1)B(k)B^T(k)\Phi^T(n_1,k+1) \ .$$

The stronger form for uniform controllability is analogous to that in Theorem 3.1.

Theorem 3.6:

$$S_3': \qquad \underline{x}(n+1) = A\underline{x}(n) + B\underline{v}(n)$$

S_3' is completely controllable if and only if the controllability matrix P_c has rank r where

$$P_c = [B \vdots AB \vdots \ \ldots \ \vdots A^{r-1}B] \ .$$

Theorem 3.7:

S_4 is completely observable if and only if all n_o there exists an n_1, $n_o < n_1 < \infty$, such that the matrix $N(n_o,n_1)$ is nonsingular where

$$N(n_o,n_1) = \sum_{k=n_o+1}^{n_1} \Phi^T(k,n_o+1)C^T(k)C(k)\Phi(k,n_o+1) .$$

The stronger form for uniform observability is analogous to that in Theorem 3.3.

Theorem 3.8:

$$S_4': \qquad x(n+1) = A\underline{x}(n)$$

$$\underline{z}(n) = C\underline{x}(n)$$

S_4' is completely observable if and only if the observability matrix P_o has rank r where

$$P_o = [C^T \vdots A^T C^T \vdots \ \ldots \vdots (A^T)^{r-1} C^T] \ .$$

The following two examples illustrate the use of the four theorems which apply to stationary systems.

Example 3.30

Consider the second-order discrete-time stationary system for $n \geq n_o = 0$ (A nonsingular):

$$\underline{x}(n+1) = A\underline{x}(n) + \underline{b}v(n) \qquad |A| \neq 0 \ .$$

$$z(n) = \underline{c}^T \underline{x}(n) \ .$$

a) Find a sequence v(n), $0 \leq n \leq n_1$, such that for $\underline{x}(0) \neq \underline{0}$, $\underline{x}(n_1) = \underline{0}$. First, if $n_1 = 1$,

$$\underline{x}(1) = \underline{0} = A\underline{x}(0) + \underline{b}v(0)$$

or

$$\underline{x}(0) = -A^{-1}\underline{b}v(0) = \underline{s}_1 v(0).$$

Thus in the special case where $\underline{x}(0)$ is colinear with $\underline{s}_1$, $n_1 = 1$ and v(0) is chosen to satisfy the equation.

Second, if $n_1 = 2$,

$$\underline{x}(2) = \underline{0} = A\underline{x}(1) + b\underline{v}(1)$$

$$= A^2\underline{x}(0) + A\underline{b}v(0) + \underline{b}v(1)$$

or

$$\underline{x}(0) = -A^{-1}\underline{b}v(0) - A^{-2}\underline{b}v(1)$$

$$= \underline{s}_1 v(0) + \underline{s}_2 v(1).$$

Now $\underline{s}_1$ and $\underline{s}_2$ are in R_2, and if they are linearly independent, they form a basis in R_2. Furthermore, $\underline{s}_1$ and $\underline{s}_2$ are linearly independent if $[\underline{s}_2 \vdots \underline{s}_1] = -[A^{-2}\underline{b} \vdots A^{-1}\underline{b}]$ is rank 2. Now $P_c = [\underline{b} \vdots A\underline{b}]$ is rank 2 if the system is controllable (Theorem 3.6). Then $-A^{-2}P_c = [\underline{s}_2 \vdots \underline{s}_1]$ and $[\underline{s}_1 \vdots \underline{s}_2] = S$ also will be rank 2. Therefore,

$$\underline{x}(0) = S \begin{bmatrix} v(0) \\ v(1) \end{bmatrix}$$

and

$$\begin{bmatrix} v(0) \\ v(1) \end{bmatrix} = S^{-1}\underline{x}(0) .$$

We see that every $\underline{x}(0) \varepsilon R_2$ can be brought to the origin in at most $n_1 = 2$ time increments if the system is controllable, and the sequence v(0), v(1) is found from the equation above.

b) From a sequence z(n), $0 \le n \le n_1$, find $\underline{x}(0)$ if it is known that v(n)=0 for all n. Now

$$z(0) = \underline{c}^T\underline{x}(0)$$

and

$$\underline{x}(1) = A\underline{x}(0)$$

and

$$z(1) = \underline{c}^T\underline{x}(1) = \underline{c}^T A\underline{x}(0),$$

therefore,

$$\begin{bmatrix} z(0) \\ z(1) \end{bmatrix} = \begin{bmatrix} \underline{c}^T \\ \cdots\cdots \\ \underline{c}^T A \end{bmatrix} \underline{x}(0) .$$

The observability matrix P_o is

$$P_o = [C^T \vdots A^T C^T] = [\underline{c} \vdots A^T \underline{c}]$$

and

$$P_o^T = \begin{bmatrix} \underline{c}^T \\ \cdots\cdots \\ \underline{c}^T A \end{bmatrix} = Q \ .$$

If the system is observable (Theorem 3.8), P_o, and consequently Q, are rank 2. We therefore know that $n_1=1$ and

$$\underline{x}(0) = Q^{-1} \begin{bmatrix} z(0) \\ z(1) \end{bmatrix} .$$

We see that $\underline{x}(0)$ can be found from two outputs if the system is observable. The case for $v(n)\neq 0$ is similar. If $v(0)\neq 0$ in this simple example,

$$\underline{x}(0) = Q^{-1} \begin{bmatrix} z(0) \\ z(1) \end{bmatrix} - Q^{-1} \begin{bmatrix} 0 \\ 1 \end{bmatrix} \underline{c}^T \underline{b} v(0) \ .$$

Example 3.31

Consider the system

$$(D^2 + 3D + 2)y(t) = (0.5D + 1)u(t).$$

a) Using Eq. 3.68, a state variable model (first primal) is:

$$\dot{\underline{x}} = \begin{bmatrix} 0 & 1 \\ -2 & -3 \end{bmatrix} \underline{x} + \begin{bmatrix} 0 \\ 1 \end{bmatrix} u$$

$$y = [1 \quad 0.5]\ \underline{x}.$$

Using Theorem 3.2,

$$P_c = [\underline{b} \vdots A\underline{b}] = \begin{bmatrix} 0 & 1 \\ 1 & -3 \end{bmatrix}$$

which has rank 2 and thus this form is controllable.

Using Theorem 3.4,

$$P_o = [\underline{c} \vdots A^T\underline{c}] = \begin{bmatrix} 1 & -1 \\ 0.5 & -0.5 \end{bmatrix}$$

which has rank 1 and thus this form is not observable.

b) Using Eq. 3.71, a state variable model (third primal) is:

$$\dot{\underline{x}}' = \begin{bmatrix} 0 & 1 \\ -2 & -3 \end{bmatrix} \underline{x}' + \begin{bmatrix} 0.5 \\ -0.5 \end{bmatrix} u$$

$$y = [1 \quad 0]\, \underline{x}' .$$

Using Theorem 3.2,

$$P_c = \begin{bmatrix} 0.5 & -0.5 \\ -0.5 & 0.5 \end{bmatrix}$$

which has rank 1 and thus this form is not controllable.

Using Theorem 3.4,

$$P_o = \begin{bmatrix} 1 & 0 \\ 0 & 1 \end{bmatrix}$$

which has rank 2 and this form is observable.

It is apparent that although the A-matrix in a) and b) are the same, the $\underline{b}$ and $\underline{c}^T$ vectors are different. These changes result in different controllability and observability properties for the two models. Inspection of the transfer function

$$H(s) = \frac{0.5s+1}{s^2 + 3s + 2} = \frac{0.5(s+2)}{(s+1)(s+2)}$$

shows that there is a pole-zero cancellation -- a condition that was not allowed in earlier discussions of rth-order forms of system models.

c) If the model in a) is converted to diagonal form, we have

$$\dot{\underline{x}}'' = \begin{bmatrix} -1 & 0 \\ 0 & -2 \end{bmatrix} \underline{x}'' + \begin{bmatrix} 1 \\ -1 \end{bmatrix} u$$

$$y = [0.5 \quad 0]\ \underline{x}''$$

and it is obvious by inspection that the mode associated with $\lambda_2=-2$ cannot be observed, but both modes for $\lambda_1=-1$ and $\lambda_2=-2$ can be controlled.

If the model is b) is converted to diagonal form, we have

$$\dot{x}''' = \begin{bmatrix} -1 & 0 \\ 0 & -2 \end{bmatrix} \underline{x}''' + \begin{bmatrix} 0.5 \\ 0 \end{bmatrix} u$$

$$y = [1 \quad 1]\ \underline{x}'''$$

and it is obvious that the mode associated with $\lambda_2=-2$ cannot be controlled, but both modes for $\lambda_1=-1$ and $\lambda_2=-2$ can be observed.

The results in Example 3.31 suggest that there are other ways of checking for controllability and observability. Two theorems, which apply in the special case of stationary continuous-time or discrete-time systems where all the characteristic values of A are of multiplicity one, are given here [5.]. For the general case of Jordan forms, see [12.].

Theorem 3.9

The pair A,B is not controllable if $B_1 = T^{-1}B$ has one or more rows with all zeros. T is such that

$$\Lambda = T^{-1}AT = \text{diag. } [\lambda_1 \ldots \lambda_j \ldots \lambda_r]$$

and all λ_j are of multiplicity one.

Theorem 3.10

The pair A,C is not observable if $C_1 = CT$ has one or more columns with all zeros. T and the condition on the λ_j of A are as in the previous theorem.

Example 3.31c illustrates these two theorems. The use of Theorems 3.9 and 3.10 immediately shows the modes of a given state model which are not controllable, not observable, or both. In fact, it now can be shown that any state model of a stationary system can be decomposed into four parts: controllable-observable, not controllable-observable, controllable-not observable, and not controllable-not observable. Furthermore, a subspace of R_r spanned by a set of vectors associated with the controllable modes is a controllable subspace and a subspace of R_r spanned by a set of vectors associated with the observable modes is an observable subspace. We consider an original stationary model

$$\dot{\underline{x}} = A\underline{x} + B\underline{u}$$

$$\underline{y} = C\underline{x} + D\underline{u}$$

with a controllability matrix P_c as in Theorem 3.2 and assume P_c has rank $m \leq r$. Then we form a transformation T_1 such that the first m columns are linearly independent columns of P_c and the remaining r-m columns are such that T_1 is nonsingular. For $\underline{x}=T_1\underline{x}'$, the resulting controllability canonical form is

$$\dot{\underline{x}}' = A'\underline{x}' + B'\underline{u} \tag{3.100}$$

$$\underline{y} = C'\underline{x}' + D\underline{u}$$

with

$$A' = T_1^{-1}AT_1 = \begin{bmatrix} A'_{11} & \vdots & A'_{12} \\ \cdots & \cdots & \cdots \\ \Theta & \vdots & A'_{22} \end{bmatrix}$$

$$B' = T_1^{-1}B = \begin{bmatrix} B'_1 \\ \cdots \\ \Theta \end{bmatrix}$$

$$C' = CT_1 = [C'_1 \vdots C'_2]$$

and $\underline{x}'$ is partitioned as

$$\underline{x}' = \begin{bmatrix} \underline{x}'_m \\ \cdots \\ \underline{x}'_{r-m} \end{bmatrix} .$$

The pair (A'_{11}, B'_1) is controllable and $\underline{x}'_m$ is a state vector of a controllable subsystem of the original model. Also, the first m columns of T_1 form a basis of a controllable subspace $R_m \subseteq R_r$. Similarly, if the original system has an observability matrix P_o as in Theorem 3.4 and P_o has rank

$k \leq r$, we form a transformation T_2 such that the first k columns are linearly independent columns of P_o and the remaining r-k columns are such that T_2 is nonsingular. For $\underline{x}=T_2\underline{x}''$, the resulting observability canonical form is

$$\dot{\underline{x}}'' = A''\underline{x}'' + B''\underline{u} \qquad (3.101)$$

$$\underline{y} = C''\underline{x}'' + D\underline{u}$$

with

$$A'' = T_2^{-1}AT_2 = \left[\begin{array}{c|c} A_{11}'' & \Theta \\ \hline A_{21}'' & A_{22}'' \end{array}\right]$$

$$B'' = T_2^{-1}B = \begin{bmatrix} B_{11}'' \\ \cdots\cdots \\ B_{22}'' \end{bmatrix}$$

$$C'' = CT_2 = [C_1'' \ \vdots \ \Theta]$$

and $\underline{x}''$ is partitioned as

$$\underline{x}'' = \begin{bmatrix} \underline{x}_k'' \\ \text{------} \\ \underline{x}_{r-k}'' \end{bmatrix} .$$

The pair (A_{11}'', C_1'') is observable and $\underline{x}_k''$ is a state vector of an observable subsystem of the original model. Also, the first k columns of T_2 form a basis of an observable subspace $R_k \subseteq R_r$. By combining the two transformations, it is possible to decompose the original system into the four subsystems described previously. For example, if the original system is first put in the form of 3.100 and two subsystems

$$\dot{\underline{x}}'_m = A'_{11}\underline{x}'_m + A'_{12}\underline{x}'_{r-m} + B'_1\,\underline{u}$$

$$\underline{y}_o = C'_1\underline{x}'_m + D\underline{u}$$

and

$$\dot{\underline{x}}'_{r-m} = A'_{22}\underline{x}'_{r-m}$$

$$\underline{y}_1 = C'_2\underline{x}'_{r-m}$$

are defined with $\underline{y}=\underline{y}_o+\underline{y}_1$, a transformation T_3 formed from an observability matrix P'_o of (A'_{11}, C'_1) reduces the controllable subsystem to observable and nonobservable parts and a transformation T_4, formed from an observability matrix P''_o of (A'_{22}, C'_2), reduces the noncontrollable subsystem to observable and nonobservable parts. The resulting canonical form is

$$\begin{bmatrix} \dot{\underline{x}}_1 \\ \dot{\underline{x}}_2 \\ \dot{\underline{x}}_3 \\ \dot{\underline{x}}_4 \end{bmatrix} = \left[\begin{array}{c|c|c|c} A_{11} & \Theta & A_{13} & \Theta \\ \hline A_{21} & A_{22} & A_{23} & A_{24} \\ \hline \Theta & \Theta & A_{33} & \Theta \\ \hline \Theta & \Theta & \Theta & A_{44} \end{array}\right] \begin{bmatrix} \underline{x}_1 \\ \underline{x}_2 \\ \underline{x}_3 \\ \underline{x}_4 \end{bmatrix} + \begin{bmatrix} B_1 \\ B_2 \\ \Theta \\ \Theta \end{bmatrix} \underline{u} \tag{3.102}$$

$$\underline{y} = [C_1 \vdots \Theta \vdots C_3 \vdots \Theta] \begin{bmatrix} \underline{x}_1 \\ \hline \underline{x}_2 \\ \hline \underline{x}_3 \\ \hline \underline{x}_4 \end{bmatrix} + D\underline{u}$$

where

$\underline{x}_1$ is the state of a controllable/observable subsystem

$\underline{x}_2$ is the state of a controllable/nonobservable subsystem

$\underline{x}_3$ is the state of a noncontrollable/observable subsystem

$\underline{x}_4$ is the state of a noncontrollable/nonobservable subsystem.

A generalization for time-varying systems is given in [12.], and the corresponding forms for discrete-time systems are obvious.

A third point of view on controllability and observability is associated with pole/zero cancellations in the transfer functions or elements of the transfer matrices defined in Section 3.3. The results are given here and are discussed further in the next section on system model conversions where the relationships between A,B,C,D and H(s) or H(z) are developed. The results are applicable to stationary systems.

<u>Theorem 3.11</u>

S_1' of Theorem 3.2 is completely controllable if and only if there are no cancellations in $\Phi(s)B$.

<u>Theorem 3.12</u>

S_2' of Theorem 3.4 is completely observable if and only if there are no cancellations in $C\Phi(s)$.

<u>Theorem 3.13</u>

S_3' of Theorem 3.6 is completely controllable if and only if there are no cancellations in $z^{-1}\Phi(z)B$.

<u>Theorem 3.14</u>

S_4' of Theorem 3.8 is completely observable if and only if there are no cancellations in $Cz^{-1}\Phi(z)$.

<u>Example 3.32</u>

The system of Example 3.31 is

$$(D^2+3D+2)y(t) = (0.5\ D+1)u(t)$$

and

$$A = \begin{bmatrix} 0 & 1 \\ -2 & -3 \end{bmatrix}$$

for both first and third primal forms. Thus we have

$$\Phi(s) = [sI-A]^{-1} = \begin{bmatrix} s & -1 \\ 2 & s+3 \end{bmatrix}^{-1}$$

$$= \frac{1}{(s+1)(s+2)} \begin{bmatrix} s+2 & 1 \\ -2 & 5 \end{bmatrix} .$$

a) For the first primal form

$$\underline{b} = \begin{bmatrix} 0 \\ 1 \end{bmatrix} , \qquad \underline{c}^T = [1 \quad 0.5] .$$

Therefore,

$$\Phi(s)\underline{b} = \frac{1}{(s+1)(s+2)} \begin{bmatrix} s+3 & 1 \\ -2 & s \end{bmatrix} \begin{bmatrix} 0 \\ 1 \end{bmatrix}$$

$$= \frac{1}{(s+1)(s+2)} \begin{bmatrix} 1 \\ s \end{bmatrix}$$

and the model is controllable. Also,

$$\underline{c}^T\Phi(s) = \frac{[1 \quad 0.5]}{(s+1)(s+2)} \begin{bmatrix} s+3 & 1 \\ -2 & s \end{bmatrix}$$

$$= \frac{[(s+2) \quad 0.5(s+2)]}{(s+1)(s+2)}$$

and the model is not observable.

b) For the third primal form

$$\underline{b} = \begin{bmatrix} 0.5 \\ -0.5 \end{bmatrix} \qquad \underline{c}^T = [1 \quad 0] .$$

Therefore,

$$\Phi(s)\underline{b} = \frac{1}{(s+1)(s+2)} \begin{bmatrix} s+3 & 1 \\ -2 & s \end{bmatrix} \begin{bmatrix} 0.5 \\ -0.5 \end{bmatrix}$$

$$= \frac{1}{(s+1)(s+2)} \begin{bmatrix} 0.5(s+2) \\ -0.5(s+2) \end{bmatrix}$$

and the model is not controllable. Also,

$$\underline{c}^T \ (s) = \frac{[1 \quad 0]}{(s+1)(s+2)} \begin{bmatrix} s+3 & 1 \\ -2 & s \end{bmatrix}$$

$$= \frac{[(s+3) \quad 1]}{(s+1)(s+2)}$$

and the model is observable.

Example 3.33

Consider the system

$$(E^2-0.75E + 0.125)z(n) = (E-0.5)v(n)$$

and

$$A = \begin{bmatrix} 0 & 1 \\ -0.125 & 0.75 \end{bmatrix}$$

for both first and third primal forms. Thus we have

$$z^{-1}\Phi(z) = [zI-A]^{-1} = \begin{bmatrix} z & -1 \\ 0.125 & z-0.75 \end{bmatrix}^{-1}$$

$$= \frac{1}{(z-0.5)(z-0.25)} \begin{bmatrix} z-0.75 & 1 \\ -0.125 & z \end{bmatrix} .$$

a) For the first primal form

$$\underline{b} = \begin{bmatrix} 0 \\ 1 \end{bmatrix} , \quad \underline{c}^T = [-0.5 \quad 1] .$$

Therefore,

$$z^{-1}\Phi(z)\underline{b} = \frac{1}{(z-0.5)(z-0.25)} \begin{bmatrix} 1 \\ z \end{bmatrix}$$

and the model is controllable. Also,

$$\underline{c}^T z^{-1} (z) = \frac{[-0.5(z-0.5) \quad (z-0.5)]}{(z-0.5)\ (z-0.25)}$$

and the model is not observable.

b) For the third primal form

$$\underline{b} = \begin{bmatrix} 1 \\ 0.25 \end{bmatrix} \qquad \underline{c}^T = [1 \quad 0] .$$

Therefore,

$$z^{-1}\Phi(z)\underline{b} = \frac{1}{(z-0.5)(z-0.25)} \begin{bmatrix} (z-0.5) \\ 0.25(z-0.5) \end{bmatrix}$$

and the model is not controllable. Also,

$$\underline{c}^T z^{-1}\Phi(z) = \frac{[(z-0.75) \quad 1]}{(z-0.5)(z-0.25)}$$

and the model is observable.

From Theorems 3.11, 3.12, 3.13, and 3.14 it now is clear that the restrictions on pole/zero cancellations used in developing the models of Eq. 3.78 and 3.81 were necessary to be sure that there were no noncontrollable or nonobservable modes. A transfer function without pole/zero cancellations or a transfer matrix with no cancellations in any of its elements represents only the controllable-observable part of a system. The other three parts, if they exist, are included in the general state variable model. It also is clear from these theorems and Examples 3.32 and 3.33 why the first and fourth primal forms are controllable and the second and third are observable. No cancellations can occur for $\underline{b}^T = [0 \ \dots \ 0 \ \ 1]$ or $\underline{c}^T = [1 \ \ 0 \ \ \dots \ \ 0]$, and, of course, no cancellations occur in $\Phi(s)$ or $\Phi(z)$ when the state variables are chosen properly as a minimal set. An alternative point of view which shows this is to consider the controllability and observability matrices of these forms. For example, for Eq. 3.68, where X indicates possible nonzero elements, we have

$$P_c = \begin{bmatrix} 0 & \dots & & & 0 & 1 \\ 0 & \dots & & 0 & 1 & X \\ 0 & \dots & 0 & 1 & X & X \\ & & \vdots & & & \\ 0 & 1 & X & \dots & X & X \\ 1 & X & X & \dots & X & X \end{bmatrix}_{r\times r}$$

and the rank of P_c is r and the single-input first primal form is controllable.

Another important comment that can be made relates to system stability. Earlier in this chapter, system stability was defined as a bounded input-bounded output condition. Although stability is considered in much greater detail in Chapter 5, it should be apparent that modes which are not observable may cause difficulty. In particular, modes associated with characteristic values with positive real parts (for continuous-time systems) or magnitudes greater than one (for discrete-time systems) may become unbounded when an input or a nonzero initial state occurs. If these modes are not observable, a test for system stability based on an examination of the system output will be misleading.

Finally, in concluding this section on controllability and observability, the concept of <u>duality</u> is introduced. Kalman noticed the duality between these two properties, and it sometimes is convenient to change a problem of one on controllability to one on observability or vice versa. A formal statement is as follows:

Theorem 3.15

a) The system of 3.93 is controllable (observable) if, and only if, the system of 3.94 is observable (controllable).
b) The system of 3.98 is controllable (observable) if, and only if, the system of 3.99 is observable (controllable).

Examination of P_c and P_o for the dual systems shows the validity of the principle.

3.6 SYSTEM MODEL CONVERSIONS

3.6.1 Summary of System Models

Several linear system models have now been examined, and this section is concerned with ways of going from one model form to another when the systems are stationary. Table 3.4 summarizes the forms and the defining equations. The two primary forms are the rth-order forms of linear differential or difference equations (as characterized by the $M(\cdot)$ and $N(\cdot)$ polynomials or the polynomial or rational matrices $M(\cdot)$, $N(\cdot)$, and $P(\cdot)$), and state variable forms (as characterized by the A, B, C, D matrices). Several response functions have been derived from the rth-order forms by choosing various elementary functions for a spectral model of the input signal and then solving for the corresponding system output. The response function for the state variable models is the transition matrix -- a zero input response.

There are three types of model conversions which are of interest here. The first is conversion from an rth-order model to a state variable model or vice versa. The second is a conversion from a state variable model to a specified standard form state variable model. Conversions are needed for single input/output and multiple input/output for both

TABLE 3.4

LINEAR SYSTEM MODELS FOR CONTINUOUS- AND DISCRETE-TIME

A. Primary Forms - Stationary Systems

Single input/output	M(D),N(D), Eq. 3.19 A,B,C,D, Table 3.2	M(E),N(E), Eq. 3.21 A,B,C,D, Table 3.2
Multiple input/output	M(D),N(D), P(D), Eq. 3.51 A,B,C,D, Eq. 3.57, Table 3.3	M(E),N(E), P(E), Eq. 3.54 A,B,C,D, Eq. 3.60, Table 3.3

B. Response Functions - Stationary Systems

Single input/output	$F_c(t,k)$, Eq. 3.22 $h(\tau)$, Eq. 3.30 $a(\tau)$, Eq. 3.34 $R(t,s)$, Eq. 3.47 $H(s)$, Eq. 3.48 $\Phi(t-t_0)$, Eq. 3.65	$F_d(n,k)$, Eq. 3.24 $h(j)$, Eq. 3.40 $a(j)$, Eq. 3.44 $R(n,z)$, Eq. 3.49 $H(z)$, Eq. 3.50 $\Phi(n-n_0)$, Eq. 3.66
Multiple input/output	$H(\tau)$, Eq. 3.53 $H(s)$, Eq. 3.52 $\Phi(t-t_o)$, Eq. 3.65	$H(j)$, Eq. 3.56 $H(z)$, Eq. 3.55 $\Phi(n-n_0)$, Eq. 3.66

continuous- and discrete-time systems and these cases are summarized in Part A of Table 3.4. Third, conversions from a continuous-time to an approximating discrete-time system (or vice versa) is useful for computer analysis of linear system response.

3.6.2 Conversion between rth-Order and State Forms

For the first type of conversion, it frequently happens

that the system transfer function or transfer matrix is given rather than an rth-order form and the A, B, C, D matrices are required (or vice versa). Although the transfer functions or matrices are related to M, N, and P by Equations 3.48, 3.49, 3.52, and 3.55, there will be differences between the two if any pole/zero cancellations occur. By Theorems 3.11, 3.12, 3.13, and 3.14, this means that the state model is not controllable or not observable or both. Thus, if H(·) is the original model (with no common poles and zeros), the state model will be controllable and observable. If the original model is a state variable form, cancellations may occur and H(·) may represent a reduced number of system response modes -- those that are controllable and observable. Examples 3.31, 3.32, and 3.33 illustrate this case.

Conversion I. r-th Order Form to State Variable (Single Input/Output).

When M(D), N(D) (or M(E), N(E)) are known, the α_i, i=1,...,r, and the β_j, j=1,...,r may be substituted directly into Equations 3.68, 3.69, 3.71, and 3.73 for the <u>primal forms</u>. By taking H(s) = N(s)/M(s) (or H(z) = N(z)/M(z)) and cancelling common poles and zeros, the part of the original system which is controllable and observable may put in <u>diagonal</u> or <u>Jordan form</u> as in Equations 3.78 and 3.80 or in <u>cascade form</u> as in Equation 3.81. An alternative for finding the diagonal and Jordan forms, which does not use the transfer function, requires finding any of the primal forms directly and then using the methods of Appendix B. The uncontrollable and/or unobservable modes of the system are included in this manner.

Conversion II. State Variable to rth-Order Form (Single input/output).

The derivations of the following results are given in Chapter 4. If A, $\underline{b}$, $\underline{c}^T$, and d are given, the unit impulse response is

$$h(\tau) = \underline{c}^T \Phi(\tau) \underline{b} u_1(\tau) + d u_o(\tau) \tag{3.103}$$

with $\Phi(\tau) = \varepsilon^{A\tau}$, and the unit pulse response

$$h(j) = \underline{c}^T \Phi(j-1) \underline{b} v_1(j-1) + d v_o(j) \tag{3.104}$$

with $\Phi(j) = A^j$. Also,

$$H(s) = \mathcal{L}\, h(\tau) = \underline{c}^T \Phi(s) \underline{b} + d \tag{3.105}$$

and

$$H(z) = \mathcal{Z}\, h(j) = \underline{c}^T z^{-1} \Phi(z) \underline{b} + d \tag{3.106}$$

with $\Phi(s) = \mathcal{L}\,\Phi(\tau) = [sI-A]^{-1}$ and $\Phi(z) = \mathcal{Z}\,\Phi(j) = z[zI-A]^{-1}$. The $M(\cdot)$ and $N(\cdot)$ polynomials can then be found from the transfer functions (common poles and zeros should not be cancelled). The meaning of Theorems 3.11 through 3.14 is also clear from these forms. An alternative to this procedure is to transform the original system into first or second primal form as shown in Conversion V. below. The coefficients α_i and β_j of $M(\cdot)$ and $N(\cdot)$ can then be found directly from the new matrices.

Conversion III. Transfer Matrix to State Variable.

This particular conversion has been the subject of much research over the past decade, and it frequently is referred to as the "realization problem". The extension of Conversion I. to this multivariable case is complicated by the fact that the state dimension of the system is not apparent by inspection of the transfer matrix H(s). It is assumed that a minimal dimension state model is required. Several methods that start with a transfer matrix H(s) have been suggested. In [25.][26.][27.], forms are developed from a partial fraction expansion and these lead to Jordan forms in analogy to Conversion I. (Eq. 3.78 and 3.80). In [28.][29.], use is made of a Smith-McMillan construction to develop phase-variable forms, and [30.] gives a method suitable for digital computation. The first of these methods [25.] is discussed here to demonstrate one way of making Conversion III. For a pxq transfer matrix H(·) with m distinct poles and with elements h_{ij} that have no multiple poles, a partial fraction expansion is obtained first. Thus, for a continuous-time system,

$$H(s) = \sum_{i=1}^{m} \frac{K_i}{s-\lambda_i} + D \tag{3.107a}$$

with

$$K_i = \lim_{s \to \lambda_i} (s-\lambda_i)H(s)$$

and

$$D = \lim_{s \to \infty} H(s) \quad ,$$

for a proper system. For r_i = rank K_i, we find

$$r = \sum_{i=1}^{m} r_i$$

is the minimal order [25.] or the minimal dimension of $\underline{x}$. To find A, B, and C in the model

$$\dot{\underline{x}} = A\underline{x} + B\underline{u}$$

$$\underline{y} = C\underline{x} + D\underline{u}$$

it is necessary to find a factorization of the K_i such that

$$K_i = E_i F_i$$

where E_i is $p x r_i$, F_i is $r_i x q$, and rank E_i = rank $F_i = r_i$. One way of doing this is to choose r_i linearly independent columns of K_i to form E_i. Then

$$F_i = (E_i^T E_i)^{-1} E_i^T K_i ,$$

and

$$K_i = \sum_{j=1}^{r_i} \underline{e}_{ji} \underline{f}_{ij}^T .$$

Substituting K_i into Eq. 3.107a,

$$H(s) = \sum_{i=1}^{m} \sum_{j=1}^{r_i} \frac{\underline{e}_{ji} \underline{f}_{ij}^T}{s-\lambda_i} + D. \tag{3.107b}$$

Now consider the forms

$$A = \begin{bmatrix} \lambda_1 I_1 & & & & 0 \\ & \lambda_2 I_2 & & & \\ & & \cdot & & \\ & 0 & & \cdot & \\ & & & & \cdot \\ & & & & \lambda_m I_m \end{bmatrix} \tag{3.108}$$

$$B = \begin{bmatrix} F_1 \\ F_2 \\ \cdot \\ \cdot \\ \cdot \\ F_m \end{bmatrix}$$

$$C = [E_1 \quad E_2 \quad \ldots \quad E_m],$$

where the submatrices are dimensioned according to the r_i, e.g., I_1 is the $r_1 x r_1$ identity matrix, F_1 is $r_1 xq$, and E_1 is pxr_1. Using 3.111 in Conversion IV,

$$H(s) = C[sI-A]^{-1}B + D$$

$$= [E_1 \dots E_m] \begin{bmatrix} (s-\lambda_1)I_1 & & & \\ & \cdot & & \\ & & \cdot & \\ & & & (s-\lambda_m)I_m \end{bmatrix}^{-1} \begin{bmatrix} F_1 \\ \cdot \\ \cdot \\ \cdot \\ F_m \end{bmatrix} + D$$

$$= E_1 \frac{I_1}{(s-\lambda_1)} F_1 + \dots + E_m \frac{I_m}{(s-\lambda_m)} F_m + D$$

which is the same as 3.107b.

Example 3.34

a) For

$$H(s) = \begin{bmatrix} \frac{4s+6}{(s+1)(s+2)} & \frac{1}{s+1} \end{bmatrix}$$

$$D = \lim_{s\to\infty} H(s) = 0$$

$$K_1 = \lim_{s\to -1} (s+1)H(s) = [2 \quad 1]$$

$$K_2 = \lim_{s\to -2} (s+2)H(s) = [2 \quad 0]$$

and $r_1 = r_2 = 1$, thus $r=2$, the minimal dimension. Now let

$$E_1 = [2], \text{ therefore, } F_1 = [1 \quad 1/2]$$

and

$$E_2 = [2], \text{ therefore, } F_2 = [1 \quad 0].$$

Using 3.108,

$$A = \begin{bmatrix} -1 & 0 \\ 0 & -2 \end{bmatrix}$$

$$B = \begin{bmatrix} F_1 \\ F_2 \end{bmatrix} = \begin{bmatrix} 1 & 1/2 \\ 1 & 0 \end{bmatrix}$$

$$C = [E_1 \quad E_2] = [2 \quad 2]$$

for a realization of H(s).

b) For (see [25.])

$$H(s) = \begin{bmatrix} \frac{1}{s+1} & \frac{2}{s+1} \\ \frac{-1}{(s+1)(s+2)} & \frac{1}{s+2} \end{bmatrix}$$

$$D = 0$$

$$K_1 = \lim_{s \to -1} (s+1)H(s) = \begin{bmatrix} 1 & 2 \\ -1 & 0 \end{bmatrix}$$

$$K_2 = \lim_{s \to -2} (s+2)H(s) = \begin{bmatrix} 0 & 0 \\ 1 & 1 \end{bmatrix}$$

and $r_1=2$, $r_2=1$, and $r=3$. Let

$$E_1 = \begin{bmatrix} 1 & 1 \\ -1 & 0 \end{bmatrix}$$

then

$$F_1 = \left[\begin{bmatrix} 1 & -1 \\ 1 & 0 \end{bmatrix} \begin{bmatrix} 1 & 1 \\ -1 & 0 \end{bmatrix}\right]^{-1} \begin{bmatrix} 1 & -1 \\ 1 & 0 \end{bmatrix} \begin{bmatrix} 1 & 2 \\ -1 & 0 \end{bmatrix}$$

$$= \begin{bmatrix} 1 & 0 \\ 0 & 2 \end{bmatrix}$$

and let

$$E_2 = \begin{bmatrix} 0 \\ 1 \end{bmatrix}$$

then

$$F_2 = [1 \quad 1] .$$

Using 3.108 we get

$$A = \begin{bmatrix} -1 & 0 & 0 \\ 0 & -1 & 0 \\ 0 & 0 & -2 \end{bmatrix}$$

$$B = \begin{bmatrix} 1 & 0 \\ 0 & 2 \\ 1 & 1 \end{bmatrix}$$

$$C = \begin{bmatrix} 1 & 1 & 0 \\ -1 & 0 & 1 \end{bmatrix}$$

for a realization of H(s). Note that $r_1=2$ implies that the multiplicity of λ_1 is 2.

There are two remaining details. If any of the poles of the elements $h_{ij}(s)$ is of multiplicity greater than one, an alternative method [26.] must be used. Also for discrete-time systems, H(z) is treated exactly as H(s) in the preceding development to find a realization as

$$\underline{x}(n+1) = A\underline{x}(n) + B\underline{v}(n)$$

$$\underline{z}(n) = C\underline{x}(n) + D\underline{v}(n)$$

Conversion IV. State Variable to Transfer Matrix.

The derivations of the following results are given in Chapter 4. If A, B, C, and D are given, the unit impulse response matrix is

$$H(\tau) = C\Phi(\tau)Bu_1(\tau) + Du_0(\tau) \tag{3.109}$$

with $\Phi(\tau) = \varepsilon^{A\tau}$ and the unit pulse response matrix is

$$H(j) = C\Phi(j-1)Bv_1(j-1) + Dv_0(j) \tag{3.110}$$

with $\Phi(j) = A^j$. Also, the transfer matrices are

$$H(s) = C\Phi(s)B + D \tag{3.111}$$

and

$$H(z) = Cz^{-1}\Phi(z)B + D\ . \tag{3.112}$$

If the analogy to the single input/single output Conversions I. and II. is to be complete, the multivariable forms must also be related to the M(·), N(·) and P(·) matrices of Equations 3.51 and 3.54. To complete the ideas of Conversion III, the conversion from M(·), N(·), and P(·) to a transfer matrix H(·) is given by Equations 3.52 and 3.55. The direct conversion from M(·), N(·), and P(·) to a state model, i.e., to A, B, C, and D, also has been studied in [31.][32.][33.], and direct construction is given in [32.]. To complete the ideas of Conversion IV, a conversion from H(·) to M(·), N(·), and P(·) is needed. A method for constructing a minimal realization, i.e., one such that the dimension of M(·) is minimal, is given in [33.]. With these additions, the possibilities are summarized in Figure 3.14 for both single input/output and multivariable models:

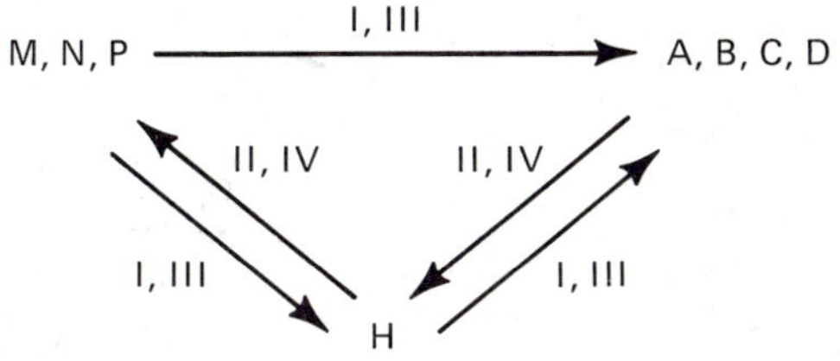

Figure 3.14

We note that the multivariable Conversion III from M(·), N(·), and P(·) to A, B, C, and D includes the conversion of sets of independent loop current equations or node voltage equations of an electric network (see Sections 3.3 and 6.3) to state variable form.

3.6.3 Conversion between State Forms

Conversion V. State Variable to a Standard Form (Single input/output).

In this case there are five standard forms of interest: Jordan form, and first, second, third, and fourth primal forms. The original model must be controllable to have a first or fourth form and observable to have a second or third form. In all five cases, the original A, $\underline{b}$, $\underline{c}^T$ and d are transformed using a linear transformation T where $\underline{x}=T\underline{x}'$ as in Eq. 3.75.

The conversion to Jordan form (including the case of a diagonal form) is done by choosing columns of T as characteristic vectors (or generalized characteristic vectors) of A as shown in Appendix B. The conversion to first primal form results from choosing

$$T = P_c M_\alpha , \tag{3.113}$$

where P_c is the controllability matrix of $(A,\underline{b})$ and

$$M_\alpha = \begin{bmatrix} \alpha_1 & \alpha_2 & \alpha_3 & \cdots & \alpha_r \\ \alpha_2 & \alpha_3 & \cdots & \cdot & \\ \alpha_3 & \vdots & \cdot & & \\ \vdots & \cdot & & & 0 \\ \alpha_r & & & & \end{bmatrix} .$$

The α_j, j=1,2,...,r, are coefficients of $p(\lambda)$, the characteristic polynomial of A. The conversion to second primal form results from choosing

$$T = (N_\alpha P_o^T)^{-1} \tag{3.114}$$

where P_o is the observability matrix of $(A,\underline{c}^T)$ and

$$N_\alpha = \begin{bmatrix} 1 & & & & & \\ \alpha_{r-1} & & \cdot & & 0 & \\ \vdots & & & \cdot & & \\ & & & & \cdot & \\ \alpha_3 & & & & & \\ \alpha_2 & \alpha_3 & & & & \\ \alpha_1 & \alpha_2 & \alpha_3 & \cdots & \alpha_{r-1} & 1 \end{bmatrix} .$$

The conversion to third primal form is made by choosing

$$T = (P_o^T)^{-1} \tag{3.115}$$

and the conversion to fourth primal form is done by choosing

$$T = P_c F \tag{3.116}$$

where F is the transformation of Eq. 3.74.

<u>Example 3.35</u>

For

$$\dot{\underline{x}} = \begin{bmatrix} -3 & -2 & -1 \\ 0 & -1 & 0 \\ 0 & 0 & -2 \end{bmatrix} \underline{x} + \begin{bmatrix} 0 \\ 1 \\ 1 \end{bmatrix} u$$

$$y = [1 \quad 2 \quad 0]\ \underline{x}$$

$$P_c = \begin{bmatrix} 0 & -3 & 13 \\ 1 & -1 & 1 \\ 1 & -2 & 4 \end{bmatrix}, \quad \det P_c = -4$$

$$P_o = \begin{bmatrix} 1 & -3 & 9 \\ 2 & -4 & 10 \\ 0 & -1 & 5 \end{bmatrix}, \quad \det P_o = 2\ .$$

The system is, therefore, controllable and observable.

a) Convert to Jordan form

$$C(\lambda) = \begin{bmatrix} (\lambda+1)(\lambda+2) & -2(\lambda+2) & -(\lambda+1) \\ 0 & (\lambda+2)(\lambda+3) & 0 \\ 0 & 0 & (\lambda+1)(\lambda+3) \end{bmatrix}$$

and

$$p(\lambda) = \lambda^3 + 6\lambda^2 + 11\lambda + 6$$

with

$$\lambda_1 = -3, \ \lambda_2 = -1, \ \lambda_3 = -2 \ .$$

Using

$$C(\lambda_1), \ C(\lambda_2), \ C(\lambda_3),$$

let

$$T = \begin{bmatrix} 2 & -2 & 1 \\ 0 & 2 & 0 \\ 0 & 0 & -1 \end{bmatrix},$$

then

$$A' = \begin{bmatrix} -3 & 0 & 0 \\ 0 & -1 & 0 \\ 0 & 0 & -2 \end{bmatrix}$$

$$\underline{b}' = \begin{bmatrix} 1 \\ \frac{1}{2} \\ -1 \end{bmatrix}$$

$$\underline{c}'^T = [2 \quad 2 \quad 1] \ .$$

b) Convert to first primal form

From $p(\lambda)$, $\alpha_o=6$, $\alpha_1=11$, $\alpha_2=6$, $\alpha_3=1$ and

$$M_\alpha = \begin{bmatrix} 11 & 6 & 1 \\ 6 & 1 & 0 \\ 1 & 0 & 0 \end{bmatrix}$$

and we obtain

$$T = P_c M_\alpha = \begin{bmatrix} -5 & -3 & 0 \\ 6 & 5 & 1 \\ 3 & 4 & 1 \end{bmatrix},$$

then

$$A' = \begin{bmatrix} 0 & 1 & 0 \\ 0 & 0 & 1 \\ -6 & -11 & -6 \end{bmatrix}$$

$$\underline{b}' = \begin{bmatrix} 0 \\ 0 \\ 1 \end{bmatrix}$$

$$\underline{c}'^T = [7 \quad 7 \quad 2] .$$

Thus we have $\beta_o=7$, $\beta_1=7$, $\beta_2=2$.

c) Convert to second primal form

$$N_\alpha = \begin{bmatrix} 1 & 0 & 0 \\ 6 & 1 & 0 \\ 11 & 6 & 1 \end{bmatrix}$$

and

$$T^{-1} = N_\alpha P_o^T = \begin{bmatrix} 1 & 2 & 0 \\ 3 & 8 & -1 \\ 2 & 8 & -1 \end{bmatrix}$$

then

$$A' = \begin{bmatrix} -6 & 1 & 0 \\ -11 & 0 & 1 \\ -6 & 0 & 0 \end{bmatrix}$$

$$\underline{b}' = \begin{bmatrix} 2 \\ 7 \\ 7 \end{bmatrix}$$

$$\underline{c}'^T = [1 \quad 0 \quad 0] \; .$$

d) Convert to third primal form

P_o is given at the beginning of the example, and

$$(P_o^T)^{-1} = T = \begin{bmatrix} -5 & -5 & -1 \\ 3 & 5/2 & 1/2 \\ 3 & 4 & 1 \end{bmatrix}$$

then

$$A' = \begin{bmatrix} 0 & 1 & 0 \\ 0 & 0 & 1 \\ -6 & -11 & -6 \end{bmatrix}$$

$$\underline{b}' = \begin{bmatrix} 2 \\ -5 \\ 15 \end{bmatrix}$$

$$\underline{c}'^T = [1 \quad 0 \quad 0]$$

Conversion VI. State Variable to a Standard Form (multiple input/output).

The construction of a transformation T to bring a multiple input/output model to a Jordan form as in 3.87 is discussed in Appendix B. The construction of T to bring the model into one of the forms 3.83, 3.84, 3.85, or 3.86 is considerably more complex than the analogous case in Conversion V. Basically, the independent columns or rows of P_c or P_o are used, but the details require a rather lengthy development. Therefore, the methods are illustrated with an example here, and the references cited should be consulted for the general constructions.

Example 3.36

For the system

$$\dot{\underline{x}} = \begin{bmatrix} -3 & -2 & -1 \\ 0 & -1 & 0 \\ 0 & 0 & -2 \end{bmatrix} \underline{x} + \begin{bmatrix} -1 & 0 \\ 2 & 0 \\ 0 & 1 \end{bmatrix} \underline{u}$$

$$\underline{y} = \begin{bmatrix} 0 & 1 & 1 \\ 1 & 1 & 1 \end{bmatrix} \underline{x}$$

$$P_c = \begin{bmatrix} -1 & 0 & -1 & -1 & 1 & 5 \\ 2 & 0 & -2 & 0 & 2 & 0 \\ 0 & 1 & 0 & -2 & 0 & 4 \end{bmatrix}$$

and

$$P_o = \begin{bmatrix} 0 & 1 & 0 & -3 & 0 & 9 \\ 1 & 1 & -1 & -3 & 1 & 9 \\ 1 & 1 & -2 & -3 & 4 & 9 \end{bmatrix} .$$

These are both rank = 3 and the system is, therefore, controllable and observable.

a) Using the method of Reference [22.] to find form 3.83

$$T = \begin{bmatrix} 0 & -5 & -1 \\ 0 & 6 & 2 \\ 1 & 0 & 0 \end{bmatrix}$$

and using this,

$$A_1 = T^{-1}AT = \begin{bmatrix} -2 & 0 & 0 \\ 0.5 & 0 & 1 \\ -0.5 & -3 & -4 \end{bmatrix}$$

and

$$B_1 = T^{-1}B = \begin{bmatrix} 0 & 1 \\ 0 & 0 \\ 1 & 0 \end{bmatrix} .$$

b) Using the method of Reference [19.] to find form 3.84,

$$T^{-1} = \begin{bmatrix} -2 & -1 & 0 \\ 1 & -2 & -8 \\ 6 & 5 & 2 \end{bmatrix}$$

and

$$T = \frac{1}{38} \begin{bmatrix} -44 & -2 & -8 \\ 50 & 4 & 16 \\ 7 & -4 & 3 \end{bmatrix}$$

and from these,

$$A_2 = T^{-1}AT = \begin{bmatrix} 0 & 0 & 1 \\ 0.97 & -1.84 & 0.13 \\ -3.37 & 0.21 & -4.16 \end{bmatrix}$$

and

$$T^{-1}B = \begin{bmatrix} 0 & 0 \\ \hline 3 & -8 \\ 4 & 2 \end{bmatrix} = \begin{bmatrix} --- \\ F \end{bmatrix}$$

$$B_2 = T^{-1}BF^{-1} = \begin{bmatrix} 0 & 0 \\ 1 & 0 \\ 0 & 1 \end{bmatrix},$$

where the new input $\underline{u}' = F\underline{u}$.

c) Using the method of Reference 20. to find form 3.85

$$T^{-1} = \begin{bmatrix} 0 & 1 & 1 \\ 0 & -1 & -2 \\ 1 & 1 & 1 \end{bmatrix}$$

$$T = \begin{bmatrix} -1 & 0 & 1 \\ 2 & 1 & 0 \\ -1 & -1 & 0 \end{bmatrix}$$

and using these,

$$A_3 = T^{-1}AT = \begin{bmatrix} 0 & 1 & 0 \\ -2 & -3 & 0 \\ 0 & 0 & -3 \end{bmatrix}$$

$$C_3 = CT = \begin{bmatrix} 1 & 0 & 0 \\ 0 & 0 & 1 \end{bmatrix}.$$

d) Using the method of Reference 21. to find form 3.86

$$T^{-1} = \begin{bmatrix} 0 & 1 & 1 \\ 1 & 1 & 1 \\ 0 & -1 & -2 \end{bmatrix}$$

$$T = \begin{bmatrix} -1 & 1 & 0 \\ 2 & 0 & 1 \\ -1 & 0 & -1 \end{bmatrix}$$

and using these,

$$A_4 = T^{-1}AT = \begin{bmatrix} 0 & 0 & 1 \\ 0 & 3 & 0 \\ -2 & 0 & -3 \end{bmatrix}$$

and

$$C_4 = CT = \begin{bmatrix} 1 & 0 & 0 \\ 0 & 1 & 0 \end{bmatrix} .$$

3.6.4 Conversion between Continuous-time and Discrete-time Forms

Conversion VII. Continuous-time to Discrete-time.

This conversion is useful for changing a differential equation to an approximating difference equation that can be solved numerically on a digital computer. The approximation results from using a piecewise constant representation of the original input signal. This also is called a sampled-data model since a zero-order hold at the input of a linear plant has been widely used in the design and analysis of linear sampled-data control systems. Starting with the stationary system

$$\dot{\underline{x}} = A\underline{x} + B\underline{u} \tag{3.117}$$

$$\underline{y} = C\underline{x} + D\underline{u}$$

a sampling period T_s is chosen. Then let

$$\underline{u}^a(t) = \underline{u}(nT_s), \qquad nT_s \leq t < (n+1)T_s ,$$

and $n=n_o, n_o+1, n_o+2, \ldots, n_o+k, \ldots, (n_o=t_o/T_s$ integer). Then using $\underline{u}^a$, the resulting state $\underline{x}(t)$ is*

$$\underline{x}(t) = \Phi(t-t_o)\underline{x}(t_o) + \int_{t_o}^{t} \Phi(t-\lambda)B\underline{u}^a(\lambda)d\lambda$$

where Φ is the transition matrix, i.e., $\Phi(\tau) = \exp A\tau$. For $t=t_o+T_s=(n_o+1)T_s$

$$\underline{x}(t_o+T_s) = \underline{x}((n_o+1)T_s) = \Phi(T_s)\underline{x}(n_oT_s) + \\ + \int_{n_oT_s}^{(n_o+1)T_s} \Phi((n_o+1)T_s-\lambda)B\underline{u}(n_oT_s)d\lambda$$

or

$$\underline{x}((n_o+1)T_s) = \Phi(T_s)\underline{x}(n_oT_s) + HB\underline{u}(n_oT_s)$$

where

$$H = \int_{n_oT_s}^{(n_o+1)T_s} \Phi((n_o+1)T_s-\lambda)d\lambda \quad ,$$

or

$$H = \int_{0}^{T_s} \Phi(\sigma)d\sigma \quad .$$

Similarly for $n=n_o+k$, we have the general result

$$\underline{x}((n+1)T_s) = \Phi(T_s)\underline{x}(nT_s) + HB\underline{u}(nT_s)$$

and with obvious changes in notation,

$$\underline{x}(n+1) = A_d\underline{x}(n) + B_d\underline{v}(n)$$

where $A_d = \Phi(T_s)$, $B_d = HB$, and $\underline{v}(n) = \underline{u}(nT_s)$. Thus a

---

*See Chapter 4.

discrete-time approximation to Equation 3.117 is (with $\underline{y}(nT_s) = \underline{z}(n)$)

$$\underline{x}(n+1) = A_d\underline{x}(n) + B_d\underline{v}(n) \tag{3.118}$$

$$\underline{z}(n) = C\underline{x}(n) + D\underline{v}(n)$$

and all results for stationary discrete-time systems apply. In particular, the transition matrix is $\Phi_d(j) = A_d^j$. The sampling time T_s is chosen so that $\underline{u}^a$ is "close" to $\underline{u}$. Also, in order to avoid loss of controllability and observability (assuming both for 3.117), a sufficient condition is that for every pair of characteristic values λ_i, λ_j of A with $[\text{Re}\lambda_i - \text{Re}\lambda_j]=0$, require $\text{Im}(\lambda_i - \lambda_j) \neq 2\pi k/T_s$, k an integer [24.]. For single input/output systems, the condition is necessary. Finally, since $A_d = \Phi(T_s)$, it always is nonsingular; this was Property I. after Eq. 3.64.

Conversion VIII. Discrete-time to Continuous-time.

A very simple version of an "inverse" of Conversion VII can be found using a first forward difference approximation to a derivative. Starting with

$$\underline{x}(n+1) = A'\underline{x}(n) + B'\underline{v}(n) \tag{3.119}$$

$$\underline{z}(n) = C'\underline{x}(n) + D'\underline{v}(n)$$

it is assumed that a continuous-time approximation to the difference equation in 3.119 exists in the form

$$\dot{\underline{x}}(t) = A_c\underline{x}(t) + B_c\underline{u}(t).$$

Let

$$\dot{\underline{x}}(t) \cong \frac{1}{T_s}(\underline{x}(t-T_s) - \underline{x}(t))$$

and then

$$\underline{x}(t+T_s) \cong (I+T_sA_c)\underline{x}(t) + T_sB_c\underline{u}(t).$$

For $t \to t_n = nT_s$,

$$\underline{x}((n+1)T_s) \cong (I+T_sA_c)\underline{x}(nT_s) + T_sB_c\underline{u}(nT_s).$$

Using the same change in notation as in Conversion VII,

$$\underline{x}(n+1) \cong (I+T_sA_c)\underline{x}(n) + T_sB_c\underline{v}(n).$$

Comparing,

$$A' = I+T_sA_c$$

$$B' = T_sB_c$$

Obviously the closeness of the approximation depends on T_s being small and on the "smoothness" of $\underline{x}(n)$. Strictly speaking, the state $\underline{x}(t)$, $\forall t$, is only inferred from a reconstruction using $\underline{x}(n)$ (see Chapter 2, Section 2.9.3 on reconstruction). Comparison with the results in Conversion VII shows that Conversion VIII is a simpler approximation. In Conversion VII we have

$$A_d = \Phi(T_s) = (\exp AT_s) = I+AT_s+ \tfrac{1}{2}A^2T_s{}^2 + \ldots$$

but in Conversion VIII,

$$A' = I + A_cT_s.$$

We see that A_d of the new discrete-time system is given as a series whereas A' of the original discrete-time system is related to A_c by the first two terms of a similar series. A simple first-order conversion for VII rather than the sampled-data model used here would produce similar forms in the two cases. Finally,

$$\dot{\underline{x}}(t) = A_c\underline{x}(t) + B_c\underline{u}(t) \tag{3.120}$$

$$\underline{y}(t) = C\underline{x}(t) + D\underline{u}(t)$$

is an approximation to 3.119 with A_c and B_c as given above and with $\underline{u}(t)$ a reconstruction of $\underline{v}(n)$ as discussed in Chapter 2.

3.7 LINEARIZATION

The system models that have been considered in this chapter have been linear, and properties and various response forms were developed as a consequence of this assumption. It is quite possible, in many practical cases of interest, that the form of the system model is known and nonlinear and that under certain conditions a useful linear approximation can be found. This is a linearization procedure, and it is entirely analytic. The use of experimental data to find the unknown parameters of an assumed linear model is discussed in the introduction to identification in the next chapter (Section 4.4).

The two primary system models in Table 3.4 were rth-order and vector-matrix or state variable forms. A general form for use in this discussion is

$$\dot{\underline{x}} = \underline{f}(x, \underline{u}, t) \tag{3.121}$$

$$\underline{y} = \underline{g}(\underline{x}, \underline{u}, t)$$

where $\underline{x}$ is an r-dimension state vector, the input $\underline{u}$ is qx1, and the output $\underline{y}$ is px1. If, on the other hand, the original model is in the form of one or more higher-order ordinary nonlinear differential equations, a form like Eq. 3.121 can be obtained first. For example, assume the inputs and outputs are related by

$$\frac{d^j y_1}{dt^j} = h_1(\dot{y}_1, y_1, \ldots, y^{(j-1)}, y_2, \dot{y}_2, \ldots, y_2^{(k-1)}, y_3, \ldots,$$

$$\ldots, y_p, \dot{y}_p, \ldots, y_p^{(m-1)}, u_1, \ldots, u_q, t)$$

$$\frac{d^k y_2}{dt^k} = h_2(\qquad)$$

$$\vdots$$

$$\frac{d^m y_p}{dt^m} = h_p(\qquad)$$

where the parentheses on the right may contain all the arguments listed for the first function. Then for $r = j+k+\ldots+m$, let

$$\begin{aligned}
\dot{x}_1 &= x_2 \\
\dot{x}_2 &= x_3 \\
&\vdots \\
\dot{x}_j &= h_1(x_1, x_2, \ldots, x_j, x_{j+1}, \ldots, x_r, u_1, \ldots, u_q, t) \\
\dot{x}_{j+1} &= x_{j+2} \\
&\vdots \\
\dot{x}_{j+k} &= h_2(\qquad) \\
&\vdots \\
\dot{x}_r &= h_p(\qquad)
\end{aligned}$$

and

$$\begin{aligned}
y_1 &= x_1 \\
y_2 &= x_{j+1} \\
&\vdots \\
y_p &= x_{r-m+1} \quad .
\end{aligned}$$

The functions $\underline{f}$ and $\underline{g}$ are then defined from these.

Now assume that a solution of Eq. 3.121 is known for a given $\underline{x}^o(t_o)$ and $\underline{u}^o(t)$, $t \geq t_o$, i.e.,

$$\dot{\underline{x}}^o = \underline{f}(\underline{x}^o, \underline{u}^o, t) \tag{3.122}$$

$$\underline{y}^o = \underline{g}(\underline{x}^o, \underline{u}^o, t)$$

for $t \geq t_o$. Next, consider solutions $\underline{x}$ of Eq. 3.121 that are "close" to $\underline{x}^o$, and let

$$\underline{\alpha} = \underline{u} - \underline{u}^o$$

$$\underline{\beta} = \underline{x} - \underline{x}^o$$

$$\underline{\delta} = \underline{y} - \underline{y}^o \ .$$

Substituting in Eq. 3.121,

$$\dot{\underline{x}}^o + \dot{\underline{\beta}} = \underline{f}(\underline{x}^o + \underline{\beta}, \underline{u}^o + \underline{\alpha}, t)$$

$$\underline{y}^o + \underline{\delta} = \underline{g}(\underline{x}^o + \underline{\beta}, \underline{u}^o + \underline{\alpha}, t)$$

with initial condition $\underline{x}(t_o) = \underline{x}^o(t_o)+\underline{\beta}(t_o)$. Assuming the necessary differentiability, the right hand sides can be expanded about $\underline{x}^o$, $\underline{u}^o$ in a Taylor Series:

$$\dot{\underline{x}}^o+\dot{\underline{\beta}} = \underline{f}(\underline{x}^o,\underline{u}^o,t) + (J_{\underline{x}}\underline{f}\Big|_{\underline{x}^o,\underline{y}^o})\underline{\beta}$$

$$+ (J_{\underline{u}}\underline{f}\Big|_{\underline{x}^o,\underline{u}^o})\underline{\alpha} + \ldots \tag{3.123}$$

$$\underline{y}^o + \underline{\delta} = \underline{g}(\underline{x}^o, \underline{y}^o, t) + (J_{\underline{x}}\underline{g}\Big|_{\underline{x}^o, \underline{u}^o})\underline{\beta}$$

$$+ (J_{\underline{u}}\underline{g}\Big|_{\underline{x}^o, \underline{u}^o})\underline{\alpha} + \ldots$$

where the Jacobian matrix is defined in Appendix B, e.g.,

$$J_{\underline{x}}\underline{f} = \begin{bmatrix} \frac{\partial f_1}{\partial x_1} & \cdots & \frac{\partial f_1}{\partial x_r} \\ \vdots & & \vdots \\ \frac{\partial f_r}{\partial x_1} & \cdots & \frac{\partial f_r}{\partial x_r} \end{bmatrix} .$$

The higher order terms in 3.123 can be neglected for the squared norms $||\underline{\beta}||^2$ and $||\underline{\alpha}||^2$ small. Then using 3.122 to eliminate the nominal solution terms in 3.123, we have

$$\dot{\underline{\beta}} = A(t)\underline{\beta} + B(t)\underline{\alpha} \qquad (3.124)$$

$$\underline{\delta} = C(t)\underline{\beta} + D(t)\underline{\alpha}$$

where

$$A(t) = J_{\underline{x}}\underline{f}\Big|_{\underline{x}^o, \underline{u}^o}$$

and B(t), C(t), D(t) are defined similarly in terms of $J_{\underline{u}}\underline{f}$, $J_{\underline{x}}\underline{g}$, and $J_{\underline{u}}\underline{g}$, respectively. This is a set of linear, time-varying equations in the form of Eq. 3.57, and these are the linearized equations that replace the nonlinear model of Eq. 3.121. This linear model is valid for small perturbations about $\underline{x}^o$, $\underline{u}^o$ for $t \geq t_o$. Although $||\underline{\alpha}||$ can be made small by proper choice of $\underline{u}$, $||\underline{\beta}||$, which depends on the system, may or may not be small depending on the stability of the system described by Eq. 3.124. This problem is discussed in Chapter 5 where conditions on the matrices are given that assure stability. If the matrices are constants (and D=0) with (A,B) controllable and (A,C) observable, it is necessary and sufficient that the real parts of the characteristic values of A be negative (see Theorem 5.12).

Example 3.37

Consider a modification of the mass-spring-dashpot system of Example 3.1c

$$\frac{W}{g}\frac{d^2y}{dt^2} + C\frac{dy}{dt} + Ky^3 = u,$$

a nonlinear differential equation. First, we write

$$\frac{d^2y}{dt^2} = -\frac{gC}{W}\frac{dy}{dt} - \frac{gK}{W}y^3 + \frac{g}{W}u$$

and then let

$$y = x_1$$

and

$$\dot{x}_1 = x_2$$

$$\dot{x}_2 = -\frac{gC}{W}x_2 - \frac{gK}{W}x_1^3 + \frac{g}{W}u\ .$$

Then for

$$\underline{x}^T = (x_1\ x_2),$$

$$\dot{\underline{x}} = \underline{f}(\underline{x}, u)$$

$$y = g(\underline{x})$$

where

$$f_1 = x_2$$

$$f_2 = -\frac{gC}{W}x_2 - \frac{gK}{W}x_1^3 + \frac{g}{W}u$$

and

$$g = x_1 .$$

Now

$$J_{\underline{x}}\underline{f} = \begin{bmatrix} 0 & 1 \\ \frac{-3gKx_1^2}{W} & \frac{-gC}{W} \end{bmatrix}$$

$$J_u\underline{f} \quad \begin{bmatrix} 0 \\ \frac{g}{W} \end{bmatrix}$$

$$J_{\underline{x}}g = [1 \quad 0]$$

$$J_u g = 0.$$

For a constant input u^o,

$$K(x_1^o)^3 = u^o$$

or

$$x_1^o = (u^o/K)^{1/3} .$$

Then

$$A = \begin{bmatrix} 0 & 1 \\ -\frac{3gK}{W}(x_1^o)^2 & -\frac{gC}{W} \end{bmatrix} = \begin{bmatrix} 0 & 1 \\ -a & -b \end{bmatrix}$$

$$\underline{b} = \begin{bmatrix} 0 \\ \frac{g}{W} \end{bmatrix}$$

$$\underline{c}^T = [1 \quad 0]$$

$$d = 0,$$

and

$$\dot{\underline{\beta}} = A\underline{\beta} + \underline{b}\alpha$$

$$\delta = \underline{c}^T\underline{\beta}$$

is the linearized model. We require

$$\alpha = u - u^o \quad \text{and} \quad \underline{\beta}(t_o) = \underline{x}(t_o) - \begin{bmatrix} x_1^o \\ 0 \end{bmatrix}$$

both small. Since the linearized model is controllable and observable, and the characteristic values of A are

$$\lambda_1, \lambda_2 = \frac{-b \pm \sqrt{b^2-4a}}{2}, \qquad b > 0,$$

the model is stable in the b.i.b.o. sense as defined in the introduction to this chapter.

A more complete example is given in Chapter 6, Section 6.4 where the longitudinal equations of motion of an aircraft are obtained.

REFERENCES

1. Bunge, M., "Causality, Chance, and Law," American Scientist, v. 49, n. 4, Dec. 1961, pp. 432-448.
2. Schwarz, R. and Friedland, B., "Linear Systems," New York: McGraw-Hill Book Co., 1965.
3. Cannon, R. H., Jr., "Dynamics of Physical Systems," New York: McGraw-Hill Book Co., 1967.
4. Trimmer, J. D., "Response of Physical Systems," New York: John Wiley & Sons, 1950.
5. De Russo, P., Roy, R., and Close, C., "State Variables for Engineers," New York: John Wiley & Sons, 1965.
6. Kaplan, W., "Operational Methods for Linear Systems," Reading, Mass.: Addison-Wesley, 1962.
7. Zadeh, L. and Desoer, C., "Linear System Theory," New York: McGraw-Hill Book Co., 1963.
8. Timothy, L and Bona, B., "State Space Analysis: An Introduction," New York: McGraw-Hill Book Co., 1968.
9. Rubio, J., "The Theory of Linear Systems," New York: Academic Press, 1971.
10. Cadzow, J. and Martens, H., "Discrete-Time and Computer Control Systems," Englewood Cliffs, New Jersey: Prentice-Hall, 1970.
11. Mastacusa, E. J., "A Method for Calculating ε^{At} Based on the Cayley-Hamilton Theorem," Proc. IEEE, v. 57, n. 7, July, 1969, pp. 1328-1329.
12. Wiberg, D., "State Space and Linear Systems," New York: McGraw-Hill Book Co., 1971.
13. Chen, C. F., and Parker, R. R., "Generalization of Heaviside's Expansion Techniques to Transition Matrix

Evaluation," IEEE Trans. on Education, v. E-9, n. 4, Dec. 1966, pp. 209-212.

14. Kalman, R. E., "Mathematical Description of Linear Dynamical Systems," S.I.A.M. Journ. Control, v. 1, n. 2, 1963, pp. 152-192.
15. Luenberger, D., "Canonical Forms for Linear Multivariable Systems," IEEE Trans. on Automatic Control, v. AC-12, n. 3, July, 1967, pp. 290-293.
16. Johnson, C. D., "A Unified Canonical Form for Controllable and Uncontrollable Linear Dynamical Systems," Int'l. Journ. Control, v. 13, n. 3, 1971, pp. 497-517.
17. Asseo, S., "Phase Variable Canonical Transformation of Multicontroller Systems," IEEE Trans. on Automatic Control, v. AC-13, n. 1, Feb. 1968, pp. 129-131.
18. Curran, R. and Franklin, G., "Comments on 'Phase Variable Canonical Transformation on Multivariable Systems,'" IEEE Trans. on Auto. Control, v. AC-16, n. 1, Feb. 1971, pp. 108-110.
19. Asseo, S., "Author's Reply," IEEE Trans. Auto. Control, v. AC-16, n. 1, Feb. 1971, pp. 110-111.
20. Bucy, R. S., "Canonical Forms for Multivariable Systems," IEEE Trans. Auto. Control, v. AC-13, n. 5, October, 1968, pp. 567-569.
21. Budin, M., "Canonical Forms for Multi-Input/Multi-Output Minimal Systems," IEEE Trans. Auto. Control, v. AC-17, n. 4, Aug. 1972, p. 554.
22. Anderson, B. and Luenberger, D., "Design of Multivariable Feedback Systems," Proc. IEE, v. 114, n. 3, March, 1967, pp. 395-399.
23. Meditch, J., "Stochastic Optimal Linear Estimation and Control," New York: McGraw-Hill Book Co., 1969.
24. Kalman, R., et al, "Controllability of Linear Dynamical Systems," Contributions to Differential Equations, v. 1, n. 2, New York: Interscience Publ. Inc., 1962, pp. 189-213.
25. Gilbert, E., "Controllability and Observability in

Multivariable Control Systems," S.I.A.M. Journ. Control, v. 1, n. 2, 1963, pp. 128-151.
26. Kalman, R., "Mathematical Description of Linear Dynamical Systems," S.I.A.M. Journ. Control, v. 1, n. 2, 1963, pp. 152-192.
27. Panda, S. and Chen, C., "Irreducible Jordan Form Realization of a Rational Matrix," IEEE Trans. Auto Control, v. AC-14, n. 1, Feb., 1969, pp. 66-69.
28. Kalman, R., "On Structural Properties of Linear, Constant, Multivariable Systems," Proc. I.F.A.C. Congress, 1966, Paper No. 6.A.
29. Kalman, R., "Irreducible Realizations and the Degree of a Rational Matrix," S.I.A.M. Journ, v. 13, n. 2, June, 1965, pp. 520-524.
30. Ho, B., and Kalman, R., "Effective Construction of Linear State Variable Models for Input/Output Data," Proc. 3rd Allerton Conf., October, 1965, pp. 449-459.
31. Polak, E., "An Algorithm for Reducing a Linear, Time-Invariant Differential System to State Form," IEEE Trans. Auto. Control, v. AC-11, n. 3, July, 1966, pp. 577-579.
32. Wolovich, W., "The Determination of State-Space Representations for Linear Multivariable Systems," Automatica, v. 9, n. 1, January, 1973, pp. 97-106.
33. Rosenbrock, H. H., "State-Space and Multivariable Theory," New York: John Wiley & Sons, 1970.
34. Kwakernaak, H. and Sivan, R., "Linear Optimal Control Systems," New York: John Wiley & Sons, 1972.

PROBLEMS

1. A simple system has an input/output relation.

$$y(t) = au(t) + b \qquad a>0,\ b>0,\ \forall t$$

Specify whether or not the system is:

a) realizable

b) stationary
c) linear
d) stable.

2. A simple system is described by:

$$\frac{dy(t)}{dt} = b\ u(t) \qquad \forall t$$

Specify whether or not the system is:
a) realizable
b) stationary
c) linear
d) stable.

3. Repeat Problem 2 for a system

$$y(t) = b\ \frac{du(t)}{dt} \qquad \forall t$$

Consider u(t) differentiable for all t first and then u(t) continuous for all t, then u(t) piecewise continuous for all t. Note that this is <u>not</u> in the class of systems considered in Section 3.3.

4. For a system

$$\frac{dy(t)}{dt} + ay(t) = bu(t), \qquad \forall t$$

specify whether the system is
a) realizable
b) stationary
c) linear.
Consider cases for a constant a and a variable a(t).

5. Generalize the methods used in Problem 4. for Equations 3.19a, 3.19b, 3.21a, and 3.21b.

6. For a system

$$\frac{dy(t)}{dt} + ay(t) = bu(t-\tau), \qquad a > 0$$

under what condition is the system realizable?

7. The unit exponentials u_e and v_e were used in Problems 33 and 34. of Chapter 2 to obtain real spectral models

for continuous and discrete-time signals. Using the methods of Section 3.2, obtain corresponding response forms. Call the responses to $u_e(t)$ and $v_e(n)$ the unit exponential responses $e(t,\lambda)$ and $e(n,k)$.

8. For a curve $u(\tau)$, $0 \le \tau \le t$, the area under the curve is

$$A = y(t) = \int_0^t u(\tau)d\tau .$$

The area A is to be approximated using a numerical integrator. The trapezoidal rule is:

$$A_n = y(n) = T[\tfrac{1}{2}(u(0)+u(n))+(u(1)+\ldots+ u(n-1))], \quad n \ge 1$$

with $nT=t$. Find $y(n+1)$ in terms of $y(n)$, and therefore, specify $M(E)$ and $N(E)$.

9. A simple linear model of a vibration isolation mounting is shown.

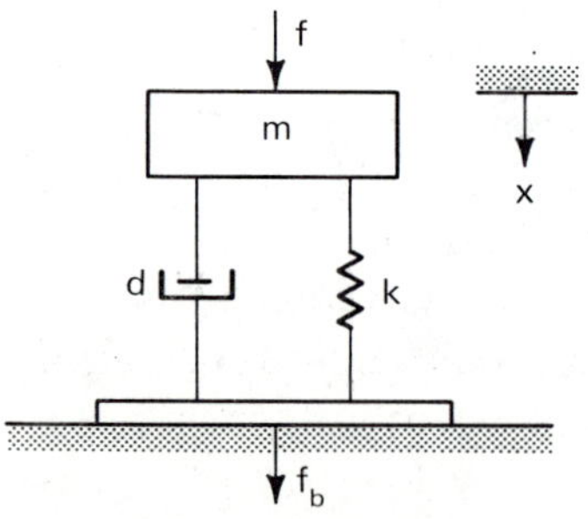

Let f be the input force and f_b be the output force transmitted to the floor. Find $M(D)$ and $N(D)$.

10. Find $M(D)$ and $N(D)$ of the circuit shown if i is the input and v_o the output.

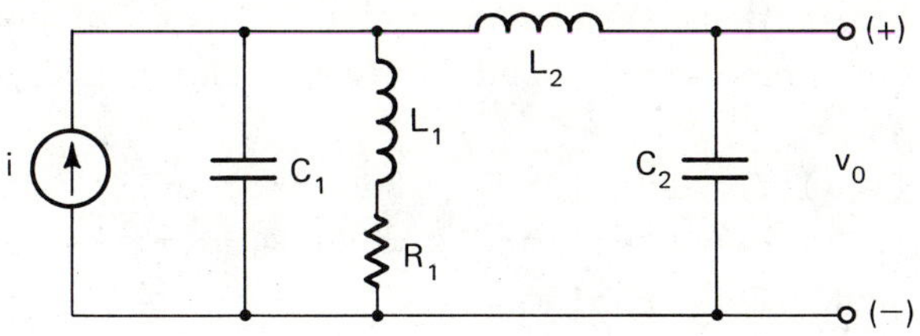

11. Find M(D) and N(D) for the parallel-T shown; e_1 is input and e_2 is output.

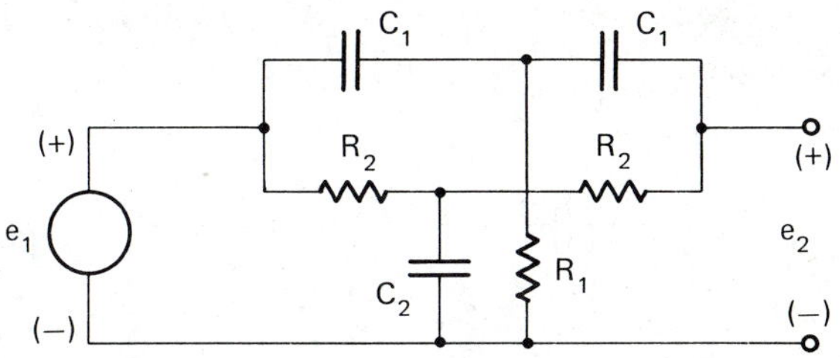

Note that the circuit is not strictly proper, i.e., r=m=3.

12. A simple model for a time payment plan is as follows.

Let z(n) = principal after nth payment

v(n) = amount of nth payment

r = interest rate (interest compounded after each payment).

Find a difference equation relating z(n) and v(n) for $n \geq 0$. What are M(E) and N(E)?

13. A D-C motor drive is shown. Find M(D) and N(D) when the input is e(t) and the output is $e_t(t)$. Let

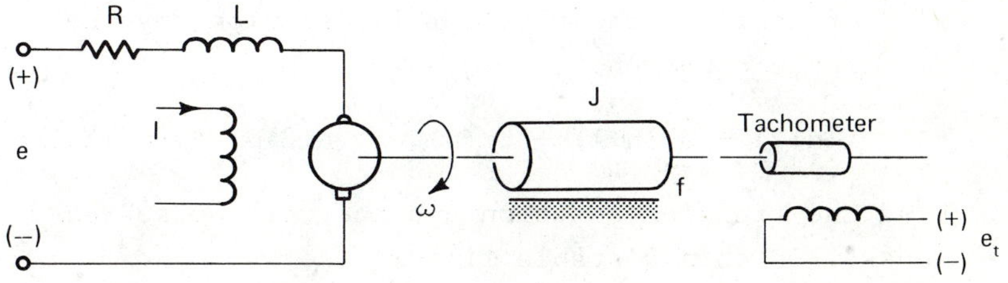

K_a be a motor voltage constant and K_t a torque constant for constant I, and let K be the tachometer constant.

14. Find $h(t,\lambda)$ for Problem 9. when m=1, d=1, k=1.
15. Find $h(t,\lambda)$ for Problem 11. when $R_1=C_1=1$ and $R_2=C_2=2$.
16. Find a(n,k) for Problem 8.
17. Using the definitions of Section 3.4 and letting x=y in Problem 2., show that the system is zero-state and zero-input linear and satisfies the decomposition property.
18. For the network shown, find M(D) and N(D) and a state variable model for $i_L=x_1$ and $v_c=x_2$.

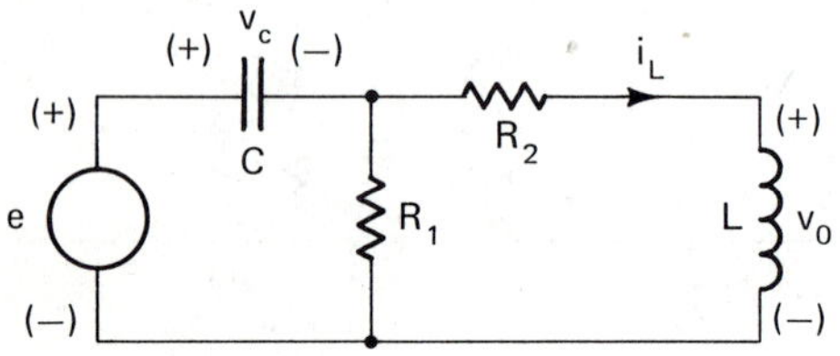

19. Find a state model and the transition matrix for the filter shown.

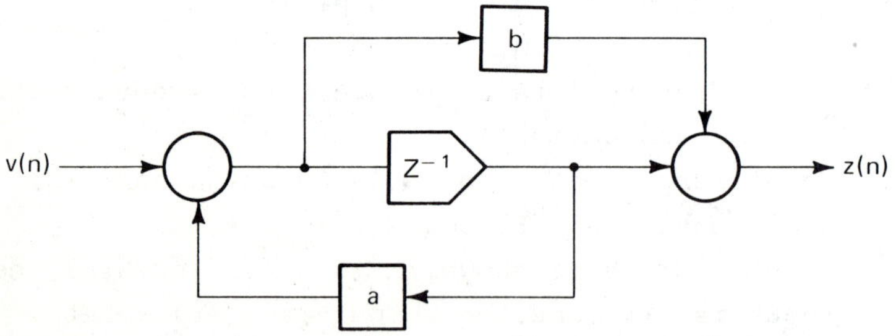

20. A discrete-time linear filter is described by the difference equation:

$$z(n+2) + 5z(n+1) + 6z(n) = v(n+2) + 4v(n+1).$$

Find the transfer function and the unit pulse response. Is this a stable filter?

21. Using the power series definition of the matrix exponential, show that

a) $[\varepsilon^{At}]^{-1} = \varepsilon^{-At}$

b) $\frac{d}{dt}\varepsilon^{At} = A\varepsilon^{At}$

c) $\varepsilon^{At}\varepsilon^{-At_0} = (\exp A(t-t_0))$.

22. For the system whose simulation diagram is shown, find A, B, C, D.

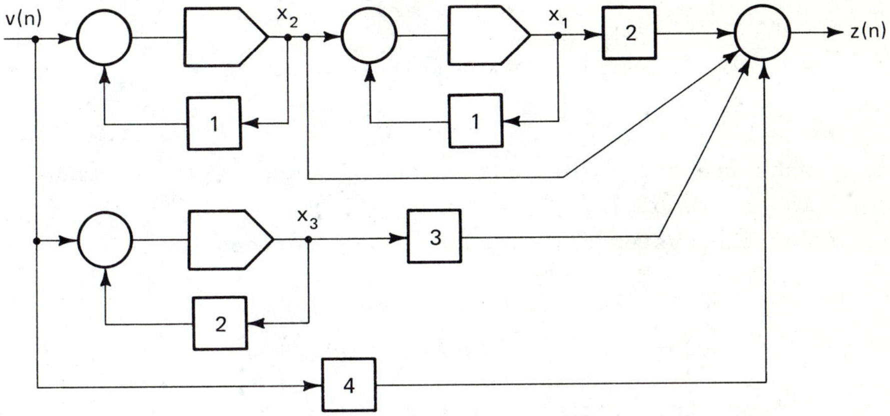

23. A single input/single output system has a transfer function

$$H(s) = \frac{5(s+2)}{(s+1)(s+3)}$$

a) Find a state variable model in first primal form.

b) Show that the model in a) is controllable and observable.

24. A motor has a transfer function

$$H(s) = \frac{10}{s(s+2)}$$

a) Find a state variable model in first primal form.

b) Using a sampling period $T_s=1$, find a suitable discrete-time model that approximates the original model.

25. A linear system is described by

$$\dot{\underline{x}} = \begin{bmatrix} a_{11} & a_{12} \\ a_{21} & a_{22} \end{bmatrix} \underline{x} + \begin{bmatrix} 0 \\ 1 \end{bmatrix} u$$

What are conditions on the a_{ij} such that the model is controllable?

26. For the system in Problem 25. with

$$y = [1 \quad 0]\underline{x} ,$$

what are the conditions on the a_{ij} such that the model is observable?

27. For the system described by the equations

$$(D^2+2D+1)y_1 + (2D+1)y_2 = u_a$$
$$(2D+1)y_1 + (D^2+2D+2)y_2 = Du_a + u_b ,$$

find a state variable model.

28. An equivalent circuit for a 3-stage transistor amplifier is shown. Find a state variable model.

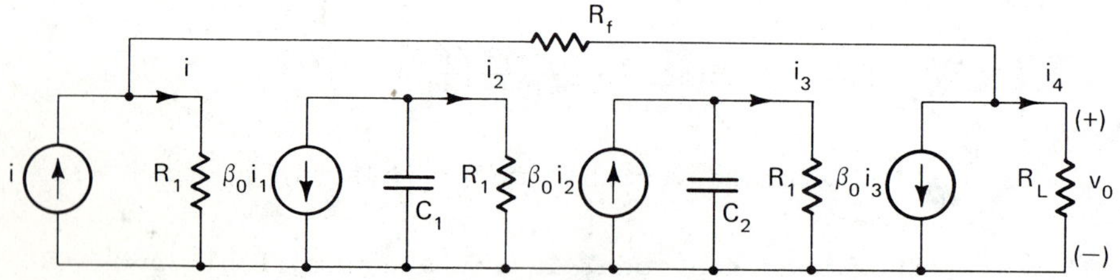

29. Verify that the first primal and second primal form models of Eq. 3.68 and Eq. 3.69 are correct by finding the original rth-order form from the state variable model.

30. Sketch $|H(j\omega)|$ and $\text{Ang}H(j\omega)$ versus ω for the transfer function of Problem 23.

31. Sketch $|H(\gamma)|$ and $\text{Ang}H(\gamma)$ versus θ for the transfer function of Problem 20.

32. A series RLC circuit is driven by a voltage source u(t) which is a square wave with amplitude ±1 and period $T_o=2\pi/\omega_o$. The output y(t) is the voltage across R. Find U_k, $F_c(t,k)$, and Y_k.

33. Show that for any linear transformation

$$\underline{x} = T\underline{x}' ,$$

and input $\underline{u}$ and output $\underline{y}$, a general result in a form like Eq. 3.76 exists.

34. Find a T such that a second-order system represented by two elements of the form in Figure 3.8a with complex $\lambda_2=\lambda_1^*$ is transformed to the form of Figure 3.8b.

35. Starting with Eq. 3.68, write the general form of the dual first primal form and show that it always is observable. Let $\beta_r=0$.

36. For a system model

$$\dot{\underline{x}} = \begin{bmatrix} 2 & 0.5 & -3 \\ 0 & -1 & 0 \\ 0 & 0.5 & -1 \end{bmatrix} \underline{x} + \begin{bmatrix} 1 & 0 \\ 0 & 2 \\ 1 & 0 \end{bmatrix} \underline{u}$$

$$\underline{y} = \begin{bmatrix} 1 & 0 & -1 \\ 1 & 0 & -1 \end{bmatrix} \underline{x}$$

a) Find the characteristic values of A and a corresponding set of characteristic vectors.

b) Check the model for controllability and observability.

c) Transform the model to one where $A' = T^{-1}AT$ is a Jordan form.

d) By observation of the transformed model, decompose the model into controllable-observable, controllable-nonobservable, noncontrollable-observable, and noncontrollable-nonobservable parts (if they exist). Sketch the simulation diagram partitioned in this way.

37. Find a state variable model for Problem 8.

38. Find a state variable model for Problem 9.

39. Find a state variable model for Problem 10.

40. Find a state variable model for Problem 13., where the state variables are i, θ, and $\dot{\theta} = \omega$. Use the same input and output. Is the model controllable? observable?

41. Use Eq. 3.113 to transform the model of Problem 22. to first primal form.

42. Assume a set of numerical values for Problem 10. and use the result in Problem 39. and Conversion VII to find a suitable discrete-time approximation for digital computation. Choose T_s relative to $2\lambda/|\lambda_i|$, i=1,2,3,4 and then compute A_d and B_d as suggested in App. D.

43. Find the transfer matrix H(s) for the system of Problem 36.

44. For a system

$$\dot{\underline{x}} = A\underline{x} + B\underline{u}$$

$$\underline{y} = C\underline{x} + D\underline{u}$$

and a transformation $\underline{x} = T\underline{x}'$, prove that:

a) The transformed model (A´,B´) is controllable if and only if (A,B) is controllable.

b) The transformed model (A´,C´) is observable if and only if (A,C) is observable.

45. For the system

$$(E^2 - E + 2/9)z(n) = (E-1/2)v(n), \qquad n \geq 0 .$$

a) Find a(n-k).

b) Using Δ_b, find h(n-k) from the answer to a). Be careful at n-k=j=0.

46. a) For the circuit shown, find a state variable model using v_2 as output.

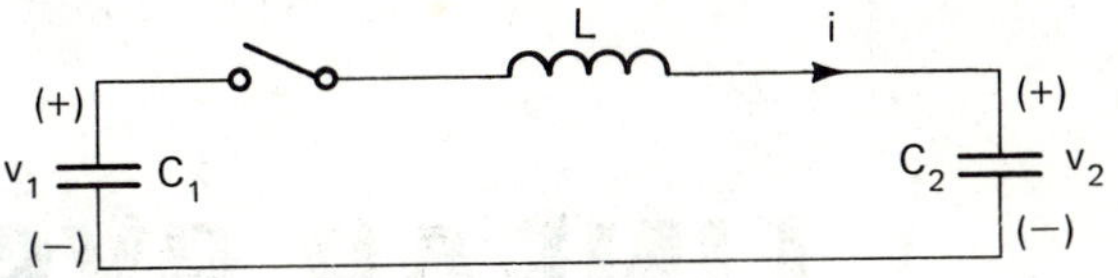

b) If $C_1=C_2=L=1$, find $\Phi(t)$.

c) If the switch closes at $t_o=0$, and $v_1(0)=1$ and $v_2(0)=0$, find an expression for $v_2(t)$, $t\geq 0$.

47. A simple low pass digital filter is described by

$$z(n) = \alpha z(n-1) + (1-\alpha)v(n), \qquad |\alpha|<1 .$$

a) Find the filter transfer function $H(\gamma)$.

b) If Θ_1 is such that

$$|H(\gamma_1)| = |H(\varepsilon^{j\Theta_1})| = 0.707,$$

find α to make $\Theta_1 = \pi/2$.

Chapter 4

LINEAR SYSTEM RESPONSE

4.1 GENERAL INPUT/OUTPUT PROBLEMS

System response is concerned with finding an output when an input and a system are given. Having developed a variety of signal models in Chapter 2 and several system models in Chapter 3, we now are prepared to find ways of solving for the response of a system. The models of the output signals will, of course, be of the same types as those for the inputs. In Chapter 3, Section 3.4, we also used the terminology of a system transforming (or mapping) a signal of an input space to another signal in an output space. In a mathematical sense, the basic problem is to solve ordinary linear differential or difference equations. Many of the methods already have been introduced in Chapter 3. Section 3.3 used classical and transform methods, and Section 3.4 introduced vector-matrix forms.

4.1.1 Summary of Signal and System Models

Examination of Table 2.6, a summary of signal models, shows a basic division between deterministic and stochastic signals. Also, the three categories of a signal as a function, as a spectral form derived from the function, or as an auxiliary function are shown. Similarly, Table 3.4, indicates a distinction between primary forms (differential or difference equations in rth-order or vector-matrix form) and response forms. In fact, using these two tables we can immediately outline a plan for this chapter. As a matter of convenience, we shall consider the deterministic cases first and then follow these with the stochastic ones. Furthermore, the parallelism between continuous-time and discrete-time results will be continued as it was in the two previous chapters. Usually we shall develop single input/output cases first and then extend these to multivariable system responses.

4.1.2 Deterministic Input/Output Problems

The signal models of Chapter 2 as noted above are of three types: u(t) or v(n) as specified function on T_{c1}, T_{c2}, T_{d1}, or T_{d2}; as a spectral function related to a given elementary or basis function; and as an auxiliary function of a time displacement variable. For deterministic signals, each of these types leads to a general method of expressing a similar form of system output in terms of the input and an appropriate system model. Thus, for example, if u(t) is given we can:

1. solve the differential equation 3.19 directly for y(t), or
2. convert u(t) to a spectral form (e.g., U(s)) and using the associated system model (H(s)), find the output spectral form (Y(s)), or

3. convert u(t) to an auxiliary function (e.g., $w_u(\tau)$) and using the appropriate system model ($h(\cdot)$), find the output auxiliary function ($w_y(\tau)$).

4.1.3 Stochastic Input/Output Problems

For stochastic signals, only the third case above is considered in this book although the spectral forms shown in Table 2.6 also are used in some cases.

3. For a given stochastic signal model (e.g., $r_U(\tau)$ of $\{U(t)\}$), use an appropriate system model ($h(\cdot)$) to find a response ($r_Y(\tau)$ of $\{Y(t)\}$).

4.1.4 Outline of Methods

In Section 4.2 on deterministic cases, after a brief review of classical methods of solving rth-order linear differential and difference equations, the forms that follow from a spectral description of the input are discussed. Many of these results already have been found in Chapter 3 in the development of the system response functions summarized in Table 3.4. Next, the system responses using auxiliary functions are given. Finally, system responses when the models are in state variable form are given. These actually are just solutions of linear vector-matrix differential or difference equations. In Section 4.3 on stochastic cases, auxiliary functions (i.e., autocorrelation functions of the stationary inputs or their transforms, the spectral density functions) are used to find the corresponding system output functions where the outputs also are stationary. This will sometimes be called the case of "infinite averaging time" or the case of $t_o \to -\infty$. Finally, when t_o is finite, a time-varying auxiliary function response, e.g., a mean square $S_Y^2(t)$, will be found. Section 4.4 provides a brief introduction to system identification where, under certain assumptions, we can develop methods for finding system models from experimental input/output data. This is, in some ways, analogous to the methods of signal identification (e.g., least squares approximations) discussed in Chapter 2. Section 4.5 develops some interesting applications of many of the system response results by using analog and digital computers to generate specified signals. These are useful in laboratory testing and in the simulation of systems.

4.2 DETERMINISTIC CASES

4.2.1 Review of Classical Methods

When the system model is in the form of Eq. 3.19a or Eq. 3.19b, rth-order ordinary linear differential equations, and the signal model is u(t), a specified function, the system response y(t) can be found by using classical methods of solving these differential equations. The complete solution y(t) is the sum of a particular solution $y_p(t)$ and a complementary solution $y_c(t)$, Therefore, we find

$$y(t) = y_p(t) + y_c(t), \qquad t \geq t_o .$$

The complementary solution $y_c(t)$ is a linear combination of r independent solutions $y_1(t),\ldots,y_r(t)$ of the homogeneous equation

$$M(D,t)y(t) = 0 \tag{4.1}$$

for time-varying systems, or

$$M(D)y(t) = 0$$

for stationary systems. Then for stationary systems, i.e., systems with constant coefficients,

$$y_c(t) = k_1\varepsilon^{\lambda_1 t} +\ldots+ k_r\varepsilon^{\lambda_r t} \tag{4.1a}$$

where the λ_j are simple zeros of $M(\lambda)$. For a zero, e.g., λ_1, of multiplicity j,

$$y_c(t) = k_1\varepsilon^{\lambda_1 t} + k_2 t\varepsilon^{\lambda_1 t} +\ldots+ k_j t^{j-1}\varepsilon^{\lambda_1 t} +$$

$$+ k_{j+1}\varepsilon^{\lambda_{j+1}} +\ldots+ k_r\varepsilon^{\lambda_r t} \tag{4.1b}$$

These forms have already occurred as solutions to Eq. 3.27a. For time-varying systems, no general method for finding $y_c(t)$ is available although many special cases can be solved, e.g., cases where the coefficients are polynomials or periodic functions of t. Power series methods often are useful. A more complete discussion of time-varying system response is given later in this chapter when state variable models are used.

To find $y_p(t)$, two methods are commonly used: the method of undetermined coefficients and the method of variation of parameters[1.][2.]. The method of undetermined coefficients is useful for stationary systems where the input u(t) has a finite number of forms resulting from $D^m u(t)$, $m=0,1,2,\ldots,$ The method of variation of parameters is applicable for finding $y_p(t)$ of both stationary and time-varying systems provided that $y_c(t)$ is known.

The final step in finding y(t) is the evaluation of the r constants $k_1,\ldots,k_r$. A set of r independent initial conditions at t_o, e.g., $y(t_o)$, $\dot{y}(t_o),\ldots,y^{(r-1)}(t_o)$, is required. These methods are well known and will not be repeated here.

When the system model is in the form of Eq. 3.21a or Eq. 3.21b, rth-order ordinary linear difference equations, and the signal model is v(n), a specified function of n, the system response z(n) can be found by using classical methods of solving difference equations. The complete solution z(n) is the sum of a particular solution $z_p(n)$ and a complementary solution $z_c(n)$. Thus we find

$$z(n) = z_p(n) + z_c(n) \qquad n \geq n_o \tag{4.2}$$

The complementary solution $z_c(n)$ is a linear combination of r independent solutions $z_1(n),\ldots,z_r(n)$ of the homogeneous equation

$$M(E,n)z(n) = 0$$

for time-varying systems, or

$$M(E)z(n) = 0$$

for stationary systems. Then for stationary systems

$$z_c(n) = k_1\rho_1^n + \ldots + k_r\rho_r^n \qquad (4.2a)$$

where the ρ_j are simple zeros of $M(\rho)$. For a zero, e.g., ρ_1, of multiplicity i, we use

$$z_c(n) = k_1\rho_1^n + k_2 n\rho_1^n + \ldots + k_i n^{i-1}\rho_i^n + $$
$$+ k_{i+1}\rho_{i+1}^n + \ldots + k_r\rho_r^n \; . \qquad (4.2b)$$

These forms have occurred as solutions of Eq. 3.37a. As in the previous case of continuous-time systems, no general method for finding $z_c(n)$ is available for time-varying discrete-time systems.

To find $z_p(n)$, the two methods cited previously are frequently used [2.][3.]. The method of undetermined coefficients is useful for stationary systems where the input v(n) has a finite number of forms resulting from $E^m v(n)$, m=0,1,2, ... The method of variation of parameters applies to both stationary and time-varying systems when $z_c(n)$ is known.

The final step in finding z(n) is the evaluation of the r constants $k_1, \ldots, k_r$; a set of r conditions for $n \geq n_0+1$, e.g., $z(n_0+1), \ldots, z(n_0+r)$, is required. $z(n_0)$ is treated as a special case as shown in Example 4.1b.

<u>Example 4.1</u>

a) For the system described by

$$(D^2+3D+2)y = (2D+3)u \; ,$$

find y(t), t>0, when $u(t) = u_2(t)$ and $y(0+) = \dot{y}(0+) = 0$. Now

$$M(\lambda) = \lambda^2+3\lambda+2$$

with zeros

$$\lambda_1, \lambda_2 = -2, \; -1$$

and therefore,

$$y_c(t) = k_1 \varepsilon^{-2t} + k_2 \varepsilon^{-t} \ .$$

Using the method of undetermined coefficients, let

$$y_1(t) = Au_2(t) + Bu_1(t)$$

$$= At + B, \qquad t > 0$$

and substituting in the system model, for t>0,

$$3A + 2At + 2B = 2 + 3t \ .$$

Then, equating coefficients of like powers of t,

$$3A + 2B = 2$$

$$2A = 3$$

and

$$A = 3/2, \qquad B = -5/4.$$

From these

$$y(t) = 3/2t - 5/4 + k_1 \varepsilon^{-2t} + k_2 \varepsilon^{-t} \ .$$

At t=0+, using the initial conditions,

$$0 = -5/4 + k_1 + k_2$$

$$0 = 3/2 - 2k_1 - k_2$$

and

$$k_1 = 1/4, \qquad k_2 = 1 \ .$$

Therefore,

$$y(t) = 3/2t - 5/4 + 1/4\varepsilon^{-2t} + \varepsilon^{-t}, \qquad t \geq 0 \ .$$

b) For the system described by

$$(E^2-0.75E+0.125)z = (E-0.4)v \ ,$$

find z(n), n>0, when $v(n) = v_2(n)$ and $z(0) = z(-1) = 0$. Now

$$M(\rho) = \rho^2-0.75\rho + 0.125$$

with zeros

$$\rho_1, \rho_2 = 0.5,\ 0.25$$

and therefore,

$$z_c(n) = k_1(0.5)^n + k_2(0.25)^n, \qquad n > 0.$$

Using the method of undetermined coefficients, let

$$z_p(n) = Av_2(n) + Bv_1(n)$$

$$= An + B, \qquad n > 0$$

and substituting in the system model, for n>0, with

$$Ez_p(n) = A(n+1) + B$$

$$E^2z_p(n) = A(n+2) + B$$

we get

$$[A(n+2)+B]-0.75[A(n+1)+B]+0.125[An+B] = (n+1)-0.4(n).$$

Therefore, equating coefficients of like powers of n,

$$A-0.75A + 0.125A = 0.6$$

$$2A+B-0.75A-0.75B+0.125B = 1.$$

Then

$$A = 1.6, \qquad B = -2.67 .$$

From these,

$$z(n) = 1.6v_2(n) - 2.67v_1(n) + k_1(0.5)^n$$
$$+ k_2(0.25)^n, \qquad n > 0 .$$

Now z(1) and z(2) are needed so that k_1 and k_2 can be found. For n=-1 in the system model

$$z(1) = 0.75z(0) - 0.125z(-1)$$
$$+ v_2(0) - 0.4v_2(-1)$$
$$= 0,$$

and for n=0 in the system model

$$z(2) = 0.75z(1) - 0.125z(0)$$
$$+ v_2(1) - 0.4v_2(0)$$
$$= 1 .$$

Using these,

$$0 = 1.6 - 2.67 + k_1(0.5) + k_2(0.25)$$

$$1 = 3.12 - 2.67 + k_1(0.5)^2 + k_2(0.25)^2$$

and

$$k_1 = 1.63, \qquad k_2 = 1.06 .$$

Therefore,

$$z(n) = 1.6n - 2.67$$
$$+ 1.6(0.5)^n + 1.07(0.25)^n, \quad n > 0$$

and this is also true for n=0 in this example.

The foregoing classical methods of finding the response for single input/single output system models can be extended to multiple input/multiple output systems such as Eq. 3.51 and Eq. 3.54. This is a problem of solving simultaneous ordinary linear differential or difference equations. Each internal variable w_i, $i=1,2,\ldots,m$, is expressed as a sum of a particular solution, w_{ip}, and a complementary solution, w_{ic}. Thus we have

$$w_i = w_{ip} + w_{ic}.$$

The w_{ip} must be found using the input function $\underline{f}$ and the w_{ic} are formed by using the zeros of the system characteristic equation

$$\det M(\cdot) = 0 .$$

For $i=1,2,\ldots,m$, we obtain

$$w_i(t) = w_{ip}(t) + k_{i1}\varepsilon^{\lambda_1 t} +\ldots+ k_{ir}\varepsilon^{\lambda_r t} \tag{4.1c}$$

for a stationary continuous-time system with simple zeros $\lambda_1,\ldots,\lambda_r$, or

$$w_i(n) = w_{ip}(n) + k_{i1}\rho_1^n +\ldots+ k_{ir}\rho_r^n \tag{4.2c}$$

for a stationary discrete-time system with simple zeros $\rho_1,\ldots,\rho_r$. The constants $k_{i1},\ldots,k_{ir}$ are evaluated from initial conditions, and the outputs $\underline{y}$ and $\underline{z}$ are found using $P(\cdot)$. The general method becomes quite cumbersome for systems where, m, the dimension of $\underline{w}$ is large, and will, therefore, not be considered further. A simple electric network example is given to illustrate the required procedure.

Example 4.2

For the two-input/two output network shown, find the outputs v_1 and v_2 when the inputs are $e_1=u_1(t)$ and $e_2=2u_1(t)$.

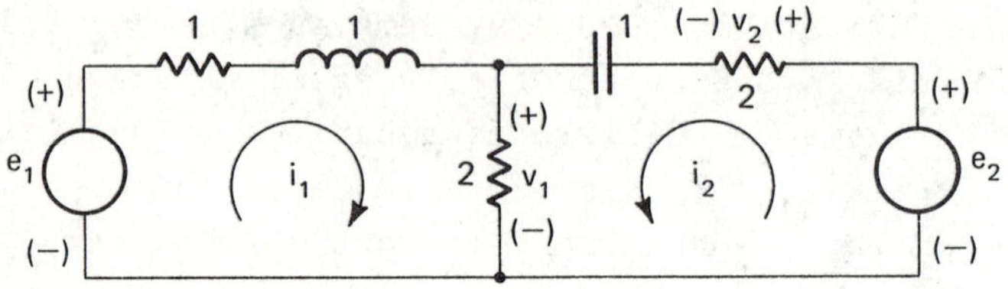

Choosing the internal variables i_1 and i_2 as shown,

$$e_1 = (1+D)i_1 + 2(i_1+i_2)$$

$$e_2 = 2(i_1+i_2) + (2+\frac{1}{D})i_2$$

or

$$(3+D)i_1 + 2i_2 = e_1$$

$$2i_1 + (4+\frac{1}{D})i_2 = e_2 .$$

Also

$$v_1 = 2(i_1+i_2)$$

$$v_2 = 2i_2.$$

In the general notation of Eq. 3.51:

$$\underline{w} = \begin{bmatrix} i_1 \\ i_2 \end{bmatrix}, \quad \underline{f} = \begin{bmatrix} e_1 \\ e_2 \end{bmatrix}, \quad \underline{y} = \begin{bmatrix} v_1 \\ v_2 \end{bmatrix}$$

$$M(D) = \begin{bmatrix} 3+D & 2 \\ 2 & 4+\frac{1}{D} \end{bmatrix}, \quad N(D) = I$$

$$P(D) = \begin{bmatrix} 2 & 2 \\ 0 & 2 \end{bmatrix}.$$

Now

$$\det M(\lambda) = (3+\lambda)(4+\frac{1}{\lambda}) - 4,$$

therefore, from this the characteristic equation is

$$4\lambda^2 + 9\lambda + 3 = 0$$

and

$$\lambda_1, \lambda_2 = \frac{-9 \pm \sqrt{33}}{8} = -0.41, -1.85 .$$

Then

$$i_{1c} = k_{11}(\exp -0.41t) + k_{12}(\exp -1.85t)$$

$$i_{2c} = k_{21}(\exp -0.41t) + k_{22}(\exp -1.85t).$$

By inspection of this simple network,

$$i_{1p} = 0.33u_1(t)$$

$$i_{2p} = 0.$$

Thus we have for $t \geq 0$,

$$i_1 = 0.33u_1(t) + k_{11}(\exp -0.41t) + k_{12}(\exp -1.85t)$$

$$i_2 = k_{21}(\exp -0.41t) + k_{22}(\exp -1.85t).$$

At t=0, the current in the inductor and the voltage across the capacitor are zero, therefore,

$$\left.\frac{di_1}{dt}\right|_{0+} + 2i_2(0+) = 1$$

$$4i_2(0+) = 2$$

and $i_1(0+) = 0$, $i_2(0+) = 0.5$, $\left.\frac{di_1}{dt}\right|_{0+} = 0$. Using the derivative of the second differential equation at 0+,

$$2\left.\frac{di_1}{dt}\right|_{0+} + 4\left.\frac{di_2}{dt}\right|_{0+} + i_2(0+) = 0$$

and from this

$$\left.\frac{di_2}{dt}\right|_{0+} = -0.125 \ .$$

The four constants can now be evaluated by using the two solutions and their derivatives at t=0+:

$$0 = 0.33 + k_{11} + k_{12}$$

$$0 = -\ 0.41k_{11} - 1.85k_{12}$$

$$0.5 = k_{21} + k_{22}$$

$$-0.125 = -\ 0.41k_{21} - 1.85k_{22}$$

and $k_{11} = -0.43$, $k_{12} = 0.095$, $k_{21} = 0.56$, $k_{22} = -0.056$.

Therefore, for $t \geq 0$,

$$i_1 = 0.33 - 0.43(\exp -0.41t) + 0.095(\exp -1.85t)$$

$$i_2 = 0.56(\exp -0.41t) - 0.056(\exp -1.85t).$$

Finally, for $t \geq 0$,

$$v_1 = 0.67 + 0.26(\exp -0.41t) + 0.08(\exp -1.85t)$$

$$v_2 = 1.12(\exp -0.41t) - 0.112(\exp -1.85t).$$

In the case of stationary systems, the determination of the system response requires finding solutions of single or simultaneous ordinary linear differential or difference equations with constant coefficients. An alternative to the classical methods in the use of transform methods, and the Laplace and Z-Transforms are particularly convenient and widely used. They have the advantage of making it relatively simple to deal with arbitrary initial conditions and to require only some algebraic manipulation and the use of one of the inversion methods of Appendix A, e.g., partial fraction expansion or an inversion integral evaluation using residue theory.

Example 4.3

a) For the problem of Ex. 4.1a, using the Laplace Transform,

$$(s^2+3s+2)Y(s) = (2s+3)\frac{1}{s^2} + (s+3)y(0+) + \dot{y}(0+) \ .$$

Since $y(0+) = \dot{y}(0+) = 0$,

$$Y(s) = \frac{(2s+3)}{s^2(s+2)(s+1)} \ .$$

Now $y(t) = \sum \text{Res}\ Y(s)\varepsilon^{ts}$, $t > 0$, at $s=0$, $s=-2$, and $s=-1$. Therefore,

$$y(t) = (3/2)t - 5/4 + 1/4\varepsilon^{-2t} + \varepsilon^{-t}, \qquad t > 0.$$

b) For the problem of Ex. 4.1b, using the Z-Transform,

$$(z^2-0.75z+0.125)Z(z) = (z-0.4)\frac{z}{(z-1)^2} + (z+1)z(0) + zz(1).$$

Since $z(0) = z(1) = 0$,

$$Z(z) = \frac{z(z-0.4)}{(z-1)^2(z-0.5)(z-0.25)} \ .$$

Now

$$z(n) = \sum \text{Res } Z(z)z^{n-1}, \qquad n \geq 0,$$

at z=1, z=0.5, and z=0.25. Therefore,

$$z(n) = 1.6n - 2.67 + 1.6(0.5)^n + 1.07(0.25)^n, \qquad n \geq 0.$$

4.2.2 Convolution Forms – Direct Input/Output

The second general way of finding the response of a linear system makes use of a spectral model of the input and an associated system response function. The most common time-domain forms in this group result from resolving the input signal in terms of $u_o(t)$, $u_1(t)$, $v_o(n)$, and $v_1(n)$. The scalar signal models are therefore given by Eq. 2.55, 2.55a, 2.56, and 2.56a, respectively, and the associated system response functions are $h(t,\lambda)$, $a(t,\lambda)$, $h(n,k)$, and $a(n,k)$. The corresponding response forms are given by Eq. 3.12, 3.13, 3.14, and 3.15, and these are repeated here:

$$y(t) = \int_{-\infty}^{\infty} u(\lambda)h(t,\lambda)d\lambda \tag{4.3}$$

$$y(t) = \int_{-\infty}^{\infty} \dot{u}(\lambda)a(t,\lambda)d\lambda \tag{4.4}$$

$$z(n) = \int_{-\infty}^{\infty} v(k)h(n,k) \tag{4.5}$$

$$z(n) = \int_{-\infty}^{\infty} \Delta_b v(k)a(n,k) \ . \tag{4.6}$$

When the input signals $u(t)=0$, $t<t_o$, or $v(n)=0$, $n<n_o$, and the linear system is realizable and stationary, the usual convolution integrals or sums follow. Thus we let

$$h(t,\lambda) \rightarrow h(t-\lambda)$$

$$a(t,\lambda) \rightarrow a(t-\lambda)$$
$$h(n,k) \rightarrow h(n-k)$$
$$a(n,k) \rightarrow a(n-k)$$

and h and a in both cases are zero for negative argument. The above equations then become for $t>t_0$ and $n>n_0$

$$y(t) = \int_{t_0}^{t} u(\lambda)h(t-\lambda)d\lambda \tag{4.7}$$

$$y(t) = u(t_0+)a(t) + \int_{t_0+}^{t} \dot{u}(\lambda)a(t-\lambda)d\lambda \tag{4.8}$$

$$z(n) = \sum_{k=n_0}^{n} v(k)h(n-k) \tag{4.9}$$

$$z(n) = v(n_0)a(n) + \sum_{k=n_0+1}^{n} \Delta_b v(k)a(n-k) \ . \tag{4.10}$$

In all of these cases, the system is at rest prior to t_0 or n_0. When the system is not initially relaxed (i.e., there is an initial storage of energy), the previous classical or transform methods or the state variable forms discussed later may be used. Multiple input/multiple output response forms follow by analogy to Eq. 4.7 through 4.10. Using the definitions for linear stationary system unit impulse or unit pulse response matrices given by Eq. 3.53 and 3.56

$$\underline{y}(t) = \int_{t_0}^{t} H(t-\lambda)\underline{u}(\lambda)d\lambda \tag{4.11}$$

and

$$\underline{z}(n) = \sum_{k=n_0}^{n} H(n-k)\underline{v}(k) \ . \tag{4.12}$$

When the spectral argument of the input signal model is complex or imaginary (s, jω, z, γ), the common frequency domain response forms occur. For stationary, realizable linear system and transformable input signals; the $\mathcal{L}$, $\mathcal{F}$, Z, and

$\mathcal{F}_d$ Transforms lead to spectral functions U(s), U(jω), V(z), and V(γ), and the associated system response functions to ε^{st}, $\varepsilon^{j\omega t}$, z^n, and γ^n are R(t,s), R(t,jω), R(n,z), and R(n,γ). Using Eq. A.5, the inverse Laplace Transform,

$$u(t) = \frac{1}{2\pi j}\int_{\sigma-j\infty}^{\sigma+j\infty} U(s)\varepsilon^{ts}ds$$

and for the system characterized by the linear operator L_c,

$$L_c u(t) = y(t) = \frac{1}{2\pi j}\int_{\sigma-j\infty}^{\sigma+j\infty} U(s)L_c\varepsilon^{ts}ds \ .$$

But $L_c\varepsilon^{ts} = R(t,s)$, a response to ε^{ts}, and using Eq. 3.47,

$$R(t,s) = (N(s)/M(s))\varepsilon^{ts}$$

$$= H(s)\varepsilon^{ts} \ .$$

Therefore,

$$y(t) = \mathcal{L}^{-1}Y(s) = \frac{1}{2\pi j}\int_{\sigma-j\infty}^{\sigma+j\infty} U(s)H(s)\varepsilon^{ts}ds$$

where

$$Y(s) = U(s)H(s) \ . \tag{4.13}$$

This is a well-known form and has been developed in this sequence to follow the plan of (1) choosing an input signal model, (2) finding the corresponding system response function, and then (3) finding the output signal. A more direct method would be to simply apply the real convolution theorem of the Laplace Transform (App. A) to Eq. 4.7 to give Eq. 4.13 immediately. Note that Eq. 4.13 has the desirable property expressed after Eq. 3.16, i.e., the spectral function of the output is obtained directly from the spectral function of the input and the appropriate system model -- in this case H(s), the transfer function. The response y(t) again is that for an initially relaxed system. A similar development using the Fourier Transform leads to

$$R(t,j\omega) = (N(j\omega)/M(j\omega))\varepsilon^{j\omega t}$$
$$= H(j\omega)\varepsilon^{j\omega t}$$

and

$$y(t) = \mathfrak{F}^{-1}Y(j\omega) = \frac{1}{2\pi}\int_{-\infty}^{\infty} U(j\omega)H(j\omega)\varepsilon^{jt\omega}d\omega$$

where

$$Y(j\omega) = U(j\omega)H(j\omega) \quad . \qquad (4.14)$$

Also, using the Z-Transform and Eq. 3.49,

$$R(n,z) = (N(z)/M(z))z^n$$
$$= H(z)z^n$$

and

$$z(n) = \mathfrak{Z}^{-1}Z(z) = \frac{1}{2\pi j}\oint V(z)H(z)z^{n-1}dz$$

where

$$Z(z) = V(z)H(z) \quad . \qquad (4.15)$$

Furthermore, using the $\mathfrak{F}_d$-Transform,

$$R(n,\gamma) = (N(\gamma)/M(\gamma))\gamma^n$$
$$= H(\gamma)\gamma^n$$

and

$$z(n) = \mathfrak{F}_d^{-1}Z(\gamma) = \frac{1}{2\pi j}\oint V(\gamma)H(\gamma)\gamma^{n-1}d\gamma$$

where

$$Z(\gamma) = V(\gamma)H(\gamma) \ . \tag{4.16}$$

Also included in this group of response forms are those resulting from the use of the Fourier Series. Thus, using Eq. 2.45 and 3.22,

$$F_c(t,k) = (N(jk\omega_o)/M(jk\omega_o))(\exp jk\omega_o t)$$

$$= H(jk\omega_o)(\exp jk\omega_o t)$$

and

$$y(t) = \sum_{k=-\infty}^{\infty} Y_k(\exp jk\omega_o t)$$

where

$$Y_k = U_k H(jk\omega_o). \tag{4.17}$$

Similarly, using Eq. 2.46 and 3.24,

$$F_d(n,k) = (N(\exp jk\lambda_o)/M(\exp jk\lambda_o))(\exp jk\lambda_o)$$

$$= H(\exp jk\lambda_o)(\exp jk\lambda_o)$$

and

$$z(n) = \sum_{k=-K}^{K} Z_k(\exp jk\lambda_o n)$$

where

$$Z_k = V_k H(\exp jk\lambda_o) \tag{4.18}$$

for N_o odd. A similar result for N_o even follows from Eq. 2.47.

Multiple input/multiple output response forms are obtained by extending the above results. For example, using the definitions of transfer matrices given by Eq. 3.52 and 3.55

and applying the real convolution theorems of Appendix A to Eq. 4.11 and 4.12,

$$\underline{Y}(s) = H(s)\underline{U}(s) \tag{4.19}$$

with

$$\underline{y}(t) = \mathcal{L}^{-1}\underline{Y}(s)$$

and

$$\underline{Z}(z) = H(z)\underline{V}(z) \tag{4.20}$$

with

$$\underline{z}(n) = \mathcal{Z}^{-1}\underline{Z}(z) .$$

<u>Example 4.4</u>

For the problem of Ex. 4.2, the transfer matrix is

$$H(s) = P\,M(s)^{-1}N$$

$$= \begin{bmatrix} 2 & 2 \\ 0 & 2 \end{bmatrix} \begin{bmatrix} s+3 & 2 \\ 2 & \frac{4s+1}{s} \end{bmatrix}^{-1} \begin{bmatrix} 1 & 0 \\ 0 & 1 \end{bmatrix}$$

$$= \frac{s}{4s^2+9s+3} \begin{bmatrix} \frac{4s+2}{s} & 2(s+1) \\ -4 & 2(s+3) \end{bmatrix}$$

Thus

$$\underline{Y}(s) = \begin{bmatrix} V_1(s) \\ V_2(s) \end{bmatrix} = H(s) \begin{bmatrix} \frac{1}{s} \\ \frac{2}{s} \end{bmatrix}$$

and

$$\underline{y}(t) = \begin{bmatrix} v_1(t) \\ \\ v_2(t) \end{bmatrix} = \mathcal{L}^{-1}\underline{Y}(s) .$$

The result is given in Ex. 4.2. Note that if there had been an initial current in the inductor or an initial voltage across the capacitor, this method would not give the complete response.

Example 4.5

Consider the numerical integrator of Ex. 3.23.

$$M(z) = z^2-1$$

$$N(z) = \frac{1}{3}T\ (z^2+4z+1)$$

thus

$$H(z) = N(z)/M(z) = \frac{1}{3}T\ (\frac{z^2+4z+1}{z^2-1})$$

The results of Ex. 3.23 can be checked for $v(n) = n^2 = 2v_3(n)$. From Chapter 2, Section 2.4,

$$V(z) = \frac{z(z+1)}{(z-1)^3} .$$

Using this (with T=1),

$$Z(z) = V(z)H(z)$$

$$= \frac{1}{3}\ \frac{z(z+1)(z^2+4z+1)}{(z-1)^3(z^2-1)}$$

$$= \frac{1}{3}\ \frac{z^3+4z^2+z}{z^4-4z^3+6z^2-4z+1}$$

and by long division

$$Z(z) = \frac{1}{3}\ [z^{-1} + 8z^{-2} + 27z^{-3} + 64z^{-4} +\ldots]$$

and z(0) = 0, z(1) = 1/3, z(2) = 8/3, z(3) = 9, z(4) = 64/3, etc. We note again as in Ex. 3.23 that the initial state in this example is zero and v(0)=0, and the solution is correct. If the Z-Transform were used to solve the difference equation in Ex. 3.23, z(1) = 1/3 and v(1) = 1 would be included (see Section A.3.3, Property 4), but there would be a cancellation which results in the same form as that found above by using H(z). See also the discussion after Eq. 3.38b.

Example 4.6

Two simple graphical examples of the use of Eq. 4.7 and 4.9 are given since it is relatively easy to see the meaning of real convolution in this way.

a) For u(·) and h(·) as shown, y(t), t>0, can be found by carrying out the four steps that are illustrated. These are folding ($h(\lambda) \to h(-\lambda)$), translating ($h(-\lambda) \to h(t-\lambda)$), multiplying ($u(\lambda)h(t-\lambda)$), and integrating.

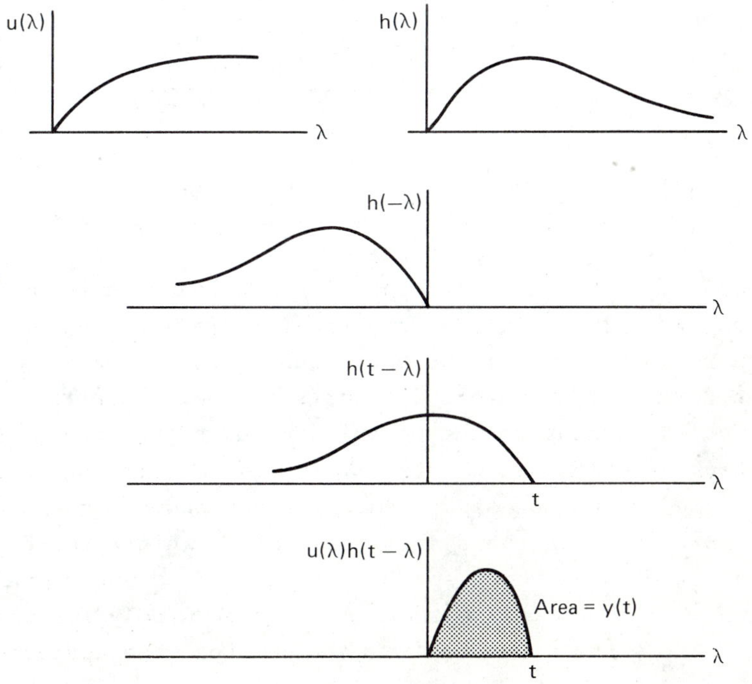

b) The same procedure as in a) applies to the discrete-time case except that the last step is summing.

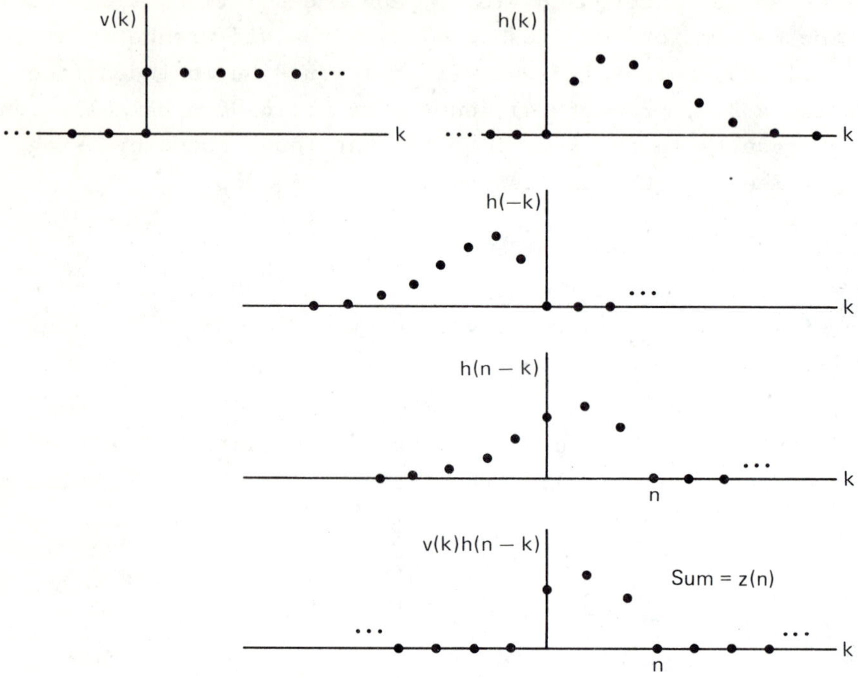

4.2.3 Use of Signal Auxiliary Functions

The third way of studying system response uses a time or frequency domain auxiliary function model of the input signal. The response is then the corresponding auxiliary function of the output. This differs from the two previous methods since in this case only input and output signal "averages" are considered rather than the signals or their spectral functions. Referring to the definitions of signal energy and translation functions in Chapter 2, it can be seen that this point of view is one of signal energy input and output, and the interpretation of the system response in this form is different from cases where specific input and output signals are known. Although this method is not used widely for the deterministic signals covered in this section, the analogous methods for stochastic signals -- the use of correlation functions -- is a very popular method. The results given here are sometimes useful in studying the integral square performance of systems, as shown in Chapter 5, and in the design of certain systems (e.g., signal filters) which minimize an integral of squared error.

From Eq. 2.69, the autotranslation function of u is

$$w_u(\tau) = \int_{-\infty}^{\infty} u(t)u(t+\tau)dt$$

and for a stationary, realizable system with $u(t) = 0$, $t<t_o$, from Eq. 4.7 we have

$$y(t) = \int_{t_o}^{t} u(\lambda)h(t-\lambda)d\lambda$$

$$= \int_{0}^{t-t_o} u(t-\lambda)h(\lambda)d\lambda$$

$$= \int_{0}^{\infty} u(t-\lambda)h(\lambda)d\lambda \ , \qquad t > t_o .$$

Also,

$$y(t+\tau) = \int_{0}^{\infty} u(t+\tau-\alpha)h(\alpha)d\alpha$$

and therefore,

$$w_y(\tau) = \int_{-\infty}^{\infty}\left\{\int_{0}^{\infty} h(\lambda)u(t-\lambda)d\lambda \int_{0}^{\infty} h(\alpha)u(t+\tau-\alpha)d\alpha\right\} dt$$

Changing the order of integration and integrating over t, we get

$$w_y(\tau) = \int_{0}^{\infty} h(\lambda)d\lambda \int_{0}^{\infty} h(\alpha)w_u(\tau-\alpha+\lambda)d\alpha \ . \qquad (4.21)$$

A similar procedure for discrete-time systems starts with Eq. 2.70 where

$$w_v(i) = \sum_{i=-\infty}^{\infty} v(n)v(n+i)$$

and for a stationary, realizable system with $v(n) = 0,\ n<n_0$, from Eq. 4.9,

$$z(n) = \sum_{j=n_0}^{n} v(j)h(n-j)$$

$$= \sum_{j=0}^{\infty} v(n-j)h(j), \qquad n > n_0$$

and

$$z(n+i) = \sum_{k=0}^{\infty} v(n+i-k)h(k) \ .$$

Then

$$w_z(i) = \sum_{j=0}^{\infty} h(j) \sum_{k=0}^{\infty} w_v(i-k+j)h(k) \ . \tag{4.22}$$

Equations 4.21 and 4.22 also can be expressed in terms of two integral or sums by using cross translation functions. The analogous forms are developed in the next section for correlation functions.

When the input and output signals are vectors,

$$W_{\underline{u}}(\tau) = \int_{-\infty}^{\infty} \underline{u}(t)\underline{u}^T(t+\tau)dt$$

as shown in Chapter 2, and following a sequence of steps similar to those above and using Eq. 4.11, we obtain

$$W_{\underline{y}}(\tau) = \int_{0}^{\infty} H(\lambda)d\lambda \int_{0}^{\infty} W_{\underline{u}}(\tau-\alpha+\lambda)H^T(\alpha)d\alpha \tag{4.23}$$

where $H(\cdot)$ is the impulse response matrix of the linear, stationary, realizable system. For discrete-time systems,

$$W_{\underline{v}}(i) = \sum_{n=-\infty}^{\infty} \underline{v}(n)\underline{v}^T(n+i)$$

and

$$W_{\underline{z}}(i) = \sum_{j=0}^{\infty} h(j) \sum_{k=0}^{\infty} W_{\underline{v}}(i-k+j)H^T(k) \qquad (4.24)$$

The time-domain results given by Eq. 4.21, 4.22, 4.23, and 4.24 may be expressed in the frequency domain by using the appropriate transform. The frequency-domain forms are convenient to use if the inversion integral for finding the final time-domain answer (i.e., the translation function or a matrix of translation functions) is evaluated by using the calculus of residues or by tables [4.][5.]. The time-domain forms can be used when a computer is to be used. Starting with Eq. 4.21,

$$W_y(s) = \mathcal{L}_2 w_y(\tau)$$

$$= \int_{-\infty}^{\infty} \varepsilon^{-s\tau} \int_{0}^{\infty} h(\lambda)d\lambda \int_{0}^{\infty} h(\alpha)w_u(\tau-\alpha+\lambda)d\alpha d\tau$$

$$= \int_{0}^{\infty} h(\lambda)\varepsilon^{s\lambda}d\lambda \int_{-\infty}^{\infty} \varepsilon^{-s\lambda}\varepsilon^{-s\tau} \int_{0}^{\infty} h(\alpha)w_u(\tau-\alpha+\lambda)d\alpha d\tau$$

by interchanging the order of integration. Now let $\sigma = \tau-\alpha+\lambda$, then

$$W_y(s) = \int_{0}^{\infty} h(\lambda)\varepsilon^{s\lambda}d\lambda \int_{-\infty}^{\infty} \varepsilon^{-s\sigma}w_u(\sigma)d\sigma \int_{0}^{\infty} h(\alpha)\varepsilon^{-s\alpha}d\alpha$$

$$= H(-s)W_u(s)H(s) \qquad (4.25)$$

where H(s) is the system transfer function. Similarly, for

$$W_z(z) = \mathfrak{Z}_2 w_z(i)$$

the result is

$$W_z(z) = H(z^{-1})W_v(z)H(z) \qquad (4.26)$$

and the corresponding forms for multiple input/multiple output are

$$W_{\underline{y}}(s) = H(-s)W_{\underline{u}}(s)H^T(s) \qquad (4.27)$$

and

$$W_{\underline{z}}(z) = H(z^{-1})W_{\underline{v}}(z)H^T(z) \ . \qquad (4.28)$$

<u>Example 4.7</u>

For the signal of Example 2.11 as input, i.e.,

$$u(t) = \varepsilon^{-at}u_1(t)$$

the autotranslation function is

$$w_u(\tau) = \frac{1}{2a}\ (\exp -a|\tau|)$$

and

$$W_u(s) = \frac{1}{a^2-s^2} \ .$$

For a simple system with impulse response

$$h(\lambda) = k\varepsilon^{-b}u_1(\lambda), \qquad b > 0, \qquad b \neq a,$$

we have

$$H(s) = \frac{k}{s+b} \ .$$

Then using Eq. 4.25,

$$W_y(s) = \frac{k}{(-s+b)} \cdot \frac{1}{(-s+a)(s+a)} \cdot \frac{k}{(s+b)}$$

$$= \frac{k^2}{(s-b)(s-a)(s+a)(s+b)} \ .$$

Then

$$w_y(\tau) = \mathcal{L}_2^{-1} W_y(s)$$

$$= \sum \text{Res } W_y(s)\varepsilon^{\tau s}\Big|_{\substack{s=-a \\ s=-b}}, \qquad \tau > 0$$

where the contour of integration encloses the left half s-plane. Carrying out the integration,

$$w_y(\tau) = \frac{k^2}{(-a-b)(-2a)(-a+b)} \varepsilon^{-a\tau}$$

$$+ \frac{k^2}{(-2b)(-b-a)(-b+a)} \varepsilon^{-b\tau}, \qquad \tau > 0$$

$$= k_1 \left[\frac{\varepsilon^{-a\tau}}{a} - \frac{\varepsilon^{-b\tau}}{b} \right], \qquad \tau > 0 .$$

where

$$k_1 = \frac{k^2}{2(b-a)(b+a)} .$$

Since $w_y(\tau)$ is an even function,

$$w_y(\tau) = k_1 \left[\frac{(\exp -a|\tau|)}{a} - \frac{(\exp -b|\tau|)}{b} \right], \quad \forall \tau.$$

The integral square system output is then

$$w_y(o) = k_1 \left[\frac{1}{a} - \frac{1}{b} \right]$$

$$= \frac{k^2}{2ab\ (a+b)} .$$

Three approaches to the calculation of system response have now been considered, and these may be classified in Table 4.1 (single input/output included as special cases).

TABLE 4.1

MULTIPLE INPUT/OUTPUT FORMS

Method	Input Signal Model	System Model
1.	Function	rth-order form
	$\underline{u}(t)$ $\underline{v}(n)$	M(D), N(D), P(D) M(E), N(E), P(E)
2.	Spectral Function	Response Function
	$\underline{u}(\lambda)$, $\dot{\underline{u}}(\lambda)$ $\underline{v}(k)$, $\Delta_b\underline{v}(k)$ $\underline{U}(s)$ $\underline{V}(z)$ U_k V_k	$H(t-\lambda)$, $A(t-\lambda)$ $H(n-k)$, $A(n-k)$ $H(s)\varepsilon^{ts} = R(t,s)$ $H(z)z^n = R(n,z)$ $F_c(t,k)$ $F_d(n,k)$
3.	Translation Function	Response Function
	$W_{\underline{u}}(\tau)$, $W_{\underline{v}}(i)$	$H(\lambda)$, $H(j)$

4.2.4 State Variable Forms

A fourth point of view follows if in the first method the system model is a state-variable form. It is necessary to solve linear vector-matrix differential or difference equations. The multiple input/output forms for stationary systems are considered first and then the more general forms for time-varying systems are given. Starting with Eq. 3.57a, i.e.,

$$\dot{\underline{x}} = A\underline{x} + B\underline{u}$$

and for $\underline{u}=\underline{0}$, the corresponding homogeneous solution is given in terms of the transition matrix of Eq. 3.65,

$$\Phi(t-t_o) = (\exp A(t-t_o)),$$

therefore

$$\underline{x}_c(t) = \Phi(t-t_o)\underline{\xi}_o ,$$

where $\underline{\xi}_o$ is an arbitrary constant vector. When $\underline{u}\neq\underline{0}$, $t\geq t_o$, let

$$\underline{x}_p(t) = \Phi(t-t_o)\underline{f}(t)$$

where $\underline{f}(t)$ is to be determined. Substituting in the system differential equation using

$$\dot{\underline{x}}_p(t) = \dot{\Phi}(t-t_o)\underline{f}(t) + \Phi(t-t_o)\dot{\underline{f}}(t) ,$$

leads to

$$\dot{\Phi}(t-t_o)\underline{f}(t) + \Phi(t-t_o)\dot{\underline{f}}(t) = A\Phi(t-t_o)\underline{f}(t) + B\underline{u}(t) .$$

Now $\dot{\Phi}(\cdot) = A\Phi(\cdot)$, therefore, require

$$\Phi(t-t_o)\dot{\underline{f}}(t) = B\underline{u}(t)$$

or

$$\dot{\underline{f}}(t) = \Phi(t_o-t)B\underline{u}(t)$$

and

$$\underline{f}(t) = \int_{t_o}^{t} \Phi(t_o-\lambda)B\underline{u}(\lambda)d\lambda .$$

Using this we have

$$\underline{x}_p(t) = \int_{t_o}^{t} \Phi(t-\lambda)B\underline{u}(\lambda)d\lambda$$

and the complete solution is then

$$\underline{x}(t) = \underline{x}_c(t) + \underline{x}_p(t)$$

$$= \Phi(t-t_o)\underline{\xi}_o + \int_{t_o}^{t} \Phi(t-\lambda)B\underline{u}(\lambda)d\lambda \quad .$$

Now at $t=t_o$, $\underline{x}(t_o)$ is known and using the general solution, we get

$$\underline{x}(t_o) = \Phi(t_o-t_o)\underline{\xi}_o = \underline{\xi}_o$$

so that

$$\underline{x}(t) = \Phi(t-t_o)\underline{x}(t_o) + \int_{t_o}^{t} \Phi(t-\lambda)B\underline{u}(\lambda)d\lambda \quad . \tag{4.29a}$$

Also, the output is

$$\underline{y}(t) = C\underline{x}(t) + D\underline{u}(t). \tag{4.29b}$$

When the system is time-varying, a similar sequence of steps gives

$$\underline{x}(t) = \Phi(t,t_o)\underline{x}(t_o) + \int_{t_o}^{t} \Phi(t,\lambda)B(\lambda)\underline{u}(\lambda)d\lambda \tag{4.30a}$$

with

$$\underline{y}(t) = C(t)\underline{x}(t) + D(t)\underline{u}(t). \tag{4.30b}$$

These general results are in terms of the transition matrix and methods for finding it have been discussed in Chapter 3. Digital computer programs for finding numerical solutions of Eq. 4.29 are common (see Appendix D), and one popular method is to use a system model as in Eq. 3.118. This requires solving linear vector-matrix difference equations as the discrete-time counterparts to Eq. 4.29.

For discrete-time systems we start with Eq. 3.60a,

$$\underline{x}(n+1) = A\underline{x}(n) + B\underline{v}(n)$$

and for $\underline{v}=\underline{0}$, the homogeneous solution is in terms of the transition matrix of Eq. 3.66,

$$\Phi(n-n_o) = A^{(n-n_o)}$$

therefore

$$\underline{x}_c(n) = \Phi(n-n_o)\underline{\xi}_o$$

where $\underline{\xi}_o$ is an arbitrary constant vector. When $\underline{v}\neq\underline{0}$, $n\geq n_o$, the particular solution $\underline{x}_p(n)$ could be found using a sequence of steps analogous to those used for $\underline{x}_p(t)$ above. Alternatively, an iterative procedure may be used to find the complete solution directly. Thus for $n=n_o$, we get

$$\underline{x}(n_o+1) = A\underline{x}(n_o) + B\underline{v}(n_o),$$

and for $n=n_o+1$,

$$\begin{aligned}\underline{x}(2+n_o) &= A\underline{x}(n_o+1) + B\underline{v}(n_o+1)\\ &= A^2\underline{x}(n_o) + AB\underline{v}(n_o) + B\underline{v}(n_o+1)\end{aligned}$$

and so on. In general, let $n=n_o+i-1$, to obtain

$$\underline{x}(i+n_o) = A^i\underline{x}(n_o) + \sum_{j=0}^{i-1} A^j B\underline{v}(n_o-j+i-1).$$

Now let $m=i+n_o$,

$$\underline{x}(m) = A^{m-n_o}\underline{x}(n_o) + \sum_{j=1}^{m-n_o-1} A^j B\underline{v}(m-j-1)$$

and then let $m\to n$ and $j=n-k-1$,

$$\underline{x}(n) = A^{(n-n_o)}\underline{x}(n_o) + \sum_{k=n_o}^{n-1} A^{n-k-1}B\underline{v}(k)$$

or for $n>n_o$,

$$\underline{x}(n) = \Phi(n-n_o)\underline{x}(n_o) + \sum_{k=n_o}^{n-1} \Phi(n-k-1)B\underline{v}(k) \quad (4.31a)$$

with

$$\underline{z}(n) = C\underline{x}(n) + D\underline{v}(n). \quad (4.31b)$$

When the system is time-varying, a similar sequence of steps leads to

$$\underline{x}(n) = \Phi(n,n_o)\underline{x}(n_o) + \sum_{k=n_o}^{n-1} \Phi(n,k+1)B(k)\underline{v}(k) \quad (4.32a)$$

for $n>n_o$, and where

$$\Phi(n,i) = \prod_{j=i}^{n-1} A(j), \qquad n > i,$$

and the output is

$$\underline{z}(n) = C(n)\underline{x}(n) + D(n)\underline{v}(n) . \quad (4.32b)$$

Thus Eq. 4.32 gives a general numerical form for finding the response of a linear time-varying system. Conversion of a time-varying continuous-time system model to an approximate discrete-time model makes it possible to compute approximate numerical results by using Eq. 4.32. (System conversions were discussed in Chapter 3; see, for example, Conversions VII and VIII and consider extensions to time-varying cases). These forms generally are more useful than those resulting from rth-order form system models, and this is why the solutions of rth-order time-varying linear differential and difference equations were not discussed in Chapter 3.

Example 4.8

Consider the system of Example 3.22 with $\alpha_1=a$, $\alpha_o=\beta_1=0$, $\beta_o=1$

$$\underline{\dot{x}} = \begin{bmatrix} 0 & 1 \\ 0 & -a \end{bmatrix} \underline{x} + \begin{bmatrix} 0 \\ 1 \end{bmatrix} u$$

$$y = [1 \quad 0]\underline{x} \quad .$$

The transition matrix was found in Example 3.28

$$\Phi(\lambda) = \varepsilon^{A\lambda} = \begin{bmatrix} 1 & \frac{1}{a}(1-\varepsilon^{-a\lambda}) \\ 0 & \varepsilon^{-a\lambda} \end{bmatrix}$$

Using Eq. 4.29a, with $t_o=0$, $u(t)=u_1(t)$, and $\underline{x}(0)=[1 \quad 0]^T$, we obtain

$$\underline{x}(t) = \begin{bmatrix} 1 & \frac{1}{a}(1-\varepsilon^{-at}) \\ 0 & \varepsilon^{-at} \end{bmatrix} \begin{bmatrix} 1 \\ 0 \end{bmatrix} +$$

$$+ \int_o^t \begin{bmatrix} 1 & \frac{1}{a}(1-\varepsilon^{-a(t-\lambda)}) \\ 0 & \varepsilon^{-a(t-\lambda)} \end{bmatrix} \begin{bmatrix} 0 \\ 1 \end{bmatrix} d\lambda$$

Thus, for $t>0$,

$$x_1(t) = 1 + \int_o^t \frac{1}{a}(1-(\exp -a(t-\lambda))d\lambda$$

$$x_2(t) = \int_o^t (\exp -a(t-\lambda))d\lambda$$

or

$$\underline{x}(t) = \begin{bmatrix} 1 + \frac{1}{a}t - \frac{1}{a}2(1-\varepsilon^{-at}) \\ \\ \frac{1}{a}(-\varepsilon^{-at}) \end{bmatrix} .$$

Also,

$$y(t) = x_1(t) .$$

Example 4.9

Consider the numerical integrator of Example 3.23

$$\underline{x}(n+1) = \begin{bmatrix} 0 & 1 \\ 1 & 0 \end{bmatrix} \underline{x}(n) + \begin{bmatrix} 4/3 \\ 2/3 \end{bmatrix} v(n)$$

$$z(n) = [1 \quad 0]\ \underline{x}(n) + [1/3]\ v(n).$$

The transition matrix was found in Example 3.28

$$\Phi(k) = \begin{bmatrix} \frac{1}{2}(1+(-1)^k) & \frac{1}{2}(1-(-1)^k) \\ \frac{1}{2}(1-(-1)^k) & \frac{1}{2}(1+(-1)^k) \end{bmatrix}$$

Using Eq. 4.31a with $n_o=0$, $v(n)=v_o(n)$, and $\underline{x}(0)=\underline{0}$,

$$\underline{x}(n) = \sum_{k=0}^{n-1} \Phi(n-k-1) \begin{bmatrix} 4/3 \\ 2/3 \end{bmatrix} v_o(k).$$

Therefore for $n>0$,

$$x_1(n) = 2/3(1+(-1)^{n-1}) + 1/3(1-(-1)^{n-1})$$

$$x_2(n) = 2/3(1-(-1)^{n-1}) + 1/3(1+(-1)^{n-1})$$

and

$$z(0) = 1/3$$

$$z(n) = x_1(n), \quad n > 0 ,$$

a response to a unit pulse input. We also could obtain this result (since the initial conditions are zero) by taking $\mathcal{Z}^{-1}H(z)$ where $H(z)$ is given in Ex. 3.19b. However, we now have $z(0)\neq 0$, a problem that was anticipated in Ex. 3.23. It is easy to see that if $x_1(0)=-1/3$, the initial value of z will be zero. By indicating the new response resulting from $\underline{x}^T(0) = [-\frac{1}{3} \quad x_2(0)]$ with primes, we have

$$\underline{x}'(n) = \Phi(n)\begin{bmatrix} -\frac{1}{3} \\ \\ x_2(0) \end{bmatrix} + \underline{x}(n), \quad n>0$$

where $\underline{x}^T(n)=[x_1(n) \;\; x_2(n)]$ is given above. Then

$$z'(n) = -\frac{1}{3}\phi_{11}(n) + \phi_{12}(n)x_2(0)+x_1(n), \quad n>0$$

with $z'(0)=0$. The value of $x_2(0)$ can be chosen to satisfy any other condition, e.g., a condition on $z'(1)$. See Problem 17 at the end of this chapter for a similar situation.

The results given in Chapter 3 on System Conversions II and IV can now be obtained by using Eq. 4.11, 4.12, 4.29, and 4.31. From Eq. 4.11, we obtain a first equation

$$\underline{y}(t) = \int_{t_o}^{t} H(t-\lambda)\underline{u}(\lambda)d\lambda$$

and from Eq. 4.29,

$$\underline{y}(t) = C\Phi(t-t_o)\underline{x}(t_o) + \int_{t_o}^{t} C\Phi(t-\lambda)B\underline{u}(\lambda)d\lambda+D\underline{u}(t).$$

For $\underline{u}$ resolved in terms of u_o, i.e.,

$$\underline{u}(t) = \int_{t_o}^{t} \underline{u}(\lambda)u_o(t-\lambda)d\lambda ,$$

the second equation for $\underline{y}(t)$ becomes

$$\underline{y}(t) = C\Phi(t-t_o)\underline{x}(t_o) + \int_{t_o}^{t} [C\Phi(t-\lambda)B + Du_o(t-\lambda)]\underline{u}(\lambda)d\lambda .$$

Now $\underline{y}(t)$ in terms of the impulse response matrix in the first equation is the zero-state response. Therefore, comparing this with the zero-state response in terms of C, Φ, B, and D in the second equation, we have

$$H(t-\lambda) = C\Phi(t-\lambda)B + Du_o(t-\lambda), \qquad t \geq \lambda$$

or

$$H(\tau) = C\Phi(\tau)Bu_1(\tau) + Du_o(\tau)$$

which is Eq. 3.109. Similarly, from Eq. 4.12,

$$\underline{z}(n) = \sum_{k=n_o}^{n} H(n-k)\underline{v}(k)$$

and from Eq. 4.31,

$$\underline{z}(n) = C\Phi(n-n_o)\underline{x}(n_o) + \sum_{k=n_o}^{n-1} C\Phi(n-k-1)B\underline{v}(k) + D\underline{v}(n)$$

$$= C\Phi(n-n_o)\underline{x}(n_o) + \sum_{k=n_o}^{n} [C\Phi(n-k-1)B\underline{v}(k)v_1(n-k-1) + D\underline{v}(k)v_o(n-k)] .$$

Thus we find

$$H(j) = C\Phi(j-1)Bv_1(j-1) + Dv_o(j)$$

which is Eq. 3.110. The other forms in Chapter 3 follow directly from obvious changes for scalar input and output and by using the Laplace and Z-Transforms.

4.3 STOCHASTIC CASES

Although it is possible to carry out a development which is analogous to that in Section 4.2 on deterministic inputs, we choose a more limited goal. Relative to the three general methods mentioned in the introduction of this chapter, the counterpart of the first method when the system input is stochastic requires the solution of stochastic differential or difference equations for the corresponding stochastic output [6.]. The second method, which uses spectral forms, involves the use of spectral functions that are random variables. These were discussed briefly in connection with Table 2.6b in Chapter 2. Both of these points of view are useful in certain types of problems, but they are not developed in this book. The third method, the use of an auxiliary function, is the usual way of studying stochastic input/output problems. As pointed out in the introduction, if the initial time t_o is in the infinite past, the initial conditions do not affect the response (assuming a stable, stationary system) and a "steady-state" response using signal correlation functions is obtained. This has a close analogy to the previous use of autotranslation functions when the stochastic signals are ergodic and time averages are used. Alternatively, if t_o is finite, initial conditions affect the response and the moments of the output process (e.g., mean and mean square) become functions of time. In the former case, the output process or sequence is stationary if the input and system are stationary. In the latter case the output process or sequence is not stationary.

4.3.1 Stationary Response

Starting with a linear, realizable, stationary system

described by h(·) and a wide-sense stationary input process {U(t)}, using Eq. 4.7 with $t_0 \to -\infty$,

$$Y(t) = \int_{-\infty}^{\infty} h(\lambda)U(t-\lambda)d\lambda \ .$$

The mean value of the output is

$$m_Y = E[Y] = E\int_{-\infty}^{\infty} h(\lambda)U(t-\lambda)d\lambda$$

and interchanging the order of operations,

$$m_Y = \int_{-\infty}^{\infty} h(\lambda)m_U d\lambda = m_U\int_{-\infty}^{\infty} h(\lambda)d\lambda \quad ,$$

and a useful form in terms of the transfer function H(s) is

$$m_Y = H(0)m_U \ . \tag{4.33}$$

To find the autocorrelation function of the output process,

$$Y(t+\tau) = \int_{-\infty}^{\infty} h(\alpha)U(t+\tau-\alpha)d\alpha$$

and

$$\begin{aligned} r_Y(\tau) &= E[Y(t)Y(t+\tau)] \\ &= E\int_{-\infty}^{\infty} h(\lambda)U(t-\lambda)d\lambda\int_{-\infty}^{\infty} h(\alpha)U(t+\tau-\alpha)d\alpha \\ &= E\int_{-\infty}^{\infty} h(\lambda)d\lambda\int_{-\infty}^{\infty} h(\alpha)U(t-\lambda)U(t+\tau-\alpha)d\alpha \\ &= \int_{-\infty}^{\infty} h(\lambda)d\lambda\int_{-\infty}^{\infty} h(\alpha)r_U(\tau-\alpha+\lambda)d\alpha \ . \end{aligned} \tag{4.34}$$

This obviously is analogous to Eq. 4.21. The double integral form can be changed to two single integrals by finding the cross correlation between the input and output processes.

$$r_{UY}(\tau) = E\,[U(t)Y(t+\tau)]$$

$$= E[U(t)\int_{-\infty}^{\infty} h(\alpha)U(t+\tau-\alpha)\,d\alpha]$$

$$= \int_{0}^{\infty} h(\alpha)r_U(\tau-\alpha)\,d\alpha \quad . \qquad (4.35a)$$

Similarly,

$$r_{YU}(\tau) = E\,[Y(t)U(t+\tau)]$$

$$= \int_{0}^{\infty} h(\alpha)r_U(\tau+\alpha)\,d\alpha \quad . \qquad (4.35b)$$

Now

$$r_{UY}(\tau+\lambda) = \int_{0}^{\infty} h(\alpha)r_U(\tau+\lambda-\alpha)\,d\alpha$$

and comparing this with Eq. 4.34,

$$r_Y(\tau) = \int_{-\infty}^{\infty} h(\lambda)r_{UY}(\tau+\lambda)\,d\lambda \quad . \qquad (4.36)$$

Also,

$$r_{YU}(-\tau-\lambda) = \int_{0}^{\infty} h(\alpha)r_U(-\tau-\lambda+\alpha)\,d\alpha$$

$$= \int_{0}^{\infty} h(\alpha)r_U(\tau+\lambda-\alpha)\,d\alpha$$

and comparing again with Eq. 4.34,

$$r_Y(\tau) = \int_{-\infty}^{\infty} h(\lambda)r_{YU}(-\tau-\lambda)\,d\lambda \quad .$$

Since r_Y is an even function,

$$r_Y(\tau) = \int_{-\infty}^{\infty} h(\lambda) r_{YU}(\tau-\lambda)\, d\lambda \quad . \tag{4.37}$$

Equations 4.33, 4.34, 4.36, and 4.37 give the output mean and autocorrelation function. The mean square output $S_Y^2 = r_Y(0)$. Frequently m_Y and S_Y^2 provide satisfactory measures of system response. It also may be seen that Eq. 4.36 and 4.37, as well as the expressions for r_{YU} and r_{UY}, are real convolution forms and r_Y may, therefore, be found by computing two real convolutions.

A similar procedure for discrete-time systems starts with Eq. 4.9 where $h(\cdot)$ describes a linear, realizable, stationary system, $\{V(n)\}$ is a wide-sense stationary input sequence, and $n_0 \to -\infty$. Then we have

$$Z(n) = \sum_{j=-\infty}^{\infty} h(j) V(n-j) \quad .$$

The mean value of the output is

$$m_Z = E[Z] = m_V \sum_{j=0}^{\infty} h(j)$$

and a useful form in terms of the transfer function H(z) is

$$m_Z = H(1) m_V \quad . \tag{4.38}$$

To find the autocorrelation function of the output sequence,

$$Z(n+k) = \sum_{i=-\infty}^{\infty} h(i) V(n+k-i)$$

and

$$r_Z(k) = E[Z(n)Z(n+k)]$$

$$= E \sum_{j=-\infty}^{\infty} h(j) V(n-j) \sum_{i=-\infty}^{\infty} h(i) V(n+k-i)$$

$$= E\sum_{j=-\infty}^{\infty} h(j) \sum_{i=-\infty}^{\infty} h(i)V(n-j)V(n+k-i)$$

$$= \sum_{j=-\infty}^{\infty} h(j) \sum_{i=-\infty}^{\infty} h(i)r_V(k-i+j) \quad . \tag{4.39}$$

This (with i and k interchanged) is analogous to Eq. 4.22. The double summation forms can be written in terms of cross correlation functions as before. We obtain

$$r_{VZ}(k) = \sum_{i=0}^{\infty} h(i)r_V(k-i) \tag{4.40a}$$

and

$$r_{ZV}(k) = \sum_{i=0}^{\infty} h(i)r_Z(k+i) \tag{4.40b}$$

and using these in 4.39,

$$r_Z(k) = \sum_{j=-\infty}^{\infty} h(j)r_{VZ}(k+j) \tag{4.41}$$

or

$$r_Z(k) = \sum_{j=-\infty}^{\infty} h(j)r_{ZV}(k-j) \quad . \tag{4.42}$$

The lower limit on the integrals and sums in Eq. 4.36, 4.37, 4.41, and 4.42 may be changed to zero since $h(\cdot)$ in both cases is realizable. When the input and output signals are vectors,* the vector means and correlation matrices can be obtained. Thus for the same assumptions,

$$\underline{m}_{\underline{Y}} = E[\underline{Y}] = E\int_{0}^{\infty} H(\lambda)\underline{U}(t-\lambda)d\lambda$$

where $H(\lambda)$ is the impulse matrix. A useful form in terms of the transfer matrix $H(s)$, $s=0$, is

---

* See paragraph after Eq. 4.60 on notation.

$$\underline{m}_{\underline{Y}} = H(0)\underline{m}_{\underline{U}} \tag{4.43}$$

Also, we have

$$R_{\underline{Y}}(\tau) = E[\underline{Y}(t)\underline{Y}^T(t+\tau)]$$
$$= \int_{-\infty}^{\infty} H(\lambda)\,d\lambda$$
$$\cdot \int_{-\infty}^{\infty} R_{\underline{U}}(\tau-\alpha+\lambda)H^T(\alpha)\,d\alpha \tag{4.44}$$

where $H(\cdot)$ is the impulse response matrix. For discrete-time systems

$$\underline{m}_{\underline{Z}} = H(1)\underline{m}_{\underline{V}} \tag{4.45}$$

where $H(1)$ is the transfer matrix $H(z)$, $z=1$, and

$$R_{\underline{Z}}(k) = \sum_{j=-\infty}^{\infty} H(j) \sum_{i=-\infty}^{\infty} R_{\underline{V}}(k-i+j)H^T(i) \tag{4.46}$$

where $H(\cdot)$ is the pulse response matrix.

Frequency domain results may be obtained by using the appropriate transform. Using Eq. 4.34, the spectral density function is

$$R_Y(s) = \mathcal{L}_2 r_Y(\tau)$$
$$= \int_{-\infty}^{\infty} \varepsilon^{-s\tau} \int_{0}^{\infty} h(\lambda)\,d\lambda \int_{0}^{\infty} h(\alpha)\,r_U(\tau-\alpha+\lambda)\,d\alpha\,d\tau$$

and following the same steps as in the derivation of Eq. 4.25

$$R_Y(s) = H(-s)R_U(s)H(s) \ . \tag{4.47}$$

Similarly, using Eq. 4.39,

$$R_Z(z) = \mathfrak{z}\ {}_2 r_Z(k)$$

$$= H(z^{-1})R_V(z)H(z) \quad . \qquad (4.48)$$

Equations 4.47 and 4.48 also can be found by using the forms in Eq. 2.86, 2.87 and Eq. 4.13, 4.15. Substituting, we obtain

$$R_Y(s) = \lim_{T\to\infty} E\left[\frac{Y_T(-s)Y_T(s)}{2T}\right]$$

$$= \lim_{T\to\infty} E\left[\frac{H(-s)U_T(-s)H(s)U_T(s)}{2T}\right]$$

$$= H(-s)R_U(s)H(s) \qquad (4.47)$$

and using cross spectra,

$$R_Y(s) = \lim_{T\to\infty} E\left[\frac{H(-s)U_T(-s)Y_T(s)}{2T}\right]$$

$$= H(-s)R_{UY}(s) \qquad (4.49)$$

or

$$R_Y(s) = \lim_{T\to\infty} E\left[\frac{Y_T(-s)H(s)U_T(s)}{2T}\right]$$

$$= R_{YU}(s)H(s) \quad . \qquad (4.50)$$

Also, in the same way we can find

$$R_{UY}(s) = H(s)R_U(s) \qquad (4.51)$$

and

$$R_{YU}(s) = H(-s)R_U(s) \tag{4.52}$$

which can be checked by taking the Laplace Transforms of Equation 4.35a and 4.35b. Similarly, for discrete-time systems,

$$R_Z(z) = \lim_{N\to\infty} E\left[\frac{Z_N(z^{-1})Z_N(z)}{2N+1}\right]$$

$$= H(z^{-1})R_V(z)H(z) \tag{4.48}$$

and using cross spectra,

$$R_Z(z) = H(z^{-1})R_{VZ}(z) \tag{4.53}$$

$$= R_{ZV}(z)H(z) \tag{4.54}$$

and

$$R_{VZ}(z) = H(z)R_V(z) \tag{4.55}$$

$$R_{ZV}(z) = H(z^{-1})R_V(z) \ . \tag{4.56}$$

The multiple input/output forms are found similarly.* The spectral density matrix obtained from Eq. 4.44 is

$$R_{\underline{Y}}(s) = \mathcal{L}_2 R_{\underline{Y}}(\tau)$$
$$= H(-s)R_{\underline{U}}(s)H^T(s) \tag{4.57}$$

---

* There is some ambiguity in the literature on vector forms that goes back to the definitions of $r_u(\tau)$ and $r_v(k)$ in 2.79. If $-\tau$ and $-k$ are used in these, a consistent development leads to an interchange of s and $-s$ and z and z^{-1} in $H(\cdot)$ and the right hand H in each case is a Hermetian.

and from Eq. 4.46,

$$R_{\underline{Z}}(z) = \mathcal{Z}_2 R_{\underline{Z}}(k)$$

$$= H(z^{-1}) R_{\underline{V}}(z) H^T(z) . \qquad (4.58)$$

Properties of these spectral density matrices shown in Chapter 2 are

$$R_{\underline{Y}}(-s) = R_{\underline{Y}}^T(s)$$

and

$$R_{\underline{Z}}(z^{-1}) = R_{\underline{Z}}^T(z) .$$

If there are no poles on the imaginary axis of the s-plane, s may be replaced by $j\omega$ and if there are no poles on the unit circle of the z-plane, z may be replaced by γ. Also

$$R_{\underline{Y}}(s) = H(-s) R_{\underline{UY}}(s)$$

$$= R_{\underline{YU}}(s) H^T(s) \qquad (4.59)$$

and

$$R_{\underline{UY}}(s) = R_{\underline{U}}(s) H^T(s)$$

$$R_{\underline{YU}}(s) = H(-s) R_{\underline{U}}(s) .$$

Similarly, for discrete-time systems,

$$R_Z(z) = H(z^{-1}) R_{\underline{VZ}}(z) \qquad (4.60)$$

$$= R_{\underline{ZV}}(z) H^T(z)$$

and

$$R_{\underline{VZ}}(z) = R_{\underline{V}}(z) H^T(z)$$

$$R_{\underline{ZV}}(z) = H(z^{-1}) R_{\underline{V}}(z) .$$

We observe again that the notational convention of H for
both continuous-time and discrete-time systems is for convenience in remembering corresponding forms.   In the vector
cases (i.e., multiple input/output), there is an additional
need for care since the time domain matrices become capital
letters.   No confusion should arise if the arguments are
checked as in:

$H(\lambda)$ - continuous-time system impulse response
matrix

$H(j)$ - discrete-time system pulse response matrix

$H(s)$ - continuous-time system transfer matrix

$H(z)$ - discrete-time system transfer matrix .

The results for the correlation matrices in Eq. 4.44 and
4.46 can be related to the covariance kernels as follows.
The covariance kernel for $\underline{Y}$, where the process $\{\underline{Y}(t)\}$ is
wide-sense stationary, is

$$P_{\underline{Y}}(\tau) = E[(\underline{Y}(t) - \underline{m}_{\underline{Y}})(\underline{Y}(t+\tau) - \underline{m}_{\underline{Y}})^T]$$

$$= R_{\underline{Y}}(\tau) - \underline{m}_{\underline{Y}}\underline{m}_{\underline{Y}}^T$$

$$= R_{\underline{Y}}(\tau) - H(0)\underline{m}_{\underline{U}}\underline{m}_{\underline{U}}^T H^T(0) \qquad (4.61)$$

where $H(0)$ is $H(s)$ for s=0.   Obviously $P_{\underline{Y}}(\tau) = R_{\underline{Y}}(\tau)$ if $\underline{m}_{\underline{U}} = \underline{0}$;
this is a common situation.   Similarly,

$$P_{\underline{Z}}(k) = R_{\underline{Z}}(k) - H(1)\underline{m}_{\underline{V}}\underline{m}_{\underline{V}}^T H^T(1) \qquad (4.62)$$

where $H(1)$ is $H(z)$ for z=1.   It is also clear that forms in
$P_{\underline{Y}}$ and $P_{\underline{Z}}$ like Eq. 4.44 and 4.46 can be written in terms of
$P_{\underline{U}}$ and $P_{\underline{V}}$.

Example 4.10

A first-order linear system is characterized by

$$h(t) = \varepsilon^{-at}u_1(t), \qquad a > 0,$$

and the input is a zero mean, wide-sense stationary process $\{U(t)\}$ with

$$r_U(\tau) = W_o u_o(\tau).$$

Find the mean, mean square, and autocorrelation function of the output process $\{Y(t)\}$. Using Eq. 4.33,

$$m_Y = H(0)m_U$$

$$= \left.\frac{1}{s+a}\right|_{s=0} \cdot 0 = 0.$$

To find $r_Y(\tau)$, start with Eq. 4.35b,

$$r_{YU}(\tau) = \int_o^\infty h(\alpha) r_U(\tau+\alpha)\, d\alpha$$

$$= \int_o^\infty \varepsilon^{-a\alpha} W_o u_o(\tau+\alpha)\, d\alpha$$

or

$$r_{YU}(\tau) = W_o \varepsilon^{a\tau}, \qquad \tau < 0$$

$$= 0 \quad , \qquad \tau > 0$$

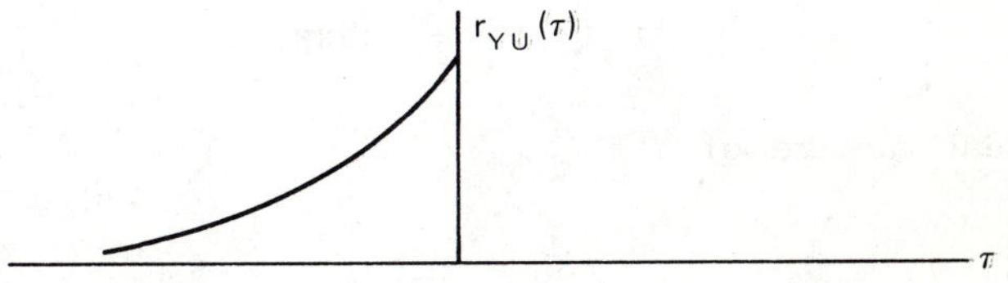

Then, using Eq. 4.37,

$$r_Y(\tau) = \int_o^\infty \varepsilon^{-a\lambda} W_o (\exp a(\tau-\lambda))\, d\lambda, \qquad \tau - \lambda < 0$$

$$= 0 \qquad , \qquad \tau - \lambda > 0.$$

Thus for $\tau>0$,

$$r_Y(\tau) = \int_{\tau}^{\infty} W_o \varepsilon^{a\tau} (\exp -2a\lambda) d\lambda \ .$$

The lower limit follows from $r_{YU}(\tau-\lambda)$, $\tau > 0$:

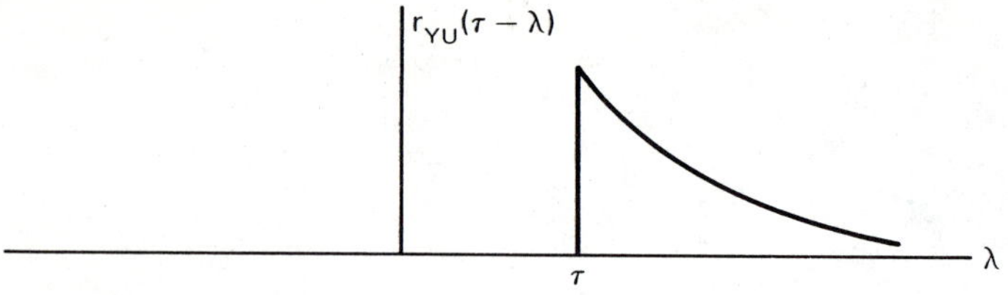

For $\tau < 0$,

$$r_Y(\tau) = \int_{o}^{\infty} W_o \varepsilon^{a\tau} (\exp -2a\lambda) d\lambda \ .$$

Therefore,

$$r_Y(\tau) = \frac{W_o}{2a} \varepsilon^{-a\tau} , \qquad \tau > 0$$

$$= \frac{W_o}{2a} \varepsilon^{a\tau} \qquad \tau < 0$$

or

$$r_Y(\tau) = \frac{W_o}{2a} (\exp -a|\tau|) \ .$$

The mean square of Y is

$$S_Y^2 = r_Y(0) = \frac{W_o}{2a} \ .$$

The output autocorrelation function also can be found by using the frequency domain form Eq. 4.47,

$$R_Y(s) = \frac{1}{-s+a} \cdot \frac{1}{s+a} \cdot W_o$$

$$= \frac{W_o}{-s^2+a^2} = \frac{-W_o}{(s+a)(s-a)} .$$

Then

$$r_Y(\tau) = \mathcal{L}_2^{-1} R_Y(s)$$

$$= \frac{1}{2\pi j} \int_{-j\infty}^{j\infty} \frac{-W_o}{(s+a)(s-a)} \varepsilon^{\tau s} ds$$

and using the calculus of residues (see App. A),

$$r_Y(\tau) = \frac{-W_o}{-2a} \varepsilon^{-a\tau} = \frac{W_o}{2a} \varepsilon^{-a\tau}, \qquad \tau > 0$$

$$= \frac{-W_o}{-2a} \varepsilon^{a\tau} = \frac{W_o}{2a} \varepsilon^{a\tau}, \qquad \tau < 0 .$$

The mean square could be found directly as

$$S_Y^2 = r_Y(0) = \frac{1}{2\pi j} \int_{-j\infty}^{j\infty} \frac{-W_o}{(s+a)(s-a)} ds$$

$$= \frac{W_o}{2a}$$

and for $s \to j\omega$, a real integral form is

$$S_Y^2 = \frac{1}{2\pi} \int_{-\infty}^{\infty} \frac{W_o}{\omega^2+a^2} d\omega .$$

Example 4.11

A first-order system is characterized by

$$H(z) = \frac{z}{z+a} , \qquad -1 < a < 1 ,$$

and a wide-sense stationary input sequence $\{V(n)\}$ has a spectral density function

$$R_V(z) = K.$$

Using Eq. 4.48

$$R_Z(z) = \frac{z}{z+a} \cdot \frac{z^{-1}}{z^{-1}+a} \cdot K$$

$$= \frac{K}{a} \frac{z}{(z+a)(z+1/a)} .$$

For $r_Z(k)$,

$$r_Z(k) = \mathcal{Z}_2^{-1} R_Z(z)$$

$$= \frac{1}{2\pi j} \oint_\Gamma \frac{K/a \; z^k}{(z+a)(z+1/a)} dz$$

and using the calculus of residues,

$$r_Z(k) = \frac{K}{a} \frac{(-a)^k}{(\frac{1}{a} - a)} , \qquad k \geq 0$$

$$= - \frac{K}{a} \frac{(-1/a)^k}{(a - \frac{1}{a})} , \qquad k \leq 0$$

or

$$r_Z(k) = \frac{K}{(1 - a^2)} (-a)^{|k|} .$$

A simulation diagram for this example is

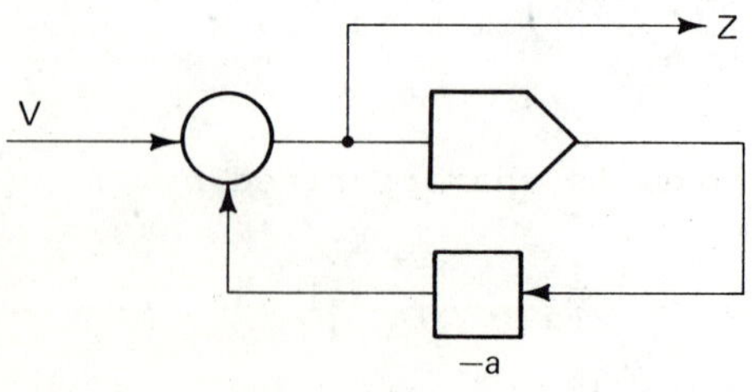

We see in this example that {Z} is a random sequence with autocorrelation $r_Z(k)$ when the input is a sequence of independent random variables V. More general cases of this are discussed later in this chapter when the generation of signals with specified characteristics using digital and analog computers is considered. Obviously V is obtained from a random number generator and Z is related to V by a simple difference equation.

4.3.2 Nonstationary Response

Equations 4.61 and 4.62 give the output covariance kernels of multiple input/multiple output stationary linear systems when the inputs are wide-sense stationary vector processes or sequences and the initial times, t_o or n_o, are in the infinite past. Equations 4.43 and 4.45 give the means. An extension to cases where t_o and n_o are finite is needed. This means that the effect of initial conditions must be included, and the resulting output processes or sequences are not stationary. To include arbitrary initial conditions and the possibility of time-varying systems, it is convenient to use state variable models of the system. Therefore, the following problem is of interest for continuous-time systems:

System: $\dot{\underline{x}}(t) = A(t)\underline{x}(t) + B(t)\underline{u}(t)$

$\underline{y}(t) = C(t)\underline{x}(t)$

Input: Process $\{\underline{U}(t)\}$ with mean

$$E[\underline{U}(t)] = \underline{m}_{\underline{U}}(t)$$

and covariance kernel

$$E[(\underline{U}(t_1)-\underline{m}_{\underline{U}}(t_1))(\underline{U}(t_2)-\underline{m}_{\underline{U}}(t_2))^T] =$$
$$= P_{\underline{U}}(t_1,t_2) \ .$$

Initial State: Random variable $\underline{X}_o$ at t_o with mean

$$E[\underline{X}_o] = \underline{m}_o$$

and covariance

$$E[(\underline{X}_o-\underline{m}_o)(\underline{X}_o-\underline{m}_o)^T] = P_o \ .$$

Find: The mean of the output

$$E[\underline{Y}(t)] = \underline{m}_Y(t)$$

and the covariance kernel of the output

$$E[(\underline{Y}(t_1)-\underline{m}_{\underline{Y}}(t_1))(\underline{Y}(t_2)-\underline{m}_{\underline{Y}}(t_2))^T] =$$

$$= P_{\underline{Y}}(t_1,t_2) \ .$$

Similarly for the discrete-time systems:

System: $\underline{x}(n+1) = A(n)\underline{x}(n) + B(n)\underline{v}(n)$

$$\underline{z}(n) = C(n)\underline{x}(n)$$

Input: Sequence $\{\underline{V}(n)\}$ with mean

$$E[\underline{V}(n)] = \underline{m}_{\underline{V}}(n)$$

and covariance kernel

$$E[(\underline{V}(n_1)-\underline{m}_{\underline{V}}(n_1))(\underline{V}(n_2)-\underline{m}_{\underline{V}}(n_2))^T] =$$

$$= P_{\underline{V}}(n_1,n_2) \ .$$

Initial State: Random variable $\underline{X}_o$ at n_o with mean

$$E[\underline{X}_o] = \underline{m}_o$$

covariance

$$E[(\underline{X}_o-\underline{m}_o)(\underline{X}_o-\underline{m}_o)^T] = P_o \ .$$

Find: The mean of the output

$$E[\underline{Z}(n)] = \underline{m}_{\underline{Z}}(n)$$

and the covariance kernel of the output

$$E[(\underline{Z}(n_1)-\underline{m}_{\underline{Z}}(n_1))(\underline{Z}(n_2)-\underline{m}_{\underline{Z}}(n_2))^T] =$$

$$= P_{\underline{Z}}(n_1,n_2) \quad .$$

Before developing the general forms, we consider a simple example that illustrates the difference between the results obtained previously and those required now.

Example 4.12

An R-L circuit has a wide-sense stationary input voltage U as shown. The switch closes at $t_o=0$ and the response Y is the current. Let L=1 and R/L=a. Find the mean square current.

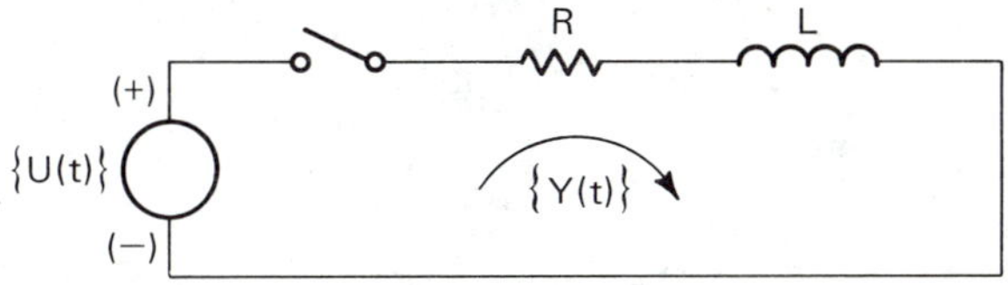

The impulse response is $h(t) = \varepsilon^{-at}u_1(t)$, a>0, and $m_U=0$ and $r_U(\tau)=K\varepsilon^{-b|\tau|}$, b≠a, b>0. Since the initial current is zero the convolution form Eq. 4.7 may be used. Therefore,

$$Y(t_1) = \int_o^{t_1} U(\lambda)h(t_1-\lambda)d\lambda$$

and

$$Y(t_2) = \int_o^{t_2} U(\sigma)h(t_2-\sigma)d\sigma \quad .$$

Then

$$r_Y(t_1,t_2) = E[Y(t_1)Y(t_2)]$$

$$= E\int_0^{t_1} h(t_1-\lambda)d\lambda \int_0^{t_2} U(\lambda)U(\sigma)h(t_2-\sigma)d\sigma$$

$$= \int_0^{t_1} h(t_1-\lambda)d\lambda \int_0^{t_2} r_U(\lambda,\sigma)h(t_2-\sigma)d\sigma \quad .$$

Now since the input is wide-sense stationary, let $\tau=\sigma-\lambda$, then

$$r_Y(t_1,t_2) = \int_0^{t_1} h(t_1-\lambda)d\lambda \int_{-\lambda}^{t_2-\lambda} r_U(\tau)h(t_2-\tau-\lambda)d\tau \quad .$$

For the mean square Y, let $t_1=t_2=t$, then

$$r_Y(t,t) = S_Y^2(t) = \int_0^{t} h(t-\lambda)d\lambda \int_{-\lambda}^{t-\lambda} r_U(\tau)h(t-\tau-\lambda)d\tau$$

and substituting r_U and h,

$$S_Y^2(t) = K\int_0^{t} (\exp -a(t-\lambda))u_1(t-\lambda)d\lambda$$

$$\int_{-\lambda}^{t-\lambda} (\exp -b|\tau|)(\exp -a(t-\tau-\lambda))u_1(t-\tau-\lambda)d\tau.$$

Now $0\leq\lambda\leq t$ and $-\infty<\tau<\infty$, so the second integral is written

$$\int_{-\lambda}^{0} \varepsilon^{b\tau}(\exp -a(t-\tau-\lambda))u_1(t-\tau-\lambda)d\tau \;+$$

$$+ \int_0^{t-\lambda} \varepsilon^{-b\tau}(\exp -a(t-\tau-\lambda))u_1(t-\tau-\lambda)d\tau \quad .$$

Performing the integrations leads to

$$S_Y^2(t) = \frac{K}{a(a+b)} - \frac{K}{a(b-a)} (\exp -2at) +$$

$$+ \frac{2K}{(b^2-a^2)} (\exp -(a+b))t, \qquad t \geq 0.$$

For t=0, $S_Y^2(0) = 0$, which is expected since all realizations of $\{Y(t)\}$, i.e., all $y^i(t)$, are zero at t=0. Also, for $t \to \infty$,

$$S_Y^2(\infty) = \frac{K}{a(a+b)} ,$$

the "steady state" or asymptotic mean square current. This can be checked by using the previous results for $t_o \to -\infty$. Using Eq. 4.47

$$R_Y(s) = \frac{1}{-s+a} \cdot \frac{1}{s+a} \cdot \frac{-2Kb}{(s+b)(s-b)} .$$

Then taking the inverse Laplace Transform and letting $\tau=0$,

$$S_Y^2(\infty) = r_Y(0) = \frac{K}{a(a+b)} .$$

Returning now to the general continuous-time problem, a first step in obtaining the general forms for the mean and covariance kernel of the output requires the following conditions:

1. The input is a white process, therefore,

$$P_{\underline{U}}(t_1,t_2) = M(t_1)u_o(t_2-t_1)$$

where $M(\cdot)$ is the intensity matrix.

2. The initial state and the input are uncorrelated for all t, i.e.,

$$E[\underline{X}_o-\underline{m}_o)(\underline{U}(t)-\underline{m}_{\underline{U}}(t))^T] = \Theta, \qquad \forall t .$$

Furthermore, the system state rather than the output is considered first and the covariance of the state is obtained before finding the covariance kernel of the state. Thus, for $t=t_1=t_2$ we use the notation

$$P_{\underline{X}}(t_1,t_2) \rightarrow P(t) \ ,$$

as the covariance matrix of the state.* This is analogous to finding $S_Y^2(t)$ in Example 4.12. The mean of the state can be found by taking the expected value of the state differential equation with $\underline{U}(t)$ as input and $\underline{X}(t)$ the random state vector. For $t \geq t_o$:

$$E[\dot{\underline{X}}(t)] = A(t)E[\underline{X}(t)] + B(t)E[\underline{U}(t)]$$

or

$$\dot{\underline{m}}_{\underline{X}}(t) = A(t)\underline{m}_{\underline{X}}(t) + B(t)\underline{m}_{\underline{U}}(t) \qquad (4.63)$$

and $\underline{m}_{\underline{X}}(t_o)=\underline{m}_o$. The solution for $\underline{m}_{\underline{X}}(t)$ is given by Eq. 4.30a with $\underline{x}(t)$ replaced by $\underline{m}_{\underline{X}}(t)$ and $\underline{u}(\lambda)$ replaced by $\underline{m}_{\underline{U}}(\lambda)$. Now using

$$\dot{\underline{X}}(t) = A(t)\underline{X}(t) + B(t)\underline{U}(t) \ ;$$

and Eq. 4.63,

$$(\dot{\underline{X}}(t)-\dot{\underline{m}}_{\underline{X}}(t)) = A(t)(\underline{X}(t)-\underline{m}_{\underline{X}}(t)) + B(t)(\underline{U}(t)-\underline{m}_{\underline{U}}(t)) \ .$$

Using this at t_1 with the indicated post multiplier at t_2, we have

$$[\dot{\underline{X}}(t_1)-\underline{m}_{\underline{X}}(t_1)][\underline{X}(t_2)-\underline{m}_{\underline{X}}(t_2)]^T =$$

---

*Note that this is a change in notation from Table 2.5 where the covariance matrix is $C_{\underline{X}}(t)$.

$$= A(t_1)[\underline{X}(t_1)-\underline{m}_{\underline{X}}(t_1)][\underline{X}(t_2)-\underline{m}_{\underline{X}}(t_2)]^T +$$

$$+ B(t_1)[\underline{U}(t_1)-\underline{m}_{\underline{U}}(t_1)][\underline{X}(t_2)-\underline{m}_{\underline{X}}(t_2)]^T \ .$$

Taking the transpose of both sides and interchanging t_1 and t_2, we have

$$[\underline{X}(t_1)-\underline{m}_{\underline{X}}(t_1)][\dot{\underline{X}}(t_2)-\dot{\underline{m}}_{\underline{X}}(t_2)]^T =$$

$$= [\underline{X}(t_1)-\underline{m}_{\underline{X}}(t_1)][\underline{X}(t_2)-\underline{m}_{\underline{X}}(t_2)]^T A^T(t_2) +$$

$$+ [\underline{X}(t_1)-\underline{m}_{\underline{X}}(t_1)][\underline{U}(t_2)-\underline{m}_{\underline{U}}(t_2)]^T B^T(t_2).$$

Now if these last two equations are added and the expected value of the sum is taken, it is obvious that the left side will contain two partial derivatives of $P_{\underline{X}}(t_1,t_2)$ and the right side will contain two terms in $P_{\underline{X}}(t_1,t_2)$ and two cross covariance terms in $\underline{X}$ and $\underline{U}$. To obtain an equation in the covariance of the state, let $t=t_1=t_2$. Then we get

$$\dot{P}(t) = A(t)P(t) + P(t)A^T(t) +$$

$$+ B(t)\ E[(\underline{U}(t)-\underline{m}_{\underline{U}}(t))(\underline{X}(t)-\underline{m}_{\underline{X}}(t))^T] +$$

$$+ E[(\underline{X}(t)-\underline{m}_{\underline{X}}(t))(\underline{U}(t)-\underline{m}_{\underline{U}}(t))^T]B^T(t) \ .$$

Using Eq. 4.30a,

$$\underline{X}(t) = \Phi(t,t_o)\underline{X}_o + \int_{t_o}^{t} \Phi(t,\lambda)B(\lambda)\underline{U}(\lambda)d\lambda$$

$$\underline{m}_{\underline{X}}(t) = \Phi(t,t_o)\underline{m}_o + \int_{t_o}^{t} \Phi(t,\lambda)B(\lambda)\underline{m}_{\underline{U}}(\lambda)d\lambda \ .$$

Therefore,

$$[\underline{X}(t)-\underline{m}_{\underline{X}}(t)] = \Phi(t,t_o)[\underline{X}_o-\underline{m}_o] +$$

$$+\int_{t_o}^{t} \Phi(t,\lambda)B(\lambda)[\underline{U}(\lambda)-\underline{m}_{\underline{U}}(\lambda)]d\lambda$$

and

$$[\underline{X}(t)-\underline{m}_{\underline{X}}(t)][\underline{U}(t)-\underline{m}_{\underline{U}}(t)]^T$$

$$= \Phi(t,t_o)[\underline{X}_o-\underline{m}_o][\underline{U}(t)-\underline{m}_{\underline{U}}(t)]^T +$$

$$+\int_{t_o}^{t} \Phi(t,\lambda)B(\lambda)[\underline{U}(\lambda)-\underline{m}_{\underline{U}}(\lambda)][\underline{U}(t)-\underline{m}_{\underline{U}}(t)]^T d\lambda .$$

Taking the expected value, we have

$$E[(\underline{X}(t)-\underline{m}_{\underline{X}}(t))(\underline{U}(t)-\underline{m}_{\underline{U}}(t))^T] =$$

$$= \int_{t_o}^{t} \Phi(t,\lambda)B(\lambda)E[(\underline{U}(\lambda)-\underline{m}_{\underline{U}}(\lambda))(\underline{U}(t)-\underline{m}_{\underline{U}}(t))^T]d\lambda$$

since the expected value of the first term on the right is zero by Condition 2. Also, from Condition 1.,

$$E[(\underline{U}(\lambda)-\underline{m}_{\underline{U}}(\lambda))(\underline{U}(t)-\underline{m}_{\underline{U}}(t))^T] = \tfrac{1}{2}M(\lambda)u_o(\lambda-t) .$$

The factor 1/2 appears since $u_o(\cdot)$ is symmetric and the impulse occurs at the end of the interval of integration. Then,

$$E[(\underline{X}(t)-\underline{m}_{\underline{X}}(t))(\underline{U}(t)-\underline{m}_{\underline{U}}(t))^T] = \tfrac{1}{2}B(t)M(t) .$$

Taking the transpose,

$$E[(\underline{U}(t)-\underline{m}_{\underline{U}}(t))(\underline{X}(t)-\underline{m}_{\underline{X}}(t))^T] = \tfrac{1}{2}M(t)B^T(t)$$

since M(t) is symmetric. Finally, the required differential equation is

$$\dot{P}(t) = A(t)P(t) + P(t)A^T(t) + B(t)M(t)B^T(t) \qquad (4.64)$$

with initial condition $P(t_o)=P_o$. The solution can be found by starting with the definition of P(t),

$$P(t) = E\{(\underline{X}(t)-\underline{m}_{\underline{X}}(t))(\underline{X}(t)-\underline{m}_{\underline{X}}(t))^T\}$$

and since $\underline{X}(t)-\underline{m}_{\underline{X}}(t)$ was found above, we substitute to obtain

$$P(t) = E\left\{[(\Phi(t,t_o)[\underline{X}_o-\underline{m}_o]+ \int_{t_o}^{t} \Phi(t,\lambda)B(\lambda)(\underline{U}(\lambda)-\underline{m}_{\underline{U}}(\lambda))d\lambda] \cdot \cdot[(\Phi(t,t_o)[\underline{X}_o-\underline{m}_o] + \int_{t_o}^{t} \Phi(t,\lambda)B(\lambda)[\underline{U}(\lambda)-\underline{m}_{\underline{U}}(\lambda)]d\lambda)]^T\right\}$$

and using Conditions 1. and 2.,

$$P(t) = \Phi(t,t_o)P_o\Phi^T(t,t_o) + \int_{t_o}^{t} \Phi(t,\lambda)B(\lambda)M(\lambda)B^T(\lambda)\Phi^T(t,\lambda)d\lambda \; . \qquad (4.65)$$

This can be verified by differentiating and substituting in Eq. 4.64. Although Eq. 4.65 is the solution of Eq. 4.64, it frequently is more convenient to compute P(t) numerically by using Eq. 4.64 directly.

To find the covariance kernel of the state,

$$P_{\underline{X}}(t_1,t_2) = E[(\underline{X}(t_1)-\underline{m}_{\underline{X}}(t_1)(\underline{X}(t_2)-\underline{m}_{\underline{X}}(t_2))^T].$$

Using the result above for $\underline{X}(t)-\underline{m}_{\underline{X}}(t)$,

$$(\underline{X}(t_2)-\underline{m}_{\underline{X}}(t_2)) = \Phi(t_2,t_1)(\underline{X}(t_1)-\underline{m}_{\underline{X}}(t_1)) + \int_{t_1}^{t_2} \Phi(t_2,\lambda)B(\lambda)[\underline{U}(\lambda)-\underline{m}_{\underline{U}}(\lambda)]d\lambda.$$

Then

$$[\underline{X}(t_1)-\underline{m}_X(t_1)][\underline{X}(t_2)-\underline{m}_X(t_2)]^T =$$

$$= [\underline{X}(t_1)-\underline{m}_X(t_1)][\underline{X}(t_1)-\underline{m}_X(t_1)]^T\Phi^T(t_2,t_1) +$$

$$+\int_{t_1}^{t_2} [\underline{X}(t_1)-\underline{m}_X(t_1)][\underline{U}(\lambda)-\underline{m}_X(\lambda)]^T B^T(\lambda)\Phi^T(t_2,\lambda)d\lambda$$

and taking the expected value,

$$P_{\underline{X}}(t_1,t_2) = P(t_1)\Phi^T(t_2,t_1), \qquad t_2 \geq t_1 \qquad (4.66a)$$

The integral term is zero since $\underline{X}(t_1)$ is not correlated with $\underline{U}(\lambda)$, $t_1 \leq \lambda \leq t_2$, under the assumption of a white input process. Similarly,

$$P_{\underline{X}}(t_1,t_2) = \Phi(t_1,t_2)P(t_2), \qquad t_1 \geq t_2 \qquad (4.66b)$$

Thus the covariance kernel of the state can be found if the covariance $P(t)$ is known. In the general case of time-varying systems, a major difficulty, as noted earlier, is finding $\Phi(t_1,t_2)$, the transition matrix of the system. For stationary systems and a wide-sense stationary input (M a constant matrix), use of Eq. 4.64 or 4.65 is relatively easy. Finally, since $\underline{Y}(t) = C(t)\underline{X}(t)$, we have

$$\underline{m}_Y(t) = C(t)\underline{m}_X(t) \qquad (4.67)$$

where $\underline{m}_X(t)$ is the solution of Eq. 4.63, and

$$P_{\underline{Y}}(t_1,t_2) = C(t_1)P_{\underline{X}}(t_1,t_2)C^T(t_2) \qquad (4.68)$$

These are the results which were sought for continuous-time systems.

For discrete-time systems, a parallel to the development above is carried out. The two conditions are:

1. The input is a white sequence with

$$P_{\underline{V}}(n_1,n_2) = M(n_1)v_o(n_2-n_1)$$

2. The initial state and the input are uncorrelated for all n, i.e.,

$$E[(\underline{X}_o-\underline{m}_o)(\underline{V}(n)-\underline{m}_{\underline{V}}(n))^T] = \Theta, \qquad \forall\ n\ .$$

From the state difference equation,

$$E[\underline{X}(n+1)] = A(n)E[\underline{X}(n)] + B(n)E[\underline{V}(n)]$$

or

$$\underline{m}_{\underline{X}}(n+1) = A(n)\underline{m}_{\underline{X}}(n) + B(n)\underline{m}_{\underline{V}}(n)$$

with $\underline{m}_{\underline{X}}(n_o)=\underline{m}_o$. The solution for $\underline{m}_{\underline{X}}(n)$ is given by Equation 4.32a with $\underline{x}(n)$ replaced by $\underline{m}_{\underline{X}}(n)$ and $\underline{v}(k)$ replaced by $\underline{m}_{\underline{V}}(k)$. Then

$$(\underline{X}(n+1)-\underline{m}_{\underline{X}}(n+1)) = A(n)(\underline{X}(n)-\underline{m}_{\underline{X}}(n))$$
$$+ B(n)(\underline{V}(n)-\underline{m}_{\underline{V}}(n)).$$

Also,

$$[\underline{X}(n_1+1)-\underline{m}_{\underline{X}}(n_1+1)][\underline{X}(n_2+1)-\underline{m}_{\underline{X}}(n_2+1)]^T =$$
$$= A(n_1)\ [\underline{X}(n_1)-\underline{m}_{\underline{X}}(n_1)][\underline{X}(n_2+1)-\underline{m}_{\underline{X}}(n_2+1)]^T$$
$$+ B(n_1)[\underline{V}(n_1)-\underline{m}_{\underline{V}}(n_1)][\underline{X}(n_2+1)-\underline{m}_{\underline{X}}(n_2+1)]^T.$$

Taking the expected value and letting $n=n_1=n_2$,

$$P(n+1) = A(n)E[\underline{X}(n)-\underline{m}_{\underline{X}}(n)][\underline{X}(n+1)-\underline{m}_{\underline{X}}(n+1)]^T$$
$$+ B(n)E[\underline{V}(n)-\underline{m}_{\underline{V}}(n)][\underline{X}(n+1)-\underline{m}_{\underline{X}}(n+1)]^T$$

$$= A(n)E[\underline{X}(n)-\underline{m}_{\underline{X}}(n)][\underline{X}(n)-\underline{m}_{\underline{X}}(n)]A^T(n)$$

$$+ A(n)E[\underline{X}(n)-\underline{m}_{\underline{X}}(n)][\underline{V}(n)-\underline{m}_{\underline{V}}(n)]^T B^T(n)$$

$$+ B(n)E[\underline{V}(n)-\underline{m}_{\underline{V}}(n)][\underline{X}(n)-\underline{m}_{\underline{X}}(n)]^T A^T(n)$$

$$+ B(n)E[\underline{V}(n)-\underline{m}_{\underline{V}}(n)][\underline{V}(n)-\underline{m}_{\underline{V}}(n)]^T B^T(n)$$

The second and third terms on the right, which are transposes of each other, are zero. To show this, using Eq. 4.32a, we have

$$\underline{X}(n)-\underline{m}_{\underline{X}}(n) = \Phi(n,n_o)[\underline{X}_o-\underline{m}_o] + \sum_{k=n_o}^{n-1} \Phi(n,k+1)B(k)(\underline{V}(k)-\underline{m}_{\underline{V}}(k)) .$$

Then

$$[\underline{X}(n)-\underline{m}_{\underline{X}}(n)][\underline{V}(n)-\underline{m}_{\underline{V}}(n)]^T =$$

$$= \Phi(n,n_o)[\underline{X}_o-\underline{m}_o][\underline{V}(n)-\underline{m}_{\underline{V}}(n)]^T + \sum_{k=n_o}^{n-1} \Phi(n,k+1)B(k)[\underline{V}(k)-\underline{m}_{\underline{V}}(k)][\underline{V}(n)-\underline{m}_{\underline{V}}(n)]^T$$

Taking the expected value, the first term is zero because of Condition 2. and the second term is zero because of Condition 1. Therefore*

$$P(n+1) = A(n)P(n)A^T(n) + B(n)M(n)B^T(n) \qquad (4.69)$$

with initial condition $P(n_o)=P_o$. The solution is obtained by using Eq. 4.69 iteratively. The result is

---

*Note that this is a change in notation from Table 2.5 where the covariance matrix is $C_{\underline{X}}(n)$.

$$P(n) = \Phi(n,n_o)P_o\Phi^T(n,n_o) + \\ + \sum_{k=n_o}^{n-1} \Phi(n,k+1)B(k)M(k)B^T(k)\Phi^T(n,k+1) \quad (4.70)$$

where

$$\Phi(n,i) = \prod_{j=i}^{n-1} A(j), \qquad n > i .$$

To find the covariance kernel of the state, recall that

$$P_{\underline{X}}(n_1,n_2) = E[(\underline{X}(n_1)-\underline{m}_{\underline{X}}(n_1))(\underline{X}(n_2)-\underline{m}_{\underline{X}}(n_2))^T] .$$

Now by using Eq. 4.32a,

$$(\underline{X}(n_2)-\underline{m}_{\underline{X}}(n_2)) = \Phi(n_2,n_1)(\underline{X}(n_1)-\underline{m}_{\underline{X}}(n_1)) + \\ + \sum_{k=n_1}^{n_2-1} \Phi(n_2,k+1)B(k)(\underline{V}(k)-\underline{m}_{\underline{V}}(k)).$$

Then we get

$$[\underline{X}(n_1)-\underline{m}_{\underline{X}}(n_1)][\underline{X}(n_2)-\underline{m}_{\underline{X}}(n_2)]^T = \\ = [\underline{X}(n_1)-\underline{m}_{\underline{X}}(n_1)][\underline{X}(n_1)-\underline{m}_{\underline{X}}(n_1)]^T\Phi^T(n_2,n_1) + \\ + \quad [\underline{X}(n_1)-\underline{m}_{\underline{X}}(n_1)][\underline{V}(k)-\underline{m}_{\underline{V}}(k)]^T B^T(k)\Phi^T(n_2,k+1)$$

and taking the expected value,

$$P_{\underline{X}}(n_1,n_2) = P(n_1)\Phi^T(n_2,n_1), \qquad n_2 \geq n_1 \quad (4.71a)$$

since $\underline{X}(n_1)$ is not correlated with $\underline{V}(k)$, $n_1 \leq k \leq n_2-1$. Similarly,

$$P_{\underline{X}}(n_1,n_2) = \Phi(n_1,n_2)P(n_2), \qquad n_1 \geq n_2. \quad (4.71b)$$

Finally, since $\underline{Z}(n) = C(n)\underline{X}(n)$,

$$\underline{m}_{\underline{Z}}(n) = C(n)\underline{m}_{\underline{X}}(n) \tag{4.72}$$

and we have

$$P_{\underline{Z}}(n_1,n_2) = C(n_1)P_{\underline{X}}(n_1,n_2)C^T(n_2). \tag{4.73}$$

In summary, Eq. 4.68 and 4.73 are the general results which were sought. In many cases, however, Eq. 4.64 and 4.69 are the ones used in the analysis of multiple input/multiple output linear systems driven by white inputs. We shall use these in Chapter 5. In the special case where the inputs and initial states are Gaussian, the outputs also are Gaussian and the means and covariances completely describe the corresponding output processes and sequences.

4.3.3 System Augmentation

The restriction to white input processes or sequences, the first of the two conditions used in obtaining the results of the previous section, can be relaxed by augmenting the original system. Any stochastic process or sequence that can be modeled as the output of a linear system driven by a white input can be used as an input to the system under study, and the above results are then used for the augmented linear system which has a white input. For Gaussian white inputs to the augmented system, the inputs to the system under study are Gauss-Markov.

The method is illustrated here for the special case where each component of $\underline{U}(t)$ or $\underline{V}(n)$ is the output of a first-order linear system driven by a white input. The components also are assumed to be independent. The more general cases are considered in the last section of this chapter, Section 4.5. Starting with the system*

$$\dot{\beta} = -\alpha\beta + w, \qquad \alpha > 0 \tag{4.74a}$$

*We are using lower case for random variables.

where w is a stationary scalar white input with zero mean and intensity w_o, Equation 4.65 leads to

$$p(t) = (\exp -2\alpha(t-t_o))p_o + w_o \int_{t_o}^{t} (\exp -2\alpha(t-\lambda)d\lambda$$

where p(t) is the variance of β and p_o is the variance of $\beta(t_o)$. Then, after integrating,

$$p(t) = (\exp -2\alpha(t-t_o))p_o + \frac{w_o}{2\alpha}(1 - (\exp -2\alpha(t-t_o))).$$

Now if $p_o = \frac{w_o}{2\alpha}$, we would have

$$p(t) = \frac{w_o}{2\alpha}, \qquad \forall\ t$$

a constant variance of β. From Equation 4.66,

$$p_\beta(t_1,t_2) = \frac{w_o}{2\alpha}(\exp -\alpha(t_2-t_1)), \qquad t_2 \geq t_1$$

$$= \frac{w_o}{2\alpha}(\exp -\alpha(t_1-t_2)), \qquad t_1 \geq t_2$$

or for $t_2-t_1=\tau$, we can write

$$p_\beta(\tau) = r_\beta(\tau) = \frac{w_o}{2\alpha}(\exp -\alpha|\tau|), \qquad (4.74b)$$

since the mean of β is zero.

Thus the first-order system of Eq. 4.74a has a stationary output process $\{\beta(t)\}$ such that the autocorrelation function $r_\beta(\tau)$ is given by Eq. 4.74b when the system is driven by a stationary white input of intensity w_o and when p_o, the variance of $\beta(t_o)$, is $w_o/2\alpha$.

To augment the system

$$\dot{\underline{x}} = A\underline{x} + B\underline{u}$$

$$\underline{y} = C\underline{x}$$

when, for example, the ith component of $\underline{u}$ is a wide-sense stationary process with

$$r_{U_i}(\tau) = K(\exp - c|\tau|), \quad c > 0$$

and all other components u_j of $\underline{u}$, $j=1,2,\ldots,q$, $j \neq i$, are white, let $u_i = x_{r+1}$ and let

$$\dot{x}_{r+1} = -cx_{r+1} + w$$

with $w_0 = 2cK$ and var $x_{r+1}(t_0) = K$. Then the augmented system is

$$\dot{\underline{x}}' = A'\underline{x}' + B'\underline{u}' \tag{4.75}$$

$$\underline{y} = C'\underline{x}'$$

where $(\underline{x}')^T = (\underline{x}^T \vdots x_{r+1})$, $(\underline{u}')^T = (u_1 u_2 \ldots 0 \ldots u_q \vdots w)$, i.e., the i-th component of $\underline{u}'$ is zero, and

$$A' = \left[\begin{array}{c|c} A & \underline{b}_i \\ \hline \underline{0}^T & -c \end{array}\right], \qquad B' = \left[\begin{array}{c|c} B & \underline{0} \\ \hline \underline{0}^T & 1 \end{array}\right]$$

$$C' = \left[\begin{array}{c|c} C & \underline{0} \end{array}\right]$$

where $\underline{b}_i$ is the i-th column of B. For the intensity matrix of $\underline{u}'$, a white vector process, we have

$$M' = \left[\begin{array}{c|c} M_0 & \underline{0} \\ \hline \underline{0}^T & 2cK \end{array}\right]$$

where $M_0 = \text{diag}\,[m_{11}, m_{22} \ldots 0 \ldots m_{qq}]$, i.e., $m_{ii} = 0$ and m_{jj}, $i \neq j$, is the intensity of u_j. Also, for the initial state covariance matrix,

$$P_o' = \begin{bmatrix} P_o & \vdots & \underline{0} \\ \cdots\cdots & \vdots & \cdots\cdots \\ \underline{0}^T & \vdots & K \end{bmatrix} .$$

The augmented system can then be analyzed as in the previous section since it is a linear system with a white input.

Example 4.13

The results of Example 4.12 can be obtained by using a system augmentation to give an equivalent problem with a white input. The original problem was for

$$\dot{x}_1 = -ax_1 + u, \qquad t \geq 0 .$$

$$y = x_1$$

with

$$r_U(\tau) = K(\exp -b|\tau|) \text{ and } p_o = 0 .$$

We therefore let

$$\dot{x}_2 = -bx_2 + w$$

with

$$r_w(\tau) = 2bKu_o(\tau) .$$

Thus $(\underline{x}')^T = (x_1 \quad x_2), (\underline{u}')^T = (0 \quad w)$, and

$$A' = \begin{bmatrix} -a & 1 \\ 0 & -b \end{bmatrix} \qquad B' = \begin{bmatrix} 1 & 0 \\ 0 & 1 \end{bmatrix}$$

$$C' = [1 \quad 0].$$

Also,

$$M' = \begin{bmatrix} 0 & 0 \\ 0 & 2bK \end{bmatrix}$$

and

$$P_o' = \begin{bmatrix} 0 & 0 \\ 0 & K \end{bmatrix}.$$

Then, using Eq. 4.65,

$$P(t) = \Phi(t)P_o'\Phi^T(t) + \int^t \Phi(t-\lambda)B'M'B'^T\Phi^T(t-\lambda)d\lambda$$

and the covariance of the output (which also is the mean square output in this example) is

$$(C')P(t)(C')^T = p_{11}(t)$$

where p_{11} is an element of P.

Now

$$\Phi(t) = \mathcal{L}^{-1}[sI-A']^{-1}$$

$$= \begin{bmatrix} \varepsilon^{-at} & \frac{1}{b-a}(\varepsilon^{-at} - \varepsilon^{-bt}) \\ 0 & \varepsilon^{-bt} \end{bmatrix}.$$

Then substituting and integrating we find that

$$p_{11}(t) = \frac{K}{(b-a)^2}(\varepsilon^{-at} - \varepsilon^{-bt})^2 +$$

$$+\int_o^t \frac{2bK}{(b-a)^2}((\exp -a(t-\lambda))-(\exp -b(t-\lambda)))^2 d\lambda$$

$$= \frac{K}{a(a+b)} - \frac{K}{a(b-a)} (\exp -2at)$$

$$+ \frac{2K}{(b^2-a^2)} (\exp -(a+b)t), \qquad t \geq 0$$

which checks the previous result for $S_Y^2(t)$.

Example 4.14

For the series RLC circuit shown, find the variance of the output v for $t \geq 0$. The switch closes at t=0.

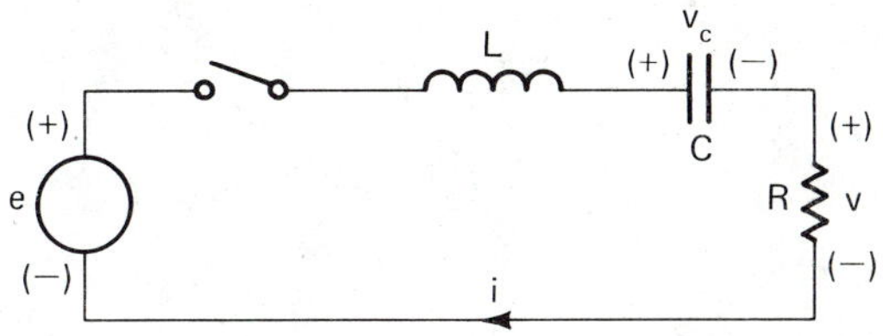

The voltage source is a wide-sense stationary process with $r_E(\tau) = 2(\exp -3|\tau|)$ and the initial charge on C is a random variable Q_o which is Gaussian with zero mean and a variance of one. R=4, L=1, C=1.

For a state variable model, let

$$x_1 = i \qquad e = u$$

$$x_2 = v_c \qquad v = y$$

and the result is

$$\dot{\underline{x}} = \begin{bmatrix} -4 & -1 \\ 1 & 0 \end{bmatrix} \underline{x} + \begin{bmatrix} 1 \\ 0 \end{bmatrix} u$$

$$y = [4 \quad 0]\underline{x}.$$

Since $x_1(0)=0$ and var $x_2(0)=\text{var}\ \frac{Q_o}{C}=1$, we have

$$P_o = \begin{bmatrix} 0 & 0 \\ 0 & 1 \end{bmatrix}.$$

Now the input is not white, therefore we augment the system. Let

$$\dot{x}_3 = -3x_3 + w, \qquad u = e = x_3$$

and

$$w_o = 12$$

and

$$\text{var}\ x_3(0) = 2\ .$$

Then

$$A' = \begin{bmatrix} -4 & -1 & 1 \\ 1 & 0 & 0 \\ 0 & 0 & -3 \end{bmatrix} \quad B' = \begin{bmatrix} 1 & 0 \\ 0 & 0 \\ 0 & 1 \end{bmatrix}$$

$$C' = [4 \quad 0 \quad 0]$$

with $(\underline{x}')^T=(x_1 \quad x_2 \quad x_3)$ and $(\underline{u}'^T)=(0 \quad w)$. Then

$$\dot{\underline{x}}' = A'\underline{x}' + B'\underline{u}'$$

$$y = C'\underline{x}'$$

and

$$P'_o = \begin{bmatrix} 0 & 0 & 0 \\ 0 & 1 & 0 \\ 0 & 0 & 2 \end{bmatrix}, \quad M' = \begin{bmatrix} 0 & 0 \\ 0 & 12 \end{bmatrix}$$

Now we use Equation 4.65. First

$$\Phi(s) = [sI-A']^{-1}$$

$$= \frac{1}{(s+3)(s-\lambda_1)(s-\lambda_2)} \begin{bmatrix} s^2+3s & -(s+3) & s \\ s+3 & s^2+7s+12 & 1 \\ 0 & 0 & s^2+4s+1 \end{bmatrix}$$

with $\lambda_1, \lambda_2 = -2 \pm \sqrt{3}$, the characteristic values of A. Substituting in Eq. 4.65, we get

$$p_{11}(t) = \phi_{12}^2(t)+2\phi_{13}^2(t)+12\int_0^t \phi_{13}^2(t-\lambda)d\lambda, \qquad t \geq 0 .$$

where

$$\phi_{12}(t) = \frac{1}{2\sqrt{3}} \left(\varepsilon^{\lambda_2 t} - \varepsilon^{\lambda_1 t}\right)$$

$$\phi_{13}(t) = \frac{\lambda_1 \varepsilon^{\lambda_1 t}}{(3+\lambda_1)(\lambda_1-\lambda_2)} + \frac{\lambda_2 \varepsilon^{\lambda_2 t}}{(3+\lambda_2)(\lambda_2-\lambda_1)} - \frac{3\varepsilon^{-3t}}{(3+\lambda_1)(3+\lambda_2)} .$$

Finally,

$$\text{var } v = (C')P(t)(C')^T = 16\, p_{11}(t) .$$

The analogous system augmentation for the discrete-time case starts with

$$\beta(n+1) = \alpha\beta(n) + w(n), \qquad 0 < \alpha < 1 \qquad (4.76)$$

where w(n) is a stationary white input with zero mean and intensity w_o. Using Equation 4.70,

$$p(n) = (\alpha)^{2(n-n_o)} p_o + w_o \sum_{k=n_o}^{n-1} (\alpha)^{2(n-k-1)}$$

$$= (\alpha)^{2(n-n_o)} p_o + w_o \left[\frac{1 - (\alpha)^{2(n-n_o)}}{1 - (\alpha)^2} \right] .$$

Now if

$$p_o = \frac{w_o}{1 - \alpha^2} ,$$

we have

$$p(n) = \frac{w_o}{1 - \alpha^2} , \qquad \forall n .$$

From Equation 4.71,

$$p_\beta(n_1,n_2) = \frac{w_o}{1-\alpha^2}(\alpha)^{(n_2-n_1)} \qquad n_2 \geq n_1$$

$$= \frac{w_o}{1-\alpha^2}(\alpha)^{(n_1-n_2)} \qquad n_1 \geq n_2$$

and for $k=n_2-n_1$, we obtain

$$p_\beta(k) = r_\beta(k) = \frac{w_o}{1-\alpha^2}(\alpha)^{|k|} , \tag{4.77}$$

Thus the first-order system of Eq. 4.76 has a stationary output sequence $\{\beta(n)\}$ such that the autocorrelation function is given by Eq. 4.77 when the system is driven by a white input of intensity w_o and when p_o, the variance of $\beta(n_o)$, is $w_o/(1-\alpha^2)$.

To augment the system

$$\underline{x}(n+1) = A\underline{x}(n) + B\underline{v}(n)$$

$$\underline{z}(n) = C\underline{x}(n).$$

when, for example, the i-th component of $\underline{v}$ is a wide-sense stationary sequence with

$$r_{v_i}(k) = K(c)^{|k|}, \qquad 0 < c < 1$$

and all other components v_j of $\underline{v}$, $j=1,2,\ldots,q$, $j \neq i$, are white let $v_i = x_{r+1}$ and

$$x_{r+1}(n+1) = cx_{r+1}(n) + w(n)$$

with $w_o = (1-c^2)K$ and var $x_{r+1}(n_o) = K$. Then the augmented system is

$$\underline{x}'(n+1) = A'\underline{x}'(n) + B'\underline{v}'(n) \qquad (4.78)$$

$$\underline{z}(n) = C'\underline{x}'(n)$$

where $(\underline{x}')^T = (\underline{x}^T(n) \vdots x_{r+1}(n))$, $(\underline{v}'(n))^T = (v_1, v_2, \ldots, 0, \ldots, v_q \vdots w)$, i.e., the i-th component of $\underline{v}'$ is zero, and

$$A' = \left[\begin{array}{c|c} A & \underline{b}_i \\ \hline \underline{0}^T & c \end{array}\right], \qquad B' = \left[\begin{array}{c|c} B & \underline{0} \\ \hline \underline{0}^T & 1 \end{array}\right]$$

$$C' = \left[\begin{array}{c|c} C & \underline{0} \end{array}\right],$$

where $\underline{b}_i$ is the i-th column of B. For the intensity matrix of $\underline{v}'$, a white vector sequence, we use

$$M' = \left[\begin{array}{c|c} M_o & \underline{0} \\ \hline \underline{0}^T & (1-c^2)K \end{array}\right]$$

where $M_o = \text{diag}[m_{11}\ m_{22}\ \ldots\ 0\ \ldots\ m_{qq}]$, $m_{ii} = 0$, and for the initial state covariance we use

$$P_o' = \left[\begin{array}{c|c} P_o & \underline{0} \\ \hline \underline{0}^T & K \end{array}\right].$$

The augmented system can be analyzed as in the previous section for the discrete-time case since it is a linear system with a white input.

Example 4.15

Consider the use of Simpson's Rule for numerical integration again. Then a state variable model is, for T=1,

$$\underline{x}(n+1) = \begin{bmatrix} 0 & 1 \\ 1 & 0 \end{bmatrix} \underline{x}(n) + \begin{bmatrix} 4/3 \\ 2/3 \end{bmatrix} v(n)$$

$$z(n) = [1 \quad 0]\ x(n) + [1/3]\ v(n) .$$

To study the effect of using a fixed word length binary representation of the input, assume m bits are used. Then the uncertainty or quantization error is bounded by $\pm\Delta/2$ where $\Delta=2^{-m}$. If the random errors are assumed to be independent and uniformly distributed, the mean of the error is zero, the variance of the error is $\Delta^2/12$, and the input sequence of random errors is white. Assume all other inputs zero. Then the problem is one of finding the variance of the output z(n) when the input is a white, zero mean sequence with intensity $\Delta^2/12$.

a) If the locations in memory holding $x_1(n)$ and $x_2(n)$ are cleared or set at known values at n=0,

$$P_o = \Theta.$$

Then using Equation 4.69,

$$P(1) = \begin{bmatrix} 4/3 \\ 2/3 \end{bmatrix} \left(\frac{\Delta^2}{12}\right) [4/3 \quad 2/3] = \frac{\Delta^2}{12} \begin{bmatrix} 16/9 & 8/9 \\ 8/9 & 4/9 \end{bmatrix}$$

$$P(2) = \begin{bmatrix} 0 & 1 \\ 1 & 0 \end{bmatrix} P(1) \begin{bmatrix} 0 & 1 \\ 1 & 0 \end{bmatrix} + P(1)$$

$$= \frac{\Delta^2}{12} \begin{bmatrix} 20/9 & 16/9 \\ 16/9 & 20/9 \end{bmatrix},$$

$$P(3) = \frac{\Delta^2}{12} \begin{bmatrix} 36/9 & 24/9 \\ 24/9 & 24/9 \end{bmatrix},$$

$$P(4) = \frac{\Delta^2}{12} \begin{bmatrix} 40/9 & 32/9 \\ 32/9 & 40/9 \end{bmatrix},$$

etc.

Thus, e.g., $p_{11}(n) = \frac{5\Delta^2}{54} n$, n even

$$= \frac{5\Delta^2}{54} n + \frac{\Delta^2}{18}, \quad n \text{ odd},$$

and if m=10 bits, $\Delta^2 \cong 10^{-6}$ and the rms <u>error in x_1</u> is approximately $3\text{x}10^{-4}\sqrt{n}$. This also is close to the rms <u>error in the output</u> where the 1/3 v(n) term must be included.

b) If the initial contents of the locations in memory holding $x_1(n)$ and $x_2(n)$ are uncertain, the initial state covariance is

$$P_o = \begin{bmatrix} p_{10} & 0 \\ 0 & p_{20} \end{bmatrix}$$

where p_{10} and p_{20} are the covariances of uncorrelated $x_1(0)$ and $x_2(0)$. The solution then is similar to that for part a) where we again use Eq. 4.69.

c) If the random quantization errors in the input are not assumed independent but are assumed to have an autocorrelation sequence

$$r(k) = \frac{\Delta^2}{12}(c)^{|k|}, \qquad 0 < c < 1\ ,$$

then

$$x_3(n+1) = cx_3(n) + w \quad , \qquad x_3 = v$$

where w has intensity $w_o = \frac{\Delta^2}{12}(1-c^2)$ and var $x_3(0) = \Delta^2/12$. The augmented system is

$$\underline{x}'(n+1) = A'\underline{x}'(n) + B'\underline{v}'(n)$$

$$z(n) = C'\underline{x}'(n) + D'\underline{v}'(n)$$

where $(\underline{x}')^T = (x_1 \quad x_2 \quad x_3)$, $(\underline{v}')^T = (0 \quad w)$ and

$$A' = \begin{bmatrix} 0 & 1 & 4/3 \\ 1 & 0 & 2/3 \\ 0 & 0 & c \end{bmatrix} \qquad B' = \begin{bmatrix} 4/3 & 0 \\ 2/3 & 0 \\ 0 & 1 \end{bmatrix}$$

$$C' = [1 \quad 0 \quad 1/3] \qquad D' = \underline{0}^T$$

and

$$P_o' = \begin{bmatrix} p_{10} & 0 & 0 \\ 0 & p_{20} & 0 \\ 0 & 0 & \frac{\Delta^2}{12} \end{bmatrix}$$

$$M' = \begin{bmatrix} 0 & 0 \\ 0 & \frac{\Delta^2}{12}(1-c^2) \end{bmatrix} .$$

The solution for the augmented state covariance matrix is then obtained by using Eq. 4.69 or Eq. 4.70.

At this point in the book we have completed a major part of the goal stated in Chapter 1, i.e., we have developed a

TABLE 4.2

TIME DOMAIN INPUT/OUTPUT FORMS

a) Single input/output		
	Continuous-time	Discrete-time
Deterministic	1. Eq. 4.1 (rth-order) 2. Eq. 4.7, 4.8 3. Eq. 4.21	1. Eq. 4.2 (rth-order) 2. Eq. 4.9, 4.10 3. Eq. 4.22
Stochastic	3. Eq. 4.34, 4.36, 4.37	3. Eq. 4.39, 4.41, 4.42

b) Multiple input/output		
	Continuous-time	
Deterministic	1. Eq. 4.1c (rth-order) 1. Eq. 4.29, 4.30 (state) 2. Eq. 4.11 3. Eq. 4.23	1. Eq. 4.2c (rth-order) 1. Eq. 4.31, 4.32 (state) 2. Eq. 4.12 3. Eq. 4.24
Stochastic	3. Eq. 4.44 3. Eq. 4.65, 4.68 (t_o finite)	3. Eq. 4.46 3. Eq. 4.70, 4.73 (n_o finite)

Numbers at left refer to cases in Sections 4.1.2 and 4.1.3

TABLE 4.3

FREQUENCY DOMAIN INPUT/OUTPUT FORMS

a) Single input/output		
	Continuous-time	Discrete-time
Deterministic	2. Eq. 4.13, 4.14, 4.17 3. Eq. 4.25	2. Eq. 4.15, 4.16, 4.18 3. Eq. 4.26
Stochastic	3. Eq. 4.47	3. Eq. 4.48

b) Multiple input/output		
	Continuous-time	Discrete-time
Deterministic	2. Eq. 4.19 3. Eq. 4.27	2. Eq. 4.20 3. Eq. 4.28
Stochastic	3. Eq. 4.57	3. Eq. 4.58

Numbers at left refer to cases in Section 4.1.2 and 4.1.3

set of signal models (Table 2.6), a set of system models (Table 3.4), and now a set of input/output results. Tables 4.2 and 4.3 summarize these results for general time domain forms and for special frequency domain forms. Much of the rest of the book is concerned with some special topics that use these results.

4.4 IDENTIFICATION

The development of signal models in Chapter 2 and system models in Chapter 3 primarily was concerned with idealizations. The use of experimental data to obtain signal models was illustrated by examples in Chapter 2 where the method of least squares and expansions in terms of sets of elementary functions were introduced. The corresponding problem of using experimental data to obtain a system model was not discussed in Chapter 3; it was deferred until results for system response were developed in this chapter. The entire problem of system identification is introduced here to show the wide range of possible approaches. First, a distinction is made between classification and identification of a system. Generally, classification has to do with selecting the form of the model from a set of classes. In this section, it is assumed that the class of linear, stationary systems is to be identified. This, of course, is a great simplication and many practical difficulties arise in attempting to justify such an assumption by using test results. Similar difficulties occur for classifying signal models, e.g., justifying an assumption of wide-sense stationarity or ergodicity. Second, identification, as discussed here, is concerned with finding specific numerical results for a model (linear and stationary) of a given system. This usually is posed as a problem of finding numerical values of the parameters of a system model. Thus, many of the methods in use today have their origins in classical methods of parameter estimation. It also must be noted that much of the research of the last two decades on system identification has been motivated by an interest in analyzing and controlling systems that are both nonstationary and nonlinear. Furthermore, the variabilities of the parameters are stochastic, i.e., the parameters are random variables or random processes or sequences. These difficult problems are beyond the scope of this book, although some of the references given at the end of this chapter consider these situations. An assumption that frequently is made in these cases is one of piecewise stationarity, i.e., the system is stationary over a finite time interval. It is understood throughout this section that the actual system being modeled has only its input and output terminals accessible and the model therefore cannot be obtained by direct measurement of the parameters and the use of physical laws as was done in Example 3.1.

The two basic approaches considered first are: (1) the use of specific test input signals to find a corresponding response function, e.g., an input $u_1(t)$ or $v_1(n)$ to find $a(\cdot)$, and (2) the use of related input and output signals, e.g., $u(t)$ ($v(n)$) and $y(t)$ ($z(n)$) to find $h(\cdot)$. No distinction is made between continuous-time and discrete-time cases although digital processing is far more common than analog. Also, only single input/single output cases are considered; the extensions to multiple inputs and outputs are straightforward in concept, but the computational difficulties grow at an enormous rate.

4.4.1 Test Inputs

A simple case of the first approach occurs if the test input to the system is u_1 or v_1. From Equations 4.8 or 4.10, the output is then $a(\cdot)$. Basically, the problem is one of recording the output data $y(t)$ or $z(n)$ and fitting a model to these as was done for signals in Chapter 2, e.g., by using a least squares fit of a polynomial. If accurate measurements were available, mathematical operations on $a(\cdot)$ would produce satisfactory results. Thus, by differentiating or differencing, $h(\cdot)$ can be found, and by transforming, $H(\cdot)$, the transfer function, can be found. The model conversions of Chapter 3, Section 3.6 also can be used to find a state model of the controllable, observable part of the system. The practical difficulties arise in obtaining accurate data, and using only a finite recording time. There also are problems (e.g., saturation) associated with subjecting the system to a step function input. A modification is discussed below.

A second simple case of the first approach occurs if the test input is sinusoidal, i.e., u_a or v_a of Chapter 2. Then from Equations 4.17 or 4.18, $H(\cdot)$ can be found. If the input frequencies are not necessarily equally-spaced, Equations 4.14 or 4.16 apply. In practice, it is necessary to obtain the magnitude and phase data where

$$H(\cdot) = |H(\cdot)| \underline{/\phi_H(\cdot)}$$

and

$$|H(\cdot)| = |Y(\cdot)| / |U(\cdot)|$$

and

$$\phi_H(\cdot) = \phi_Y(\cdot) - \phi_U(\cdot) .$$

The methods of Chapter 2, Section 2.4 for fitting straight-line asymptotes can be used to find a model for the experimental magnitude data; Reference [7.] also describes this method. If the system is known to be minimum phase, the phase function is directly related to the slope of the magnitude function; otherwise, a separate fit for the phase data is required. For discrete-time systems, the bilinear transformation of Chapter 2, Section 2.4. can be used and the curve fitting for the magnitude data is similar to that for continuous-time systems. Generally, the phase data must be known too since the transfer functions of w are not minimum phase.

An alternative to using deterministic test inputs such as step functions or sinusoids is to use a specified type of random input. In the development of Eq. 4.36, we found

$$r_{UY}(\tau) = \int_0^\infty h(\alpha) r_U(\tau-\alpha)\, d\alpha \ .$$

If the test input is a white process, we have

$$r_U(\tau) = w_o u_o(\tau),$$

and therefore,

$$\begin{aligned} r_{UY}(\tau) &= w_o \int_0^\infty h(\alpha) u_o(\tau-\alpha)\, d\alpha \\ &= w_o h(\tau), \qquad \tau \geq 0 \\ &= 0 \qquad\qquad \tau < 0 \ . \end{aligned}$$

Thus if a single input/single output linear, stationary system is driven by a white process, the crosscorrelation function $r_{UY}(\tau)$ is proportional to the required impulse response. In the application of this method, time averaging over a finite period must be done, i.e., we use the approximation

$$r_{UY}(\tau) \cong \frac{1}{T} \int_0^T u^i(t-\tau) y^i(t)\, dt, \qquad \tau \geq 0$$

for large T. The method has at least two advantages over the previous ones in that the test can be done while the system is in normal operation if w_o is kept small, and the effects of measurement errors can be reduced by using T sufficiently large [8.]. The actual input need only be a process that is wide band relative to the system band width, and with obvious changes in notation, a discrete-time (digital) rather than continuous-time (analog) implementation can be used:

$$r_{VZ}(k) \cong \frac{1}{N+1} \sum_{n=0}^{N} v^i(n-k) z^i(n)$$

and

$$r_{VZ}(k) = w_o h(k) \quad .$$

An interesting example of the use of this method is given in [22.] where the impulse response of an attemporator loop of a superheated steam supply is measured.

4.4.2 Input/Output Analysis

The second approach to system identification uses related pairs of input and output signals rather than requiring special test inputs. These signals may be obtained, for example, while the system is in normal operation, and this has a clear practical advantage, i.e., "on-line" identification is possible. A numerical form of deconvolution can be found by starting with Eq. 4.9:

$$z(n) = \sum_{k=n_o}^{n} v(k)h(n-k) \quad .$$

With $z(\cdot)$ and $v(\cdot)$ known over a finite interval $[n_o, N]$, it is possible to find $h(\cdot)$. It is assumed that $v(n)=0$, $n<n_o$, so that the system has no stored energy at $n=n_o$. For $n=0, 1,\ldots,N$ with $n_o=0$,

$$z(0) = v(0)h(0)$$

$$z(1) = v(0)h(1) + v(1)h(0)$$

$$\vdots$$

$$z(N) = \sum_{k=0}^{N} v(k)h(N-k) \quad .$$

In vector-matrix form

$$\underline{z} = V\underline{h}$$

where

$$\underline{z}^T = (z(0) \ \ldots \ z(N))$$

$$\underline{h}^T = (h(0) \ \ldots \ h(N))$$

and

$$V = \begin{bmatrix} v(0) & & & 0 \\ v(1) & \ddots & & \\ \vdots & \ddots & \ddots & \\ v(N) & \ldots & v(1) & v(0) \end{bmatrix} .$$

For $v(0) \neq 0$, V^{-1} exists and

$$\underline{h} = V^{-1}\underline{z} . \tag{4.79}$$

A similar result for a numerical version of Equation 4.7 also can be developed, and this gives N+1 values of the impulse response which can be used for finding a fit to obtain $h(\tau)$. It is necessary that the system under test be stable and N must be chosen large enough to include all "significant" variations in the impulse response curve. The effects of measurement errors are made small in the curve fitting procedure.

A reformulation of this problem leads directly to a "best" identification in the least squares sense [9.]. Thus if $h(\cdot)$ is modeled by any suitable form (such as one of those given in Chapter 2) with parameters $p_1, p_2, \ldots, p_K$, i.e., $\underline{p}^T = (p_1 \ \ldots \ p_K)$, the required response function is

$$h(n;\underline{p})$$

where the parameter vector $\underline{p}$ that minimizes

$$S = \sum_{n=N_a}^{N_b} (z(n) - f(n))^2$$

is to be found. $z(n)$ is the measured output over $[N_a, N_b]$ and

$$f(n) = \sum_{k=0}^{N_s-1} h(k;\underline{p})v(n-k) \equiv f(n;\underline{p}) \ .$$

This follows from Eq. 4.5 if the system is assumed to have a finite settling time, i.e., $h(n;\underline{p})=0$, $n \geq N_s$, as well as being stationary and realizable. The required number of output values is $N_b-N_a+1 \geq K$ and the required number of input values is the number of output values plus N_s-1. Thus the minimum number of values $M=K+N_s-1$. Note that this formulation does not require that the input be zero before n_o as was the case for the simple deconvolution of Eq. 4.79. If the number of values is greater than M, the redundant data reduces the effect of measurement error as was discussed in Chapter 2.

Necessary conditions for minimum S are

$$\frac{\partial S}{\partial p_i} = 0, \qquad i = 1, 2, \ldots, K \ .$$

Usually, $h(n;\underline{p})$ is a transcendental function of the p_i (e.g., if the model of Eq. 2.40 is used) and an iterative solution of the K equations is necessary. One such solution [10.] uses an initial guess $\underline{p}_o$; then we let

$$\hat{f}(n;\underline{p}_o) = f(n;\underline{p}_o) + \left.\frac{\partial f}{\partial p_1}\right|_{\underline{p}_o} \Delta p_{1o} + \ldots +$$

$$+ \left.\frac{\partial f}{\partial p_K}\right|_{\underline{p}_o} \Delta p_{Ko} + \ldots$$

Using only the first-order terms, the "best" corrections Δp_{io} are found by requiring

$$\frac{\partial \hat{S}}{\partial \Delta p_{io}} = 0, \qquad i = 1, 2, \ldots, K \ ,$$

where $\hat{S}$ is obtained by replacing $f(\cdot)$ with $\hat{f}(\cdot)$ in S. The resulting linear equations are solved for the Δp_{io}, and then $\underline{p}_1 = \underline{p}_o + \lambda_o \underline{\Delta p}_o$ where $(\underline{\Delta p}_o)^T = (\Delta p_{1o} \ \dots \ \Delta p_{Ko})$ and λ_o is a scalar, $0 < \lambda_o < 1$, chosen to obtain convergence. The iteration is continued until some appropriate stopping condition is satisfied, e.g., the difference between $\hat{S}$ on one iteration and $\hat{S}$ on the next, is less than a specified number. A computational procedure is as follows: Let

$$a^j(n) = f(n;\underline{p}_j)$$

and

$$b_i^j(n) = \left.\frac{\partial f}{\partial p_i}\right|_{\underline{p}_j} ,$$

then

$$\underline{\Delta p}_j = G_j^{-1}\underline{q}_j \tag{4.80}$$

where the components of $\underline{q}_j$ are

$$q_{ji} = \sum_{m=N_a}^{N_b} (z(m) - a^j(m))b_i^j(m), \qquad i = 1, \dots, K$$

and the components of the symmetric matrix G_j are

$$g_{jrs} = \sum_{m=N_a}^{N_b} b_r^j(m)b_s^j(m), \qquad r,s = 1, 2, \dots, K .$$

In all of these, j is the iteration number, thus $j = 0, 1, 2, \dots$. At each step

$$\underline{p}_{j+1} = \underline{p}_j + \lambda_j \underline{\Delta p}_j, \qquad 0 < \lambda_j < 1$$

and the computation is stopped when

$$|\hat{S}_{j+1} - \hat{S}_j| < \delta \quad .$$

The values of the λ_j and δ usually are arrived at by experiment to obtain a good compromise between accuracy and speed of convergence. The following simple example illustrates the procedure.

Example 4.16

For the configuration shown, the input and output data are given

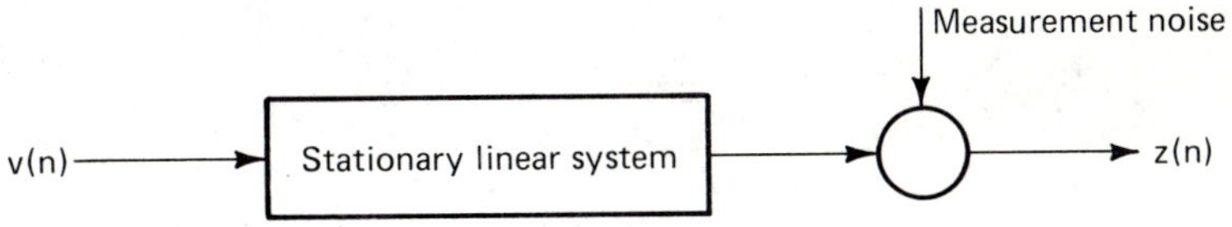

n	v(n)	z(n)
0	1	3.02
1	0	1.47
2	1	3.74
3	-1	-1.10
4	0	-0.58
5	1	2.75
6	1	4.32
7	-1	-0.85
8	0	-0.43
9	1	2.78

Assume the unit pulse response can be approximated by

$$h(n) = \alpha\beta^n, \quad n \geq 0, \quad \alpha > 0 \quad 0 < \beta < 1,$$

$$= 0 \quad , \quad n < 0 .$$

Then $\underline{p}^T = (\alpha \;\; \beta)$ and K=2. Also assume N_s=6. Then M=2+6-1=7 and there is redundant data in the 10 values given. Also, let $N_b=9$ and let $N_a=5$, the smallest value consistent with the assumed N_s.

For j=0, assume $\alpha^o=2$ and $\beta^o=0.4$. First, compute

$$a^o(n) = \sum_{k=0}^{5} 2(0.4)^k v(n-k)$$

$$b_1^o(n) = \sum_{k=0}^{5} (0.4)^k v(n-k)$$

$$b_2^o(n) = \sum_{k=0}^{5} 2k(0.4)^{k-1} v(n-k) \quad .$$

Second, compute

$$q_{01} = \sum_{m=5}^{9} (z(m)-a^o(m))b_1^o(m)$$

$$q_{02} = \sum_{m=5}^{9} (z(m)-a^o(m))b_2^o(m)$$

$$g_{011} = \sum_{m=5}^{9} (b_1^o(m))^2 \quad .$$

$$g_{022} = \sum_{m=5}^{9} (b_2^o(m))^2$$

$$g_{012} = g_{021} = \sum_{m=5}^{9} b_1^o(m)b_2^o(m) \quad .$$

After this first iteration,

$$\underline{\Delta p}_o = G_o^{-1}\,\underline{q}_o = \begin{bmatrix} 3.80 & 0.06 \\ 0.06 & 13.85 \end{bmatrix}^{-1} \begin{bmatrix} 3.86 \\ 2.19 \end{bmatrix}$$

$$= \begin{bmatrix} 1.01 \\ 0.12 \end{bmatrix} .$$

If $\lambda_o = 0.8$,

$$\underline{p}_1 = \begin{bmatrix} 2 \\ 0.4 \end{bmatrix} + 0.8 \begin{bmatrix} 1.01 \\ 0.12 \end{bmatrix} = \begin{bmatrix} 2.81 \\ 0.52 \end{bmatrix} .$$

The true values of α and β used to generate the noisy data z(n) were 3 and 0.5. Also,

$$\hat{S}_o = \sum_{n=5}^{9} (z(n)-\hat{f}(n;\underline{p}_o))^2 = 4.26$$

and

$$\hat{S}_1 = 0.011 .$$

Repeated iteration gave the following results using $\lambda = 0.8$

j	α^j	β^j	$\hat{S}_j$	α Error	β Error
0	2.00	0.40	4.26	1.00	0.10
1	2.81	0.52	0.011	0.19	-0.02
2	2.96	0.50	0.004	0.04	0.0
3	2.99	0.50	0.004	0.01	0.0
4	3.00	0.50	0.004	0.00	0.0

The stopping condition in this very simple case was a specified j.

Stochastic analogs of this least squares method could be used if a model for the measurement noise as a stochastic sequence is assumed. In this case we could find $\underline{p}$ such that

$$S_1 = E[(z(n)-f(n))^2]$$

is minimal or

$$S_2 = E \sum_{n=N_a}^{N_b} (z(n)-f(n))^2$$

is minimal where

$$f(n) = \sum_{k=0}^{N_s-1} h(k;\underline{p})v(n-k)$$

as before. The use of S_1, a mean square error, implies that $\underline{p}$ is chosen to minimize "error power" and for S_2, "error energy" is minimized (See Tables 2.7 and 2.8). An identification of a state-variable model using a form like S_1 is given later in this section.

The transfer function H(z) also can be found directly by using least squares or minimum mean square methods similar to those above [11.][12.]. From Equation 3.20,

$$(E^r+\alpha_{r-1}E^{r-1}+\ldots+\alpha_0)z(n) = (\beta_m E^m+\ldots+\beta_0)v(n), \qquad m \leq r$$

and from Equation 3.50 then

$$H(z) = \frac{\beta_m z^m + \ldots + \beta_0}{z^r + \alpha_{r-1}z^{r-1} + \ldots + \alpha_0}$$

$$= \frac{\beta_m z^{m-r} + \ldots + \beta_0 z^{-r}}{1 + \alpha_{r-1}z^{-1} + \ldots + \alpha_0 z^{-r}}$$

and there are r+m+1 parameters to be found. The estimates of the α_j, j=0,...,r-1 and β_j, j=0,...,m are $\hat{\alpha}_j$ and $\hat{\beta}_j$, and we let

$$S = \sum_{n=N_a}^{N_b} (z(n)-g(n))^2$$

where z(n) is again the measured output data, and we have

$$g(n) = -\hat{\alpha}_{r-1}z(n-1) - \ldots - \hat{\alpha}_o z(n-r) + $$
$$+ \hat{\beta}_m v(n+m-r) + \ldots + \hat{\beta}_o v(n-r) \ .$$

Minimizing S provides the best estimates of the α_j and β_j. When g(n) is substituted in S, it is clear that there are terms involving products $z(\cdot)z(\cdot)$, $z(\cdot)v(\cdot)$, and $v(\cdot)v(\cdot)$. This leads to making definitions of "pseudocorrelation" (or discrete pseudotranslation) functions as

$$\theta_{zz}(k;N,I) = \sum_{i=I-N}^{I} z(i)z(i+k)$$

$$\theta_{zv}(k;N,I) = \sum_{i=I-N}^{I} z(i)v(i+k)$$

$$\theta_{vv}(k;N,I) = \sum_{i=I-N}^{I} v(i)v(i+k)$$

when $N=N_b-N_a$ and $I=N_b$. Using the r+m+1 necessary conditions

$$\frac{\partial S}{\partial \hat{\alpha}_j} = 0, \qquad \frac{\partial S}{\partial \hat{\beta}_j} = 0$$

leads to r+m+1 linear algebraic equations in the $\hat{\alpha}_j$ and $\hat{\beta}_j$. These equations easily can be solved by using the iterative procedure in [11.]. Alternatively, we can start with

$$\hat{H}(z) = \frac{\hat{\beta}_m z^{m-r} + \ldots + \hat{\beta}_o z^{-r}}{1+\hat{\alpha}_{r-1}z^{-1} + \ldots + \hat{\alpha}_o z^{-r}} = \frac{\theta_{zz}(z;N,I)}{\theta_{zv}(z;N,I)} \ . \tag{4.81}$$

This equality on the right can be shown by considering the form

$$w_z(k) = \sum_{j=-\infty}^{\infty} h(j) w_{zv}(k-j) \ ,$$

the deterministic analogy to Equation 4.42. Its Z-Transform is

$$W_z(z) = H(z) W_{zv}(z)$$

and modification of this for pseudotranslation functions gives the result above. Then for 4.81 we have

$$\hat{H}(z) = \frac{\sum_{j=r-m}^{r} \hat{\beta}_{r-j} z^{-j}}{1 + \sum_{j=1}^{r} \hat{\alpha}_{r-j} z^{-j}}$$

$$= \frac{\sum_{k=-\infty}^{\infty} \theta_{zz}(k;N,I) z^{-k}}{\sum_{k=-\infty}^{\infty} \theta_{zv}(k;N,I) z^{-k}} \ . \qquad (4.82)$$

For p=j+k, cross multiplying and equating coefficients of like powers of z leads to

$$\sum_{j=r-m}^{r} \hat{\beta}_{r-j} \theta_{zv}(p-j) - \sum_{j=1}^{r} \hat{\alpha}_{r-j} \theta_{zz}(p-j) = \theta_{zz}(p) \qquad (4.83)$$

where the N and I dependence has been suppressed, and it is assumed that all translation functions are computed for the same N. By choosing r+m+1 values for p, e.g., p=0, -1,..., -(r+m), r+m+1 linear algebraic equations are obtained and these can be solved for the $\hat{\alpha}_j$ and $\hat{\beta}_j$. It is desirable to have N>>r+m for good accuracy when measurement errors are present.

Example 4.17

Assume that the required input and output data are available and a model of the form

$$H(z) = \frac{\hat{\beta}_1 z}{z+\hat{\alpha}_o} = \frac{\hat{\beta}_1}{1 + \hat{\alpha}_o z^{-1}}$$

is used. Obtain equations in $\hat{\beta}_1$ and $\hat{\alpha}_o$. Now for minimizing S, we have

$$g(n) = \hat{\beta}_1 v(n) - \hat{\alpha}_o z(n-1)$$

and

$$S = \sum_{n=I-N}^{I} (z(n)-\hat{\beta}_1 v(n)+\hat{\alpha}_o z(n-1))^2$$

$$= \sum_{n=I-N}^{I} (z(n)^2+\hat{\beta}_1^2 v(n)^2+\hat{\alpha}_o^2 z(n-1)^2$$

$$-2\hat{\beta}_1 z(n)v(n)-2\hat{\beta}_1\hat{\alpha}_o z(n-1)v(n) + 2\hat{\alpha}_o z(n)z(n-1)).$$

Taking

$$\frac{\partial S}{\partial \hat{\alpha}_o} = 0 \quad \text{and} \quad \frac{\partial S}{\partial \hat{\beta}_1} = 0,$$

we obtain

$$0 = \sum_{n=I-N}^{I} (\hat{\alpha}_o z(n-1)^2-\hat{\beta}_1 z(n-1)v(n)+z(n)z(n-1))$$

$$0 = \sum_{n=I-N}^{I} (-\hat{\alpha}_o z(n-1)v(n)+\hat{\beta}_1 v(n)^2-z(n)v(n)) .$$

Using the definitions of the pseudotranslation functions,

$$0 = \hat{\alpha}_o \theta_{zz}(0) - \hat{\beta}_1 \theta_{zv}(1) + \theta_{zz}(1)$$

$$0 = \hat{\alpha}_o \theta_{zv}(1) + \hat{\beta}_1 \theta_{vv}(0) - \theta_{zv}(0)$$

and these can be put in the form

$$\begin{bmatrix} \theta_{zz}(0) & -\theta_{zv}(1) \\ -\theta_{zv}(1) & \theta_{vv}(0) \end{bmatrix} \begin{bmatrix} \hat{\alpha}_o \\ \hat{\beta}_1 \end{bmatrix} = \begin{bmatrix} -\theta_{z\dot{z}}(1) \\ \theta_{zv}(0) \end{bmatrix}$$

To show that these two equations obtained by minimizing S are equivalent to 4.83, we can start with p=1 and $\hat{\beta}_o=0$. We obtain

$$\hat{\beta}_1 \theta_{zv}(1) - \hat{\alpha}_o \theta_{zz}(0) = \theta_{zz}(1)$$

which is the same as the first equation. Similarly, for p=2,

$$\hat{\beta}_1 \theta_{zv}(2) - \hat{\alpha}_o \theta_{zz}(1) = \theta_{zz}(2)$$

which can be related to the second equation by multiplying the equation for $\frac{\partial S}{\partial \hat{\beta}_1} = 0$ by

$$\frac{z(n-2)}{v(n)}$$

and then using the definitions of the autotranslation functions. Finally, the required equations as found from 4.83 with p=1 and p=2 are:

$$\begin{bmatrix} \theta_{zz}(0) & -\theta_{zv}(1) \\ -\theta_{zz}(1) & \theta_{zv}(2) \end{bmatrix} \begin{bmatrix} \hat{\alpha}_o \\ \hat{\beta}_1 \end{bmatrix} = \begin{bmatrix} -\theta_{zz}(1) \\ \theta_{zz}(2) \end{bmatrix}$$

We note that although r=m=1, only two values of p were used since $\beta_o=0$.

Example 4.18

A simple numerical example of the use of Eq. 4.83 is as follows. With a model

$$H(z) = \frac{3z}{z-0.5} = \frac{Z(z)}{V(z)} ,$$

a data record of v(n) and z(n) was generated on a digital computer for v(n) a white Gaussian sequence with an intensity of 1 (see Section 4.5.2 for the method). This record was then used to compute the required "pseudocorrelation" functions in Eq. 4.83. The values of p were 0 and -1. Two cases were tried:

a) No measurement noise added to z(n)

N+1	$\hat{\alpha}_o$	$\hat{\beta}_1$
50	-0.5376	2.9890
500	-0.4960	2.9997
1000	-0.5026	3.0001

b) Measurement noise added to z(n)

The noise added to z(n) was white Gaussian with an intensity of 1. Since the mean square z(n) without noise is 12, as found by using the inverse of 4.48 with k=0, the signal/noise ratio at the output was 12. For N+1=1000,

$$\hat{\alpha}_o = -0.38$$

$$\hat{\beta}_1 = 4.02$$

Obviously the accuracy of the identification depends on N and on the measurement noise. Small differences also occur for other values of p, e.g., p=1 and 2, since the required correlation functions are different. Finally, we note that in this example, the order of the system was assumed known, i.e., m=r=1 as in Ex. 4.17. In practical applications, the order is not known, and we must assume m and r. This may contribute to further inaccuracy. See Problem 30 at the end of this chapter.

4.4.3 State Model

The second identification approach, methods which use general input/output data, has now been shown for finding h(n), the unit pulse response, and H(z), the transfer function. It also is possible to identify a single input/single output system as a state variable model. The form of the model chosen here is a third primal form, a discrete version of Eq. 3.71 with $f_o=0$. Thus the parameters $\alpha_o,\ldots,\alpha_{r-1}$ and $b_1,\ldots,b_r$ are required in

$$\underline{x}(n+1) = \left[\begin{array}{c|c} \underline{0} & I_{r-1} \\ \hline \multicolumn{2}{c}{-\underline{\alpha}^T} \end{array}\right] \underline{x}(n) + \begin{bmatrix} b_1 \\ \vdots \\ b_r \end{bmatrix} v(n)$$

or

$$\underline{x}(n+1) = A\underline{x}(n) + \underline{b}v(n) \tag{4.84a}$$

and

$$z(n) = [1 \quad 0 \quad \ldots \quad 0]\underline{x}(n) = \underline{c}^T\underline{x}(n) \tag{4.84b}$$

where $\underline{\alpha}^T=(\alpha_o\ \alpha_1\ \ldots\ \alpha_{r-1})$. A simple case occurs if v(n)=0, $\forall$ n, and z(n) is measured without error for z(1), z(2),..., z(2r). The third primal form is observable, and if the initial state and A are such that the identifiability matrix

$$P_I = [\underline{x}(0) \vdots A\underline{x}(0) \vdots \; \dots \; \vdots A^{r-1}\underline{x}(0)]$$

is rank r, the required result is [13.]

$$A = S_{2r} S_{2r-1}^{-1} \tag{4.85}$$

where

$$S_{2r} = \begin{bmatrix} z(2) & z(3) & \dots & z(r+1) \\ z(3) & & & \cdot \\ \vdots & & \cdot & \\ & \cdot & & \\ z(r+1) & & \dots & z(2r) \end{bmatrix}$$

and

$$S_{2r-1} = \begin{bmatrix} z(1) & z(2) & \dots & z(r) \\ z(2) & & \cdot & \vdots \\ & \cdot & & \\ \cdot & & & \\ z(r) & & \dots & z(2r-1) \end{bmatrix}.$$

This is, of course, a very simple situation and a more realistic problem would include a nonzero input and an output with errors in the measurements. In these cases, least squares or minimum mean square estimates are required [13.] [14.]. Of the many possibilities, only one [14.] is given here. The model is

$$\underline{x}(n+1) = A\underline{x}(n) + \underline{b}v(n) + \underline{\gamma}w(n) \tag{4.86a}$$

$$z(n) = \underline{c}^T\underline{x}(n) + m(n) \tag{4.86b}$$

where A, $\underline{b}$, and $\underline{c}^T$ are as in Eq. 4.84 and w(n) and m(n) are white (zero mean) sequences that are not correlated. v(n) is measured without error and the measured z(n) contains errors due to m(n). Now let

$$\underline{\theta} = \begin{bmatrix} \underline{c}^T A^{2r-1} \underline{b} \\ \vdots \\ \underline{c}^T \underline{b} \end{bmatrix} = \begin{bmatrix} h(2r) \\ \vdots \\ h(1) \end{bmatrix} = \begin{bmatrix} \theta_1 \\ \theta_2 \\ \vdots \\ \theta_{2r} \end{bmatrix}$$

where the second equality follows from Eq. 3.104. (Note that h(0)=0 since f_o was assumed zero). Also, let

$$S = E[(z(n) - y(n))^2]$$

where

$$y(n) = \sum_{j=1}^{2r} h(j)v(n-j+1)$$

which is similar to f(n) and g(n) in the previous methods. Now by using the definition of $\underline{\theta}$, we can show that

$$\underline{b} = \begin{bmatrix} \theta_{2r} \\ \theta_{2r-1} \\ \vdots \\ \theta_{r+1} \end{bmatrix} \tag{4.87a}$$

and

$$\underline{\alpha} = -\begin{bmatrix} \theta_{2r} & \theta_{2r-1} & \cdots & \theta_{r+1} \\ \theta_{2r-1} & & & \theta_r \\ \vdots & & & \vdots \\ \theta_{r+1} & \theta_r & \cdots & \theta_2 \end{bmatrix}^{-1} \begin{bmatrix} \theta_r \\ \theta_{r-1} \\ \vdots \\ \theta_1 \end{bmatrix}, \tag{4.87b}$$

and if estimates of $\underline{\theta}$, i.e., $\hat{\underline{\theta}}$, can be found, estimates $\hat{\underline{\alpha}}$ and $\hat{\underline{b}}$ also can be found. An iterative (stochastic approximation) algorithm that converges to estimates such that the mean square error S is minimal is as follows:

$$\hat{\underline{\theta}}(k+2r+1) = \hat{\underline{\theta}}(k) + K(k)\underline{v}(k+2r) \cdot \cdot [z(k+2r+1) - \underline{v}^T(k+2r)\hat{\underline{\theta}}(k)] \quad (4.88)$$

k=0, 2r+1, 4r+2, ..., where

$$\underline{v}(n) = \begin{bmatrix} v(n-2r+1) \\ \cdot \\ \cdot \\ \cdot \\ v(n) \end{bmatrix}, \qquad \underline{z}(n) = \begin{bmatrix} z(n-2r+1) \\ \cdot \\ \cdot \\ \cdot \\ z(n) \end{bmatrix},$$

$$K(k) = \frac{P(k+2r+1)}{1 + \underline{z}^T(k+2r)P(k)\underline{z}(k+2r)},$$

and

$$P(k+2r+1) = P(k) - P(k)\underline{z}(k+2r) \cdot \cdot [\underline{z}^T(k+2r)P(k)\underline{z}(k+2r)+1]^{-1}\underline{z}^T(k+2r)P(k).$$

The computation starts with assumed $\hat{\underline{\theta}}(0)$ and P(0) using the first 2r inputs and outputs, and is repeated for k=2r+1, 4r+2,... until a suitable stopping condition is met. The estimates $\hat{\underline{\alpha}}$ and $\hat{\underline{b}}$ are then found from Eq. 4.87 using the final $\hat{\underline{\theta}}$.

These methods for identifying h(·), H(·), or the pair A, $\underline{b}$ illustrate some of the many possibilities. The use of other models for h(·), e.g., in terms of the moments of h(·) [15.], have been tried. Also, special signal models for the inputs and outputs, e.g., the use of expansions in terms of sets of orthonormal functions [16.][17.], have been used. Many other methods (e.g., gradient techniques, stochastic approximation, and quasilinearization), which can be applied in more general cases where a given functional is to be minimized, have been developed [14.][18.][19.][23.]. However, the identification of complicated, time-varying systems is still a very difficult problem.

4.5 COMPUTER GENERATION OF SIGNALS

An interesting application of the results on signal and linear system models is the generation of specified signals using analog and digital computers. These generated signals are used in many ways in simulation studies, e.g., in studies of transients or in Monte Carlo analyses. Two classes of signals are considered: (1) deterministic signals ($u(t), t \in T_{c1}$ or $v(n)$, $n \in T_{d1}$) which are the zero-input responses of stationary linear systems, and (2) stochastic signals ($\{U(t)\}$ or $\{V(n)\}$) which are the outputs of stationary linear systems with white inputs. Thus, the deterministic signals are one-sided, e.g., singularity functions and exponentials, and the stochastic signals are Markov processes or sequences. Only scalar signal generators are developed although extensions to vector signals (with possible correlation between components) can be done.

4.5.1 Deterministic Signals

From Equation 4.29, with zero input and scalar output, the signal generator in Figure 4.1 is described by

$$\underline{x}(t) = (\exp A(t-t_o))\underline{x}(t_o)$$

$$u(t) = \underline{c}^T \underline{x}(t) \qquad\qquad t \geq t_o$$

with $\underline{x}$ the state of the generator and $u(t)$ the required signal. Any generator requires a choice of a state vector $\underline{x}$ as well as A, $\underline{c}^T$, and $\underline{x}(t_o)$. The pair $(A, \underline{c}^T)$ is required to be observable. There are many ways for programming an analog computer to generate such signals; the methods

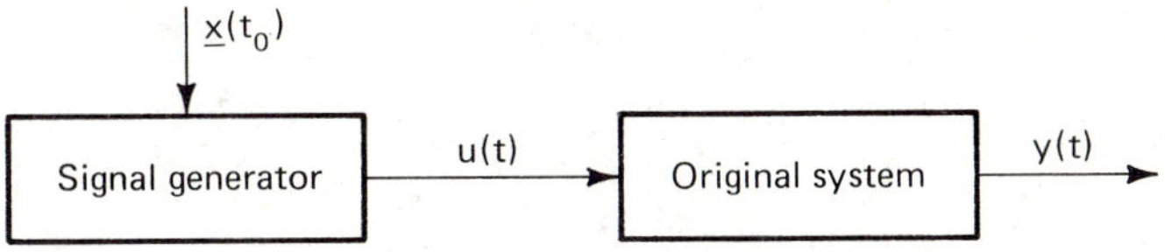

Figure 4.1 Signal generator model

suggested in this brief introduction are intended primarily to generate the signals of Table 2.1 by using some of the state models discussed in Chapter 3. An orderly procedure is as follows:

1. For the required $u(t)$, $t \in T_{cl}$, find $U(s) = \mathcal{L}\, u(t)$, a rational function.
2. For an observable system model, $U(s)=\underline{c}^T\Phi(s)\underline{x}(t_o)$, and r, the dimension of $\underline{x}$, equals the number of poles of $U(s)$.
3. Choose a matrix A_{rxr} such that its characteristic values are the same as the poles of $U(s)$. Two relatively simple forms are the diagonal form and phase-variable form of Chapter 3.
4. Find $\Phi(s) = [sI-A]^{-1}$.
5. Choose $\underline{c}^T$ and $\underline{x}(t_o)$ such that $U(s) = \underline{c}^T\Phi(s)\underline{x}(t_o)$.

The method is illustrated with some simple examples.

<u>Example 4.19</u>

a) For $u(t) = K\varepsilon^{-at}u_1(t)$, find suitable A, $\underline{c}^T$, and $\underline{x}(0)$.

$$U(s) = \frac{K}{s+a}, \quad \text{and} \quad r=1.$$

Let

$$A_{1x1} = -a,$$

then

$$\Phi(s) = [s-(-a)]^{-1} = \frac{1}{s+a}$$

and require

$$c\Phi(s)\ x(0) = \frac{K}{s+a}.$$

Therefore, we let c=1 and x(0) = K, and the model is

$$\dot{x}(t) = -a\ x(t), \quad x(0) = K, \qquad t \geq 0 .$$

$$u(t) = x(t).$$

b) Form an augmented system from a system

$$\underline{\dot{x}}'(t) = A'\underline{x}'(t) + \underline{b}'u(t),\ \underline{x}'(0) = \underline{x}'_o \quad .$$

$$y(t) = (\underline{c}')^T\underline{x}'(t)$$

where u(t) is given in a). Let

$$\underline{x}'' = \begin{bmatrix} \underline{x}' \\ x \end{bmatrix}, \qquad A'' = \begin{bmatrix} A' & \underline{b}' \\ \underline{0}^T & -a \end{bmatrix}$$

$$(\underline{c}'')^T = \left[(\underline{c}')^T \ \vdots \ 1 \right],$$

then

$$\underline{\dot{x}}''(t) = A''\underline{x}''(t) \qquad \underline{x}''(0) = \begin{bmatrix} \underline{x}'_o \\ K \end{bmatrix}.$$

$$y(t) = (\underline{c}'')^T\underline{x}''(t)$$

c) For $u(t) = (K_1\varepsilon^{-at}+K_2\varepsilon^{-bt})u_1(t)$, $a\neq b$, find suitable A, $\underline{c}^T$, $\underline{x}(0)$. Now

$$U(s) = \frac{K_1}{s+a} + \frac{K_2}{s+b} = \frac{(K_1+K_2)s+(K_1b+K_2a)}{s^2+(a+b)s+ab} .$$

Since r=2, a simple choice for A is

$$A_1 = \begin{bmatrix} -a & 0 \\ 0 & -b \end{bmatrix}$$

and for $\underline{c}_1^T$ and $\underline{x}_1(0)$,

$$\underline{c}_1^T = [1 \quad 1], \qquad \underline{x}_1^T(0) = [K_1 \quad K_2].$$

If the second form on the right of U(s) is used, another choice for A is

$$A_2 = \begin{bmatrix} 0 & 1 \\ -ab & -(a+b) \end{bmatrix} .$$

After computing $\Phi_2(s) = [sI-A_2]^{-1}$, require

$$U(s) = \underline{c}_2^T\Phi_2(s)\underline{x}_2(0) .$$

Therefore, by equating the numerators, require

$$(K_1+K_2)s+(K_1b+K_2a) = [c_1 \quad c_2] \begin{bmatrix} s+(a+b) & 1 \\ -ab & s \end{bmatrix} \begin{bmatrix} x_{10} \\ x_{20} \end{bmatrix}$$

$$= c_1[s+(a+b)]x_{10}+c_1x_{20}$$

$$-c_2abx_{10}+c_2sx_{20} .$$

From the two equations obtained by equating coefficients of

like powers of s, suitable choices for $\underline{c}_2^T$ and $\underline{x}_2(0)$ can be made. For example, for $\underline{c}_2^T = [1 \quad 0]$, we get

$$\underline{x}_2(0) = \begin{bmatrix} (K_1+K_2) \\ -(K_1 a+K_2 b) \end{bmatrix} .$$

d) For $u(t) = a_0 u_1(t)+a_1 u_2(t)$ as in Eq. 2.16, find suitable A, $\underline{c}^T$, and $\underline{x}(0)$. Now

$$U(s) = \frac{a_0}{s} + \frac{a_1}{s^2} = \frac{a_0 s+a_1}{s^2} ,$$

and r=2. Let

$$A = \begin{bmatrix} 0 & 1 \\ 0 & 0 \end{bmatrix} ,$$

a Jordan form with $\lambda_1=\lambda_2=0$. Then using the same technique as in c),

$$a_0 = c_1 x_{10}+c_2 x_{20}$$

$$a_1 = c_1 x_{20}$$

so that one solution is

$$\underline{c}^T = [1 \quad 0] \quad \text{and} \quad \underline{x}^T(0) = [a_0 \quad a_1] .$$

e) For $u(t) = [b_1 \cos\omega_0 t+a_1 \sin\omega_0 t+a_2 \sin 3\omega_0 t]u_1(t)$ find suitable A, $\underline{c}^T$, $\underline{x}(0)$. Now

$$U(s) = \frac{b_1 s}{s^2+\omega_0^2} + \frac{a_1\omega_0}{s^2+\omega_0^2} + \frac{3a_2\omega_0}{s^2+9\omega_0^2} = \frac{b_1 s+a_1\omega_0}{s^2+\omega_0^2} + \frac{3a_2\omega_0}{s^2+9\omega_0^2}$$

$$= \frac{(b_1 s + a_1 \omega_o)(s^2+9\omega_o^2) + 3a_2\omega_o(s^2+\omega_o^2)}{(s^2+\omega_o^2)(s^2+9\omega_o^2)}$$

and therefore r=4. Let

$$A = \left[\begin{array}{cc|cc} 0 & \omega_o & & \\ -\omega_o & 0 & \multicolumn{2}{c}{\Theta} \\ \hline & & 0 & 3\omega_o \\ \multicolumn{2}{c|}{\Theta} & -3\omega_o & 0 \end{array}\right]$$

if the second form of U(s) is used. The two uncoupled subsystems can be used to simplify the algebra. Thus for

$$\underline{c}^T = [1 \quad 1 \quad 1 \quad 0] \quad ,$$

we obtain

$$\underline{x}^T(0) = \left[\frac{b_1+a_1}{2} \quad \frac{b_1-a_1}{2} \quad 0 \quad a_2 \right] .$$

Discrete-time deterministic signals can be generated similarly. From Equation 4.31, with zero input and scalar output,

$$\begin{aligned} \underline{x}(n) &= A^{(n-n_o)}\underline{x}(n_o) \\ v(n) &= \underline{c}^T\underline{x}(n) \end{aligned} \qquad n \geq n_o$$

and for the required signal v(n), any generator requires a choice of a state vector $\underline{x}$ as well as A, $\underline{c}^T$, and $\underline{x}(n_o)$. The pair $(A, \underline{c}^T)$ is required to be observable. A procedure similar to the previous case is:

1. For the required v(n), $n \in T_{d1}$, find

$$V(z) = \mathcal{Z}\, v(n) .$$

2. For an observable system model

$$V(z) = \underline{c}^T \Phi(z) \underline{x}(n_o)$$

and r, the dimension of $\underline{x}$, equals the number of poles of V(z).

3. Choose a matrix A_{rxr} such that its characteristic values are the same as the poles of V(z).
4. Find $\Phi(z) = z[zI-A]^{-1}$.
5. Choose $\underline{c}^T$ and $\underline{x}(n_o)$ such that

$$V(z) = \underline{c}^T \Phi(z) \underline{x}(n_o) .$$

An alternative to this method is to find a generator for a u(t) whose samples, $u(nT_s)$, equal the required v(n) for all $n \geq n_o$. The continuous-time model can then be converted to the required discrete-time model using Conversion VII of Chapter 3.

Example 4.20

a) For $v(n) = Kb^n v_1(n)$, find suitable A, $\underline{c}^T$, and $\underline{x}(0)$.

Now

$$V(z) = \frac{Kz}{z-b} , \quad \text{and} \quad r=1.$$

We let

$$A_{1x1} = b,$$

then

$$\Phi(z) = z[z-b]^{-1} = \frac{z}{z-b} ,$$

and require

$$c\Phi(z)x(0) = \frac{Kz}{z-b} .$$

Thus we let c=1 and x(0)=K, and the model is

$$x(n+1) = bx(n)\,, \qquad x(0) = K, \quad n \geq 0.$$

$$v(n) = x(n)\,.$$

b) For $v(n) = a_0 v_1(n)+a_1 v_2(n)$ as in Eq. 2.36, find suitable A, $\underline{c}^T$, and $\underline{x}(0)$. Now

$$V(z) = \frac{a_0 z}{z-1} + \frac{a_1 z}{(z-1)^2} = \frac{a_0 z(z-1)+a_1 z}{(z-1)^2}\,,$$

and r=2. Let

$$A = \begin{bmatrix} 1 & 1 \\ 0 & 1 \end{bmatrix},$$

a Jordan form with $\lambda_1=\lambda_2=1$. Then

$$\Phi(z) = \frac{z}{(z-1)^2} \begin{bmatrix} z-1 & 1 \\ 0 & z-1 \end{bmatrix}.$$

Equating the numerator of the right-hand form of V(z) to

$$[c_1 \quad c_2]^T \, \Phi(z) \begin{bmatrix} x_{10} \\ x_{20} \end{bmatrix}$$

and then equating coefficients of like powers of z, we require

$$a_o = c_1x_{10}+c_2x_{20}$$

and

$$a_1 = c_1x_{20} \quad .$$

Thus a solution is

$$\underline{c}^T = [1 \quad 0], \qquad \underline{x}^T(0) = [a_o \quad a_1].$$

If Conversion VII were used on Example 4.19d with $T_s=1$, and

$$A_c = \begin{bmatrix} 0 & 1 \\ 0 & 0 \end{bmatrix} ,$$

and

$$\Phi(s) = [sI-A_c]^{-1} = \frac{1}{s^2} \begin{bmatrix} s & 1 \\ 0 & s \end{bmatrix}$$

we have

$$\Phi(t) = \begin{bmatrix} 1 & t \\ 0 & 1 \end{bmatrix} u_1(t)$$

so that

$$\Phi(1) = \begin{bmatrix} 1 & 1 \\ 0 & 1 \end{bmatrix} = A_d$$

which is the same as A above. $\underline{c}^T$ and $\underline{x}(0)$ also are the same.

4.5.2 Stochastic Signals

When the required signal is a zero mean wide-sense stationary process $\{U(t)\}$, it is assumed that the autocorrelation or spectral density function is given. Then, using Equation 4.47 for a white input $\{W(t)\}$ with spectral density M and output $\{U(t)\}$ with $R_U(s)$, we have

$$R_U(s) = H(-s)M\,H(s),$$

where H(s) is the transfer function of the required linear system. See Figure 4.2. Since $R_U(s)$ is even, i.e., $R_U(s) = R_U(-s)$ (see Eq. 2.84), it may be factored into two terms such that one has right-half plane poles and zeros and the other has left-half plane poles and zeros. Therefore,

$$R_U(s) = R_U^L(s)R_U^L(-s) \tag{4.89}$$

where $R_U^L(s)$ has left-half plane poles and zeros. A convenient choice of H(s) is

$$H(s) = \frac{1}{\sqrt{M}}\ R_U^L(s). \tag{4.90}$$

This means that the required system has all finite poles and zeros in the left-half s-plane, and is therefore stable in the b.i.b.o. sense. The required state variable model is

$$\dot{\underline{X}}(t) = A\underline{X}(t) + \underline{b}W(t)$$

$$U(t) = \underline{c}^T\underline{X}(t)$$

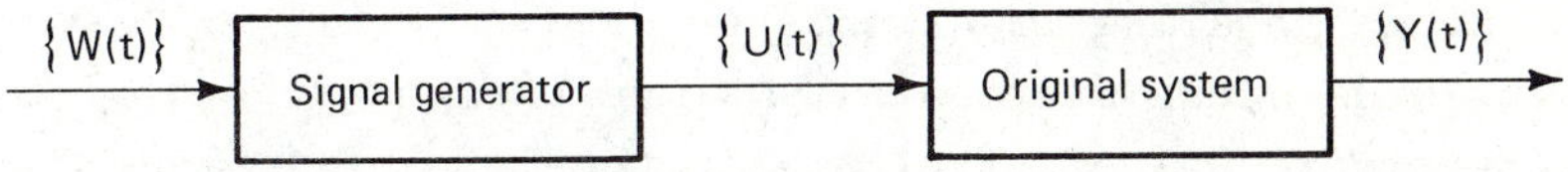

Figure 4.2 Signal generator model

where a set of A, $\underline{b}$ and $\underline{c}^T$ can be found from H(s) by using Conversion I. of Chapter 3. The initial state $\underline{X}(t_o)$ also must be considered. If {U(t)} is to be wide-sense stationary for all $t \geq t_o$, $\underline{X}(t_o)$, a random initial state, must be specified in terms of its covariance matrix P_o as in Section 4.3.2. The covariance matrix of $\underline{X}$ must therefore be constant, i.e., $P(t) = P_o$, $t \geq t_o$, and from Equation 4.64., P_o must satisfy

$$\Theta = AP_o + P_oA^T + \underline{b}M\underline{b}^T. \tag{4.91}$$

Equations of this type have been studied rather extensively, and several methods of solution are given in [20.]. (See also Chapter 5, Theorem 5.14a) In an actual simulation, then, the initial state for each run must be drawn from a probabilistic model determined by P_o. In terms of the state variable model, the original $r_U(\tau)$ can be found by starting with Eq. 4.66 for $\tau = t_2 - t_1$. Then

$$P_{\underline{X}}(\tau) = P_o\Phi^T(\tau), \qquad \tau \geq 0$$

$$= \Phi(-\tau)P_o, \qquad \tau \leq 0$$

and

$$P_U(\tau) = r_U(\tau) = \underline{c}^T P_{\underline{X}}(\tau)\underline{c} \; .$$

The simple system augmentation of Eq. 4.74 is an example. Frequently {W(t)} and $\underline{X}(t_o)$ are Gaussian and {U(t)} is then a Gauss-Markov process. As a practical matter, it is easier to specify $\underline{X}(t_o)=\underline{0}$ and let the simulation run for some time long enough to allow transients to decay. {U(t)} will tend towards wide-sense stationarity since the poles of H(s) (or characteristic values of A) are in the left-half plane. An orderly procedure for this case is:

1. Find $R_U(s) = \mathcal{L}\, r_U(\tau)$
2. Factor $R_U(s) = R_U^L(s) R_U^L(-s)$
3. For

$$H(s) = \frac{1}{\sqrt{M}} R_U^L(s) \ ,$$

find

$$A,\ \underline{b},\ \underline{c}^T.$$

4. Find P_o from Eq. 4.91.

<u>Example 4.21</u>

A zero-mean wide-sense stationary process $\{U(t)\}$ is required such that

$$r_U(\tau) = K_1(\exp -a|\tau|)+K_2(\exp -b|\tau|), \qquad a>0,\ b>0, \quad a \neq b.$$

Find suitable A, $\underline{b}$, $\underline{c}^T$, and P_o. Now

$$R_U(s) = \mathcal{L}\, r_U(\tau)$$

$$= \frac{2K_1 a}{a^2-s^2} + \frac{2K_2 b}{b^2-s^2}$$

$$= \frac{2K_1 a}{(-s+a)(s+a)} + \frac{2K_2 b}{(-s+b)(s+b)}$$

and after some manipulation,

$$R_U(s) = \beta \frac{(-s+\alpha)(s+\alpha)}{(-s+a)(s+a)(-s+b)(s+b)}$$

where

$$\beta = 2(K_1 a+K_2 b)$$

$$\alpha = \frac{2}{\beta}(K_1 ab^2+K_2 ba^2).$$

Therefore,

$$R_U^L(s) = \frac{\sqrt{\beta}(s+\alpha)}{(s+a)(s+b)}$$

and

$$H(s) = \frac{\sqrt{\beta/M}(s+\alpha)}{(s+a)(s+b)} = \frac{N(s)}{M(s)} \quad .$$

Using Eq. 3.68,

$$A = \begin{bmatrix} 0 & 1 \\ -ab & -(a+b) \end{bmatrix}$$

$$\underline{b} = \begin{bmatrix} 0 \\ 1 \end{bmatrix}, \qquad \underline{c}^T = \sqrt{\beta/M} \quad [\alpha \quad 1].$$

Substituting in Eq. 4.91,

$$\Theta = \begin{bmatrix} 0 & 1 \\ -ab & -(a+b) \end{bmatrix} \begin{bmatrix} p_{011} & p_{012} \\ p_{021} & p_{022} \end{bmatrix} +$$

$$+ \begin{bmatrix} p_{011} & p_{012} \\ p_{021} & p_{022} \end{bmatrix} \begin{bmatrix} 0 & -ab \\ 1 & -(a+b) \end{bmatrix}$$

$$+ \begin{bmatrix} 0 & 0 \\ 0 & M \end{bmatrix}.$$

Using $p_{012}=p_{021}$ and writing out the scalar equations leads to

$$p_{011} = \frac{M}{2ab(a+b)} = \text{Var } X_{10}$$

$$p_{022} = \frac{M}{2(a+b)} = \text{Var } X_{20}$$

$$p_{012} = p_{021} = 0,$$

and

$$P_o = \frac{M}{2(a+b)} \begin{bmatrix} \frac{1}{ab} & 0 \\ 0 & 1 \end{bmatrix}$$

M is the intensity of {W(t)}.

Finally, when the required signal is a zero-mean, wide-sense stationary sequence {V(n)}, it is assumed that the autocorrelation function or spectral density function is given. Then, using Equation 4.48 for a white input {W(n)} with spectral density M and output {V(n)} with $R_V(z)$, we have

$$R_V(z) = H(z^{-1})MH(z)$$

where H(z) is the transfer function of the required linear system. Since $R_V(z)=R_V(z^{-1})$ (see Eq. 2.84), it may be factored into two terms such that one has poles and zeros inside the unit circle and the other has poles and zeros outside the unit circle. Therefore we obtain

$$R_V(z) = R_V^I(z)R_V^I(z^{-1}) \tag{4.92}$$

and a convenient choice of H(z) is

$$H(z) = \frac{1}{\sqrt{M}} R_V^I(z) \ . \tag{4.93}$$

The required state variable model is

$$\underline{X}(n+1) = A\underline{X}(n) + \underline{b}W(n)$$

$$V(n) = \underline{c}^T\underline{X}(n) + dW(n)$$

where a set of A, $\underline{b}$, $\underline{c}^T$, and d can be found from H(z) by Conversion I. of Chapter 3. To consider the initial state $\underline{X}(n_o)$, a similar procedure to that for $\underline{X}(t_o)$ is used. Using Equation 4.69 with P(·) a constant P_o, $n \geq n_o$, we have

$$P_o = AP_oA^T + \underline{b}M\underline{b}^T \tag{4.94}$$

which must be solved for P_o. The original $r_V(k)$ is related to these results by starting with Eq. 4.71 with $k=n_2-n_1$. Then

$$P_{\underline{X}}(k) = P_o\Phi^T(k) \qquad k \geq 0$$

$$= \Phi(-k)P_o \qquad k \leq 0$$

and $P_V(k)=r_V(k)=\underline{c}^TP_{\underline{X}}(k)\underline{c}$. The simple system augmentation of Eq. 4.77 is an example. A procedure is:

1. Find $R_V(z) = \mathcal{Z}\, r_V(k)$
2. Factor $R_V(z) = R_V^I(z)R_V^I(z^{-1})$
3. For

$$H(z) = \frac{1}{\sqrt{M}} R_V^I(z) ,$$

   find A, $\underline{b}$, $\underline{c}^T$, d.
4. Find P_o from Eq. 4.94.

Example 4.22

We wish to generate a stationary Gauss-Markov sequence $\{V(n)\}$ with

$$r_V(k) = \frac{K}{(1-b^2)} (b)^{|k|} , \qquad 0 < |b| < 1$$

(see Example 4.11). We have

$$R_V(z) = -\frac{K}{b} \frac{z}{(z-b)(z - \frac{1}{b})} .$$

To factor, we assume

$$R_V^I(z) = K_1 \frac{z}{z-b}$$

then

$$R_V^I(z^{-1}) = \frac{K_1}{b} \frac{1}{(z - \frac{1}{b})}$$

and

$$R_V^I(z) R_V^I(z^{-1}) = -\frac{K_1^{\,2}}{b} \frac{z}{(z-b)(z - \frac{1}{b})} .$$

Therefore, comparing to $R_V(z)$, let $K_1 = \sqrt{K}$. For a white, zero mean Gaussian input $\{W(n)\}$ with intensity (mean square) of M, we have

$$H(z) = \sqrt{\frac{K}{M}} \left(\frac{z}{z-b}\right) = \frac{V(z)}{W(z)} = \frac{N(z)}{M(z)} .$$

In the notation of Eq. 3.68 (for r=1),

$$\beta_1 = \sqrt{\frac{K}{M}} , \qquad \beta_o = 0, \qquad \alpha_o = -b ,$$

and a state variable model is:

$$X(n+1) = bX(n) + W(n)$$

$$V(n) = b\sqrt{\frac{K}{M}}\ X(n) + \sqrt{\frac{K}{M}}\ W(n).$$

This may be programmed for digital computation quite easily. W(n) is obtained from a Gaussian random number generator with mean=0 and standard deviation=$\sqrt{M}$. As an example, if b=0.1 and $K/(1-b^2)=1$, we have $K=1-b^2=0.99$. A convenient choice then is to let M=K=0.99. The figure shows a result for this case with 200 samples of {V(n)} after initial transients have decayed. The plot is $v^i(n)$ where the digital plotting program reconstructs by connecting adjacent samples.

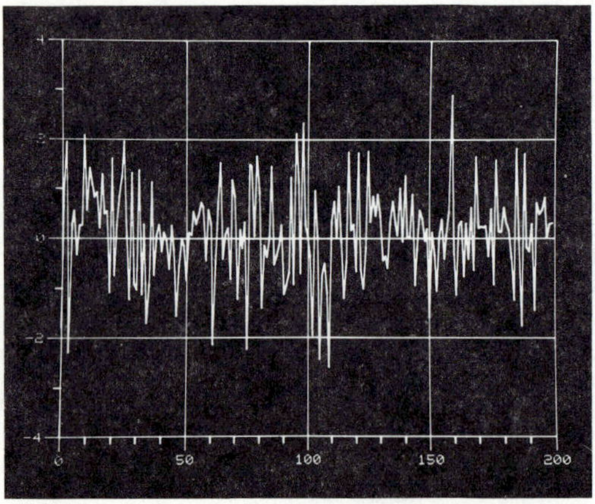

b = 0.1

A second plot for b=0.9 is shown below, and comparison with that above shows the wider frequency band character for b=0.1 and the stronger correlation between samples (i.e., low frequency character)for b=0.9.

The results of this section provide useful ways of programming analog or digital computers to generate both deterministic and stochastic signals. In the stochastic cases a white-noise (or very wide band) generator is needed

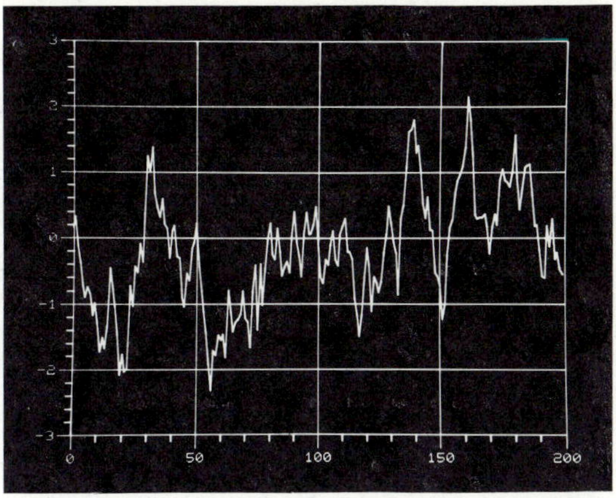

b = 0.9

for the analog input and a random number generator (usually uniform or Gaussian) is needed for the digital program. Some results for multidimensional, time-varying simulation are given in [21.].

REFERENCES

1. Kreyszig, E., "Advanced Engineering Mathematics," 2nd Ed., New York: John Wiley & Sons, 1967.
2. DeRusso, P., Roy, R., and Close, C., "State Variables for Engineers," New York: John Wiley & Sons, 1965.
3. Hildebrand, F., "Methods of Applied Mathematics," New York: Prentice-Hall, 1952.
4. Newton, G., Gould, L., and Kaiser, J., "Analytical Design of Linear Feedback Controls," New York: John Wiley & Sons, 1957.
5. Greaves, C., Gagne, G., and Bordner, G., "Evaluation of Integrals Appearing in Minimization Problems of Discrete-Data Systems," IEEE Trans. on Automatic Control, v. AC-11, n. 1, January, 1966, pp. 145-148.
6. Wong, E., "Stochastic Processes in Information and Dynamical Systems," New York: McGraw-Hill Book Co., 1971.

7. Melsa, J. and Schultz, D., "Linear Control Systems," New York: McGraw-Hill Book Co., 1969.
8. Anderson, G., Buland, N., and Cooper, G., "Use of Cross-correlation in an Adaptive Control System," Proc. Nat. Electronics Conf., v. XV, 1959, pp. 34-45.
9. Kerr, R. and Surber, W., "Precision of Impulse-Response Identification Based on Short Normal Operating Records," IRE Trans. on Automatic Control, v. AC-6, n. 2, May, 1961, pp. 173-182.
10. Kerr, R., "The Identification of Linear System Parameters," IEEE Conference Paper CP 62-84, January, 1962.
11. Kalman, R., "Design of a Self-Optimizing Control System," Trans. ASME, February, 1958, pp. 468-478.
12. Joseph, P., Lewis, J., and Tou, J., "Plant Identification in the Presence of Disturbances and Application to Digital Adaptive Systems," Trans. AIEE, Applications and Industry, March, 1961, pp. 18-24.
13. Lee, R., "Optimal Estimation, Identification, and Control," Cambridge, Massachusetts: M.I.T. Press, 1964.
14. Sage, A. and Melsa, J., "System Identification," New York: Academic Press, 1971.
15. Goodman, T. and Hillsley, R., "Continuous Measurement of Characteristics of Systems with Random Inputs: A Step Toward Self-Optimizing Control," Trans. ASME, November, 1958, pp. 1839-1848.
16. Liu, B. and Meadows, H., "A Method of Time Domain Approximation for Arbitrary Inputs," IEEE Convention Record, Part I, 1964, pp. 338-343.
17. Thorp, J. and Mintz, M., "A Method of System Identification with Random Inputs," Proc. IEEE, February, 1966, pp. 290-291.
18. Mendel, J., "Discrete Techniques of Parameter Estimation," New York: Marcel Dekker, Inc., 1973.
19. Eykhoff, P., "System Identification," New York: John Wiley & Sons, 1974.

20. Hagander, P., "Numerical Solution of $A^TS+SA+Q=0$," Information Science, v. 4, 1972, pp. 35-50.
21. Mehra, R., "Digital Simulation of Multidimensional Gauss-Markov Random Processes," IEEE Trans. Auto. Control, v. AC-14, n. 1, February, 1969, pp. 112-113.
22. Dack, D., "System Identification by On-Line Correlation," Control Engineering, March 1970, pp. 64-70.
23. - Special Issue on System Identification and Time-Series Analysis, IEEE Trans. Auto. Control, v. AC-19, n. 6, December 1974.

PROBLEMS

1. Using the results of Problem 13, Chapter 3, find $e_t(t)$, $t \geq 0$ when $e(t) = u_1(t)$ by using a classical method. Assume simple zeros of the characteristic equation.
2. Find a general solution for Problem 1 when $e(t) = \varepsilon^{-t}$, $t \geq 0$, in terms of the initial conditions $i(0)$ and $\omega(0)$. Assume simple zeros of the characteristic equation, all of which are not -1.
3. For the discrete-time filter of Problem 20, Chapter 3, find $z(n)$, $n \geq 0$ when $v(n) = v_1(n)$ by using a classical method.
4. Find a general solution for Problem 3 when $v(n) = (a)^n$, $n \geq 0$, in terms of the initial conditions $z(1)$ and $z(2)$. Assume $|a|<1$.
5. Repeat Problem 1 using a transform method.
6. Repeat Problem 3 using a transform method.
7. In Problem 12, Chapter 3 assume equal payments p, i.e., $v(n) = pv_1(n-1)$.
   a) Find $z(n)$, $n \geq 1$ when the loan is $y(0) = d$.
   b) If the loan is to be paid in N equal payments, i.e., $y(N) = 0$, find p.
8. From the result of Problem 14, Chapter 3, find $f_b(t)$, $t \geq 0$ when $f(t) = u_1(t)$ by using Eq. 4.7.

9. From the result of Problem 16, Chapter 3, find y(n), n>0, when $u(t) = u_1(t)$ by using Eq. 4.10.

10. From the result of Problem 43, Chapter 3, find $\underline{y}(t)$, $t \geq 0$ when both components of $\underline{u}$ are unit step functions.

11. For the system of Problem 28, Chapter 3, find the transfer function and then a response $v_o(t)$, $t \geq 0$, when $i(t) = 10^{-4} u_1(t)$. Assume a set of numerical values.

12. For the system of Problem 20, Chapter 3, use the transfer function to find z(n), $n \geq 0$, when $v(n) = v_1(n)$.

13. A linear system is described by:

$$\dot{\underline{x}} = \begin{bmatrix} 0 & 1 \\ -6 & -5 \end{bmatrix} \underline{x} + \begin{bmatrix} 0 \\ 3 \end{bmatrix} u$$

$$y = [1 \quad 0]\underline{x}$$

a) Find y(t), $t \geq 0$, if $u(t) = u_1(t)$ and $\underline{x}^T(0) = [0 \quad 0]$.

b) Find the mean square output if the input (applied at $t_o \to -\infty$) is a wide-sense stationary process $\{U(t)\}$ with $r_U(\tau) = 2u_o(\tau)$.

14. For the circuit,

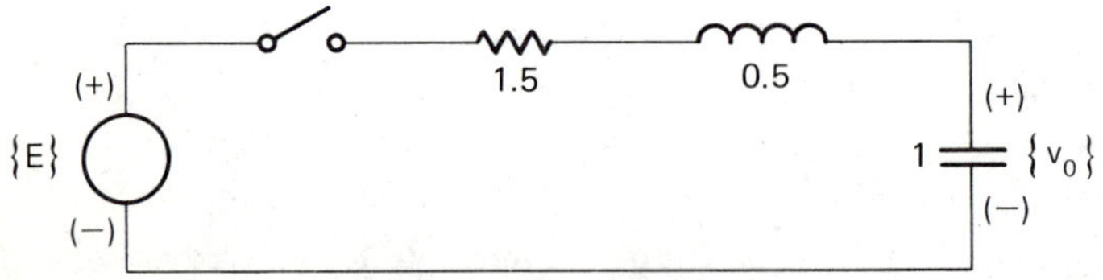

the input is a wide-sense stationary process with $r_E(\tau) = 5u_o(\tau)$. What is the mean square output? Switch closes at t=0 and the initial charge on the capacitor is zero.

15. For the filter,

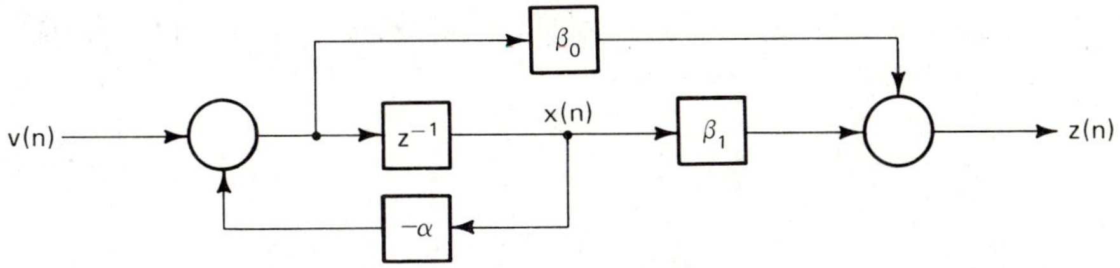

a) Find a difference equation relating v(n) and z(n).

b) Find H(z), the transfer function.

c) If x(0) = 0 and $v(n) = 0.7^n v_1(n)$, find z(n), $n \geq 0$. Let $\alpha = \beta_0 = \beta_1 = 0.8$.

16. An R-C filter is shown.

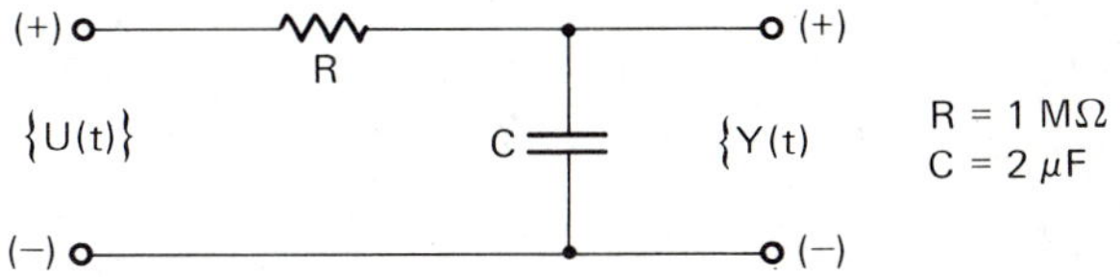

a) If {U(t)} is a wide-sense stationary process with $r_U(\tau) = (\exp -|\tau|)$, what is the ratio of output mean square to input mean square?

b) Sketch the magnitudes of the input and output spectral density functions versus ω.

c) What is the effect of the filter in terms of mean square and bandwidth?

17. The numerical integrator of Example 3.23 with T=1 is used with an input $v(n) = v_1(n)$. Show that by proper choice of $\underline{x}(0)$, $z(n) = v_2(n)$.

18. The filter shown is used to suppress stationary white noise with intensity W_o which is added to a signal process whose sample functions are of the form

$$s^i(t) = a\cos(\omega_o t + \theta^i)$$

where a and ω_o are constants and Θ is a random variable uniformly distributed on $[0,2\pi]$. The signal and noise are uncorrelated.

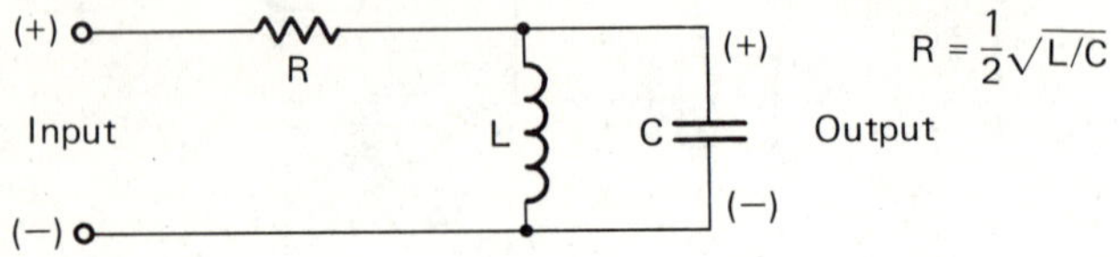

a) Find the output signal-to-noise ratio.

b) Consider the special cases

i) $\omega_o = 1/\sqrt{LC}$

ii) $\omega_o >> 1/\sqrt{LC}$.

19. A simple feedback system is shown where the input is a signal plus noise.

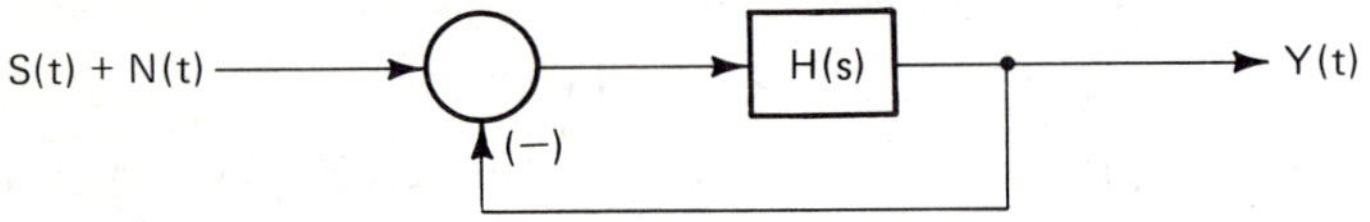

$$H(s) = \frac{K}{s(1 + T_o s)}$$

For the signal,

$$r_S(\tau) = a(\exp -|\tau|)$$

and for the noise,

$$r_N(\tau) = b\, u_o(\tau)$$

and the signal and noise are uncorrelated processes.

a) For error $E(t) = S(t) - Y(t)$, find the mean and mean square error for steady-state (i.e., $t_o \to -\infty$) conditions.

b) (Optional-see Chapter 5.) If the input is applied at t_o=0, find an expression for the mean square error $S_E^2(t)$, $t \geq 0$. Assume initial conditions are zero.

(If you have access to a digital computer, this is a problem where for numerical values of K, T_o, a, and b, a discretized version can be programmed.)

20. Derive Equations 4.57 and 4.58. Be careful to start with a definition used in this book, e.g., Eq. 4.44 (some books use $-\tau$ and this changes the form of the result).

21. Two cascaded linear systems are characterized by the impulse response matrices $H_1(t,\lambda)$ and $H_2(t,\lambda)$.

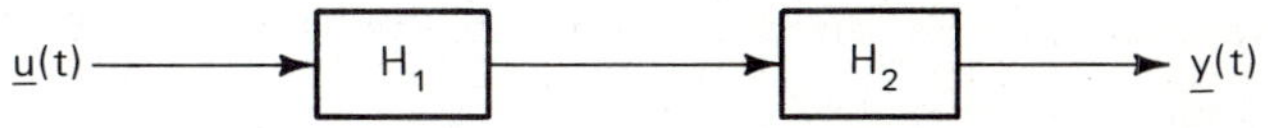

What is the impulse response matrix $H(t,\lambda)$ for the overall system in terms of H_1 and H_2?

22. A digital filter is described by

$$(E^2 - 0.75E + 0.125)z(n) = (E-a)v(n).$$

Obtain a simulation diagram of a controllable model when a = 0.6 and when a = 0.5.

23. In Problem 22 with a = 0.6, if the input $\{V(n)\}$ is a sequence of independent random values of +1 and -1 (with equal probability), find the mean square $\{Z(n)\}$, i.e., S_Z^2.

24. The system of Problem 13 is driven by an input process $\{U(t)\}$ with

$$r_U(\tau) = 2(\exp -|\tau|) .$$

Find an augmented system with a white input such that the output is the same as that of the original system.

25. The filter of Problem 22 (with a = 0.6) is driven by an input sequence {V(n)} with

$$r_V(k) = 2(0.7)^{|k|} ,$$

find an augmented system with a white input such that the output is the same as that for the original system.

26. In Problem 25, obtain numerical forms that can be solved for the mean state, covariance of the state, mean output, and covariance of the output when the input is applied at $n_o=0$. Assume that the mean and covariance of the initial state are known.

27. Find the response v_o of the circuit of Problem 18, Chapter 3 to the input e = u of Problem 12, Chapter 2. Let $R_1=R_2=1$, L=2.62, C=0.38.

28. Find the response z(n) of the filter of Problem 20, Chapter 3, to the input v(n) of Problem 18, Chapter 2.

The following two problems can be done if a digital computer with a FORTRAN compiler is available.

29. a) Prepare a FORTRAN program that generates a zero-mean Gauss-Markov sequence {Z(n)} with

$$r_Z(k) = \frac{K}{1-b^2}(b)^{|k|}$$

where K and b are parameters of the program. Use the GAUSS random number generator.

b) Obtain two plots of $z^i(n)$ for 100 samples for

$$\frac{K}{1-b^2} = 1$$

with b = 0.1 and b = 0.9. Compare the two plots. Which is the wider bandwidth signal?

30. Equation 4.83 is to be used in a simple identification problem.

a) Using a model

$$H(z) = \frac{3z}{z - 0.5} = \frac{Z(z)}{V(z)}$$

obtain a record of v(n) and z(n) for 1000 samples. Let v(n) be a white Gaussian input with intensity 1.

b) Using the records of v(n) and z(n), obtain estimates $\hat{\beta}_1$ and $\hat{\alpha}_0$ in the model

$$H(z) = \frac{\beta_1 z}{z+\alpha_0}$$

for p=0, -1.

c) Obtain a record of noisy output data

$$z^1(n) = z(n) + w(n)$$

where z(n) is the record in Part a) and w(n) is a white Gaussian signal with intensity 1. Repeat Part b) using v(n) and $z^1(n)$.

d) To illustrate the effect of choosing a model different from the true (ordinarily unknown!) model, let

$$H(z) = \frac{\beta_1 z + \beta_0}{z + \alpha_0}$$

and use v(n) and $z^1(n)$ of Part c) to find estimates $\hat{\beta}_1$, $\hat{\beta}_0$, and $\hat{\alpha}_0$ using p = 0, -1, -2.

Chapter 5

SYSTEM PERFORMANCE

System design is concerned with bringing together a number of components or subsystems to perform a specified function. Analysis of a given system is concerned with how well it performs the specified function. System analysis, therefore, is essential in the design process. Analysis may be used as part of a "trial and error" procedure where each tentative design is alternately analyzed and adjusted until an acceptable design is obtained. System evaluation also uses the techniques of system analysis. A question which arises immediately concerns a measure of how well the system performs or how well the specifications are met. In an ideal situation we would prefer to have models of the signals to which the system is subjected and a model of the system itself. Many signal and system models were discussed in Chapters 2 and 3, and in both cases, a few methods of using experimental data to find numerical forms for the models were introduced. These are the methods of signal and system identification. Once the signal and system models are known, the results of Chapter 4 can be used to find the response of the system to a specified input. This chapter is concerned with examining those responses relative to certain performance measures.

The first of these is system stability. Although this is a property of the system and logically could have been discussed in Chapter 3, it was deferred until the material in Chapter 4 had been completed. Obviously, a designer is concerned first with <u>stability</u> and second with performance relative to other criteria. Consequently this chapter is organized so that stability requirements are considered first and then additional performance measures are introduced. These are: <u>fidelity</u> or <u>accuracy</u>, a measure of how small an error or deviation from a desired performance can be achieved; <u>sensitivity</u>, a measure of how changes in signal or system parameters affect performance; and <u>reliability</u>, an estimate of the probability that the system will not fail.

5.1 STABILITY

The stability of differential and difference equations has been a subject of great mathematical interest for over a century, and there is a very large amount of available literature. Most of the results are given in the rather formal manner of definitions and theorems. Results are required here for <u>linear</u> continuous-time and discrete-time systems, and most of the theorems are given in terms of state-variable models. This concise summary of linear system stability theory is intended to provide the main ideas in the context of the goal of this book. Details concerning special cases and proofs of the theorems can be found in the references that are cited. Continuous-time systems are considered first.

5.1.1 Zero-Input Stability

For the model

$$\dot{\underline{x}}(t) = A(t)\underline{x}(t) + B(t)\underline{u}(t), \qquad t \in T_t,$$

the transition function for $\underline{u}(t) = \underline{0}$ is given by Eq. 3.58 as

$$\underline{x}(t) = \underline{f}_c(t; \underline{x}(t_o), \underline{0}), \qquad t \geq t_o, \qquad t_o \in T_t,$$

and an equilibrium state $\underline{\lambda}$ is any state that satisfies

$$\underline{\lambda} = \underline{f}_c(t; \underline{\lambda}, \underline{0}).$$

From the definition of the transition matrix,

$$\underline{x}(t) = \Phi(t,t_o)\underline{x}(t_o) = \underline{f}_c(t;\underline{x}(t_o),\underline{0})$$

and $\underline{\lambda}$ must, therefore, satisfy

$$\underline{\lambda} = \Phi(t,t_o)\underline{\lambda} . \tag{5.1}$$

Also, a constant $\underline{\lambda}$ substituted in $\dot{\underline{x}}(t) = A(t)\underline{x}(t)$ gives

$$\dot{\underline{\lambda}} = \underline{0} = A(t)\underline{\lambda} . \tag{5.2}$$

From this, an equilibrium state $\underline{\lambda}$ is unique if A(t) is nonsingular for all $t \geq t_o$. It is clear that $\underline{\lambda} = \underline{0}$ is always an equilibrium state, and for A(t) nonsingular, it is the unique equilibrium state.

Example 5.1

a) Consider a system whose transfer function is $H(s) = \frac{1}{s^2}$. A state-variable model gives

$$\underline{\dot{x}} = \begin{bmatrix} 0 & 1 \\ 0 & 0 \end{bmatrix} \underline{x} = A_1 \underline{x} .$$

For equilibrium,

$$\underline{0} = \begin{bmatrix} 0 & 1 \\ 0 & 0 \end{bmatrix} \underline{\lambda} = \begin{bmatrix} 0 & 1 \\ 0 & 0 \end{bmatrix} \begin{bmatrix} \lambda_1 \\ \lambda_2 \end{bmatrix}$$

and $\lambda_2=0$, but λ_1 can be any real constant.* A_1 is singular, and $\underline{\lambda}$ is not unique.

b) Consider $H(s) = \frac{1}{s(s+1)}$. A state-variable model gives

$$\underline{\dot{x}} = \begin{bmatrix} 0 & 1 \\ 0 & -1 \end{bmatrix} \underline{x} = A_2 \underline{x}.$$

For equilibrium,

$$\underline{0} = \begin{bmatrix} 0 & 1 \\ 0 & -1 \end{bmatrix} \underline{\lambda} = \begin{bmatrix} 0 & 1 \\ 0 & -1 \end{bmatrix} \begin{bmatrix} \lambda_1 \\ \lambda_2 \end{bmatrix}$$

and $\lambda_2 = 0$, but λ_1 = any real constant. A_2 is singular.

c) Consider $H(s) = \frac{1}{(s+1)(s+2)} = \frac{1}{s^2+3s+2}$. A state variable model gives

$$\underline{\dot{x}} = \begin{bmatrix} 0 & 1 \\ -2 & -3 \end{bmatrix} \underline{x} = A_3 \underline{x}.$$

---

*Note that in this context, λ_1 and λ_2 are components of $\underline{\lambda}$ and not characteristic values of A.

For equilibrium,

$$\underline{0} = \begin{bmatrix} 0 & 1 \\ -2 & -3 \end{bmatrix} \underline{\lambda} = \begin{bmatrix} 0 & 1 \\ -2 & -3 \end{bmatrix} \begin{bmatrix} \lambda_1 \\ \lambda_2 \end{bmatrix}$$

and $\lambda_1 = \lambda_2 = 0$, i.e., $\underline{\lambda} = \underline{0}$, is the unique equilibrium state. A_3 is nonsingular.

Now for the model

$$\dot{\underline{x}}(t) = A(t)\underline{x}(t) + B(t)\underline{u}(t),$$

let $\underline{x}'(t)$ be a known solution, $t \geq t_o$, for a given $\underline{u}(t)$ and $\underline{x}'(t_o)$. Then $\underline{x}'$ satisfies

$$\dot{\underline{x}}'(t) = A(t)\underline{x}'(t) + B(t)\underline{u}(t), \qquad t \geq t_o$$

Also, for a solution $\underline{x}''(t)=\underline{q}(t)+\underline{x}'(t)$ with initial condition $\underline{x}''(t_o)$ and the same $\underline{u}(t)$,

$$\dot{\underline{q}}(t) + \dot{\underline{x}}'(t) = A(t)(\underline{q}(t)+\underline{x}'(t)) + B(t)\underline{u}(t), \qquad t \geq t_o$$

i.e., $\underline{q}(t)$ is the deviation of $\underline{x}''(t)$ from the known solution $\underline{x}'(t)$. Since

$$\dot{\underline{x}}'(t) = A(t)\underline{x}'(t) + B(t)\underline{u}(t) ,$$

the deviation satisfies

$$\dot{\underline{q}}(t) = A(t)\underline{q}(t), \qquad t \geq t_o$$

with $\underline{q}(t_o)=\underline{x}''(t_o)-\underline{x}'(t_o)$. This is the <u>free system</u> corresponding to the original system, and $\underline{q} = \underline{\lambda} = \underline{0}$ is an equilibrium state. It follows then that an examination of the deviation $\underline{q}(t)$ from a known solution $\underline{x}'(t)$ is equivalent to the examination of the original model with $\underline{u} = \underline{0}$, i.e.,

$$\dot{\underline{x}}(t) = A(t)\underline{x}(t), \quad t \geq t_o$$

with $\underline{x}(t_o) = \underline{x}_o$. The stability of the origin $\underline{q} = \underline{0}$ of the free system is therefore of major interest. If an equilibrium state of $\dot{\underline{x}} = A\underline{x}$ is $\underline{\lambda} \neq \underline{0}$, we let $q(t) = \underline{x}(t)-\underline{\lambda}$, then

$$\begin{aligned}\dot{\underline{q}}(t) = \dot{\underline{x}}(t) &= A(t)\underline{x}(t) \\ &= A(t)(\underline{q}(t)+\underline{\lambda}) \\ &= A(t)\underline{q}(t) .\end{aligned}$$

In the following, in the usual notation, it is the stability of $\underline{x} = \underline{0}$ of $\dot{\underline{x}}(t) = A(t)\underline{x}(t)$ that is discussed first, and then the general case for $\underline{u} \neq \underline{0}$ is considered.

Definition 5.1 Zero-Input Stability

For $\dot{\underline{x}}(t) = A(t)\underline{x}(t)$, $\underline{x} = \underline{0}$ is

a) stable (in the sense of Liapunov) if for any $t \epsilon T_t$ and any $\varepsilon > 0$, there exists a $\delta > 0$ (which may depend on ε and t_o) such that $||\underline{x}(t_o)|| < \delta$ implies $||\underline{f}_c(t;\underline{x}(t_o),\underline{0}|| < \varepsilon$ for all $t \geq t_o$.

b) uniformly stable if stable and δ does not depend on t_o.

c) asymptotically stable if stable and $\underline{x}(t) \to \underline{0}$ as $t \to \infty$ for $\underline{x}(t_o)$ sufficiently close to $\underline{0}$.

d) globally asymptotically stable if stable and $\underline{x}(t) \to \underline{0}$ as $t \to \infty$ for all $\underline{x}(t_o)$ such that $||x(t_o)|| \leq \rho < \infty$.

Basically, the origin $\underline{x}=\underline{0}$ is stable if for any given ε, one can find a δ such that when the "distance" of $\underline{x}(t_o)$ from the origin is less than δ, the "distance" of $\underline{x}(t)$ from the origin is less than ε for all $t \geq t_o$. (Definitions of $||\underline{x}||$, norm of $\underline{x}$, are given in App. B.) Limit cycles and the response of various types of oscillators are examples of this behavior. The stronger requirement, asymptotic stability, implies stability plus the return of $\underline{x}(t)$ to the origin as t becomes large. The following theorem gives some general results which can be proved using the properties of vector and matrix norms (see Appendix B) [2.].

Theorem 5.1

a) The equilibrium state $\underline{x}=\underline{0}$ of $\dot{\underline{x}}(t)=A(t)\underline{x}(t)$ is stable
$\Leftrightarrow$
there exists a constant m such that

$$||\Phi(t,t_o)|| \leq m < \infty; \qquad \forall\ t \geq t_o .$$

b) The equilibrium state $\underline{x}=\underline{0}$ of $\dot{\underline{x}}(t)=A(t)\underline{x}(t)$ is asymptotically stable
$\Leftrightarrow$
it is stable and

$$\lim_{t\to\infty} ||\Phi(t,t_o)||=0, \qquad \forall\ t_o .$$

c) The equilibrium state $\underline{x}=\underline{0}$ of $\dot{\underline{x}}(t)=A(t)\underline{x}(t)$ is globally asymptotically stable
$\Leftrightarrow$
$\underline{x}=\underline{0}$ is asymptotically stable, and $\underline{x}=\underline{0}$ is the unique equilibrium state.

Generally, these results are difficult to apply since as shown in Chapter 3, finding the transition matrix $\Phi(t,t_o)$ for time-varying systems is difficult. The ideas are illustrated for the simple cases of Example 5.1.

Example 5.2

a) From Example 5.1a, for $t_o=0$,

$$\Phi(t) = \begin{bmatrix} 1 & t \\ 0 & 1 \end{bmatrix} u_1(t) .$$

Using Theorem 5.1, part a), it is clear that $\underline{x}=\underline{0}$ is not stable since $||\Phi(t)||$ is proportional to t for t large. This also can be illustrated geometrically in the 2-dimensional state space or phase plane. For the given ε,

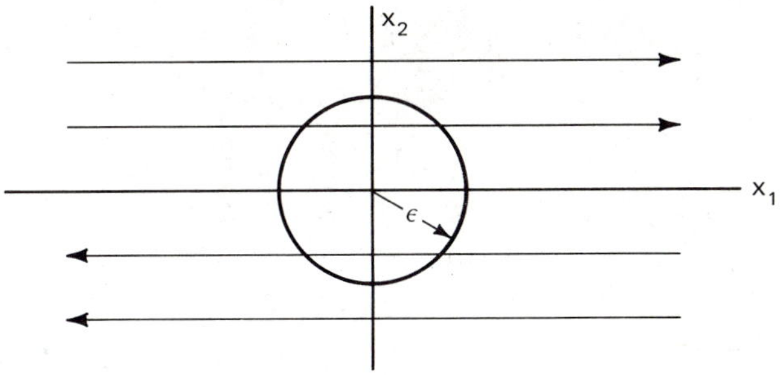

it is obvious that a δ, $\varepsilon>\delta>0$, cannot be found since for any $x_2(t_o)\neq 0$, $x_1(t)$ is unbounded.

b) From Example 5.1b, for $t_o=0$,

$$\Phi(t) = \begin{bmatrix} 1 & (1-\varepsilon^{-t}) \\ 0 & \varepsilon^{-t} \end{bmatrix} u_1(t)$$

Again using Theorem 5.1, part a) $\underline{x}=\underline{0}$ is stable since $||\Phi(t)||\leq\sqrt{2}$. From part b) of the theorem, $\underline{x}=\underline{0}$ is not asymptotically stable, since $||\Phi(t)||\to\sqrt{2}$ for large t. This also can be seen by inspection of the trajectories in the phase plane.

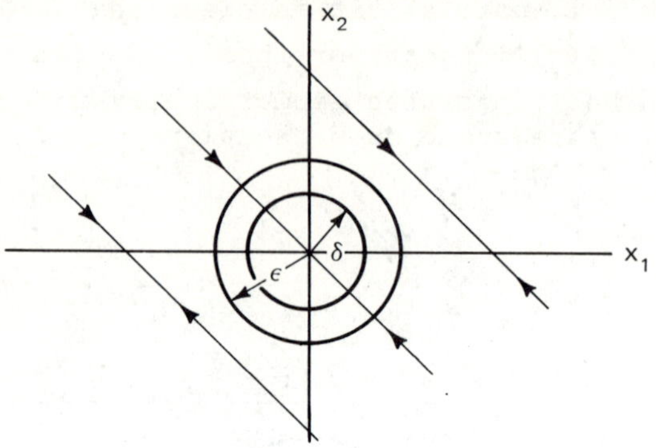

For the given ε, any initial state within the circle of radius δ leads to $\underline{x}(t)$ such that $||\underline{x}(t)||<\varepsilon$ for all t. However, for $x_2(0)\neq 0$, $x_1(t)$ approaches a nonzero constant.

c) From Example 5.1c, for $t_o=0$,

$$\Phi(t) = \begin{bmatrix} 2\varepsilon^{-t}-\varepsilon^{-2t} & \varepsilon^{-t}-\varepsilon^{-2t} \\ -2\varepsilon^{-t}+\varepsilon^{-2t} & -\varepsilon^{-t}+2\varepsilon^{-2t} \end{bmatrix} u_1(t)$$

Here, $||\Phi(t)|| \leq 1$ for all $t\geq 0$ and $||\Phi(t)|| \to 0$ for $t\to\infty$. Therefore, $\underline{x}=\underline{0}$ is globally asymptotically stable. This can be seen from the trajectories.

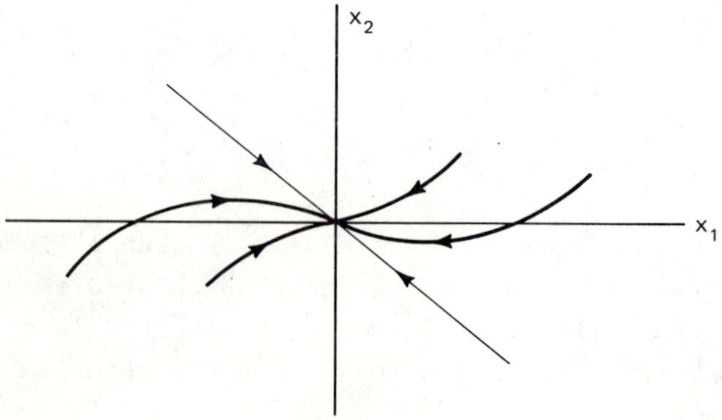

Example 5.3

A very simple time-varying system is described by

$$\dot{x}(t) = -\frac{1}{t}x(t) = A(t)x(t)$$

and

$$\Phi(t,t_o) = \frac{t_o}{t}$$

since

$$\frac{\partial\Phi(t,t_o)}{\partial t} = -\frac{t_o}{t^2} = -\frac{1}{t}\left(\frac{t_o}{t}\right) = A(t)\Phi(t,t_o)$$

and

$$\Phi(t_o,t_o) = 1 .$$

Also,

$$x(t) = \Phi(t,t_o)x(t_o), \qquad t \geq t_o.$$

Using Theorem 5.1, for

$t_o < 0$, $x=0$ is not stable

$t_o > 0$, $x=0$ is globally asymtotically stable.

Here the asymptotic stability is not uniform since it depends on t_o.

When the system is stationary, an alternative theorem which is easier to apply, can be proved.

Theorem 5.2

a) The equilibrium state $\underline{x}=\underline{0}$ of $\dot{\underline{x}}(t)=A\underline{x}(t)$, A constant is stable

$\Leftrightarrow$

1. all characteristic values λ_i of A have non-positive real parts:

$$\mathrm{Re}\lambda_i \leq 0$$

and

2. all characteristic values for which $\mathrm{Re}\lambda_i = 0$ are simple zeros of the minimal polynomial of A.

b) The equilibrium state $\underline{x}=\underline{0}$ of $\dot{\underline{x}}(t)=A\underline{x}(t)$, A constant, is <u>asymptotically stable</u>

$\Leftrightarrow$

all characteristic values λ_i of A have negative real parts:

$$\mathrm{Re}\lambda_i < 0$$

Thus, for stationary systems it only is necessary to find the characteristic values of A to test for the stability and asymptotic stability of the origin.

Example 5.4

Consider the three systems of Example 5.1.

a)

$$A = \begin{bmatrix} 0 & 1 \\ 0 & 0 \end{bmatrix}$$

$$p(\lambda) = \lambda^2 \quad \text{and} \quad \lambda_1 = \lambda_2 = 0 .$$

Since

$$C(\lambda) = \mathrm{adj}\ (\lambda I - A) = \begin{bmatrix} \lambda & 1 \\ 0 & \lambda \end{bmatrix},$$

$p(\lambda)$ is the minimal polynomial of A and $\lambda_1 = \lambda_2$ are not simple zeros. Thus $\underline{x}=\underline{0}$ is not stable.

b)

$$A = \begin{bmatrix} 0 & 1 \\ 0 & -1 \end{bmatrix}$$

$$p(\lambda) = \lambda^2 + \lambda \quad \text{and} \quad \lambda_1 = 0,\ \lambda_2 = -1 .$$

Then $\underline{x}=\underline{0}$ is stable, but not asymptotically stable.

c)

$$A = \begin{bmatrix} 0 & 1 \\ -2 & -3 \end{bmatrix}$$

$$p(\lambda) = \lambda^2 + 3\lambda + 2 \quad \text{and} \quad \lambda_1 = -1,\ \lambda_2 = -2 .$$

Therefore, $\underline{x}=\underline{0}$ is asymptotically stable.

Up to this point in the discussion of zero-input stability, it is the stability of the equilibrium state $\underline{x}=\underline{0}$ of the free system $\dot{\underline{x}}(t)=A(t)\underline{x}(t)$ that has been considered. Since it was shown that the stability of the solution $\underline{x}''(t)$ is directly related to the stability of $\underline{x}=\underline{0}$ of the free system, it frequently is convenient to use the term system stability. Therefore, the following definition is given:

<u>Definition 5.2</u> System Stability (Zero-Input)

The system $\dot{\underline{x}}(t) = A(t)\underline{x}(t)$ is stable (in a given sense) if $\underline{x}=\underline{0}$ is stable (in the same sense).

The definition of asymptotic stability requires convergence of $\underline{x}(t)$ to $\underline{0}$ for t increasing without limit. However, the manner of convergence is not specified, and in some studies of system stability, a measure of the rate of convergence is useful. Although the rate of convergence is apparent for stationary systems, it usually is not for time-varying systems, and the concept of exponential stability has been developed [3.].

Definition 5.3 Exponential Stability

The system $\dot{\underline{x}}(t) = A(t)\,\underline{x}(t)$ is exponentially stable if constants $\alpha>0$ and $\beta>0$ exist such that

$$||\underline{x}(t)|| \leq \alpha(\exp -\beta(t-t_o))||\underline{x}(t_o)||, \qquad t \geq t_o$$

for all $\underline{x}(t_o)\varepsilon R_r$.

One test for exponential stability is contained in the following theorem which requires that A(t) be bounded. Bounded vectors and matrices are defined as follows: A vector $\underline{q}(t)$, $t\geq t_o$, is bounded if there is a constant k such that $||\underline{q}(t)|| \leq k < \infty$ for all $t\geq t_o$. A matrix Q(t), $t\geq t_o$, is bounded if there is a constant k such that $||Q(t)|| \leq k < \infty$ for all $t\geq t_o$.

Theorem 5.3

The system $\dot{\underline{x}}(t) = A(t)\underline{x}(t)$, with A(t) bounded, $t\geq t_o$, is exponentially stable

$\Leftrightarrow$

there exists a constant m independent of t_o such that

$$\int_{t_o}^{t} ||\Phi(t,\tau)||\, d\tau \leq m \qquad \forall\ t \geq t_o .$$

Example 5.5

From Example 5.3 where the system was asymptotically stable (but not uniformly) for $t_o>0$ and $\Phi(t,t_o)= \frac{t_o}{t}$, we examine

$$\frac{1}{t}\int_{t_o}^{t} \tau d\tau = \frac{1}{2t}(t^2 - t_o^2) = \frac{1}{2}(t - \frac{t_o^2}{t})$$

It is clear that the system is not exponentially stable. Therefore, this is an example of a case where $x(t) \to 0$ as $t \to \infty$, $(t_o > 0)$ but the system is not exponentially stable. It is, of course, obvious in this simple example that $x(t) = \frac{t_o}{t} x(t_o)$ does not converge as fast as an exponential.

Generally, the concepts of uniform asymptotic stability and exponential stability are equivalent. In particular, for stationary systems, the following relates the two.

<u>Theorem 5.4</u>

The system $\dot{\underline{x}}(t) = A\underline{x}(t)$, A constant, is exponentially stable

$\Leftrightarrow$

it is asymptotically stable.

A number of other forms of equivalence also can be found [3.].

5.1.2 Zero-State Stability (B.I.B.O.)

The results obtained to this point concern the stability of $\dot{\underline{x}}(t) = A(t)\underline{x}(t)$, i.e., the input is zero. Theorems 5.1 and 5.3 apply to general time-varying systems and Theorems 5.2 and 5.4 to stationary systems. The next step is the investigation of stability when $\underline{u}(t) \neq \underline{0}$ but $\underline{x}(t_o) = \underline{0}$. This is a zero-state stability.

<u>Definition 5.4</u> Bounded Input - Bounded Output Stability

System S: $\dot{\underline{x}}(t) = A(t)\underline{x}(t) + B(t)\underline{u}(t)$, $\underline{x}(t_o) = \underline{0}$

$\underline{y}(t) = C(t)\underline{x}(t)$

with all elements of A, B, C continuous functions of $t \in T_t$.

System S is stable in the bounded input-bounded output (b.i.b.o.) sense if for all bounded $\underline{u}(t)$, $t \geq t_o$, and all t_o, $\underline{y}(t)$ is bounded.

The important change from the previous forms of stability is that the conditions are now on the system input and output and not explicitly on the state. This type of stability is sometimes referred to as <u>external stability</u>. The requirement of a zero initial state is not always included; the nonzero initial state is introduced later in the definition of total stability. For the general case of time-varying systems, the two following theorems apply.

<u>Theorem 5.5</u>

System S of Definition 5.4 is b.i.b.o. stable there exists a constant k such that

$$\int_{t_o}^{t} ||C(t)\Phi(t,\tau)B(\tau)||\,d\tau \leq k < \infty$$

for all t_o and $t \geq t_o$.

An equivalent form in terms of the elements of the impulse response matrix is:

<u>Theorem 5.6</u>

System S is b.i.b.o. stable

$\Leftrightarrow$

there exist constants k_{ij} such that

$$\int_{t_o}^{t} |h_{ij}(t,\tau)|\,d\tau \leq k_{ij} < \infty \qquad \begin{array}{l} i = 1, 2, \ldots, p \\ j = 1, 2, \ldots, q \end{array}$$

for all t_o and $t \geq t_o$.

When the system is stationary, a theorem which is easier to apply is as follows:

Theorem 5.7

System S_1: $\dot{\underline{x}} = A\underline{x} + B\underline{u}$, $\underline{x}(t_o) = \underline{0}$

$\underline{y} = C\underline{x}$

A, B, C constant.

System S_1 is b.i.b.o. stable
$\Leftrightarrow$
all poles of every element $H_{ij}(s)$ of the transfer matrix $H(s) = C\Phi(s)B$ have negative real parts.

It is possible that pole/zero cancellations will occur, and recalling Theorems 3.11 and 3.12, this means that S_1 may not be completely controllable and/or observable. There is the possibility, then, that some components of $\underline{x}$ may be very large or unbounded even though the output is bounded. Also, it may not be possible to keep these components of $\underline{x}$ bounded by using a bounded input. These possibilities lead immediately to the question of how the previous zero-input conditions of stability and the present zero-state forms are related. In a general way, zero-input stability is related to the internal structure that can be inferred from the A-matrix of a state model. On the other hand, b.i.b.o. stability includes the input and output matrices B and C as well as A. The following theorem gives the additional conditions of B and C that are required in order that asymptotic stability of $\underline{x}=\underline{0}$ imply b.i.b.o. stability.

Theorem 5.8

If

a) $\underline{x}=\underline{0}$ of $\dot{\underline{x}}(t) = A(t)\underline{x}(t)$ is asymptotically stable,

and

b) B(t) and C(t) are bounded for all $t \geq t_o$, then system S of Definition 5.4 is b.i.b.o. stable.

For the converse, it is necessary to recall the concepts of uniform controllability and uniform observability of Theorems 3.1 and 3.3. It is not surprising that some conditions on controllability and observability are needed to be able to infer asymptotic stability from b.i.b.o. stability.

Theorem 5.9

For system S of Definition 5.4 with

a) A(t), B(t), C(t) bounded for all $t \geq t_o$

b) A(t), B(t) uniformly completely controllable

c) A(t), C(t) uniformly completely observable,

$\underline{x}=\underline{0}$ of $\dot{\underline{x}}(t) = A(t)\underline{x}(t)$ is asymptotically stable

$\Leftrightarrow$

system S is b.i.b.o. stable.

This theorem, therefore, establishes a link between the two stability conditions.

Example 5.6

Examine the following state variable model of a system for asymptotic stability and b.i.b.o. stability.

$$\dot{\underline{x}} = \begin{bmatrix} 0 & 1 \\ 2 & 1 \end{bmatrix}\underline{x} + \begin{bmatrix} 0 \\ 1 \end{bmatrix}u$$

$$y = [-2 \quad 1]\underline{x}$$

a) For $p(\lambda) = \det[\lambda I - A]$

$$= \lambda^2 - \lambda - 2$$

$$\lambda_1 = -1, \quad \lambda_2 = 2$$

and by Theorem 5.2, the system is not asymptotically stable.

b) For $H(s) = \underline{c}^T \Phi(s) \underline{b}$

$$= \frac{1}{s+1},$$

by Theorems 5.7, the system is b.i.b.o. stable. Although the model is a controllable form, examination of the observability matrix P_o shows that it is not observable. Thus the c) requirement of Theorem 5.9 is not met. The mode associated with ε^{2t} is not observable although it can be controlled.

5.1.3 Total Stability

The combination of $\underline{u} \neq \underline{0}$ and $\underline{x}(t_o) \neq \underline{0}$ can now be considered by introducing the concept of total stability [2.].

Definition 5.5 Total Stability

System S_2: $\dot{\underline{x}}(t) = A(t)\underline{x}(t) + B(t)\underline{u}(t) \quad \underline{x}(t_o) = \underline{x}_o$

$$\underline{y}(t) = C(t)\underline{x}(t)$$

System S_2 is totally stable if for any $\underline{x}_o \; \varepsilon \; R_r$ and any bounded input $\underline{u}(t)$, $t \geq t_o$, $\underline{x}(t)$ and $\underline{y}(t)$ are bounded.

Two theorems similar to Theorems 5.8 and 5.9 follow immediately.

Theorem 5.10

If

a) $\underline{x} = \underline{0}$ of $\dot{\underline{x}}(t) = A(t)\underline{x}(t)$ is asymptotically stable, and

b) $B(t)$ and $C(t)$ are bounded for all $t \geq t_o$,

then system S_2 of Definition 5.5 is totally stable.

Theorem 5.11

For system S_2 of Definition 5.5 with

a) A(t), B(t), C(t) bounded for all $t \geq t_o$
b) A(t), B(t) uniformly completely controllable
c) A(t), C(t) uniformly completely observable,
$\underline{x} = \underline{0}$ of $\dot{\underline{x}}(t) = A(t)\underline{x}(t)$ is asymptotically stable
$\Leftrightarrow$
system S_2 is totally stable.

This theorem is a general result which combines many of the previous cases and emphasizes the conditions of A(t), B(t), and C(t). Thus, in addition to boundedness, uniform complete controllability for A(t), B(t) (Theorem 3.1), uniform complete observability for A(t), C(t) (Theorem 3.3), and asymptotic stability for A(t) (Theorem 5.1) are required for total stability.

5.1.4 Stability Tests – Stationary Systems

In addition to the decomposition of the state space R_r of stationary systems into controllable and uncontrollable subspaces (which depends on A,B) and observable and unobservable subspaces (which depends on A,C) discussed in Chapter 3, a decomposition into stable and unstable subspaces (which depends on A) can be done. The stable subspace is spanned by a set of characteristic vectors associated with those characteristic values λ_i which have negative real parts. From these decompositions, two other system properties are sometimes defined. If the unstable subspace is contained in the controllable subspace, the stationary system is stabilizable. Example 5.6 illustrates this. If the unobservable subspace is contained in the stable subspace, the stationary system is detectable. These concepts are useful in designing controllers and filters for linear systems. A theorem which summarizes many of the foregoing ideas for stationary systems also can be proved.

Theorem 5.12

Stationary System S_3: $\dot{\underline{x}} = A\underline{x} + B\underline{u}$, $\underline{x}(t_o) = \underline{x}_o$

$\underline{y} = C\underline{x}$

with (A,B) controllable and (A,C) observable.

The condition that the characteristic values λ_i of A have $Re\lambda_i<0$ is necessary and sufficient for any of the equivalent conditions:

1. S_3 is totally stable

2. S_3 is b.i.b.o. stable

3. S_3 is asymptotically stable.

From this theorem it is clear that one way of testing for the stability of a stationary system is to find the characteristic values of A, and there are computer algorithms for doing this. Also, from Theorem 5.7, it is the zeros of $Q_{ij}(s)$, where $H_{ij}(s) = P_{ij}(s)/Q_{ij}(s)$, that are required in a test for b.i.b.o. stability alone. In either event, the zeros of polynomials such as $p(\lambda)$ of A or $Q_{ij}(s)$ are needed. Rather than using a computer to solve for the zeros, the polynomials can be tested to see if all zeros have negative real parts. A <u>necessary</u> condition that the zeros of a polynomial $f(\lambda)$ have all zeros with negative real parts is that all coefficients be positive (or all coefficients be negative), thus for

$$f(\lambda) = \lambda^m + a_{m-1}\lambda^{m-1} + \ldots + a_1\lambda + a_o ,$$

require $a_o, a_1, \ldots, a_{m-1} > 0$. Polynomials whose zeros have negative real parts are called Hurwitz polynomials, and three common tests which are sufficient are the Hurwitz test, and Routh test [2.], and the Lienard-Chipart test [2.]. Only the first is discussed here. For the m-th order polynomial $f(\lambda)$, the mxm matrix H_p is formed as

$$H_p = \begin{bmatrix} a_{m-1} & a_{m-3} & \cdots & a_{1-m} \\ 1 & a_{m-2} & \cdots & a_{2-m} \\ 0 & a_{m-1} & \cdots & a_{3-m} \\ 0 & 1 & \cdots & a_{4-m} \\ & & \vdots & \\ 0 & 0 & \cdots & a_o \end{bmatrix}$$

and $a_k=0$, $k<0$. For all cases, the last column has all zeros except for the mm element. The last row is a shifted form of the first or second row according to whether m is odd or even. The following theorem then provides a test.

Theorem 5.13 Hurwitz Test

$f(\lambda)$ is Hurwitz

$\Leftrightarrow$

all principal minors of H_p are positive, i.e.,

$$D_1 = a_{m-1} > 0, \quad D_2 = \det \begin{bmatrix} a_{m-1} & a_{m-3} \\ 1 & a_{m-2} \end{bmatrix} > 0, \ldots,$$

$$D_m = \det H_p > 0.$$

The theorem is useful for stability tests as well as for establishing conditions on numerical values of some of the coefficients when they are free parameters in a system design.

5.1.5 Liapunov Theorems

An alternative point of view on stability testing involves the use of a function of the state $V(\underline{x}(t),t)$, and is usually called the second method of Liapunov [4.]. The method has been applied to a variety of nonlinear systems, and a special form can be proved for stationary linear systems. The general theorem as applied to linear systems is given first.

Theorem 5.14

For a function $V(\underline{x}(t),t)$ which is positive definite (see Appendix B for the case where V is a quadratic form), $\underline{x}=\underline{0}$ of $\dot{\underline{x}}(t) = A(t)\underline{x}(t)$ is

a) stable if $\frac{dV}{dt} \leq 0$, $t \geq t_o$ and

b) asymptotically stable if stable and if $\frac{dV}{dt} \neq 0$ along any trajectory.

The problem in applying this theorem is in finding the function V, although for linear systems, a quadratic form such as $\underline{x}^T(t)Q(t)\underline{x}(t)$, where Q(t) is a real, symmetric, positive definite matrix, is a possible choice [5.]. When the linear system is stationary, another useful theorem follows from Theorem 5.14.

Theorem 5.15

The origin $\underline{x}=\underline{0}$ of $\dot{\underline{x}}(t) = A\underline{x}(t)$, A constant, is asymptotically stable

$\Leftrightarrow$

for any positive definite, symmetric matrix Q there exists a positive definite matrix M such that

$$A^TM + MA = -Q, \qquad M^T = M.$$

The linear algebraic equation in M, called a Liapunov equation, cannot be solved directly. Numerical methods are reviewed in [6.].

Example 5.7

a) Consider the system of Example 5.2b.

$$A = \begin{bmatrix} 0 & 1 \\ 0 & -1 \end{bmatrix} \quad \text{and let } M = \begin{bmatrix} m_{11} & m_{12} \\ m_{12} & m_{22} \end{bmatrix}$$

and let Q=I. Then require

$$\begin{bmatrix} 0 & 0 \\ 1 & -1 \end{bmatrix}\begin{bmatrix} m_{11} & m_{12} \\ m_{12} & m_{22} \end{bmatrix} + \begin{bmatrix} m_{11} & m_{12} \\ m_{12} & m_{22} \end{bmatrix}\begin{bmatrix} 0 & 1 \\ 0 & -1 \end{bmatrix} = \begin{bmatrix} -1 & 0 \\ 0 & -1 \end{bmatrix}$$

For the 11 element,

$$0 + 0 \neq -1$$

and in fact for any positive definite Q, there is a contradiction. Therefore, by Theorem 5.15, the system is not asymptotically stable (as has been shown previously).

b) For the system of Example 5.2c,

$$A = \begin{bmatrix} 0 & 1 \\ -2 & -3 \end{bmatrix}$$

and using M and Q as above,

$$M = \begin{bmatrix} 5/4 & 1/4 \\ 1/4 & 1/4 \end{bmatrix}$$

which is positive definite. Therefore, the system is asymptotically stable.

5.1.6 Discrete-Time Systems

A similar sequence of results can be developed for linear discrete-time systems. In many cases, it is a simple matter of an obvious change in notation. Therefore, only those cases where differences occur are done in detail. Other results are implied by simple substitutions of n for t, n_o for t_o, and sums for integrals. For the model

$$\underline{x}(n+1) = A(n)\underline{x}(n) + B(n)\underline{v}(n), \qquad n \varepsilon T_n ,$$

the transition function for $\underline{v}(n) = \underline{0}$ is given by Eq. 3.61 as

$$\underline{x}(n) = \underline{f}_d(n;\underline{x}(n_o),\underline{0}), \qquad n \geq n_o, \qquad n_o \varepsilon T_n ,$$

and an equilibrium state $\underline{\lambda}$ is any state that satisfies

$$\underline{\lambda} = \underline{f}_d(n; \underline{\lambda}, \underline{0}) .$$

From the definition of the transition matrix,

$$\underline{x}(n) = \Phi(n,n_o)\underline{x}(n_o) = \underline{f}_d(n;\underline{x}(n_o),\underline{0})$$

and $\underline{\lambda}$ must satisfy

$$\underline{\lambda} = \Phi(n,n_o)\underline{\lambda} . \tag{5.3}$$

Also, a constant $\underline{\lambda}$ substituted in $\underline{x}(n+1) = A(n)\underline{x}(n)$ gives

$$\underline{\lambda} = A(n)\underline{\lambda}$$

and therefore,

$$\underline{0} = (A(n)-I)\underline{\lambda} . \tag{5.4}$$

Then, an equilibrium state $\underline{\lambda}$ is unique if (A(n)-I) is non-singular for $n \geq n_o$, and $\underline{\lambda} = \underline{0}$ is that unique equilibrium state. This is slightly different from Eq. 5.2. Recall, too, that in Chapter 3, A(n) was assumed nonsingular for a unique solution of Eq. 3.64, but (A(n)-I) is not necessarily nonsingular. As before, $\underline{\lambda} = \underline{0}$ is always an equilibrium state.

Example 5.8

a) From Example 3.23, a state variable model of Simpson's Rule,

$$A = \begin{bmatrix} 0 & 1 \\ 1 & 0 \end{bmatrix}$$

which is nonsingular. For equilibrium states,

$$\underline{0} = (A-I)\underline{\lambda} = \begin{bmatrix} -1 & 1 \\ 1 & -1 \end{bmatrix} \underline{\lambda} .$$

Since (A-I) is singular, $\underline{\lambda}$ is not unique. In fact any state such that $\lambda_1 = \lambda_2$ is an equilibrium state.* This also can be seen by inspection of the simulation diagram.

b) From Example 3.33,

$$A = \begin{bmatrix} 0 & 1 \\ -0.125 & 0.75 \end{bmatrix}$$

---

*Note that λ_1 and λ_2 are components of $\underline{\lambda}$, <u>not</u> characteristic values of A.

and

$$\underline{0} = (A-I)\underline{\lambda} = \begin{bmatrix} -1 & 1 \\ -0.125 & -0.25 \end{bmatrix} \underline{\lambda}$$

Here $\underline{\lambda}$ is unique since (A-I) is nonsingular; $\underline{\lambda} = \underline{0}$ is the only equilibrium state.

Following a similar argument to that for continuous-time systems, the free system is

$$\underline{q}(n+1) = A(n)\underline{q}(n) \quad .$$

The definition for zero-input stability and the corresponding theorem are analogous to Definition 5.1 and Theorem 5.1.

Example 5.9

a) From Example 5.8a (and Example 3.28), for $n_o=0$,

$$\Phi(n) = \begin{bmatrix} \frac{1}{2}(1+(-1)^n) & \frac{1}{2}(1-(-1)^n) \\ \frac{1}{2}(1-(-1)^n) & \frac{1}{2}(1+(-1)^n) \end{bmatrix} v_1(n)$$

Using this, $||\Phi(n)|| = 1$, and $\underline{x} = \underline{0}$ is stable, but not asymptotically stable.

b) From Example 5.8b,

$$A = \begin{bmatrix} 0 & 1 \\ -0.125 & 0.75 \end{bmatrix}$$

and

$$\Phi(n) = \begin{bmatrix} -(0.5)^n+2(0.25)^n & 4(0.5)^n-4(0.25)^n \\ -0.5(0.5)^n+0.5(0.5)^n & 2(0.5)^n-(0.25)^n \end{bmatrix} v_1(n)$$

and $||\Phi(n)|| \leq 1$. Also,

$$\lim_{n\to\infty} \Phi(n) = 0 ,$$

and

$$\underline{x} = \underline{0}$$

is stable and globally asymptotically stable. For a stationary discrete-time system, a theorem analogous to Theorem 5.2 can be proved.

Theorem 5.16

a) The equilibrium state $\underline{x} = \underline{0}$ of $\underline{x}(n+1) = A\underline{x}(n)$, A constant, is stable

$\Leftrightarrow$

1. all characteristic values λ_i of A have magnitudes no greater than one:

$$|\lambda_i| \leq 1$$

and

2. all characteristic values for which $|\lambda_i| = 1$ are simple zeros of the minimal polynomial of A.

b) The equilibrium state $\underline{x} = \underline{0}$ of $\underline{x}(n+1) = A\underline{x}(n)$, A constant, is asymptotically stable

$\Leftrightarrow$

all characteristic values λ_i of A have magnitudes less than one:

$$|\lambda_i| < 1 .$$

Example 5.10

From Example 5.8a and 5.9,

$$A = \begin{bmatrix} 0 & 1 \\ 1 & 0 \end{bmatrix}$$

and $p(\lambda) = \lambda^2 - 1$ and $\lambda_1, \lambda_2 = 1, -1$. $p(\lambda)$ is the minimal polynomial and using Theorem 5.16, $\underline{x} = \underline{0}$ is stable, but not asymptotically stable.

Again, the definition of zero-input system stability can be given in terms of the stability of $\underline{x} = \underline{0}$.

Definition 5.6 System Stability (Zero-Input)

The system $\underline{x}(n+1) = A(n)\underline{x}(n)$ is stable (in a given sense) if $\underline{x} = \underline{0}$ is stable (in the same sense).

The analog to exponential stability for continuous-time systems is geometric stability for discrete-time systems. The rate of convergence of $\underline{x}(n)$ to $\underline{0}$ is compared to the form $\alpha(\beta)^n$, $\alpha>0$, $0<\beta<1$, and with obvious changes in notation, results similar to Definition 5.3, Theorem 5.3, and Theorem 5.4 can be found.

For the bounded input-bounded output stability of a discrete system the following definition is analogous to Definition 5.4.

Definition 5.7 Bounded Input-Bounded Output

System S: $\underline{x}(n+1) = A(n)\underline{x}(n) + B(n)\underline{v}(n)$, $\underline{x}(n_o) = \underline{0}$

$$z(n) = C(n)\underline{x}(n)$$

with $A(n)$ nonsingular, $n \in T_n$.

System S is stable in the bounded input-bounded output (b.i.b.o.) sense if for all bounded $\underline{v}(n)$, $n \geq n_o$, and all n_o, $\underline{z}(n)$ is bounded.

Analogous to Theorem 5.5 is:

Theorem 5.17

System S of Definition 5.7 is b.i.b.o. stable

⇔

there exists a constant k such that

$$\sum_{m=n_o}^{n-1} ||C(n)\ \Phi(n,m+1)B(m)|| \leq k < \infty$$

for all n_o and $n \geq n_o$.

An equivalent form in terms of the elements $h_{ij}(n,m)$ of the pulse response matrix is analogous to Theorem 5.6. When the system is stationary, the analog of Theorem 5.7 is:

Theorem 5.18

System S_1:
$$\underline{x}(n+1) = A\underline{x}(n) + B\underline{v}(n),\ x(n_o) = \underline{0}$$
$$\underline{z}(n) = C\underline{x}(n)$$
A, B, C constant.

System S_1 is b.i.b.o. stable

⇔

all poles of every element $H_{ij}(z)$ of the transfer matrix $H(z) = Cz^{-1}\Phi(z)B$ have magnitudes less than one.

The same remarks as those made after Theorem 5.7 can be made regarding pole/zero cancellations and the relationship between asymptotic stability and b.i.b.o. stability of a discrete-time system. Using definitions of uniform complete controllability and uniform complete observability, a discrete-time system theorem analogous to Theorem 5.9 can be shown. Total stability of a discrete-time system can be defined similarly to Definition 5.5, and theorems like Theorems 5.10 and 5.11 then follow. The decomposition of the state space R_r of a stationary discrete-time system into controllable and uncontrollable subspaces, observable, and unobservable subspaces, and stable and unstable subspaces can be done. The stable subspace is spanned by a set of characteristic vectors associated with characteristic values λ_i with magnitudes less than one. A theorem which summarizes the results for stationary discrete-time systems is:

Theorem 5.19

Stationary System S_3:

$$\underline{x}(n+1) = Ax(n) + B\underline{v}(n),$$
$$\underline{x}(n_o) = \underline{x}_o$$
$$\underline{z}(n) = Cx(n)$$

with (A,B) controllable and (A,C) observable.

The condition that the characteristic values λ_i of A have $|\lambda_i| < 1$ is necessary and sufficient for any of the equivalent conditions:

1. S_3 is totally stable

2. S_3 is b.i.b.o. stable

3. S_3 is asymptotically stable.

This is analogous to Theorem 5.12 and again it is obvious that the characteristic values λ_i of A are required. The zeros of polynomials $p(\lambda)$ of A or $Q_{ij}(z)$ of $H_{ij}(z)=P_{ij}(z)/Q_{ij}(z)$ (see Theorem 5.18) need not be found, however. Rather, tests to see whether their magnitudes are less than one can be used. One method which makes use of previous results for continuous-time systems uses a bilinear transformation. Thus for $p(\alpha)$ of A or $Q_{ij}(\alpha)$ of Theorem 5.18, let

$$\alpha = \frac{1+\lambda}{1-\lambda} \tag{5.5}$$

to obtain a $p'(\lambda)$ or $Q'_{ij}(\lambda)$, the numerator polynomials after substitution. The Hurwitz test of Theorem 5.13 can then be used on p' and Q'_{ij}. This is true because the transformation of Eq. 5.5 maps the inside of the unit circle of the α-plane into the left half of the λ-plane. Thus any test to determine if all zeros of $p'(\lambda)$ or $Q'_{ij}(\lambda)$ are in the left half of the λ-plane is equivalent to a test to determine whether all zeros of $p(\alpha)$ or $Q_{ij}(\alpha)$ are inside the unit circle of the α-plane.

Example 5.11

From Example 5.9b,

$$A = \begin{bmatrix} 0 & 1 \\ -0.125 & 0.75 \end{bmatrix}$$

$$p(\alpha) = \det(\alpha I - A) = \det \begin{bmatrix} \alpha & -1 \\ 0.125 & \alpha-0.75 \end{bmatrix}$$

or

$$p(\alpha) = \alpha(\alpha-0.75) + 0.125$$

$$= \alpha^2 - 0.75\alpha + 0.125$$

For

$$\alpha = (1+\lambda)/(1-\lambda) ,$$

$$p(\frac{1+\lambda}{1-\lambda}) = \frac{(1+\lambda)^2}{(1-\lambda)^2} - 0.75 \frac{(1+\lambda)}{(1-\lambda)} + 0.125$$

and

$$p'(\lambda) = (1+\lambda)^2 - 0.75(1+\lambda)(1-\lambda) + 0.125(1-\lambda)^2$$

$$= 1.875 (\lambda^2 + 0.934\lambda + 0.200).$$

Then

$$H_p = \begin{bmatrix} 0.934 & 0 \\ 1 & 0.200 \end{bmatrix}$$

from Theorem 5.13, and

$$D_1 = 0.934 > 0 \quad \text{and} \quad D_2 = 0.167 > 0$$

and therefore all zeros of p' are in the left half λ-plane and all zeros of $p(\alpha)$ are inside the unit circle of the α-plane. The system is, therefore, asymptotically stable.

The coefficients of $p'(\lambda)$ can be found directly from the coefficients of $p(\alpha)$ for higher order polynomials (up to 9th-order) by the use of computed transformations [7.]. Note that the matrix H_p is obtained from a normalized form of $p'(\lambda)$ in agreement with Theorem 5.13. Alternatives to the use of bilinear transformation are a method similar to the Routh Test [8.] and the Schur-Cohn test [9.], which is similar to the Hurwitz test. The Schur-Cohn test starts with a polynomial

$$p(\alpha) = a_m\alpha^m + a_{m-1}\alpha^{m-1} + \ldots + a_o .$$

A set of matrices H_k is formed as

$$H_k = \begin{bmatrix} A_k & B_k \\ B_k^T & A_k^T \end{bmatrix},$$

a 2k x 2k matrix, where

$$A_k = \begin{bmatrix} a_o & 0 & \cdots & 0 \\ a_1 & a_o & \cdots & 0 \\ \vdots & & & \\ a_{k-1} & a_{k-2} & \cdots & a_o \end{bmatrix}$$

$$B_k = \begin{bmatrix} a_m & a_{m-1} & \cdots & a_{m-k+1} \\ 0 & a_m & \cdots & a_{m-k+2} \\ \vdots & & & \\ 0 & & \cdots & a_m \end{bmatrix}$$

with k = 1, 2,...,m. From these H_k, the following theorem can be proved.

<u>Theorem 5.20</u> Schur-Cohn Test

The zeros of $p(\lambda)$ lie inside the unit circle of the λ-plane

$\Leftrightarrow$

det $H_k<0$, k odd

det $H_k>0$, k even.

<u>Example 5.12</u>

Consider a system whose transfer function is

$$H(z) = \frac{K}{6z^2+z-1},$$

then

$$p(\alpha) = 6\alpha^2 + \alpha - 1$$

and

$$H_1 = \begin{bmatrix} -1 & 6 \\ 6 & -1 \end{bmatrix}$$

$$H_2 = \begin{bmatrix} -1 & 0 & 6 & 1 \\ 1 & -1 & 0 & 6 \\ 6 & 0 & -1 & 1 \\ 1 & 6 & 0 & -1 \end{bmatrix}$$

Now

$$\det H_1 = -35 < 0$$

and

$$\det H_2 = 161 > 0 ,$$

therefore, the poles of H(z) are inside the unit circle and the system is asymptotically stable.

Corresponding to Theorem 5.14, the second method of Liapunov, is a theorem for discrete-time systems.

<u>Theorem 5.21</u>

For a function $V(\underline{x}(n),n)$ which is positive definite, $\underline{x} = \underline{0}$ of $\underline{x}(n+1) = A(n)\underline{x}(n)$ is

a) stable if $\Delta_b V(\underline{x}(n),n) \leq 0,\ n \geq n_o$,

and

b) asymptotically stable if stable and if

$$\Delta_b V(\underline{x}(n),n) \neq 0 \text{ along any trajectory.}$$

As in the previous case for continuous-time systems, finding a suitable V-function may be difficult although a good possibility is again a quadratic form. When the linear system is stationary, another theorem similar to Theorem 5.15 can be proved.

Theorem 5.22

The origin $\underline{x} = \underline{0}$ of $\underline{x}(n+1) = A\underline{x}(n)$, A constant, is asymptotically stable

$\Leftrightarrow$

for any positive definite, symmetric matrix Q, there exists a positive definite matrix M such that

$$A^T MA - M = -Q, \quad M = M^T .$$

In some cases, a procedure similar to that of Example 5.7 can be used to find M when Q is chosen.

The following table summarizes the results of this section on stability for stationary systems.

TABLE 5.1

STABILITY THEOREMS FOR STATIONARY SYSTEMS

	Continuous-Time	Discrete-Time
Equilibrium	Equation 5.1, 5.2	Equation 5.3, 5.4
Zero-Input Stability	Theorem 5.2	Theorem 5.16
Zero-State Stability (b.i.b.o.)	Theorem 5.7	Theorem 5.18
Total Stability	Theorem 5.12	Theorem 5.19
Polynomial Test	Theorem 5.13 Hurwitz	Theorem 5.20 Schur-Cohn
Liapunov 2nd Method	Theorem 5.15	Theorem 5.22

5.2 FIDELITY OR ACCURACY

The evaluation of the dynamic performance of a system can be done in many ways. Generally, it starts with a system model and various input signal models. The methods of Chapter 4 can be used to find the response for many specified signals. Some of the well-known methods use elementary signals for inputs and the quality of the response is given in terms of standard parameters. For example, it is common to use the response to a unit step function ($u_1(t)$ or $v_1(n)$) to find rise time (10% to 90%), per cent overshoot, and settling time [10.]. These are time-domain measures. Also, it is common to use variable-frequency sinusoidal inputs to find bandwidth and resonance peak [10.]. Other periodic signals such as square waves, triangular waves, pulse trains, etc., as well as stochastic signals such as white inputs or band-limited white inputs are used. In all of these cases there is a general agreement as to the interpretation of the various measures. For example, short rise time implies a "fast" response and large bandwidth implies "good" fidelity. In fact, specifications on a system frequently are given in terms of these measures. The concept of defining an error -- the difference between a desired response and an actual response -- also is very useful. In simple cases an error at one point in time or frequency can be used. For example, steady-state error, an asymptotic error for very large time, is common. When the error at several points in time or frequency is to be evaluated, a summation of a nonnegative function of the error is useful. Examples of this were seen in Chapter 2 on signal identification (e.g., least squares) and in Chapter 4 on system identification. The summations provide a single number as a measure of quality.

Although the "point type" measures of error can be found directly by using the methods of Chapter 4, "summary type" measures using sums or integrals of errors usually are easier to find in other ways. This section is concerned with developing methods to obtain generalized sum and integral error measures. Forms for continuous-time and discrete-time and scalar and vector cases will be developed. Two groups of error measures are used:

1. Those which are related to the work of Wiener where terminal time (t_1 or n_1) is infinite, the system is stationary, initial times (t_o or n_o) are in the infinite past so that initial conditions are not considered, and the deterministic signals are finite energy signals or the stochastic signals are ergodic. Frequency domain methods will be used.
2. Those which consider finite terminal times, time-varying systems, nonzero initial conditions at finite t_o or n_o, and signals (deterministic and stochastic) which have finite energy on a finite interval. Time domain methods are required.

5.2.1 Infinite Time Intervals

Figure 5.1 shows a model for a continuous-time, single input/single output linear system with deterministic inputs.

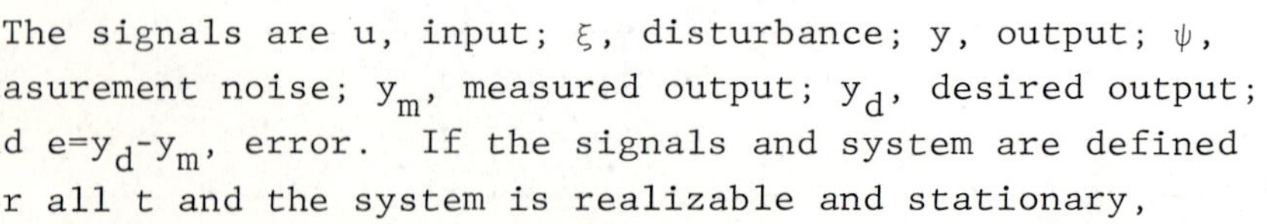

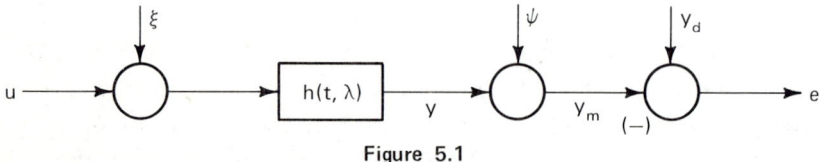

Figure 5.1

The signals are u, input; ξ, disturbance; y, output; ψ, measurement noise; y_m, measured output; y_d, desired output; and $e=y_d-y_m$, error. If the signals and system are defined for all t and the system is realizable and stationary,

$$y(t) = \int_{-\infty}^{t} h(t-\lambda)u_a(\lambda)d\lambda$$

$$u_a(t) = u(t) + \xi(t)$$

$$y_m(t) = y(t) + \psi(t)$$

and

$$e(t) = y_d(t) - y_m(t) \; . \qquad (5.6)$$

A measure of accuracy -- the integral of the squared error -- is then chosen as an index of system performance. Thus we let

$$J = \int_{-\infty}^{\infty} e^2(t)dt \; . \qquad (5.7)$$

From the Integral Square Theorem of Appendix A,

$$J = \frac{1}{2\pi j}\int_{-j\infty}^{j\infty} E(-s)E(s)ds \; ,$$

where $E(s) = \mathcal{L}\; e(t)$. Using Eq. 5.6 and the real convolution theorem,

$$Y(s) = H(s)U_a(s)$$

$$= H(s)U(s) + H(s)\xi(s)$$

and

$$E(s) = Y_d(s) - Y_m(s)$$

$$= Y_d(s) - H(s)U(s) - H(s)\xi(s) - \psi(s) \; .$$

Using the translation functions and their transforms from Chapter 2,

$$E(-s)\; E(s) = W_e(s) = \mathcal{L}\; w_e(\tau)$$

and

$$J = \frac{1}{2\pi j}\int_{-j\infty}^{j\infty} W_e(s)ds \ . \qquad (5.8)$$

Also,

$$\begin{aligned} W_e(s) = {} & [W_d(s) + W_\psi(s) - W_{d\psi}(-s)] \\ & + H(-s)H(s)[W_u(s) + W_\xi(s) + W_{u\xi}(s) + W_{u\xi}(-s)] \\ & - [H(-s)W_{du}(-s) + H(s)W_{du}(s)] \\ & - [H(-s)W_{d\xi}(-s) + H(s)W_{d\xi}(s)] \\ & + [H(-s)W_{\psi u}(-s) + H(s)W_{\psi u}(s)] \\ & + [H(-s)W_{\psi\xi}(-s) + H(s)W_{\psi\xi}(s)] \end{aligned} \qquad (5.9)$$

where

$$W_d(s) = \mathcal{L}\ w_d(\tau) = \mathcal{L}\int_{-\infty}^{\infty} y_d(t)y_d(t+\tau)dt$$

as in Eq. 2.69 and 2.71 and similar definitions are used for the other auto- and crosstranslation functions. $W_e(s)$ and the six bracketed terms on the right are even functions of s and use has been made of the property $W_{pq}(s) = W_{qp}(-s)$.

There are several ways of evaluating J. In the time domain, calculation of e(t) as $\mathcal{L}^{-1}E(s)$ and use of Eq. 5.7 is possible. If there are no poles of $W_e(s)$ on the imaginary axis of the s-plane, substitution of $j\omega$ for s gives

$$J = \frac{1}{2\pi}\int_{-\infty}^{\infty} W_e(\omega)d\omega$$

which also is a real integral. These time and frequency domain real integrals can be evaluated using numerical integration. Also, Eq. 5.8 can be evaluated using the calculus of residues where a closed contour around the left half s-plane can be used. If $W_e(s)$ approaches zero as $R \to \infty$, $s = R\varepsilon^{j\theta}$, the Cauchy Residue Theorem implies

$$J = \sum \text{Res } W_e(s) \Big|_{\text{poles in LHP}} \tag{5.10}$$

where the form assumes the singularities of $W_e(s)$ are poles. If the system is asymptotically stable and the signals are energy signals, there are no poles on the imaginary axis. Finally, tables for the evaluation of J have been published [11.] which use the coefficients of the numerator and denominator polynomials of $W_e(s)$.

Equation 5.9 is a general result which requires a large amount of calculation if all the terms are used. Two special cases of interest are considerably simpler. For

$$\xi = \psi = 0 \quad \text{and} \quad y_d = u,$$

$$W_e'(s) = [1-H(-s)-H(s)+H(-s)H(s)]W_u(s)$$

or

$$W_e'(s) = W_u(s)[H(-s)-1][H(s)-1)] \ . \tag{5.11}$$

This form, when used in Eq. 5.8, gives a measure of simple dynamic error, i.e., a measure of how well the system output matches the specified input. Also,

$$W_e'(\omega) = W_u(\omega)|H(j\omega)-1|^2 \ .$$

Second, for

$$\xi = 0, \quad \psi \neq 0, \quad y_d = u$$

$$W_e''(s) = W_e'(s)$$

$$+ W_\psi(s)+(H(-s)-1)W_{\psi u}(-s)+(H(s)-1)W_{\psi u}(s) \tag{5.12}$$

and the terms other than W_e' show the added effect of measurement error.

Example 5.13

For a simple system with

$$H(s) = \frac{5}{s+2}$$

and

$$y_d = u$$

and

$$u(t)\ \varepsilon^{-at}u_1(t). \qquad a \neq 2,\ a > 0,$$

with

$$\xi = \psi = 0,$$

find J of Eq. 5.7. Using Eq. 5.11,

$$W_e'(s) = \frac{1}{-s+a} \cdot \frac{1}{s+a} \left[\frac{5}{-s+2} - 1\right] \left[\frac{5}{s+2} - 1\right]$$

$$= - \frac{1}{(s-a)(s+a)} \left[\frac{(s+3)}{(s-2)}\right] \left[\frac{(s-3)}{(s+2)}\right]$$

and then Using Eq. 5.10,

$$J = \text{Res } W_e'(s)\Big|_{s=-a} + \text{Res } W_e'(s)\Big|_{s=-2}$$

$$= \frac{2a+9}{4a(a+2)} .$$

Note that for $a \to 0$, $J \to \infty$. The case of a=0 means that u(t) = $u_1(t)$ which is not an energy signal. Also, the simple first-order system has a steady-state response to $u_1(t)$ which is not one and it is obvious then that the integral error would be unbounded.

The vector input/output case for continuous-time, deterministic signals is similar. For

$$\underline{E}(s) = \underline{Y}_d(s) - H(s)\underline{U}(s) - H(s)\underline{\xi}(s) - \underline{\psi}(s) ,$$

$$W_{\underline{e}}(s) = \underline{E}(-s)\underline{E}(s)^T$$

as defined in Chapter 2. Now

$$J = \int_{-\infty}^{\infty} \underline{e}^T(t)\underline{e}(t)dt \qquad (5.13)$$

or

$$J = \int_{-\infty}^{\infty} (e_1^2(t) + \ldots + e_p^2(t))dt$$

and using the Integral Square Theorem,

$$J = \frac{1}{2\pi j}\int_{-j\infty}^{j\infty} (E_1(-s)E_1(s) + \ldots + E_p(-s)E_p(s)ds$$

or

$$J = \frac{1}{2\pi j}\int_{-j\infty}^{j\infty} \underline{E}^T(-s)\underline{E}(s)ds .$$

To put this in terms of $W_{\underline{e}}(s)$, we use the trace (see Appendix A), therefore,

$$J = \mathrm{tr}\ \frac{1}{2\pi j}\int_{-j\infty}^{j\infty} W_{\underline{e}}(s)ds . \qquad (5.14)$$

For the special case of $\underline{\xi} = \underline{\psi} = \underline{0}$ and $\underline{y}_d = \underline{u}$, the transfer matrix H is square and

$$W_{\underline{e}}'(s) = [H(-s)-I]\ W_{\underline{u}}(s)\ [H(s)-I]^T .$$

Also, since tr A B = tr B A, A and B square,

$$\text{tr } W'_{\underline{e}}(s) = \text{tr } \{W_{\underline{u}}(s) \; [H(s)-I]^T \; [H(-s)-I]\}$$

and the bracketed matrix on the right is a convenient form for the integrand of Eq. 5.14. Other special cases are developed similarly. The evaluation of J is similar to the single-output case except that p^2 terms in Eq. 5.14 (p is dim $\underline{y}$) are required.

The corresponding cases for discrete-time signals and systems are found similarly. Thus from Figure 5.2, for a

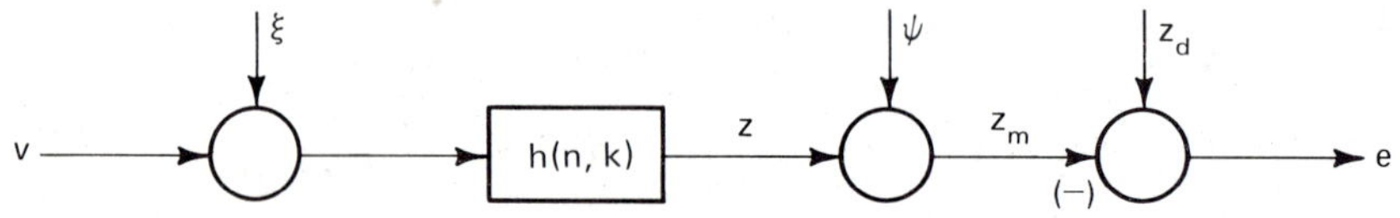

Figure 5.2

realizable and stationary system,

$$z(n) = \sum_{k=-\infty}^{n} h(n-k) \; v_a(k)$$

$$v_a(n) = v(n) + \xi(n)$$

$$z_m(n) = z(n) + \psi(n)$$

and

$$e(n) = z_d(n) - z_m(n) \; . \tag{5.15}$$

For this case a sum of the squared error is used as a measure of performance. Therefore, we use

$$J = \sum_{n=-\infty}^{\infty} e^2(n) \tag{5.16}$$

and using the Summed Square Theorem of Appendix A,

$$J = \frac{1}{2\pi j} \oint E(z^{-1})E(z)z^{-1}dz$$

where $E(z) = \mathcal{Z}\, e(n)$. Transforming,

$$\begin{aligned} Z(z) &= H(z)\, V_a(z) \\ &= H(z)\, V(z) + H(z)\xi(z) \end{aligned}$$

and

$$\begin{aligned} E(z) &= Z_d(z) - Z_m(z) \\ &= Z_d(z) - H(z)V(z) - H(z)\xi(z) - \psi(z). \end{aligned}$$

Using the discrete-time translation functions of Chapter 2,

$$E(z^{-1})E(z) = W_e(z) = \mathcal{Z}\, w_e(i)$$

and

$$J = \frac{1}{2\pi j} \oint W_e(z)z^{-1}dz. \tag{5.17}$$

Also,

$$\begin{aligned} W_e(z) = {} & [W_d(z) + W_\psi(z) - W_{d\psi}(z) - W_{d\psi}(z^{-1})] \\ & + H(z^{-1})H(z)[W_v(z)+W_\xi(z) + W_{v\xi}(z) + W_{v\xi}(z^{-1})] \\ & - [H(z^{-1})W_{dv}(z^{-1}) + H(z)\, W_{dv}(z)] \\ & - [H(z^{-1})W_{d\xi}(z^{-1}) + H(z)W_{d\xi}(z)] \\ & + [H(z^{-1})W_{\psi v}(z^{-1}) + H(z)W_{\psi v}(z)] \\ & + [H(z^{-1})W_{\psi\xi}(z^{-1}) + H(z)W_{\psi\xi}(z)] \end{aligned} \tag{5.18}$$

where

$$W_d(z) = \mathfrak{z}\; w_d(i) = \mathfrak{z} \sum_{n=-\infty}^{\infty} z_d(n) z_d(n+i) .$$

The property $W_{pq}(z) = W_{qp}(z^{-1})$ has been used in Eq. 5.18 and the six bracketed terms on the right and $W_e(z)$ are "even" in the sense of pole/zero patterns inside and outside the unit circle of the z-plane.

As in the continuous-time case, there are several ways of evaluating J. Thus, calculation of e(n) as $\mathfrak{z}^{-1}E(z)$ and use of Eq. 5.16 is possible. If there are no poles on the unit circle, substitution of $\gamma = \varepsilon^{j\theta}$ for z gives

$$J = \frac{1}{2\pi} \int_{-\pi}^{\pi} W_e(\varepsilon^{j\theta}) d\theta ,$$

a real integral. Also, Eq. 5.17 can be evaluated using the calculus of residues where the closed contour is the unit circle Γ. Then

$$J = \sum \text{Res}\; W_e(z) z^{-1} \Big|_{\text{poles inside } \Gamma} \qquad (5.19)$$

As before, the singularities of $W_e(z)z^{-1}$ are poles and the system is assumed asymptotically stable and the signals are energy signals. Also, tables in [7.] can be used with the tables in [11.] to evaluate J of Eq. 5.17. As before, special cases can be considered. Thus, for example, for

$$\xi = \psi = 0 \quad \text{and} \quad z_d = v,$$

we have

$$W_e'(z) = W_v(z)\; [H(z^{-1})-1][H(z)-1] , \qquad (5.20)$$

a measure of dynamic error.

The vector input/output case for discrete-time deterministic signals also follows. We obtain

$$\underline{E}(z) = \underline{Z}_d(z) - H(z)\underline{V}(z) - H(z)\underline{\xi}(z) - \underline{\psi}(z)$$

$$W_{\underline{e}}(z) = \underline{E}(z^{-1})\underline{E}(z)^T .$$

Now

$$J = \sum_{n=-\infty}^{\infty} \underline{e}^T(n)\underline{e}(n) \tag{5.21}$$

and as in the continuous-time case,

$$J = \frac{1}{2\pi j} \oint_\Gamma \underline{E}^T(z^{-1})\underline{E}(z)z^{-1}dz$$

$$= \text{tr}\ \frac{1}{2\pi j} \oint_\Gamma W_{\underline{e}}(z)z^{-1}dz . \tag{5.22}$$

The special case $\underline{\xi} = \underline{\psi} = \underline{0}$ and $\underline{z}_d = \underline{v}$ has

$$W'_{\underline{e}}(z) = [H(z^{-1})-I]W_{\underline{v}}(z)[H(z)-I]^T$$

and

$$\text{tr}\ W'_{\underline{e}}(z) = \text{tr}\ \{W_{\underline{v}}(z)[H(z)-I]^T\ [H(z^{-1})-I]\}$$

so that the bracketed matrix on the right is a convenient integrand (with the z^{-1} multiplier) for Eq. 5.22.

A set of forms for stochastic signals can be developed too. Again using Figure 5.1, where the signals are zero mean, wide-sense stationary processes,

$$Y(t) = \int_{-\infty}^{t} h(t-\lambda)U_a(\lambda)d\lambda$$

$$U_a(t) = U(t) + \xi(t)$$

$$Y_m(t) = Y(t) + \psi(t)$$

and

$$E(t) = Y_d(t) - Y_m(t). \qquad (5.23)$$

The performance measure is

$$J = E[E(t)^2] \, , \qquad (5.24)$$

the mean square error. If the processes are ergodic

$$J = \lim_{T\to\infty} \frac{1}{2T}\int_{-T}^{T} e^i(t)^2 dt$$

which is similar to Eq. 5.7, the integral square error. However, the J for the stochastic case (mean square error) is "average error power" whereas J for the deterministic case is "error energy" (see Tables 2.7 and 2.8). Now

$$J = E[E(t)^2] = r_E(0) = \frac{1}{2\pi j}\int_{-j\infty}^{j\infty} R_E(s)ds \qquad (5.25)$$

where $R_E(s) = \mathcal{L}\ r_E(\tau)$ is the spectral density function of error. From Eq. 2.86, we have

$$R_E(s) = \lim_{T\to\infty} E\left[\frac{E_T(-s)E_T(s)}{2T}\right],$$

and using Eq. 5.23 to find $E(s) = \mathcal{L}\ E(t)$, the spectral density function is

$$R_E(s) = [R_d(s) + R_\psi(s) - R_{d\psi}(s) - R_{d\psi}(-s)]$$

$$+ H(-s)H(s)[R_U(s) + R_\xi(s) + R_{U\xi}(s) + R_{U\xi}(-s)]$$

$$- [H(-s)R_{dU}(-s) + H(s)R_{dU}(s)]$$

$$- [H(-s)R_{d\xi}(-s) + H(s)R_{d\xi}(s)]$$

$$+ [H(-s)R_{\psi U}(-s) + H(s)R_{\psi U}(s)]$$

$$+ [H(-s)R_{\psi\xi}(-s) + H(s)R_{\psi\xi}(s)] \quad (5.26)$$

where similar definitions for all the spectral density functions are used. Eq. 5.26 is analogous to Eq. 5.9. J of Eq. 5.25 can be evaluated by using the calculus of residues or tables [11.] as for Eq. 5.8.

Example 5.14

A simple example analogous to Example 5.13 can be done where

$$H(s) = \frac{5}{s+2}$$

and

$$Y_d = U$$

and

$$r_U(\tau) = (\exp -a|\tau|), \quad a \neq 2, \quad a > 0,$$

with

$$\xi = \psi = 0.$$

Find J of Eq. 5.24. Now

$$R_{\acute{E}}(s) = R_U(s) \; [H(-s)-1][H(s)-1]$$

with

$$R_U(s) = \frac{2a}{(-s+a)(s+a)} .$$

Comparing $R'_E(s)$ with $W'_e(s)$ of Example 5.13, we see that $2aW'_e(s) = R'_E(s)$, therefore,

$$J = \frac{2a+9}{2(a+2)} .$$

The corresponding vector input/output result is obtained similarly. Thus we start with

$$\underline{E}(s) = \underline{Y}_d(s) - H(s)\underline{U}(s) - H(s)\underline{\xi}(s) - \underline{\psi}(s),$$

and

$$R_{\underline{E}}(s) = \lim_{T\to\infty} E\left[\frac{\underline{E}_T(-s)\,\underline{E}_T(s)^T}{2T}\right]$$

and let

$$J = E[\underline{E}(t)^T\ \underline{E}(t)] \tag{5.27}$$

or

$$J = \text{tr}\ \frac{1}{2\pi j}\int_{-j\infty}^{j\infty} R_{\underline{E}}(s)ds . \tag{5.28}$$

The discrete-time forms for stochastic signals also can be found similarly. Using Figure 5.2, where the signals are now zero-mean, wide-sense stationary sequences, we have

$$Z(n) = \sum_{k=-\infty}^{n} h(n-k)V_a(k)$$

$$V_a(n) = V(n) + \xi(n)$$

$$Z_m(n) = Z(n) + \psi(n)$$

and

$$E(n) = Z_d(n) - Z_m(n) \ . \tag{5.29}$$

The performance measure is

$$J = E[E(n)^2] \ . \tag{5.30}$$

If the sequences are ergodic, we also have

$$J = \lim_{N\to\infty} \frac{1}{2N+1} \sum_{n=-N}^{N} (e^i(n))^2.$$

and

$$J = E[E(n)^2] = r_E(0) = \frac{1}{2\pi j} \oint_\Gamma R_E(z) z^{-1} dz \tag{5.31}$$

when $R_E(z) = \mathfrak{z}\, r_E(k)$ is the spectral density function of error. From Eq. 2.87,

$$R_E(z) = \lim_{N\to\infty} E\left[\frac{V_N(z^{-1})V_N(z)}{2N+1}\right],$$

and using Eq. 5.29 to find $E(z) = \mathfrak{z}\, E(n)$, the spectral density function is

$$\begin{aligned} R_E(z) = {} & [R_d(z) + R_\psi(z) - R_{d\psi}(z) - R_{d\psi}(z^{-1})] \\ & + H(z^{-1})H(z)[R_V(z) + R_\xi(z) + R_{V\xi}(z) + R_{V\xi}(z^{-1})] \\ & - [H(z^{-1})R_{dV}(z^{-1}) + H(z)R_{dV}(z)] \\ & - [H(z^{-1})R_{d\xi}(z^{-1}) + H(z)R_{d\xi}(z)] \\ & + [H(z^{-1})R_{\psi V}(z^{-1}) + H(z)R_{\psi V}(z)] \\ & + [H(z^{-1})R_{\psi\xi}(z^{-1}) + H(z)R_{\psi\xi}(z)] \end{aligned} \tag{5.32}$$

where similar definitions for all the spectral density functions are used. Eq. 5.32 is analogous to Eq. 5.18. J of Eq. 5.31 can be evaluated by using the calculus of residues or tables [11.][7.] as for Eq. 5.17. The corresponding vector input/output case is similar where

$$\underline{E}(z) = \underline{Z}_d(z) - H(z)\underline{V}(z) - H(z)\underline{\xi}(z) - \underline{\psi}(z),$$

$$R_{\underline{E}}(z) = \lim_{N\to\infty} E\left[\frac{\underline{E}_N(z^{-1})\underline{E}_N(z)^T}{2N+1}\right]$$

and

$$J = E[\underline{E}(n)^T\,\underline{E}(n)] \tag{5.33}$$

or

$$J = \mathrm{tr}\,\frac{1}{2\pi j}\oint_\Gamma R_{\underline{E}}(z)z^{-1}dz\,. \tag{5.34}$$

This completes the set of forms in the first group of error measures specified at the beginning of this section. Table 5.2 summarizes the equation numbers.

5.2.2 Finite Time Intervals

The second group of performance measured involves finite terminal times and the possibility of time-varying linear systems. Figure 5.3 shows the continuous-time model with deterministic signals. From this

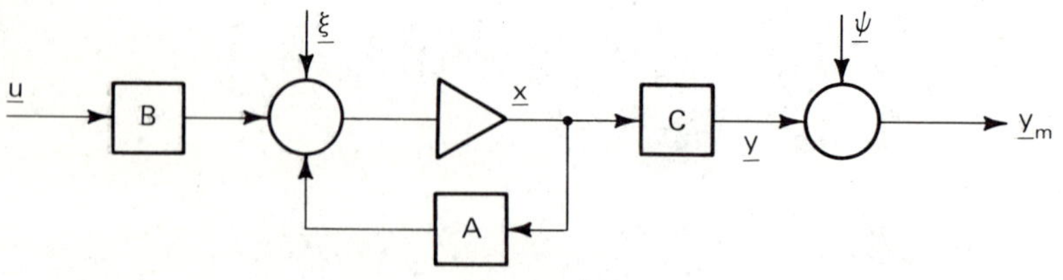

Figure 5.3

TABLE 5.2

PERFORMANCE MEASURES—GROUP 1

		Continuous-Time	Discrete-Time
deterministic	single i/0	Eq. 5.8	Eq. 5.17
	vector i/0	Eq. 5.14	Eq. 5.22
stochastic	single i/0	Eq. 5.25	Eq. 5.31
	vector i/0	Eq. 5.28	Eq. 5.34

$$\dot{\underline{x}}(t) = A(t)\underline{x}(t) + B(t)\underline{u}(t) + \underline{\xi}(t)$$

$$\underline{y}(t) = C(t)\underline{x}(t)$$

$$\underline{y}_m(t) = \underline{y}(t) + \underline{\psi}(t)$$

$$\underline{e}(t) = \underline{y}_d(t) - \underline{y}_m(t) \ , \qquad (5.35)$$

and using Eq. 4.30a and 4.30b, we have

$$\underline{y}(t) = C(t)\Phi(t,t_o)\underline{x}(t_o) +$$

$$+\int_{t}^{t_o} C(t)\Phi(t,\lambda)[B(\lambda)\underline{u}(\lambda)+\underline{\xi}(\lambda)]d\lambda \ . \qquad (5.36)$$

In analogy to the correlation matrices of Chapter 2, for any two deterministic signals $\underline{f}(t)$ and $\underline{g}(t)$, we define

$$M_{\underline{fg}}(\alpha,\lambda) = \underline{f}(\alpha)\underline{g}^T(\lambda) \qquad (5.37)$$

and assume that all of these matrices are known on a time interval $[t_o,t_1]$ for $\underline{u}$, $\underline{\xi}$, $\underline{\psi}$ as well as their cross products. As a measure of performance, we let

$$J = \int_{t_o}^{t_1} \underline{e}^T(t)Q\underline{e}(t)dt + \underline{e}^T(t_1)Q_1\underline{e}(t_1) \qquad (5.38)$$

where Q and Q_1 are symmetric positive semidefinite weighting matrices. This choice of J includes a measure of accuracy over the basic time interval of interest as well as a "point measure" at t_1. It is a quadratic form and special cases are included where either Q or Q_1 is a null matrix. To evaluate J, we start with Eq. 5.35 and Eq. 5.36. Then we obtain

$$\underline{e}(t) = \underline{y}_d(t) - \underline{\psi}(t) - C(t)\Phi(t,t_o)\underline{x}(t_o) -$$

$$- \int_{t_o}^{t_1} C(t)\Phi(t,\lambda)[B(\lambda)\underline{u}(\lambda) + \underline{\xi}(\lambda)]d\lambda] \ . \qquad (5.39)$$

Substitution of Eq. 5.39 into Eq. 5.35 gives the complete expression for J, but this result, which involves thirty-two terms, is too complicated to be of interest here. Instead of developing this, a special case with the following conditions is of interest:

$$\underline{\xi} = \underline{\psi} = \underline{0}$$

$$C(t) = I$$

$$\underline{y}_d = \underline{0}$$

This is sometimes called a deterministic, noise-free regulator problem since for $\underline{x}(t_o) \neq \underline{0}$, an input $\underline{u}(t)$, $t_o \le t \le t_1$, may be chosen to bring $\underline{x}(t_1)$ close to $\underline{0}$. J is a measure of how close $\underline{x}(t)$ is to zero for $t_o \le t \le t_1$. For this special case, the performance measure Eq. 5.38 is

$$
\begin{aligned}
J = \int_{t_o}^{t_1} \{ & \underline{x}^T(t_o)\Phi^T(t,t_o)Q\int_{t_o}^{t} \Phi(t,\lambda)B(\lambda)\underline{u}(\lambda)d\lambda \\
& + \int_{t_o}^{t} \underline{u}^T(\lambda)B^T(\lambda)\Phi^T(t,\lambda)d\lambda Q\Phi(t,t_o)\underline{x}(t_o) \\
& + \int_{t_o}^{t} \underline{u}^T(\lambda)B^T(\lambda)\Phi^T(t,\lambda)d\lambda Q\int_{t_o}^{t} \Phi(t,\alpha)B(\alpha)\underline{u}(\alpha)d\alpha \\
& + \underline{x}^T(t_o)\Phi^T(t,t_o)Q\Phi(t,t_o)\underline{x}(t_o)\} \, dt \\
& + \underline{x}^T(t_o)\Phi^T(t_1,t_o)Q_1\int_{t_o}^{t_1} \Phi(t_1,\lambda)B(\lambda)\underline{u}(\lambda)d\lambda \\
& + \int_{t_o}^{t_1} \underline{u}^T(\lambda)B^T(\lambda)\Phi^T(t_1,\lambda)d\lambda \; Q_1\Phi(t_1,t_o)\underline{x}(t_o) \\
& + \int_{t_o}^{t_1} \underline{u}^T(\lambda)B^T(\lambda)\Phi^T(t_1,\lambda)d\lambda Q_1 \int \Phi(t_1,\lambda)B(\lambda)\underline{u}(\lambda)d\lambda \\
& + \underline{x}^T(t_o)\Phi^T(t_1,t_o)Q_1 \, \Phi(t_1,t_o)\underline{x}(t_o). \qquad (5.40)
\end{aligned}
$$

First consider the third term under the first integral; this is

$$
\int_{t_o}^{t_1} \{ \int_{t_o}^{t}\int_{t_o}^{t} \underline{u}^T(\lambda)B^T(\lambda)\Phi^T(t,\lambda)Q\Phi(t,\alpha)B(\alpha)\underline{u}(\alpha)d\alpha d\lambda\} \, dt
$$

and using the identity $\underline{a}^T A\underline{a} = \mathrm{tr}[(\underline{a}\,\underline{a}^T)A]$, this term is

$$\operatorname{tr}\int_{t_o}^{t_1}\left\{\int_{t_o}^{t}\int_{t_o}^{t} \underline{u}(\alpha)\underline{u}^T(\lambda)B^T(\lambda)\Phi^T(t,\lambda)\cdot \cdot Q\Phi(t,\alpha)B(\alpha)\,d\alpha\, d\lambda\right\} dt \ .$$

Using tr AB = tr BA, the term is

$$\operatorname{tr}\int_{t_o}^{t_1}\left\{\int_{t_o}^{t}\left[\int_{t_o}^{t} B(\alpha)\underline{u}(\alpha)\underline{u}^T(\lambda)B^T(\lambda)\Phi^T(t,\lambda)\cdot \cdot Q\Phi(t,\alpha)\,d\alpha\right] d\lambda\right\} dt \ .$$

Interchanging the order of integration on λ and t first and then on α and t leads to

$$\operatorname{tr}\int_{t_o}^{t_1}\int_{t_o}^{t_1} B(\alpha)M_{\underline{u}}(\alpha,\lambda)B^T(\lambda)\left\{\int_{\text{Max }\lambda,\alpha}^{t_1} \Phi^T(t,\lambda)Q\Phi(t,\alpha)\,dt\right\} d\alpha\, d\lambda .$$

Second, consider the next to last term in 5.40; this is similar to the above without an integration with respect to t. Therefore, that term is

$$\operatorname{tr}\int_{t_o}^{t_1}\int_{t_o}^{t_1} B(\alpha)M_{\underline{u}}(\alpha,\lambda)B^T(\lambda)\Phi^T(t_1,\lambda)Q_1\cdot\Phi(t_1,\alpha)\,d\alpha\, d\lambda \ .$$

Now we let

$$F(\alpha,\lambda) = \int_{\text{Max }\alpha,\lambda}^{t_1} \Phi^T(t,\lambda)Q\Phi(t,\alpha)\,dt + \Phi^T(t_1,\lambda)Q_1\Phi(t_1,\alpha) \ .$$

Then the sum of these two terms can be written

$$\operatorname{tr}\int_{t_o}^{t_1}\int_{t_o}^{t_1} B(\alpha)M_{\underline{u}}(\alpha,\lambda)B^T(\lambda)F(\alpha,\lambda)\,d\alpha\, d\lambda \quad .$$

Similar manipulations of the other terms leads to the result

$$J = \mathrm{tr}\left\{ M_{\underline{x}_o} F(t_o,t_o) + \int_{t_o}^{t_1}\int_{t_o}^{t_1} B(\alpha)M_{\underline{u}}(\alpha,\lambda)B^T(\lambda)\cdot \cdot F(\alpha,\lambda)\,d\alpha\, d\lambda \right\}$$

$$+ 2\mathrm{tr}\left\{\int_{t_o}^{t_1} M_{\underline{x}_o\underline{u}}(t_o,\lambda)B^T(\lambda)F(t_o,\lambda)\,d\lambda\right\}. \quad (5.41)$$

This rather cumbersome result for the relatively simple special case would be difficult to evaluate for a time-varying system. A stochastic counterpart to this problem has vector stochastic inputs $\underline{U}$, $\underline{\xi}$, and $\underline{\psi}$ in Figure 5.3. These are assumed to have zero means. Also, the initial state is a random variable $\underline{X}_o$ with second moment $E[\underline{X}_o\ \underline{X}_o^T]=R_o$. The measure of performance is

$$J = E\left\{\int_{t_o}^{t_1} \underline{E}(t)^T Q\underline{E}(t)\,dt + \underline{E}^T(t_1)Q_1\underline{E}(t_1)\right\} \quad (5.42)$$

where

$$E(t) = \underline{Y}_d(t) - \underline{Y}(t) - \underline{\psi}(t)$$

and $\underline{Y}(t)$ is given as in Eq. 5.36 with $\underline{X}_o$, $\underline{U}(\lambda)$ and $\underline{\xi}(\lambda)$ substituted. Again, consider the special case:

$$\underline{\xi} = \underline{\psi} = \underline{0}$$

$$C(t) = I$$

$$\underline{Y}_d(t) = \underline{0}\ .$$

The manipulations for this case are quite similar to those above except that the expectation operation is required. The result is that the matrices R_o, $R_{\underline{U}}$, and the crosscorrelation $R_{o\underline{U}}$ between $\underline{X}_o$ and $\underline{U}$ replace the M-matrices in Eq. 5.41. Now $\overline{R}_{o\underline{U}} = E[\underline{X}_o \, \underline{U}^T] = \Theta$ since $\underline{U}$ is zero mean and $\underline{X}_o$ is independent of $\underline{U}$. Thus the result is

$$J = \mathrm{tr}\left\{R_o F(t_o, t_o) + \int_{t_o}^{t_1}\int_{t_o}^{t_1} B(\alpha) R_{\underline{U}}(\alpha,\lambda) B^T(\lambda) F(\alpha,\lambda)\, d\alpha\, d\lambda\right\} \quad (5.43)$$

A further specialization which is of interest is the case where $\{\underline{U}\}$ is a white, wide-sense stationary process. Then

$$R_{\underline{U}}(\alpha,\lambda) = M_1 u_o(\lambda-\alpha)$$

where M_1 is a constant intensity matrix. If the definition $F(\lambda,\lambda) = P(\lambda)$ is used, Eq. 5.43 becomes

$$J = \mathrm{tr}\left\{R_o P(t_o) + \int_{t_o}^{t_1} B(\lambda) M_1 B^T(\lambda) P(\lambda)\, d\lambda\right\} \quad (5.44)$$

where

$$P(\lambda) = \int_{\lambda}^{t_1} \Phi^T(t,\lambda) Q\, \Phi(t,\lambda)\, dt + \Phi^T(t_1,\lambda) Q_1 \Phi(t_1,\lambda).$$

Differentiating P with respect to λ and using properties of the transition matrix $\Phi(t,\lambda)$, the differential equation that P satisfies is

$$-\frac{dP}{d\lambda} = A^T(\lambda)P(\lambda) + P(\lambda)A(\lambda) + Q$$

with

$$P(t_1) = Q_1.$$

This form for finding P usually is easier to use when a numerical approximation is required.

The discrete-time counterparts to these results are found similarly. Figure 5.4 shows the model.

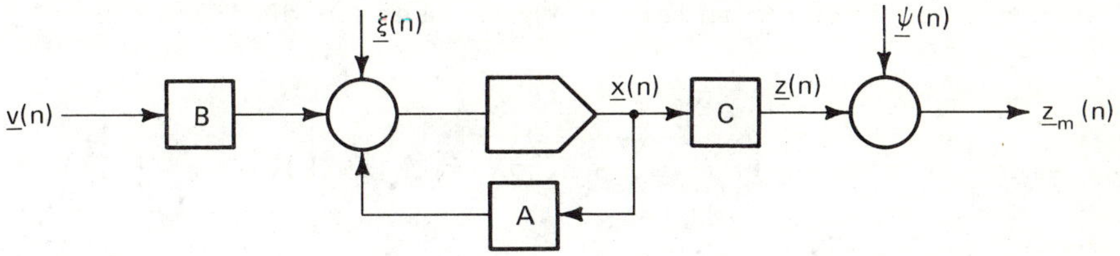

Figure 5.4

$$\underline{x}(n+1) = A(n)\underline{x}(n) + B(n)\underline{v}(n) + \underline{\xi}(n)$$

$$\underline{z}(n) = C(n)\underline{x}(n)$$

$$\underline{z}_m(n) = \underline{z}(n) + \underline{\psi}(n)$$

$$e(n) = \underline{z}_d(n) - \underline{z}_m(n) \tag{5.45}$$

and using Eq. 4.32a and Eq. 4.32b,

$$\underline{z}(n) = C(n)\Phi(n,n_o)\underline{x}(n_o) + \sum_{k=n_o}^{n-1} C(n)\Phi(n,k+1)[B(k)\underline{v}(k)+\underline{\xi}(k)]. \tag{5.46}$$

The deterministic signal models are in terms of $M_{\underline{fg}}(j,k) = \underline{f}(j)\underline{g}^T(k)$. A performance measure is

$$J = \sum_{n=n_o}^{n_1} \underline{e}^T(n)Q\underline{e}(n) + \underline{e}^T(n_1)Q_1\underline{e}(n_1). \tag{5.47}$$

The general case again includes thirty-two terms. For the special case:

$$\underline{\xi} = \underline{\psi} = \underline{0}$$

$$C(n) = I$$

$$\underline{z}_d = \underline{0}$$

a sequence of steps similar to those used in deriving Eq. 5.41 gives

$$J = \mathrm{tr}\left\{M_{\underline{x}_o} F(n_o,n_o)+\sum_{k,j=n_o}^{n_1-1}\sum B(j)M_{\underline{V}}(j,k)B^T(k)F(j+1,k+1)\right\}$$

$$+\ 2\mathrm{tr}\left\{\sum_{k=n_o}^{n_1-1} M_{\underline{x}_o\underline{V}}(n_o,k)B^T(k)F(n_o,k+1)\right\} \qquad (5.48)$$

where

$$F(j+1,k+1) = \sum_{n=\mathrm{Max}\ j+1,k+1}^{n_1} \Phi^T(n,k+1)Q\Phi(n,j+1)$$

$$+\ \Phi^T(n_1,k+1)Q_1\Phi(n_1,j+1).$$

The stochastic counterpart can be done where $\underline{V}$, $\underline{\xi}$, and $\underline{\psi}$ in Figure 5.4 are zero mean stochastic sequences and $\underline{X}_o$ is the random initial state. A performance measure is

$$J = E\left\{\sum_{n=n_o}^{n_1} \underline{E}^T(n)Q\underline{E}(n) + E^T(n_1)Q_1\underline{E}(n_1)\right\}$$

where $\underline{E}(n) = \underline{Z}_d(n) - \underline{Z}(n) - \underline{\psi}(n)$ which can be found by using Eq. 5.46. For the special case

$$\underline{\xi} = \underline{\psi} = \underline{0}$$

$$C(n) = I$$

$$\underline{Z}_d(n) = \underline{0},$$

a result similar to Eq. 5.43 is

$$J = \mathrm{tr}\left\{R_oF(n_o,n_o)+ \sum_{k,j=n_o}^{n_1-1}\sum B(j)R_{\underline{V}}(j,k)B^T(k)F(j+1,k+1)\right\} \qquad (5.49)$$

If $\{V\}$ is a wide-sense stationary white sequence, $R_{\underline{V}}(j,k) = M_1 v_0(k-j)$, and

$$J = \text{tr}\left\{R_0 P(n_0) + \sum_{k=n_0}^{n_1-1} B(k) M_1 B^T(k) P(k+1)\right\} \tag{5.50}$$

where $F(k+1,k+1) = P(k+1)$ and

$$P(k) = \sum_{n=k}^{n_1} \Phi^T(n,k) Q \Phi(n,k) + \Phi^T(n_1,k) Q_1 \Phi(n_1,k) .$$

It also can be shown that P satisfies the difference equation

$$P(k) = Q + A^T(k) P(k+1) A(k)$$

with

$$P(n_1) = Q + Q_1 .$$

In summary, the four vector input/vector output cases have been set up; explicit forms for special cases are given by Eq. 5.41, 5.43, 5.48, and 5.38. Two other special forms are given for white stochastic inputs by Eq. 5.44 and by 5.50. Obviously, the evaluation of performance measures for general time-varying, finite interval cases is difficult. However, measures of performance such as those discussed in this section are very useful in both system evaluation and system design.

5.3 SENSITIVITY

System sensitivity analysis is concerned with the effects on performance of deviations of the system parameters from their design values. The deviations may result from environmental influences such as temperature, humidity, or vibration, from aging, from inaccuracies in measurement, or from manufacturing tolerances. Previous discussions in Chapters 2 and 4 have indicated the difficulties in knowing exact signal and system parameters; the idealized signal and system models in Chapters 2 and 3 have always included parameters which were known exactly. In general, the deviations of the parameters may be described deterministically or probabilistically; only the former is considered in this introductory discussion. The latter case frequently is called the study of random parameter systems, and such systems have been investigated rather extensively -- particularly with regard to stability.

One early example of a problem resulting from a parameter variation was the effect of the variation of the amplification factor, μ, of a triode on amplifier gain. Investigation of this led to Black's Patent [12.] which describes the use of feedback to reduce the effect. Bode [13.] further developed feedback principles and Bykhovskiy [14.] published results concerning feedback control systems. Much of the research of the last three decades has been concerned with the interaction of the concept of sensitivity and the design and analysis of control systems [15., Part X]. The goal here is much more limited; it is to show some of the basic ideas of a sensitivity analysis of a linear system.

5.3.1 Definitions

Several definitions of sensitivity have been used, and one of the early ones is in terms of the change in a system transfer function relative to a change in the parameter. For example, for the feedback configuration in Figure 5.5,

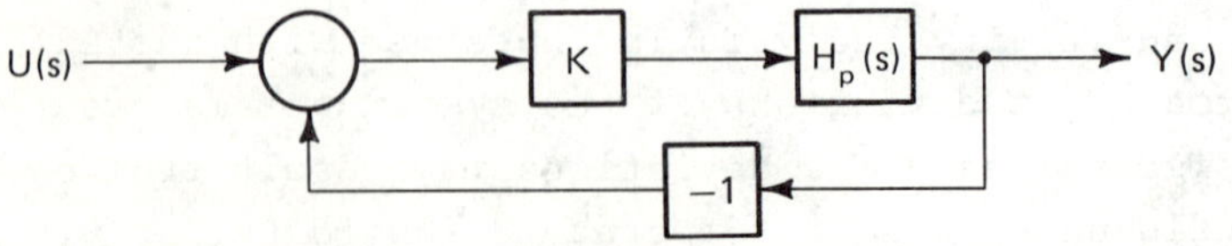

Figure 5.5

the transfer function is

$$\frac{Y(s)}{U(s)} = H(s) = \frac{KH_p(s)}{1 + KH_p(s)} \; .$$

A sensitivity function to show the effect of small changes in the gain K on H(s) is the normalized form

$$S_K^H = \frac{K}{H(s)} \frac{\partial H(s)}{\partial K} = \frac{1}{1 + KH_p(s)} \; .$$

This function of frequency can be used as a measure of the effects of small changes in K on H(s) and the designer would hope to make $|S_K^H(j\omega)|$ small over the frequency range where $|H(j\omega)|$ is large. A modification to this definition called comparison sensitivity [15., Part V], allows large deviations in the parameters, and is useful in comparing open and closed-loop systems. More recently [16.], a distinction has been between performance sensitivity and trajectory sensitivity. Performance sensitivity concerns the effect of changes in parameters on a performance functional J such as those in Section 5.2. Trajectory sensitivity shows the effect of parameter changes on a trajectory $\underline{x}(t)$ or $\underline{x}(n)$. Also, eigenvalue sensitivity [15., Part III] considers the effect of parameter changes on the characteristic values of a system. This is of importance in analog or digital realizations (e.g., filters) and in obtaining accurate simulations (see, for example Chapter 4, Section 4.5). Trajectory sensitivity is considered in the following development.

5.3.2 Trajectory Sensitivity

A stationary, single-input, continuous-time system model with m parameters is

$$\dot{\underline{x}} = A(\underline{p})\underline{x} + \underline{b}(\underline{p})u$$

$$\underline{y} = C(\underline{p})\underline{x}$$

with the parameter vector $\underline{p}^T = (p_1 \; \dots \; p_m)$. It is convenient to transform this model to first primal form (Eq. 3.68)) by using Equation 3.113, and the transformed model is

$$\dot{\underline{q}} = A_1(\underline{\gamma})\underline{q} + \underline{b}_1 u \qquad (5.51)$$

$$\underline{y} = C_1(\underline{\gamma})\underline{q}$$

where

$$\underline{x} = T\underline{q}, \quad \underline{b}_1^T = (0 \quad \dots \quad 0 \quad 1), \quad C_1 = CT ,$$

$$A_1 = T^{-1}AT = \begin{bmatrix} 0 & 1 & 0 & \dots & & 0 \\ 0 & 0 & 1 & 0 & \dots & 0 \\ \cdot & & & & & \\ \cdot & & & & & \\ \cdot & & & & & \\ -\gamma_1 & & & \dots & & -\gamma_r \end{bmatrix} ,$$

and $\underline{\gamma}^T = [\gamma_1(\underline{p}) \; \dots \; \gamma_1(\underline{p})]$. T also is a function of $\underline{p}$. Note that the numbering of the γ_i, is changed from the convention on the α_i used in Equation 3.68, i.e., $\gamma_i = \alpha_{i-1}$, $i=1, \dots, r$. An rxr sensitivity matrix $\sum$ is defined with elements

$$\sigma_{ij}(t) = \frac{\partial q_i(t)}{\partial \gamma_j} , \quad i, j = 1,2,\dots,r . \qquad (5.52)$$

The jth column of $\sum$ is the sensitivity function for $\underline{q}$ with respect to γ_j. The matrix $\sum$ has the important property that

$$\sigma_{ij} = \sigma_{i+1,j-1}, \quad i = 1, \dots, r-1 \qquad (5.53)$$

$$j = 2, \dots, r \quad .$$

This is the total symmetry property [17.]; the elements along all "antidiagonals" are equal. Thus there are 2r-1 sensitivity functions for the state $\underline{q}$ with respect to $\underline{\gamma}$. A second important property is the complete simultaneity property [17.], and this can be shown from the condition

$$\sigma_{ij} = D^{i+j-2}\sigma_{11} \qquad i,j=1, 2, \ldots, r \tag{5.54}$$

where $D \equiv \frac{d}{dt}$. This implies that all 2r-1 of the σ_{ij} can be obtained from the model 5.51 plus a second rth-order model of the same form. To obtain these results, we start with Eq. 5.51 in rth-order form

$$M(D)q_1 = u$$

where $M(D) = D^r + \gamma_r D^{r-1} + \ldots + \gamma_1$. Since

$$q_{i+1} = Dq_i \ , \qquad i = 2, \ldots, r,$$

we have

$$M(D)\,\frac{1}{D^{i-1}}\,q_i = u, \qquad i = 1,2, \ldots, r$$

and

$$q_i = D^{i-1}\left[\frac{1}{M(D)}\,u\right]. \tag{5.55}$$

Now using this with i=1 we have

$$\sigma_{11} = \frac{\partial q_1}{\partial \gamma_1} = \frac{\partial}{\partial \gamma_1}\left[\frac{1}{M(D)}\,u\right]$$

and differentiating, we get

$$\sigma_{11} = -\,\frac{1}{M(D)}\,q_1 \quad .$$

Also,

$$\sigma_{12} = \frac{\partial q_1}{\partial \gamma_2} = \frac{\partial}{\partial \gamma_2}\left[\frac{1}{M(D)}\ u\right]$$

$$= -\frac{D}{M(D)}\, q_1 = D\sigma_{11}$$

and using Eq. 5.55 with i=2, we obtain

$$\sigma_{21} = \frac{\partial q_2}{\partial \gamma_1} = \frac{\partial}{\partial \gamma_1}\left[\frac{D}{M(D)}\ u\right] = -\frac{D}{M(D)}\ q_1$$

$$= D\sigma_{11} = \sigma_{12}\ .$$

Thus Eq. 5.53 and 5.54 are verified for the first step. This procedure is continued by starting with σ_{13}, $\sigma_{14}, \ldots,$ σ_{1r}. The final set of equalities is

$$\sigma_{1r} = \frac{\partial q_1}{\partial \gamma_r} = D^{r-1}\sigma_{11}$$

$$\sigma_{2,r-1} = \frac{\partial q_2}{\partial \gamma_{r-1}} = D^{r-1}\sigma_{11}$$

$$\vdots$$

$$\sigma_{r1} = \frac{\partial q_r}{\partial \gamma_1} = D^{r-1}\sigma_{11}\ .$$

This completes the "upper left triangle" of $\sum$. Then, continuing down the last column,

$$\sigma_{2r} = \frac{\partial q_2}{\partial \gamma_r} = \frac{\partial}{\partial \gamma_r}\left[\frac{D}{M(D)}\ u\right] = D^r\sigma_{11}$$

$$\vdots \qquad\qquad \vdots$$

$$\sigma_{r2} = \frac{\partial q_r}{\partial \gamma_2} = \frac{\partial}{\partial \gamma_2}\left[\frac{D^{r-1}}{M(D)}\ u\right] = D^r\sigma_{11}$$

and so forth until we reach

$$\sigma_{rr} = \frac{\partial q_r}{\partial \gamma_r} = \frac{\partial}{\partial \gamma_r} \left[\frac{D^{r-1}}{M(D)} \quad u \right] = D^{2r-2}\sigma_{11} .$$

Thus Equations 5.53 and 5.54 are verified and

$$\sum = \begin{bmatrix} \sigma_{11} & \sigma_{12} & \sigma_{13} & \cdots & \sigma_{1r} \\ \sigma_{12} & \sigma_{13} & & & \sigma_{2r} \\ \sigma_{13} & & & \cdot & \\ \vdots & & \cdot & & \vdots \\ & \cdot & & & \\ \sigma_{1r} & \sigma_{2r} & & \cdots & \sigma_{rr} \end{bmatrix} . \qquad (5.56)$$

Now we consider the elements of the first row of $\sum$. Using Eq. 5.54,

$$D\sigma_{1j} = \dot{\sigma}_{1j} = D^j\sigma_{11} = \sigma_{1,j+1} ,$$

for j = 1,2,...,r-1, and since

$$\sigma_{1r} = \frac{\partial q_1}{\partial \gamma_r} = \frac{\partial q_r}{\partial \gamma_1} ,$$

we have for j=r,

$$D\sigma_{1r} = \dot{\sigma}_{1r} = D\,\frac{\partial q_r}{\partial \gamma_1} = \frac{\partial}{\partial \gamma_1}\,(\dot{q}_r) .$$

Using 5.51 for $\dot{q}_r$ and differentiating, we get for j=r,

$$\dot{\sigma}_{1r} = -q_1 - \sum_{i=1}^{r} \gamma_i\sigma_{1i} .$$

Using these r expressions for $\dot{\sigma}_{1j}$, j = 1,...,r a simulation model is shown in Figure 5.6. In view of 5.53 this includes all elements of the "upper left triangle" of $\sum$.

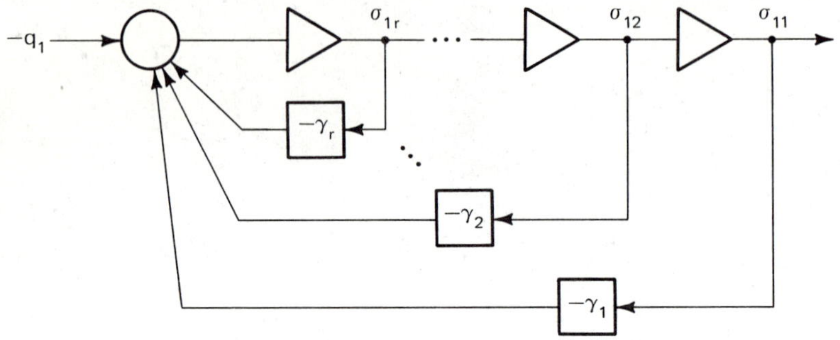

Figure 5.6 Partial sensitivity model

For i+j-2 $\geq$ r, the elements σ_{ij} are not on this diagram, and a further step is necessary. Starting with Eq. 5.54, we let i+j-2 = r+k, then we have

$$D^{r+k}\sigma_{11} = D^k D^r \sigma_{11}$$

$$= D^k D\sigma_{1r}$$

$$= D^k \dot{\sigma}_{1r}$$

or using $\dot{\sigma}_{1r}$ from above,

$$D^{r+k}\sigma_{11} = -D^k q_1 - \sum_{i=1}^{r} \gamma_i D^k \sigma_{1i}$$

and since r-1 elements of $\sum$ still are required, this is used for $0 \leq k \leq r-2$. Then for k=i+j-2-r as above, we get

$$\sigma_{ij} = -D^k q_1 - \sum_{i=1}^{r} \gamma_i D^k \sigma_{1i} \ .$$

To use this to find the elements of the last row of $\sum$ (except σ_{1r}), we let j=r, then i=k+2, and using $D^{i-1}\sigma_{11}=\sigma_{1i}$, we find that

$$\sigma_{k+2,r} = -D^k_{q1} - \sum_{i=1}^{r} \gamma_i D^{k+i-1}\sigma_{11} . \qquad (5.57)$$

Then starting with k=0,

$$\sigma_{2r} = -q_1 - \sum_{i=1}^{r} \gamma_i D^{i-1}\sigma_{11}$$

$$= -q_1 - \sum_{i=1}^{r} \gamma_i \sigma_{1i} = \dot{\sigma}_{1r} .$$

Therefore σ_{2r} is obtained from a linear combination of q_1 and σ_{1i} of Figure 5.6. For k=1,

$$\sigma_{3r} = - Dq_1 - \sum_{i=1}^{r} \gamma_i D^i \sigma_{11}$$

or

$$\sigma_{3r} = -q_2 - \sum_{i=1}^{r-1} \gamma_i \sigma_{1,i+1} - \gamma_r D^r \sigma_{11} .$$

The last term on the right is $\gamma_r \dot{\sigma}_{1r} = \gamma_r \sigma_{2r}$, therefore

$$\sigma_{3r} = -q_2 - \sum_{i=1}^{r-1} \gamma_i \sigma_{1,i+1} - \gamma_r \sigma_{2r},$$

and σ_{3r} in this form is found from q_2, the $\sigma_{1,i+1}$ of Figure 5.6, and σ_{2r} found in the preceding step. This process is repeated to give in general,

$$\sigma_{k+2,r} = -q_{k+1} - \sum_{i=1}^{r-k} \gamma_i \sigma_{1,i+k} - \sum_{i=r-k+1}^{r} \gamma_i \sigma_{i+k+1-r,r} \qquad (5.58)$$

which is true for $1 \le k \le r-2$, and σ_{2r} for k=0 is given above. This result verifies the complete simultaneity property, i.e., all the elements of $\sum$ can be obtained from combinations of signals in the original system model (Eq. 5.51) and those in a similar sensitivity model. Figure 5.7 shows a complete simulation diagram where r=3.

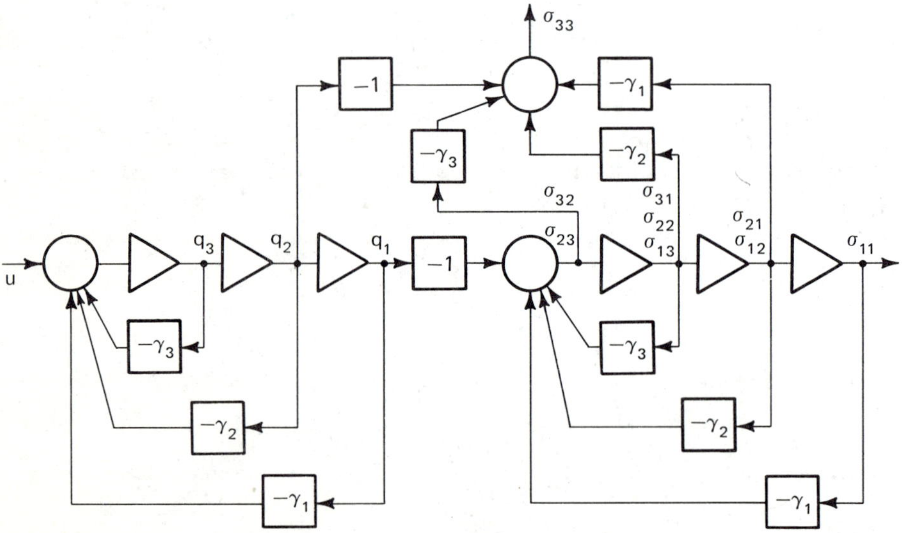

Figure 5.7 Complete sensitivity model r = 3

Finally, it is necessary to use $\sum$ to find the trajectory sensitivity functions of the system $(A,\underline{b},C)$. Starting with

$$\frac{\partial \underline{x}}{\partial p_i} = \frac{\partial}{\partial p_i} [T(\underline{p})\underline{q}(\underline{\gamma}(\underline{p}))],$$

$$= \frac{\partial T}{\partial p_i} \underline{q} + T \frac{\partial \underline{q}}{\partial \underline{\gamma}} \frac{\partial \underline{\gamma}}{\partial p_i} , \quad i = 1,2,\ldots,m .$$

and $\frac{\partial \underline{q}}{\partial \underline{\gamma}} = \sum = J_{\underline{\gamma}} \underline{q}$ ($J_{\underline{\gamma}}$ is the Jacobian of Appendix B).

Also,

$$\frac{\partial \underline{y}}{\partial p_i} = \frac{\partial}{\partial p_i} [C(\underline{p})\underline{x}] = \frac{\partial C}{\partial p_i} \underline{x} + C \frac{\partial \underline{x}}{\partial p_i} ,$$

and a schematic implementation of these two equations is shown in Figure 5.8 where $\underline{\sigma}_1$ is the first column of $\sum$ and $\underline{f}^T = [1 \ 0 \ \ldots \ 0]$. Some of the computational aspects of using these equations are discussed in [18.] where the Leverrier algorithon of Chapter 3, Section 3.4 is used to find

$$T, \ \gamma_i, \ \frac{\partial T}{\partial p_i}, \text{ and } \frac{\partial \underline{\gamma}}{\partial p_i} .$$

$$\underline{f}^T = [1 \ 0 \ \ldots \ 0]$$

$$\underline{\sigma}_1 = \text{1st col.} \sum$$

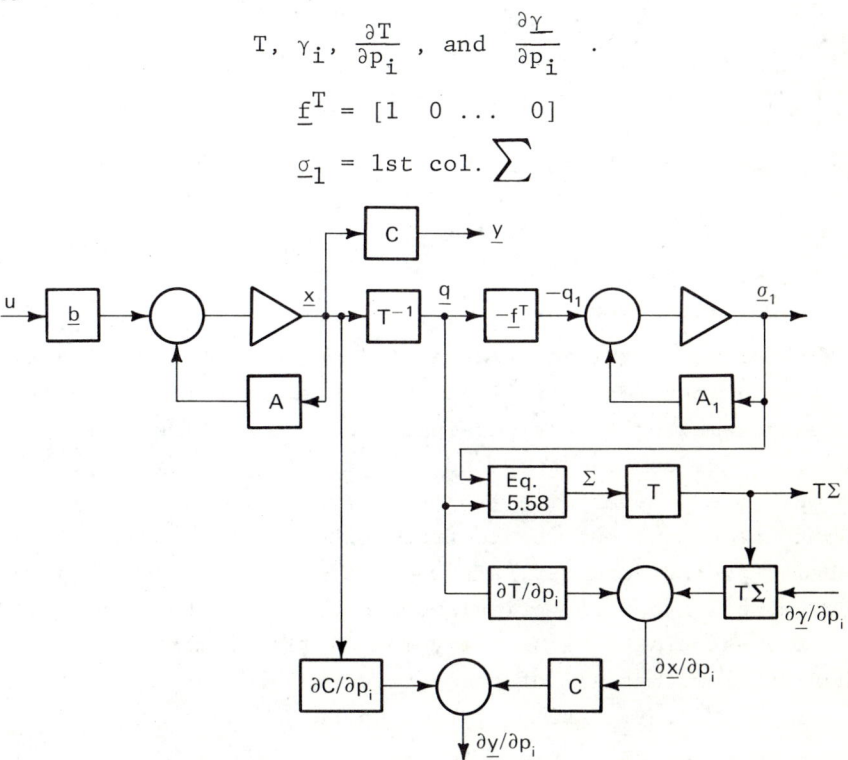

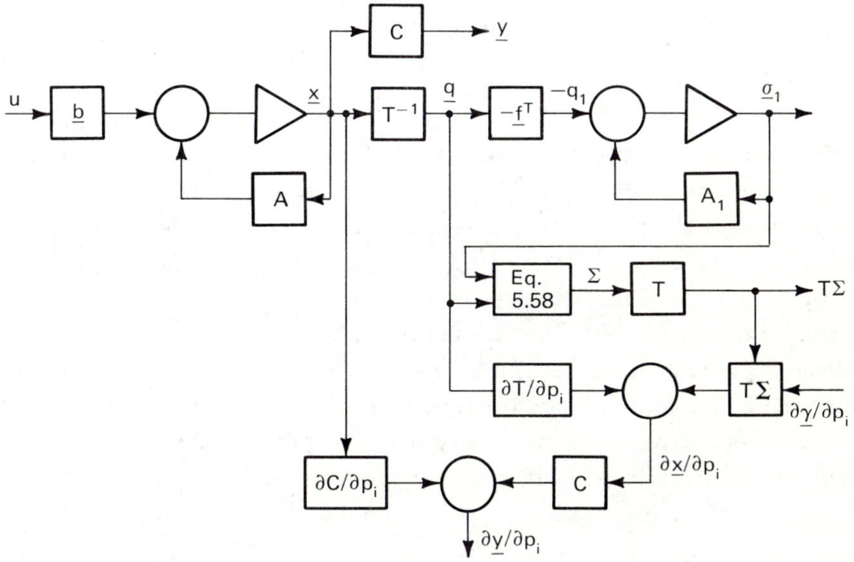

Figure 5.8

One remaining detail concerns the initial state of the sensitivity model. The initial state $\underline{x}(t_o) = \underline{x}_o$ is known, and $\underline{q}(t_o) = T^{-1}\underline{x}_o$. Also,

$$\underline{\sigma}_1(t_o) = \begin{bmatrix} \frac{\partial q_1}{\partial \gamma_1} \\ \frac{\partial q_1}{\partial \gamma_2} \\ \vdots \\ \frac{\partial q_1}{\partial \gamma_r} \end{bmatrix}_{t=t_o} = \begin{bmatrix} \frac{\partial q_1}{\partial \gamma_1} \\ \frac{\partial q_2}{\partial \gamma_1} \\ \vdots \\ \frac{\partial q_r}{\partial \gamma_1} \end{bmatrix}_{t=t_o} = \left[\frac{\partial}{\partial \gamma_1}\underline{q}(t)\right]_{t=t_o},$$

or in general,

$$\underline{\sigma}_i(t_o) = \left[\frac{\partial}{\partial \gamma_i}\underline{q}(t)\right]_{t=t_o}, \quad i = 1,2,\ldots,r .$$

From this it can be shown that

$$\underline{\sigma}_1(t_o) = \underline{0} ,$$

and usually, $\underline{x}_o=\underline{0}$ also.

Extensions of this single-input, continuous-time model can be carried out for multiple-input systems and for discrete-time systems. For multiple-input continuous-time systems, the restrictions are $\underline{x}_o=\underline{0}$ and the pair $(A(\underline{p}),B(\underline{p}))$ must be normal, i.e., each pair $(A(\underline{p}),\underline{b}_i(\underline{p}))$, $i=1,\ldots,q$ (q is the input dimension) must be controllable. In this case, 2q models, each of order r, are required. One is the original system, q-1 are first primal models of the system, and q are the associated sensitivity models [17.]. It also is easy to show that all of these results hold for linear, stationary discrete-time systems where the E-operator replaces D and unit delays replace integrators in the simulation diagrams.

<u>Example 5.15</u>

For the mass-spring-dashpot system of Example 3.1c, find the sensitivity of the displacement to the spring constant K for input $f = u_1(t)$ and zero initial displacement and velocity. The original equation is

$$\frac{W}{g}\ddot{y} + C\dot{y} + Ky = f .$$

There are three parameters W/g, C, and K and $\dot{y}(0) = y(0) = 0$, $t_o=0$. A first primal state model is obtained from

$$\ddot{y} + C\frac{g}{W}\dot{y} + K\frac{g}{W}y = \frac{g}{W}u_1(t)$$

as

$$\dot{\underline{q}} = \begin{bmatrix} 0 & 1 \\ -K\frac{g}{W} & -C\frac{g}{W} \end{bmatrix}\underline{q} + \begin{bmatrix} 0 \\ 1 \end{bmatrix} u_1(t)$$

$$y = \begin{bmatrix} \frac{g}{W} & 0 \end{bmatrix}\underline{q} , \qquad \underline{q}(0) = \underline{0} .$$

which defines A_1, $\underline{b}_1$, and $\underline{c}_1^T$ of Eq. 5.51. Since this form was written directly, T=I. Also $\gamma_1 = K\frac{g}{W}$, $\gamma_2 = C\frac{g}{W}$, and

$$\Sigma = \begin{bmatrix} \sigma_{11} & \sigma_{12} \\ \sigma_{12} & \sigma_{22} \end{bmatrix} .$$

Using Figure 5.8, the following are true:

$$p_1 = K$$

$$\frac{\partial T}{\partial p_1} = 0$$

$$\frac{\partial C}{\partial p_1} = 0$$

$$\frac{\partial \underline{\gamma}}{\partial K} = \frac{\partial}{\partial K}\begin{bmatrix} \gamma_1 \\ \\ \gamma_2 \end{bmatrix} = \begin{bmatrix} \frac{g}{W} \\ \\ 0 \end{bmatrix}$$

$$\frac{\partial \underline{q}}{\partial K} = \frac{\partial \underline{x}}{\partial K} = \sum \begin{bmatrix} \frac{g}{W} \\ \\ 0 \end{bmatrix}$$

and

$$\frac{\partial \underline{y}}{\partial K} = \underline{c}_1^T \sum \begin{bmatrix} \frac{g}{W} \\ \\ 0 \end{bmatrix} = \left(\frac{g}{W}\right)^2 \sigma_{11} \quad .$$

A state model for the complete sensitivity model is

$$\begin{bmatrix} \dot{q}_1 \\ \dot{q}_2 \\ \dot{\sigma}_{11} \\ \dot{\sigma}_{12} \end{bmatrix} = \begin{bmatrix} \dot{\underline{q}} \\ \hline \dot{\underline{\sigma}}_1 \end{bmatrix}$$

$$= \left[\begin{array}{cc|cc} 0 & 1 & & \\ -\gamma_1 & -\gamma_2 & \multicolumn{2}{c}{\Theta} \\ \hline 0 & 0 & 0 & 1 \\ -1 & 0 & -\gamma_1 & -\gamma_2 \end{array}\right] \begin{bmatrix} \underline{q} \\ \hline \underline{\sigma}_1 \end{bmatrix} + \begin{bmatrix} 0 \\ 1 \\ 0 \\ 0 \end{bmatrix} u_1(t)$$

$$\left[\frac{\partial y}{\partial K}\right] = \left[0 \quad 0 \quad \left(\frac{g}{W}\right)^2 \quad 0\right] \begin{bmatrix} \underline{q} \\ \hline \underline{\sigma}_1 \end{bmatrix}$$

with

$$\begin{bmatrix} \underline{q}(0) \\ \hline \underline{\sigma}_1(0) \end{bmatrix} = \underline{0} \quad .$$

A standard digital simulation (see App. D) of this for the nominal values W/g = 1, C=2.4 and K=4 was done and y(t) and $\frac{\partial y(t)}{\partial K}$ are shown.

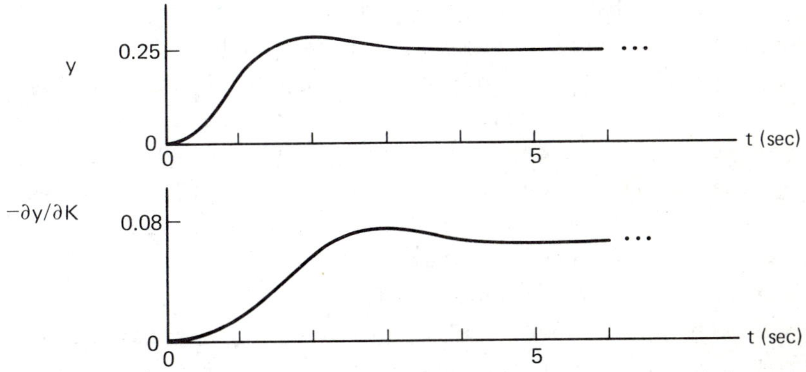

It is clear that the maximum sensitivity of y to K occurs about one second after the peak overshoot, and that the steady-state value of y is relatively quite sensitive to K as would be expected in this simple mass-spring-dashpot system.

5.4 RELIABILITY

Reliability is the probability that a component or an entire system will operate satisfactorily (or will not fail) during a prescribed operating period. A general definition is

$$R(t) = \Pr\ (\text{no failure in } (0,t]).$$

The start of the operating period is $t_o=0$ and it assumed that the system is operating properly at that time. There are, of course, several qualifying conditions on environment, stress, etc., needed before a practical application of this definition can be made. Although one goal of the equipment designer has always been to minimize the chance of failure, the development of complex military systems since World War II and the need for highly reliable equipment for space applications in the last two decades has stimulated the growth of both the mathematics and physics of reliability. In addition, the use of these theories has brought about wide application of many of the techniques of the statistician such as experimental design and parameter estimation. Basically, one is interested in <u>reliability prediction</u> during the early stages of design so that costly failures can be avoided. Also, one is interested in <u>reliability evaluation</u> of operating equipment when further improvement is needed. In practice these predictions or evaluations must be made on the basis of rather limited statistical information, and a partial answer to this difficulty has sometimes been the use of high speed computers in Monte Carlo studies.

5.4.1 Definitions

The concept of reliability adds another dimension to the ways of assessing the quality of a system. Thus, after developing signal and system models in Chapters 2 and 3, system response to various inputs was discussed in Chapter 4 as a measure of performance. In this chapter, general views of accuracy and sensitivity have been introduced. In these two cases, as well as in previous chapters (except where signal and system identification were discussed), the parameters of the models were assumed to be known exactly. The consideration of sensitivity, of course, follows, at least in part, from a concern that the parameters may not be exactly at their design values. If the parameters change in an unknown way but according to some assumed underlying probabilistic structure, prediction of system behavior in a probabilistic sense is possible. This situation obviously is related to the random parameter problem mentioned in the previous section on sensitivity. Here, however, other possibilities occur and a simple illustration is provided in the manufacture and use of a wire spring:

1. The spring is manufactured with a nominal stiffness K and K is within specified tolerance $\pm\Delta K$. The maximum allowed stretching force is F_m. The choice of material, wire size, etc., to obtain K subject to F_m depends on the properties of materials. Manufacturing to within tolerance is a question of quality control, and the effect of a deviation from K in an application is a question of sensitivity as discussed in the previous section (see Example 5.15).

2. The spring is used in an application where the applied force varies stochastically; thus the force is modeled as a process $\{F(t)\}$. If there are occasional times when $F(t)>F_m$, permanent damage may occur and the actual stiffness changes. It may eventually be outside acceptable limits, and this can be called a failure.

3. The damage in 2. may be such that when the very unlikely event that $F(t)\gg F_m$ (for a short time) occurs, the spring breaks. This is a catastrophic failure, i.e., the model of an approximate linear spring is no longer valid.

The situation in 2. could also arise if a parameter varies with an environmental effect, e.g., a change in a transistor parameter due to temperature changes modeled as a stochastic process. In such cases the parameter itself may be modeled as a stochastic process and the analysis is a kind of stochastic sensitivity analysis. The error analyses done on complex aerospace systems is a similar point of view.

These ideas then lead to the common distinction between catastrophic failure and drift failure, and it is the former that has most often been studied. A further distinction is made between component (element, unit, subsystem) failure and system failure where the system contains several components.

There are great practical difficulties in applying the mathematical theories of reliability to be discussed below. First, unambiguous definitions of failure are required as part of a test program. Environmental conditions must be controlled, adequate statistical information must be obtained, and complete system reliability flow diagrams or graphs must be developed. In addition, it frequently is difficult to obtain any verification of a prediction R(t). Nevertheless, the effort on the part of the designer to estimate reliability or to achieve a specified reliability should result in an improvement in system performance because of the attention given to the possible failure modes. A survey of the many ways of studying reliability up to 1962 is given in [19.] and rather broad and current coverage of the subject is given in [20.], [21.], and [22.].

5.4.2 Component Reliability

To study catastrophic component failure, we let the random variable T be the time of failure. Then the probability of failure in (0,t] is

$$F_T(t) = \Pr(T \le t)$$

and

$$R_c(t) = 1 - F_T(t) = \Pr(T > t) \ .$$

$F_T(t)$ is the <u>failure distribution function</u> and

$$f_T(t) = \frac{dF_T(t)}{dt}$$

is the failure probability density function. In a very small interval Δt,

$$\Pr\ (\text{failure in } (t, t+\Delta t]) \cong f_T(t)\Delta t\ .$$

A related conditional failure statement is

$$\Pr\ (\text{failure in } (t, t+\Delta t] \mid \text{no failure in } (0,t])$$

$$\cong h_c(t)\Delta t$$

and this defines $h_c(t)$, the <u>hazard function</u>. The condition in the definition of $h_c(t)$ is $R_c(t)$ and in the limiting case as $\Delta t \to 0$,

$$\begin{aligned} f_T(t) &= h_c(t)R_c(t) \\ &= h_c(t)(1-F_T(t)). \end{aligned} \qquad (5.59)$$

Using these relationships,

$$\begin{aligned} \frac{dF_T(t)}{dt} &= \frac{d}{dt}(1-R_c(t)) \\ &= -\frac{dR_c}{dt} \\ &= f_T(t) = h_c(t)R_c(t) \end{aligned} \qquad (5.60)$$

and a differential equation in $R_c(t)$ is, therefore

$$\frac{dR_c}{dt} + h_c(t)R_c = 0$$

subject to $R_c(0) = 1$. From this, we get

$$R_c(t) = (\exp - \int_0^t h_c(\tau)d\tau). \tag{5.61}$$

It is apparent that if the hazard function is known or is estimated from experimental data, the reliability of the component can be found. To relate these definitions to failure data, we assume independent failures in intervals $(t, t+\Delta t]$ and let

N_0 = original number of components

$N_s(t)$ = number of components surviving at t

$N_f(t)$ = number of components failing in $(0,t]$,

and define a failure rate F.R. at t as

$$\text{F.R.} = \frac{1}{N_s(t)} \left[\frac{\text{Number failing in } (t,\ t+\Delta t]}{\Delta t} \right] \tag{5.62}$$

$$= \frac{1}{N_s(t)} \left[\frac{N_f(t+\Delta) - N_f(t)}{\Delta t} \right]$$

$$= \frac{1}{N_s(t)} \dot{N}_f(t) \text{ as } \Delta t \text{ approaches zero.}$$

Now using a frequency ratio as a probability (see Appendix C),

$$R_c(t) = \frac{N_s(t)}{N_0} = \frac{N_0 - N_f(t)}{N_0} = 1 - \frac{N_f(t)}{N_0}$$

and

$$\frac{dR_c(t)}{dt} = - \frac{1}{N_0} \dot{N}_f(t),$$

therefore as $\Delta t \to 0$,

$$F.R. = -\frac{1}{N_s(t)}\left(N_o \frac{dR_c(t)}{dt}\right)$$

or

$$F.R. = -\frac{1}{R_c(t)} \frac{dR_c(t)}{dt} .$$

Recalling the differential equation in $R_c(t)$ above, we have

$$h_c(t) = F.R. ,$$

i.e., the hazard function $h_c(t)$ is the instantaneous failure rate which can be estimated using Equation 5.62. Also, $f_T(t)$ can be related to the experimental data since

$$f_T(t) = h_c(t)R_c(t) .$$

Use of the definition of F.R. leads to

$$f_T(t) = \frac{1}{N_o}\left[\frac{\text{Number failing in } (t, t+\Delta t]}{\Delta t}\right] \qquad (5.63)$$

and comparing Eq. 5.62 and 5.63 shows that the hazard function is failures per unit time normalized to N_s and the failure density function is failures per unit time normalized to N_o. In practice, it usually is most convenient to estimate $h_c(t)$; e.g., by using a least squares fit (see Chapter 2).

In many cases a satisfactory reliability model (over the period after "burn in" and before "wear out" occurs) is one where $h_c(t)$ is constant. In this case, for $h_c(t) = r$, a constant failure rate, we get

$$R_c(t) = \varepsilon^{-rt} , \qquad t \geq 0 \qquad (5.64)$$

and

$$f_T(t) = r\varepsilon^{-rt}, \qquad t \geq 0$$
$$= 0 \quad , \qquad t < 0$$

the well-known exponential distribution. In this case the mean time to failure is

$$\text{MTTF} = E[T] = \int_0^\infty \sigma f_T(\sigma)\,d\sigma = \frac{1}{r} \ .$$

Tabulated MTTF data is available for many common components [25.].

Example 5.16

The data on the observations of 14 light bulbs is given below in the first two columns. Calculate F.R. using Eq. 5.62 and estimate $h_c(t)=r$. From this constant failure rate model, find $R_c(100)$.

Failure No.	Time (hrs.)	F.R.
1	86	0.83×10^{-3}
2	104	4.28×10^{-3}
3	122	4.60×10^{-3}
4	148	3.50×10^{-3}
5	223	1.33×10^{-3}
6	247	4.61×10^{-3}
7	343	1.30×10^{-3}
8	358	9.50×10^{-3}
9	560	0.83×10^{-3}
10	1100	0.37×10^{-3}
11	1151	4.90×10^{-3}
12	1238	3.84×10^{-3}
13	2833	0.31×10^{-3}
14	4200	0.73×10^{-3}

The first entry for F.R. is (for t=0, Δt=86)

$$\frac{1}{14x86} = 0.83 \times 10^{-3}$$

and the second is (for t=86, Δt=18)

$$\frac{1}{13x18} = 4.28 \times 10^{-3}$$

and the rest are found similarly. For an assumed constant failure rate, the average F.R. is the least squares estimate of r, therefore, the estimated r is:

$$r = 2.91 \times 10^{-3} .$$

The estimated reliability, using Eq. 5.64, is

$$R_c(t) = (\exp -(2.91x10^{-3})t), \qquad t \geq 0$$

and MTTF = 343 hrs. Finally

$$R_c(100) = \Pr \text{ (no failure in 100 hrs.)} = (\exp -0.291)=0.75.$$

Many forms for $h_c(t)$ and $f_T(t)$ have been used [20.] (e.g., linearly varying $h_c(t)$ and the Weibull form $h_c(t) = at^b$, b>-1,) and in some cases, considerable improvement in reliability prediction has resulted.

Drift failures for components can be modeled by assuming that a parameter q varies stochastically, i.e., realizations $q^i(t)$ arise from a process $\{Q(t)\}$. Failures of this type are conditional on the assumption that no catastrophic failure (n.c.f.) occurs. Thus if

$$R_c(t) = \Pr \text{ (no catastrophic failure in } (0,t]) ,$$

we let $R_d(t) = \Pr$ (no drift failure in $(0,t]|$n.c.f.$(0,t]$) and the combined reliability is

$$R(t) = R_d(t)R_c(t) .$$

To define $R_d(t)$ it is necessary to give a specific meaning to failure. A simple choice is that the random variable $Q(t)$ not exceed some specified tolerance, i.e., one has a failure if

$$|Q(t)-q_o| \geq \alpha$$

where q_o is the nominal value and α is the maximum allowable deviation. If it is known that $\{Q(t)\}$ is such that all $q^i(t)$ are monotonic nondecreasing or monotonic nonincreasing and all $|q^i(0)-q_o|<\alpha$, this is a sensible definition [23.]. Otherwise, some function of $Q(t)$ can be defined to provide a satisfactory failure definition [24.]. Problems of this type are called level crossing problems. One wishes to compute

$$R_d(t;\alpha) = \Pr(|Q(t)-q_o|<\alpha) ,$$

when the monotonicity condition is assumed, or in normalized form

$$R_d(t;\alpha) = \Pr(|Q_n(t)|<\alpha), \; Q_n = Q-q_o .$$

Basically, a probability density function $p_{Q_n}(\lambda;t)$ is required and from this,

$$R_d(t;\alpha) = \int_{-\alpha}^{\alpha} p_{Q_n}(\lambda;t)\,d\lambda . \tag{5.65}$$

Also, letting the random variable T_d be the time of drift failure,

$$f_{T_d}(t) = -\frac{dR_d}{dt}$$

$$= -\int_{-\alpha}^{\alpha} \frac{\partial p_{Q_n}(\lambda;t)}{\partial t}\,d\lambda \tag{5.66}$$

is the drift failure density function. A drift failure hazard function also can be defined using R_d and f_{T_d} as in Eq. 5.60. The practical difficulties occur in justifying the montonicity assumption and in estimating the density function $p_{Q_n}(\lambda;t)$. Usually, sufficient test data are not available.

<u>Example 5.17</u>

Consider $Q(t) = q(0) + Vt$, $t \geq 0$, where V is a Gaussian or normal random variable such that

$$p_V(\lambda) = \frac{1}{\sqrt{2\pi}\,\sigma} \exp\left(\frac{-\lambda^2}{2\sigma^2}\right) = N(0,\sigma) \ .$$

Let $Q_n(t) = Q(t)-q(0) = Vt$, and define drift failure for $|Q_n(t)| \geq \alpha$. Then (see Appendix C on change of random variable), we obtain p_{Q_n} from p_V as

$$p_{Q_n}(\tau;t) = \frac{1}{t}\, p_V(\lambda = \frac{\tau}{t})$$

$$= \frac{1}{\sqrt{2\pi}\,\sigma t} \left(\exp - \frac{\tau^2}{2\sigma^2 t^2}\right) \qquad t > 0.$$

which is $N(0,\sigma t)$, i.e., the standard deviation is increasing linearly. From this,

$$R_d(t;\alpha) = \int_{-\alpha}^{\alpha} p_{Q_n}(\tau;t)\,d\tau \ , \qquad t > 0$$

$$= 1-2\Phi\left(\frac{-\alpha}{\sigma t}\right) = 2\Phi\left(\frac{\alpha}{\sigma t}\right) - 1$$

where $\Phi(\cdot)$ is the tabulated standardized normal distribution function [26.].

5.4.3 System Reliability

The estimation of the reliability of a system which is an interconnection of components with known reliabilities is the main problem here. To illustrate, we consider the simple system shown schematically in Figure 5.9; there are three components. The reliabilities R_1, R_2, and R_3 may be

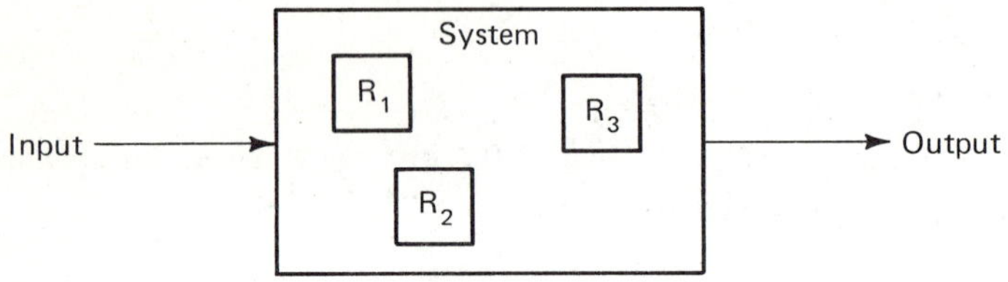

Figure 5.9

for catastrophic failures, drift failures, or both (the time argument has been suppressed). The important step here is to determine the effect of a failure of each component on the system operation. This leads to a reliability flow graph or reliability diagram. For example, if the failure of any one (or more) of the components means a failure of the system, the reliability diagram is shown in Figure 5.10a; this is "series" type. If the system operates satisfactorily when at least one component has not failed, the reliability diagram is as shown in Figure 5.10b; this is a "parallel" type. Intermediate situations occur for other combinations; for example, if the system operates satisfactorily when component 1 has not failed and either 2 or 3 (or both) has not failed, the diagram is shown in Figure 5.10c. There must be at least one path from the left to the right node if the system is to operate. Then, if <u>failures of the components are assumed to be independent</u>, the system reliabilities are

$$R_{sa} = R_1R_2R_3$$

$$R_{sb} = 1-(1-R_1)(1-R_2)(1-R_3)$$

$$R_{sc} = R_1[1-(1-R_2)(1-R_3)]$$

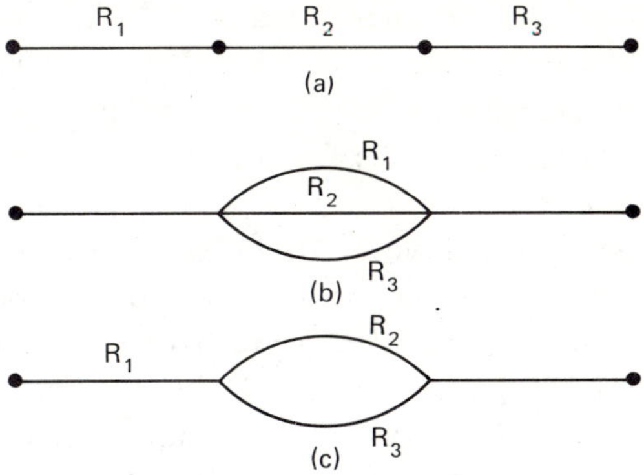

Figure 5.10

When the failures are dependent, the problem is more complicated since the estimation of R_1, R_2, and R_3 must be based on tests that are properly controlled [24.] and the resulting conditional probabilities must be combined properly. It is obvious that parallelism in these diagrams has the effect of improving reliability. Thus, if $R_1=R_2=R_3=0.9$,

$$R_{sa} = 0.729$$

$$R_{sb} = 0.999$$

$$R_{sc} = 0.891.$$

The preparation of the reliability diagram is a crucial step in system reliability estimation. The system analyst must have a thorough knowledge of the system since he must not only understand its operation in the sense of input/output behavior (as in Chapter 4 and preceding sections of Chapter 5), but he also must anticipate the effect of the failure of any component on the operation of the system. Although all the previous system models for linear systems, including sensitivity, were obtained by following specified rules, the derivation of a reliability diagram is a far less formal procedure. Decisions on how to collect components into subsystems or how to incorporate the effects of interdependence of components as well as the basic problem of anticipating the effects of component failures require great care. Once the reliability diagram is prepared, however, there is a formal procedure for computing the system reliability as illustrated above. It is, in practical systems with tens or hundreds of components, a large combinatorial problem and various methods for searching for all paths in the reliability diagram have been devised. Two useful ones are the path tracing and cut-set, tie-set methods [20.]. Algorithms for carrying out the analysis of a digital computer also have been developed [27.] when the component failures are independent. A simple illustration follows from Figure 5.10c where two tie sets[*] are 12 and 13, then

$$R_{sc} = Pr(12 + 13)$$

$$= Pr(12)+Pr(13)-Pr(12 + 13)$$

or

$$R_{sc} = R_1R_2+R_1R_3-R_1R_2R_3$$

which is the same as above. One simple result which has been used in a wide variety of applications is based on the "series" configuration with independent failures and constant hazard functions. For Figure 5.10a, we obtain

$$R_{sa} = (\exp -r_1t)(\exp -r_2t)(\exp -r_3t), \qquad t > 0$$

$$= (\exp -(r_1+r_2+r_3)t)$$

$$= (\exp -r_st)$$

---

*See Chapter 6, Section 6.3.

and the system MTTF is $1/r_s$.

<u>Example 5.18</u>

Consider the 400 Hz position servo shown. The major components and the failure rates (constant) are listed.

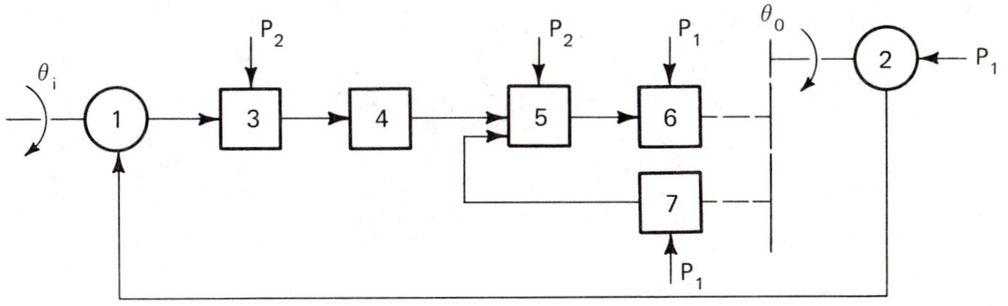

	Component	Failure Rate (per hour)
P_1	400 Hz supply	5.0×10^{-6}
P_2	d-c supply	10.0×10^{-6}
1,2	synchros	0.2×10^{-6}
3	preamplifier	3.0×10^{-6}
4	compensating network	0.1×10^{-6}
5	power amplifier	5.0×10^{-6}
6	2-phase motor	1.5×10^{-6}
7	a-c tachometer	0.3×10^{-6}

Neglect failures in the gear train, wiring, and connectors. Assume all failures are independent.

a) If the system is operated in a laboratory environment and P_1 and P_2 are not included in the calculation, find the MTTF. The reliability diagram is

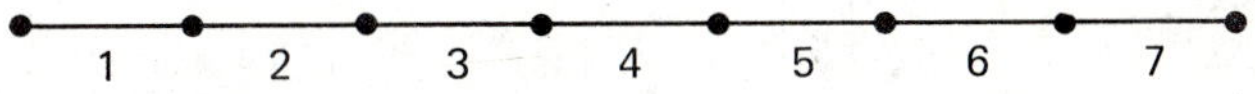

and

$$r_{sa} = (0.4+3.0+0.1+5.0+1.5+0.3)\text{x}10^{6}$$

$$= 10.3\text{x}10^{-6} \ .$$

Then

$$\text{MTTF})_a = \frac{1}{r_{sa}} = 9.7\text{x}10^{4} \text{ hrs.}$$

b) If the system is operated in an aircraft where P_1 and P_2 are generators, find the MTTF. Assume a failure rate modifier of 20 for the more severe environment. The reliability diagram is

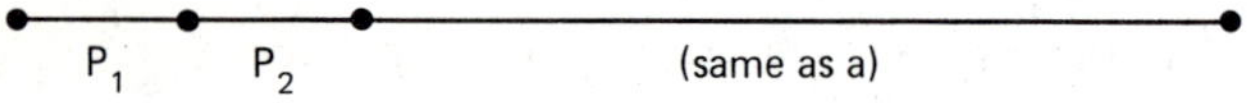

and

$$r_{sb} = 20(15+10.3)\text{x}10^{-6}$$

$$= 5.06\text{x}10^{-4}.$$

Then

$$\text{MTTF})_b = \frac{1}{r_{sb}} = 1.03\text{x}10^{3} \text{ hrs.}$$

c) Assume that dual generators are used on busses (with suitable "perfect" switching in case of failure) in the aircraft. The reliability diagram is

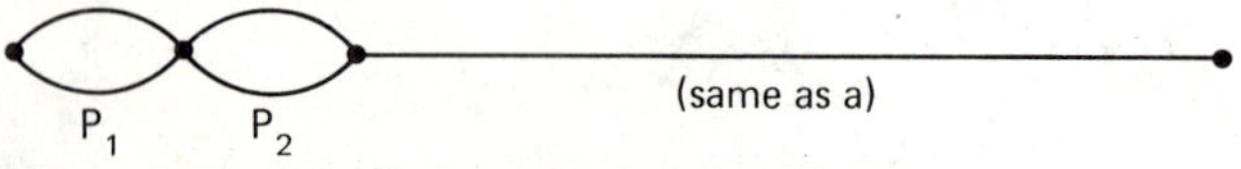

For the a-c power supply with each rate $r_1 = 5\text{x}10^{-6}$, and

reliability R_1, the reliability of the parallel configuration for P_1 is

$$R = 1-(1-R_1)^2 = 2R_1-R_1^2$$

$$= 2(\exp -r_1t)-(\exp -2r_1t).$$

Then since

$$\text{MTTF} = \int_0^\infty \sigma f_T(\sigma)\,d\sigma$$

$$= \int_0^\infty R(\sigma)\,d\sigma$$

$$= \frac{2}{r_1} - \frac{1}{2r_1} = \frac{3}{2r_1}$$

the failure rate is $\frac{2}{3}r_1 = 3.33\text{x}10^{-6}$. Similarly for the d-c supply, the failure rate of the dual system is

$$\tfrac{2}{3}(10\text{x}10^{-6}) = 6.67\text{x}10^{-6}\ .$$

Then

$$r_{sc} = 20(10+10.3)\text{x}10^{-6}$$

$$= 4.06\text{x}10^{-4}$$

and

$$\text{MTTF})_c = 2.47\text{x}10^3 \text{ hrs.}$$

Further developments in system reliability include the use of Markov models [10.], the introduction of replacements for parts that fail, i.e., the use of renewal theory [28.], and the concept of optimal system reliability [29.]. An interesting combination of optimization and renewal theory is given in [30.].

REFERENCES

1. Director, S. and Rohrer, R., "Introduction to System Theory", New York: McGraw-Hill Book Co., 1972.
2. Chen, C. T., "Introduction to Linear System Theory", New York: Holt, Rinehart, Winston, Inc., 1970.
3. Brockett, R., "Finite Dimensional Linear Systems", New York: John Wiley & Sons, Inc., 1970.
4. Schwarz, R. and Friedland, B., "Linear Systems," New York: McGraw-Hill Book Co., 1965.
5. Wiberg, D., "State Space and Linear Systems", New York: McGraw-Hill Book Co., 1971.
6. Hagander, P., "Numerical Solution of $A^TS + SA + Q = 0$", Information Science, v. 4, 1972, pp. 35-50.
7. Greaves, C., et al., "Evaluation of Integrals Appearing in Minimization Problems of Discrete-Data Systems", IEEE Trans. Auto Control, v.-AC11, n. 1, January, 1966, pp. 145-148.
8. Jury, E., and Blanchard, J., "A Stability Test for Linear Discrete Systems in Table Form", IRE Proc., v. 49, December, 1961, pp. 1947-1948.
9. Tou, J., "Digital and Sampled-Data Control Systems", New York: McGraw-Hill Book Co., 1959.
10. Eveleigh, V., "Control Systems Design", New York: McGraw-Hill Book Co., 1972.
11. Newton, G., et al., "Analytical Design of Linear Feedback Controls", New York: John Wiley and Sons, 1957.
12. Black, H. S., "Wave Translation System", U.S. Patent No. 2,102,671, Dec., 1937.
13. Bode, H., "Network Analysis and Feedback Amplifier Design", New York: Van Nostrond, 1945.
14. Bykhovskiy, M., "The Accuracy of Mechanisms in Which the States of the Components are Described by Differential Equations", Izv. AN SSR, Otd. Tekhn, Nauk., n. 11, 1947.
15. Cruz, J. (Ed.), "System Sensitivity Analysis",

Stroudsburg, Pa.: Dowden, Hutchinson, & Ross, Inc., 1973.

16. Sobral, M., "Sensitivity in Optimal Control Systems", IEEE Proc., v. 56, n. 10, Oct. 1968, pp. 1644-1652.
17. Wilkie, D. and Perkins, W., "Generation of Sensitivity Functions for Linear Systems Using Low Order Models", IEEE Trans. Auto. Control, v. AC-14, n. 2, April 1969, pp. 123-130.
18. Perkins, W., in "Feedback Systems", (J. Cruz, Ed.), New York: McGraw-Hill Book Co., 1972.
19. Blanton, H. and Jacobs, R., "A Survey of Techniques for Analysis and Prediction of Equipment Reliability", IRE Trans. Rel. Qual. Control, v. RQC-11, n. 2, July, 1962, pp. 18-35.
20. Shooman, M., "Probabilistic Reliability: An Engineering Approach", New York: McGraw-Hill Book Co., 1968.
21. Amstadter, B., "Reliability Mathematics: Fundamentals: Practices: Procedures", New York: McGraw-Hill Book Co., 1971.
22. Green, A. and Bourne, A., "Reliability Technology", London: Wiley-Interscience, 1972.
23. Drenick, R., "Mathematical Aspects of the Reliability Problem", SIAM Journal, v. 8, n. 1, March, 1962, pp. 125-149.
24. Lewis, J. and Wells, W., "Probabilistic Models for System Reliability", Proc. Aerospace Reliability and Maintainability Conference, May, 1963, pp. 133-143.
25. Earles, D. and Eddins, M., "Reliability Engineering Data Series", AVCO Res. and Dev., April, 1962.
26. Gibra, I., "Probability and Statistical Inference for Scientists and Engineers", Englewood Cliffs, N.J.: Prentice-Hall, Inc., 1973.
27. Kim, Y., et al., "A Method for Computing Complex System Reliability", IEEE Trans. Rel., v. R-21, n. 4, Nov. 1972, pp. 215-219.
28. Cox, D., "Renewal Theory", New York: John Wiley & Sons, 1962.

29. Misra, K. and Ljubojevic, M., "Optimal Reliability Design: A New Look", IEEE Trans. Rel., v. R-62, n. 5, Dec. 1973, pp. 255-258.

30. Ratner, R. and Luenberger, D., "Performance -- Adaptive Renewal Policies for Linear Systems", IEEE Trans. Auto. Control., v. AC-14, n. 4, Aug. 1969, pp. 344-351.

PROBLEMS

1. Discuss the several definitions of stability of
   a) $\dot{y} = ty + u$
   b) $\dot{y} = -ty + u$
2. Determine the range of K for which the system is totally stable.

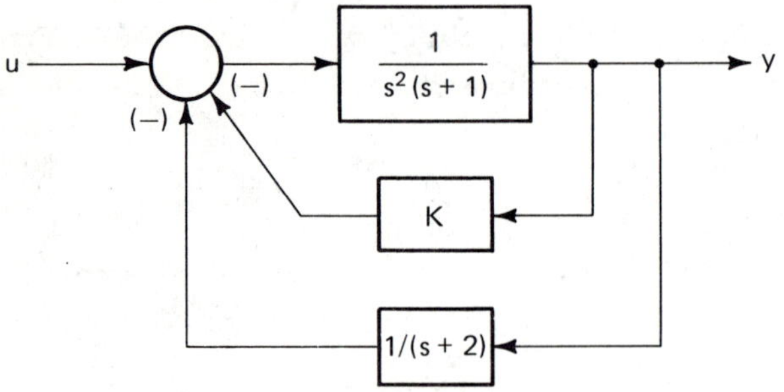

3. For

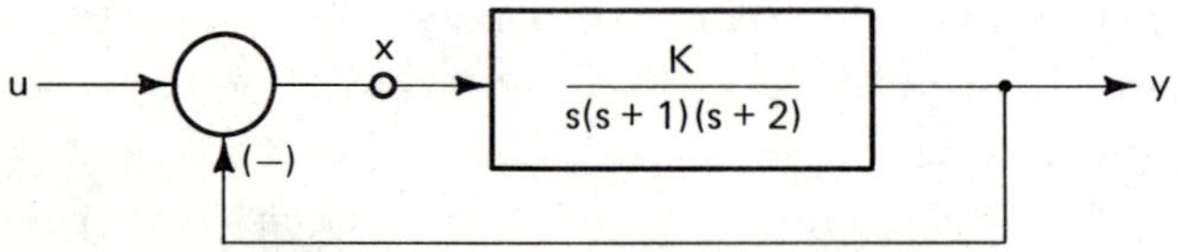

   a) Determine an allowable range of K if the system is to remain totally stable.
   b) If a uniform impulse sampler with period T_s=1 sec. is inserted at x, what is the new range of K?

4. Examine the model of Problem 36., Chapter 3 for total stability.
5. Examine a state variable model of the circuit of Problem 46., Chapter 3 for stability and asymptotic stability.
6. In Problem 28., Chapter 3, determine whether or not there are values of R_f such that the amplifier is not totally stable.
7. For the model of Problem 24b), Chapter 3, use the Schur-Cohn test (Theorem 5.20) to determine whether or not it is totally stable.
8. For the system

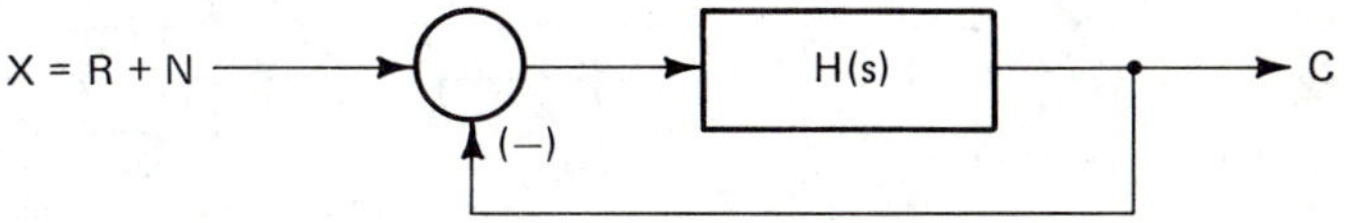

$$H(s) = \frac{\omega^2}{s(s + 2\xi\omega)}$$

ξ is fixed by the form of a desired response to a step function and ω is chosen so that $J = e_R^2 + J_1$ is minimal. e_R is the steady-state error when $R(s) = A/s^2$ acts alone. N is a white zero-mean process with intensity W, and J_1 is the mean square output when N acts alone.

a) For A = 1/2 and W = 2, obtain data to plot e_R^2, J_1, and J versus ω/ξ.

b) Specify the "best" value of ω/ξ.

9. For the filter of Problem 47., Chapter 3, define error $e(n) = v(n) - z(n)$. Find the mean square error as a function of α when the input is a zero-mean wide-sense stationary sequence with

$$r_V(k) = (0.8)^{|k|} .$$

10. For the simple feedback system shown, the wide-sense stationary input has

$$R_U(s) = \frac{1}{4-s^2}$$

and error E = U-Y. If the maximum allowable mean square error is 0.1, find the necessary positive gain K.

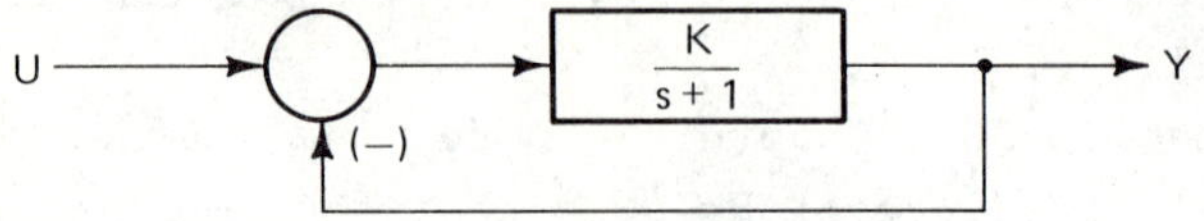

11. A single-axis gyro-stabilized inertial reference system is shown.

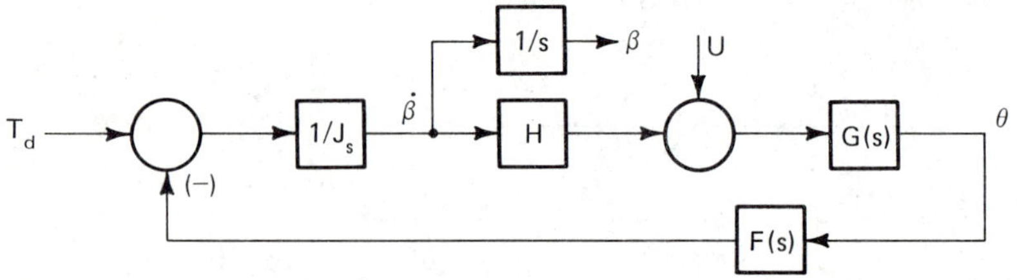

T_d = disturbance torque about gimbal axis
U = disturbance torque about input axis
J = servo motor moment of inertia
H = gyro angular momentum
I = gyro gimbal system moment of inertia
D = viscous damping of gimbal system
θ = gimbal angle
β = stabilized reference angle

$$\text{Servo } G(s) = \frac{1}{Is^2+Ds}$$

$$\text{Compensator } F(s) = K\left(\frac{1+T_1 s}{1+T_2 s}\right), \qquad T_1 > T_2$$

Consider $\dot{\beta}$ as output and U as input

a) Find the transfer function

$$\frac{\dot{\beta}(s)}{U(s)} = Q(s)$$

b) Are there values of K, T_1, and T_2 such that the closed-loop system is totally stable?

c) For $T_d=0$, assume that measurements on the drift rate $\dot{\beta}$ due to U over a time period $[0,T_m]$, $T_m >> \frac{1}{c}$, suggest an approximate autocorrelation function of $\{\dot{\beta}\}$ as

$$r_{\dot{\beta}}(\tau) = A^2(\exp -c|\tau|) .$$

Define drift angle $\beta(T)$ as

$$\beta(T) = \int_0^T \dot{\beta}(\sigma)\,d\sigma$$

and find an expression for the mean square drift angle due to U:

$$s_\beta^2(T) = E[\beta(T)^2] .$$

d) Show that for large T,

$$s_\beta^2(T) \to R_{\dot{\beta}}(0)\ T$$

$$R_{\dot{\beta}}(s) = \mathcal{L}\ r_{\dot{\beta}}(\tau) .$$

e) Is $\{\dot{\beta}\}$ wide-sense stationary?

f) Is $\{\beta\}$ wide-sense stationary?

12. An equivalent for a simple ringing circuit is shown. The sensitivity of the output v_o to R is of interest. Let L = 10 mh., c = 2.53 μf.

a) Find a sensitivity model

b) For a nominal R = 6.28 KΩ, find $\frac{\partial vo}{\partial R}(t)$ when $i = 10^{-3}u_1(t)$ amp.

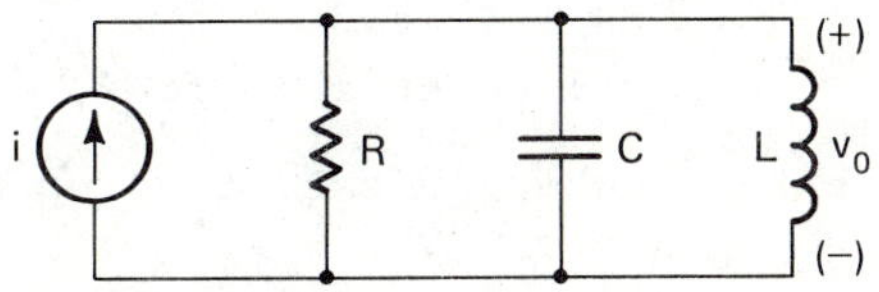

13. a) A simple digital filter is described by

$$z(n+1) = a\, z(n) + b\, v(n).$$

Find a model to study the sensitivity of the output $z(n)$ to variations in a and b.

b) What changes are made in Figure 5.6, 5.7, and 5.8 for discrete-time models?

c) Find $\frac{\partial z}{\partial a}(n)$ for $v(n) = v_1(n)$.

14. Ten servo power amplifiers are tested, and the failure times are: 90, 195, 265, 370, 395, 460, 510, 540, 600, and 670 hours.

a) Estimate $f_T(t)$ from the data and plot.

b) Estimate $h_c(t)$ from the data and plot.

c) From the results in a) and b), plot estimated $R_c(t)$.

d) If a constant failure rate model is assumed, estimate r.

e) Plot $R_c(t)$ obtained in d) and compare with the result in c).

f) Is the model of d) a reasonable one? Can you suggest better models?

15. Consider Example 5.18. Assume that two possible improved systems are being compared:

S_1: single P_1
dual P_2
improved power amplifier with
F.R. = $2\text{x}10^{-6}$

S_2: single P_1
dual P_2 with F.R. = $5\text{x}10^{-6}$
original power amplifier.

All failure rates unchanged except as noted. Which is the better choice?

Chapter 6

APPLICATIONS

6.1 INTRODUCTION

Previous chapters have been concerned with two general topics: modelling and analysis. Many models of signals and linear systems were developed in Chapters 2 and 3. Analysis of signals -- particularly with regard to spectral content -- and analysis of linear systems -- i.e., response to specified signals, stability accuracy, sensitivity, and reliability -- were covered primarily in Chapters 4 and 5. The examples used were short and frequently trivial as far as any practical applications are concerned. Consequently, this broad background is now supplemented by a series of longer examples to show the application of these ideas. There are four specific applications:

Signal Processing	- particularly discrete-time processing
Network Analysis	- linear R,L,C networks
Tracking Systems	- aircraft tracking
Power Systems	- modelling

In some of these the goal is analysis, and in others, an introduction to the design of a system is emphasized. The evaluation of the design makes use of several analysis techniques.

6.2 SIGNAL PROCESSING

6.2.1 Introduction

It was pointed out in the introduction to Chapter 2 that a signal contains information. Signal processing is done to modify this information in some way. It may be done to extract the information from a signal that also contains noise, to obtain some special part of the information (e.g., in a certain frequency band), or to convert the information to a different form (e.g., analog to digital). In addition, when experimental signal data are available, the processing may be concerned with developing a signal model, i.e., signal identification. These ideas were introduced in Chapter 2 where, for example, methods of obtaining least squares polynomial approximations were given.

Two specific signal processing operations are considered in this section; data filtering and spectral estimation. In both cases, it is assumed that a general-purpose digital computer is to be used, and this immediately suggests the use of the discrete Fourier Transform (DFT). High-speed algorithms for programming the DFT -- fast Fourier Transforms (FFT) -- have made digital signal processing a very useful alternative to analog techniques. In fact, entirely new approaches to signal processing are now possible because of the flexibility, the large storage capacity, and the precision of a modern digital computer. Realizations using microprocessors have become quite inexpensive too.

In the following, the DFT is developed by extending the Fourier Transforms ($\mathfrak{F}$ and $\mathfrak{F}_d$) of Appendix A. The use of the DFT is then considered in designing and realizing digital filters. Finally, the DFT is used in the estimation of a power spectrum (approximations to $P_{ds}(\theta)$ for v(n), a deterministic signal, or $R_V(\varepsilon^{j\theta})$ for {V(n)}, a stochastic sequence) from discrete-time experimental data $V_X(n)$ or $v_X^i(n)$. The subscript X used for experimental data in Chapter 2 will be omitted.

6.2.2 Discrete Fourier Transform

The DFT can be developed as follows. Starting with Eq. A.13,

$$F(\gamma) = \mathfrak{F}_d f(n) = \sum_{n=-\infty}^{\infty} f(n)\gamma^{-n}, \qquad \gamma = \varepsilon^{j\theta} .$$

As an approximation, we use only M values for f(n), and let

$$F_a(\gamma) = \sum_{n=0}^{M-1} f(n)\gamma^{-n} \tag{6.1}$$

Using Equation A.14, we obtain

$$f_a(n) = \frac{1}{2\pi}\int_{-\pi}^{\pi} F_a(\varepsilon^{j\theta})\varepsilon^{jn\theta}d\theta .$$

Now we convert to a discrete "frequency" variable by letting

$$\theta_k = k\Delta\theta , \qquad k = 0, \ldots, M-1.$$

with

$$\Delta\theta = \frac{2\pi}{M} .$$

Thus there are M time domain samples $f_a(n)$ and M frequency domain samples $F_a(\varepsilon^{j\theta_k})$. If an f(t) with $\mathfrak{F}\, f(t) = F(j\omega)$ is actually sampled at a rate $1/T_s$, the discrete-time variable is

$$t_n = nT_s \,, \qquad n = 0,1 \ldots, M-1$$

and the discrete "frequency" variable is

$$\omega_k = k\Omega$$

or the dimensionless "frequency" variable is

$$\theta_k = \omega_k T_s = k\Omega T_s$$

with

$$\Omega = \frac{\Delta\theta}{T_s} = \frac{2\pi}{MT_s} \quad .$$

The time data record is MT_s seconds long and the frequency record is $2\pi/T_s$ radians per second long. Finally, in view of the discretization of θ, we have an approximate $f_a(n)$ as a sum,

$$f_a(n) \cong \frac{1}{2\pi} \sum_{k=0}^{M-1} F_a(\varepsilon^{j\theta_k})\varepsilon^{jn\theta_k} \Delta\theta$$

or

$$f_a(n) \cong \frac{1}{M} \sum_{k=0}^{M-1} F_a(\varepsilon^{j\theta_k})(\exp j\tfrac{2\pi}{M}kn) \qquad (6.2)$$

and in Eq. 6.1, for $\gamma=\varepsilon^{j\theta}\to\varepsilon^{j\theta_k}$,

$$F_a(\exp j\theta_k) \cong \sum_{n=0}^{M-1} f_a(n)(\exp -jn\theta_k)$$

or

$$F_a(\varepsilon^{j\theta_k}) \cong \sum_{n=0}^{M-1} f_a(n)(\exp -j\tfrac{2\pi}{M}kn) \ . \qquad (6.3)$$

Equations 6.2 and 6.3 are now used to define the transform pair:

$$V(k) = \text{DFT}v(n) = \sum_{n=0}^{M-1} v(n)(\exp -j\beta_o kn) \qquad (6.4)$$

and

$$v(n) = \text{DFT}^{-1}V(k) = \frac{1}{M}\sum_{k=0}^{M-1} V(k)(\exp j\beta_o nk) \qquad (6.5)$$

where $\beta_o = \Delta\theta = \frac{2\pi}{M}$. For convenience, we choose M even, and, in fact, in the "fast Fourier Transform" (FFT) algorithm for high speed computation with this transform pair, it is common to let $M = 2^Q$, Q a positive integer [16.].

It is interesting to compare the DFT with Eq. 2.47 where

$$V_k' = \frac{1}{N_o}\sum_{n=0}^{N_o-1} v(n)(\exp -j\lambda_o kn) \qquad (6.6)$$

and

$$v(n) = \sum_{k=-K'}^{K'+1} V_k'(\exp j\lambda_o nk)$$

with N_o even and $N_o=2K'+2$. Using the fact that V_k' is periodic, we also can write

$$v(n) = \sum_{k=0}^{N_o-1} V_k'(\exp j\lambda_o nk) \qquad (6.7)$$

and it is obvious that except for the position of N_o and M, the DFT and the Fourier Series for periodic discrete-time data are the same (e.g., see the program HARM in the IBM Scientific Subroutine Package). The DFT treats the data record as if it were one cycle of a periodic function.

The DFT has several properties that are similar to those of the Fourier Transforms ($\mathcal{F}$ and $\mathcal{F}_d$) of Appendix A. In particular, the linearity, time shift, and frequency shift properties are true. Furthermore, forms of real and complex convolution theorems can be proved, e.g., a real convolution form can be proved for

$$f(n)*g(n) = \sum_{i=0}^{M-1} f(n-i)g(i) \tag{6.8}$$

$$= \sum_{i=0}^{M-1} f(i)g(n-i)$$

We find that

$$\text{DFT}(f(n)*g(n)) = F(k)G(k).$$

However, this differs from 2. of Section A.3.4 in terms of the summation range and in the way the shifted signal is used. Here, the variable (n-i) is modulo M so that data shifted out one end of the interval appears as data shifted into the other end. (We recall the comment after Eq. 6.7). For this reason, 6.8 is called "circular convolution".

A further question concerns the relationship between the spectral samples V(k) and the spectrum U(jω) of an original continuous-time signal u(t) which was sampled to give $u(nT_s) = v(n)$. From Eq. 2.120, with $\omega_s \geq 2\omega_o$,

$$u(t) = \sum_{n=-\infty}^{\infty} v(n) \frac{\sin \frac{\omega_s}{2}(t-nT_s)}{\frac{\omega_s}{2}(t-nT_s)} . \tag{6.9}$$

A dual form for sampling in the frequency domain can be found with the changes

$$T_s \to \Omega$$

$$\omega_s \to \frac{2\pi}{\Omega} = MT_s$$

$$t \to \omega$$

$$v(n) \to V(k) \;,$$

to give, by using Eq. 6.9,

$$U(j\omega) = \sum_{n=-\infty}^{\infty} V(k) \frac{\sin \frac{MT_s}{2} (\omega - k\Omega)}{\frac{MT_s}{2} (\omega - k\Omega)}$$

with $MT_s \geq 2t_o$. This means that u(t) should be a time-limited signal, i.e., $u(t) = 0$, $|t| > t_o$ for exact reconstruction in the frequency domain. A real signal cannot be both time limited and band limited (and quite likely is neither), and it is obvious then that the DFT can only provide approximate, although very useful, results. This result also can be seen by starting with Eq. 6.9 and taking the Fourier Transform.

$$\mathfrak{F}\, u(t) = U(j\omega) = \sum_{n=-\infty}^{\infty} v(n) \;\; \mathfrak{F}\; Sa(\frac{\omega_s}{2}(t-nT_s))$$

$$= \frac{2\pi}{\omega_s} \sum_{n=-\infty}^{\infty} v(n)\,[u_1(\omega+\frac{\omega_s}{2})-u_1(\omega-\frac{\omega_s}{2})] \cdot$$

$$\cdot (\exp -j\omega T_s)$$

$$= T_s \sum_{n=-\infty}^{\infty} v(n)(\exp -jn\omega T_s),\;\; |\omega| \leq \frac{\omega_s}{2} \;\; .$$

Then we have

$$U(jk\Omega) = T_s \sum_{n=-\infty}^{\infty} v(n)(\exp -jnk\Omega T_s), \qquad |k\Omega| \leq \frac{\omega_s}{2} \ .$$

Now

$$V(k) = \text{DFT } v(n) = \sum_{n=0}^{M-1} v(n)(\exp -j\frac{2\pi}{M}nk)$$

with $\frac{2\pi}{M} = \beta_o = \Omega T_s$. Therefore, we have $U(jk\Omega) = T_s V(k)$ if $v(n) = 0$ outside the time interval $0 \leq n \leq M$ or $0 \leq t \leq MT_s$, i.e., $2t_o \leq MT_s$. Note that the total time interval for which $u(t) \neq 0$ is $2t_o$ seconds. (Other books use a $T_o = 2t_o$ for the interval) In practical cases, then, v(n) and V(k) will characterize a signal reasonably well if the original signal can be treated as approximately band limited and time limited. The two conditions $\frac{2\pi}{T_s} \geq 2\omega_o$ and $MT_s \geq 2t_o$ imply choices of T_s and M when appropriate ω_o and t_o are assumed (or vice versa). This is discussed in Section 6.2.4. As noted earlier, the data record is MT_s seconds long and the frequency increment (frequency resolution) is $1/MT_s$ Hz. (or $\Omega = \frac{2\pi}{MT_s}$ radians/sec.). Also, it was noted that the frequency record is $1/T_s$ Hz. (or $\omega_s = \frac{2\pi}{T_s}$ radians/sec.) long. Since we usually are interested in $|V(k)|$, $|V(k)|^2$, or $C_V(k)$ (even functions), the frequency range of characterization is $-\frac{1}{2T_s} \leq f \leq \frac{1}{2T_s}$, i.e., the highest frequency is $1/2T_s$ Hz.

For deterministic signals, $|V(k)|$ is the amplitude spectrum and $\frac{1}{M}|V(k)|^2$ is the power spectrum (approximate P_{ds}) computed using the DFT. Then the signal power is

$$P_d = \frac{1}{M} \sum_{M-1}^{M-1} v(n)^2 = \frac{1}{M} \sum_{k=0}^{n=0} (\frac{|V(k)|}{M})^2$$

where the right hand equality can be shown by using Eq. 6.5 in the left hand sum. If the data record is assumed to be a section of an ergodic signal, a pseudo-correlation function is

$$c_V(n) = \frac{1}{M} \sum_{m=0}^{M-1} v(m)v(m+n)$$

and a corresponding pseudo-spectral density (approximate R_V) is

$$C_V(k) = \text{DFT } c_V(n).$$

We also can show that

$$C_V(k) = \frac{1}{M} |V(k)|^2$$

which is the same as the power spectrum for a deterministic signal. This is not surprising since only a single record v(n) is used; no expectation (as in Eq. 2.87) occurs. In actually computing $C_V(k)$, the two forms are quite different both in terms of accuracy and computer running time. These differences are illustrated in the processing of a sample function from a Gauss-Markov sequence later in this section. There are many computational details to consider in using these forms. For this elementary introduction, we shall skip several questions on errors (e.g., bias and variance) and refer to [67.] and [68.].

6.2.3 Filtering

Methods for designing continuous-time (analog) filters, i.e., two-port devices that are frequency selective, have been developed over several decades. The well-known image parameter methods for designing constant-k, m-derived, and composite passive filters are examples where inductors and capacitors are used to obtain low pass, high pass, and band pass or band reject characteristics using T and Π forms [1.] [2.]. Other structures such as lattice networks and other element pairs, particularly resistors and capacitors, also have been widely used. Approximation of an ideal filter transfer function to specify H(s) is another approach, and the Butterworth and Chebyshev forms are common choices for low pass filters. Design tables are given in [3.]. More recently, the availability of relatively inexpensive integrated circuit operational amplifiers has stimulated the development of many types of active analog filters [4.] [5.]. One simple design procedure is given in [6.], and a review of sensitivity problems is given in [7.]. In general, the various single input/single output continuous-time state variable simulation diagrams (which can be realized using R's, C's, and operational amplifiers) are in this class. Some configurations may have undesirable sensitivity problems and some may require more components than others. All of these methods basically are concerned with a realization of a transfer function H(s) under various constraints on configurations, component availability, and pole/zero locations.

For discrete-time (digital) filters, we would expect on the basis of results in Chapters 3 and 4 that there are analogies between H(s) and H(z). In particular, for any H(s), we could use Conversion VII of Section 3.6 or Eq. A.15 to find one possible H(z) from an H(s). The realization of H(z) is relatively simple if the signal processing system includes a digital computer. This is at least partly responsible for the rapid rise in interest in digital signal processing. However, it happens that in addition to the simplicity and versatility of digital realization, there is a great variety of possible forms for H(z). One classification of digital filters is according to whether the unit pulse response h(n) is of finite duration or not. Thus the terms infinite "impulse" response (IIR) and finite "impulse" response (FIR) filters are used. The IIR filters correspond roughly to the various forms of H(s) above, and, in fact, many of the design methods for H(s) carry over to the design of H(z). It is the FIR filters that are of primary interest in this section, and one method of design and a popular method of realization are discussed.

To show the differences between IIR and FIR filters, we consider

$$H(z) = \frac{N(z)}{M(z)} = \frac{\beta_o + \beta_1 z + \ldots + \beta_r z^r}{\alpha_o + \alpha_1 z + \ldots + z^r}$$

from Eq. 3.50. As an example, for r=2,

$$H(z) = \frac{\beta_o + \beta_1 z + \beta_2 z^2}{\alpha_o + \alpha_1 z + z^2} = \frac{\beta_o + \beta_1 z + \beta_2 z^2}{(z-p_1)(z-p_2)} .$$

We assume α_o and α_1 are such that $|p_1|, |p_2|<1$ where

$$p_1, p_2 = \frac{-\alpha_1 \pm \sqrt{\alpha_1^2 - 4\alpha_o}}{2} .$$

1. If α_1 and α_o are not zero, there are two cases:

$$p_1 = p_2 \quad \text{or} \quad p_1 \neq p_2 .$$

For $p_1 \neq p_2$,

$$\begin{aligned} h(n) &= 0 && n < 0 \\ &= \beta_2 && n = 0 \\ &= \left[\frac{\beta_o + \beta_1 p_1 + \beta_2 p_1^2}{p_1 - p_2}\right](p_1)^{n-1} \\ &\quad + \left[\frac{\beta_o + \beta_1 p_2 + \beta_2 p_2^2}{p_2 - p_1}\right](p_2)^{n-1} && n \geq 1 \end{aligned}$$

and for $p_1 = p_2$,

$$h(n) = 0 \qquad n < 0$$

$$= \beta_2 \qquad n = 0$$

$$= (\beta_1 + 2\beta_2 p_1)(p_1)^{n-1}$$

$$+ (\beta_0 + \beta_1 p_1 + \beta_2 p_1^2)(n-1)(p_1)^{n-2} \qquad n \geq 1$$

In both cases since p_1 and p_2 are inside the unit circle, h(n) will approach zero for large n, but will not go exactly to zero for finite n. Thus these forms are for IIR filters. (We neglect special cases $\alpha_1=0$, $\alpha_0 \neq 0$, and $\alpha_1 \neq 0$, $\alpha_0=0$).

2. If $\alpha_1 = \alpha_0 = 0$, $p_1 = p_2 = 0$ and

$$H(z) = \beta_2 + \beta_1 z^{-1} + \beta_0 z^{-2}$$

so that

$$h(n) = 0 \qquad n < 0$$

$$= \beta_2 \qquad n = 0$$

$$= \beta_1 \qquad n = 1$$

$$= \beta_0 \qquad n = 2$$

$$= 0 \qquad n \geq 3$$

which obviously is of finite duration and is therefore an FIR filter.

In Case 1, a time domain realization is

$$z(n) = \beta_2 v(n) + \beta_1 v(n-1) + \beta_0 v(n-2)$$

$$- \alpha_1 z(n-1) - \alpha_0 z(n-2) ,$$

a recursive form commonly associated with IIR filters. In

Case 2, a time domain realization is

$$z(n) = \beta_2 v(n) + \beta_1 v(n-1) + \beta_o v(n-2) ,$$

a <u>nonrecursive</u> form commonly associated with FIR filters. The recursive form uses present and past inputs and past outputs to form the present output; the nonrecursive form uses only present and past inputs. In terms of poles, the IIR filter may have poles anywhere inside the unit circle, but the FIR filter has only a multiple order pole at the origin.

Methods for designing IIR filters include:

1. Impulse invariance [8.] where h(n) is obtained as the samples of an $h^1(t) = \mathcal{L}^{-1}H^1(s)$ and forms for $H^1(s)$ are given.
2. Magnitude squared [8.] which uses Butterworth and Chebyshev forms.
3. Bilinear transformation [8.] [9.] where

$$s = \frac{z-1}{z+1} \qquad (\text{or} \quad s = \frac{2}{T_s}\ \frac{z-1}{z+1})$$

   and the design of an H(s) (by methods suggested above) is converted to an H(z).
4. Least squares [10.] where H(z) is obtained by fitting to a specified set of $H(\gamma_i)$, $i = 1,\ldots,r$.

Methods for designing FIR filters include:

1. Classical numerical methods, e.g., moving average filters and differentiating filters [11.].
2. Weighted least squares [12.] where H(z) of a specified form is obtained by fitting to a specified H(s).
3. Frequency sampling [13.] where $H(\gamma)$ is obtained by interpolating from samples $H(\gamma_k) = H(\exp j\theta_k)$.
4. Equiripple [14.] [15.] where H(z) is obtained by

specifying a maximum allowable deviation from a
desired form at a finite number of points.

Rather complete expositions of modern digital signal processing methods are given in [65.] and [66.].

The frequency sampling method of designing an FIR filter
is developed as follows. For a filter $H(\gamma) = H(\varepsilon^{j\theta})$,
frequency domain samples are

$$H_k = H(\varepsilon^{j\theta_k}) = |H_k|\varepsilon^{j\phi_k}$$

for $k = 0, 1, 2, \ldots, N-1$ and N even. This is Case A of [13.].
Using the DFT, the corresponding $h(n)$ is

$$h(n) = DFT^{-1}H_k = \frac{1}{N}\sum_{k=0}^{N-1} H_k(\exp j\beta_o nk)$$

for $n = 0, 1, 2, \ldots, N-1$. Also

$$H(z) = \mathfrak{z}\ h(n) = \sum_{n=0}^{N-1} h(n)z^{-n}$$

$$= \frac{1}{N}\sum_{n=0}^{N-1}\sum_{k=0}^{N-1} H_k(\exp j\beta_o nk)z^{-n}\quad,$$

which can be put in the form

$$H(z) = \frac{1}{N}\sum_{k=0}^{N-1} H_k\left[\frac{1-((\exp j\beta_o k)z^{-1})^N}{1-((\exp j\beta_o k)z^{-1})}\right]z^{-1}$$

$$= \frac{1-z^{-N}}{N}\sum_{k=0}^{N-1}\frac{H_k}{1-((\exp j\beta_o k)z^{-1})}\quad.$$

On the unit circle,

$$H(\gamma) = H(\varepsilon^{j\theta}) = \frac{1 - (\exp -jN\theta)}{N} \sum_{k=0}^{N-1} \frac{H_k}{1 - (\exp j\beta_o k)(\exp -j\theta)}$$

and after some manipulation

$$H(\gamma) = H(\varepsilon^{j\theta})$$
$$= \frac{(\exp -j\frac{\theta}{2}(N-1))}{N} \sum_{k=0}^{N-1} H_k \frac{(\exp -j\frac{\beta_o}{2}k)\sin(\frac{N\theta}{2})}{\sin(\frac{\theta}{2} - \frac{\beta_o}{2}k)} \quad (6.10)$$

In this form, $H(\gamma)$ is specified from the samples H_k. The function

$$\frac{\sin(\frac{N\theta}{2})}{\sin(\frac{\theta}{2} - \frac{\beta_o}{2}k)}$$

is an interpolating function. The design of the filter is now a matter of choosing N and the samples H_k such that $H(\gamma)$ as given by Eq. 6.10 has the required properties.

A procedure suggested in [13.] is as follows:

1. Choose a set of H_k such that each sample is real and the set is symmetric, i.e., $H_k = H_{N-k}$, $k = 1, 2, \ldots, \frac{N}{2}$. H_o, which is in the set, also is specified and equals H_N, which is not in the set.
2. Find $h(n) = DFT^{-1}H_k$.
3. Fold h(n) such that $h(n) = h(-n)$,

$$n = 1, 2, \ldots, \frac{N}{2}.$$

4. Add $\frac{15N}{2}$ zero values to h(n) for both positive and negative n where $|n| > \frac{N}{2}$. This means that 16N sample values for h(n) are used, and the consequent step size in θ is one-sixteenth of $\Delta\theta = \frac{2\pi}{N}$. Call the augmented pulse response $h_a(n)$.
5. Take DFT $h_a(n) = H(\exp j\theta_i)$, $i = 0, 1, \ldots, 16N-1$.

The increment in θ is $\frac{2\pi}{16N}$ and this gives a good approximation, i.e., $H(\exp j\theta_i) \cong H(\exp j\theta) = H(\gamma)$, for $0 \leq \theta \leq 2\pi$.

The choice of values for H_k is: $H_k = 1$ for the pass band, $H_k = 0$ for the stop band, and $0 < H_k < 1$ in the transition regions. For a given number of values H_k is the transition regions, optimal values can be found that minimize the maximum sidelobe level in the stop band, and extensive computations have been carried out for a number of cases. For <u>low pass</u> filters, let

N = number of samples H_k
$N-1$ = number of poles and zeros of $H(z)$
M_o = number of values of H_k in a transition band
BW = number of values for $H_k = 1$ when k goes from 0 to the transition band.

These are illustrated in Figure 6.1.

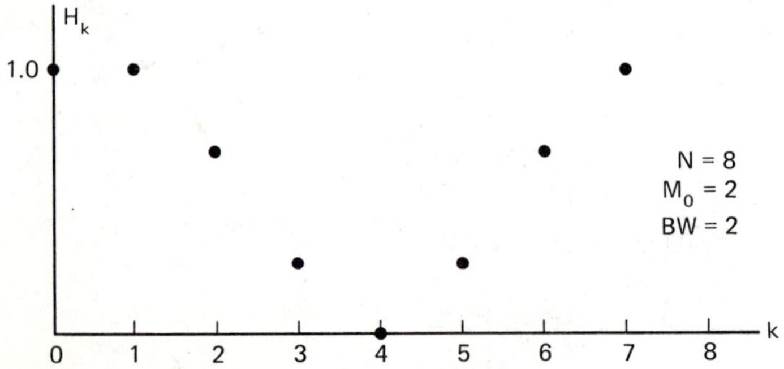

Figure 6.1 Specification of samples H_k

Tables have been prepared for low pass filters in [13.] for many values of N, M_o, BW, and parts of two of these are given in Table 6.1. As an example, if $N=64$, $BW=3$, and $M_o=2$, the values for H_k are:

TABLE 6.1*

TRANSITION VALUES FOR LOW PASS FILTERS

$M_o = 2$

BW			
	N=32		
1	-67.37020397	0.09610596	0.59045212
2	-63.93104696	0.11263428	0.60560235
3	-62.49787903	0.11931763	0.61192546
5	-61.28204536	0.12541504	0.61824023
7	-60.82049131	0.12907715	0.62307031
9	-59.74928617	0.12068481	0.60685586
11	-62.48683357	0.13004150	0.62821502
13	-70.64571857	0.11017914	0.60670943
	N=64		
1	-70.26372528	0.09376831	0.58789222
2	-67.20729542	0.10411987	0.59421778
3	-65.80684280	0.10850220	0.59666158
4	-64.95227051	0.11038818	0.59730067
5	-64.42742348	0.11113281	0.59698496
9	-63.41714096	0.10936890	0.59088884
13	-62.72142410	0.10828857	0.58738641
17	-62.37051868	0.11031494	0.58968142
21	-62.04848146	0.11254273	0.59249461
25	-61.88074064	0.11994629	0.60564501
29	-70.05681992	0.10717773	0.59842159

$M_o = 3$

BW				
	N=32			
1	-93.11873436	0.01735230	0.20052231	0.68302930
2	-89.36560249	0.02354126	0.23959557	0.71593525
4	-86.97191620	0.02770996	0.26135787	0.73350248
6	-86.69376850	0.02871094	0.26670884	0.73796855
8	-88.41957283	0.02705688	0.26084303	0.73367810
10	-92.86338043	0.02340698	0.24691517	0.72345444
12=	-115.13009739	0.01320190	0.20258948	0.68939742
	N=64			
1	-94.71326447	0.01544800	0.19125221	0.67535861
2	-89.93906212	0.02057495	0.22634942	0.70507613
3	-87.99901295	0.02438354	0.24519492	0.72204262
4	-87.48722935	0.02581177	0.25236063	0.72570913
8	-85.52755356	0.03010864	0.27213724	0.74181077
12	-85.35023785	0.02996826	0.27101526	0.74040381
16	-85.01383400	0.03095703	0.27556998	0.74434815
20	-85.68937778	0.02975464	0.27059980	0.74029023
24	-86.59522438	0.02882080	0.26751875	0.73850916
28	-115.01544189	0.01322021	0.20274639	0.68953717

(First column gives minimax values)

$$H_k = 1 \qquad k = 0,\ 1,\ 2,\ 62,\ 63$$

$$H_3 = H_{61} = 0.59666158$$

$$H_4 = H_{60} = 0.10850220$$

$$H_k = 0 \qquad k = 5,\ \ldots,\ 59$$

and the minimax value is -65.8 db. At $k = 2$, $\theta = 2\Delta\theta = 2 \cdot \frac{2\pi}{64} = \frac{\pi}{16}$ radians. The ripple amplitude in the pass band generally is less than 0.15 db [13.]. We shall use this filter in the example given after the next section.

To realize the filter using the DFT (or an FFT algorithm), for input v(n) and output z(n), perform three operations:

1. $V(k) = \text{DFT } v(n), \qquad n = 0, 1, \ldots, N-1$
2. $Z(k) = H_k V(k), \qquad k = 0, 1, \ldots, N-1$
3. $z(n) = \text{DFT}^{-1} Z(k), \qquad n = 0, 1, \ldots, N-1$.

Reference [13.] also gives design tables for band pass filters and differentiators. Furthermore, some effects of quantization (finite register lengths) and nonuniform spacing of the samples H_k are given.

6.2.4 Spectral Estimation

Returning to the two forms for $C_V(k)$ in Section 6.2.2, i.e.,

$$C_V(k) = \text{DFT } c_V(n)$$

and

$$C_V(k) = \frac{1}{M} |V(k)|^2 ,$$

we see that the first involves an averaging in the time domain while the second does not involve any averaging. Also, in a typical case, M, the number of samples of v(n) would be large (e.g., 2^{10} or more) but $c_V(n)$ would be approximately time-limited over a much smaller range of n. Thus when the DFT is taken in the first case, over say M´ values of $c_V(n)$, M´<<M, the corresponding frequency spacing for $C_V(k)$ is $\frac{2\pi}{M'T_s}$. However, when V(k) = DFT v(n) is taken for M values of v(n), the frequency spacing is $\frac{2\pi}{MT_s} << \frac{2\pi}{M'T_s}$. Furthermore, the computing time for $c_v(n)$ generally is very large. Thus, both the appearance of the results in the two cases, and the computing times will be different. One method of processing using the second form, which keeps the frequency spacing larger, is to divide the record of v(n) into a number of segments, i.e., use an M´<<M and compute a $C_V^i(k)$ for each segment. The final $C_V(k)$ is then taken as an average of the several $C_V^i(k)$ [23.].

A further refinement concerns the use of lag windows [17.]. In using Eq. 6.4, the data v(n), n = 0,1,...,M-1, are weighted equally. This is called a rectangular lag window as shown in Figure 6.15.

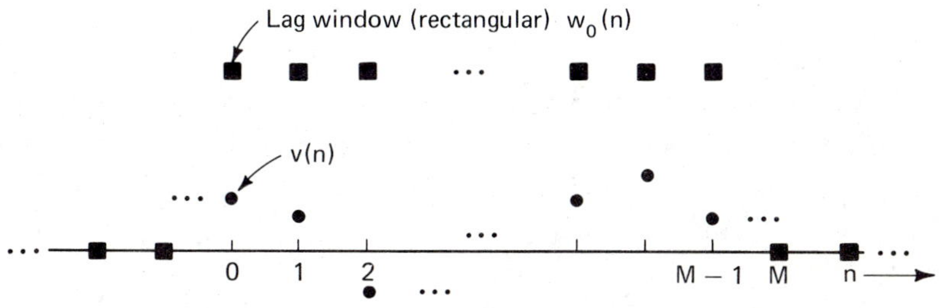

Figure 6.15

Multiplication of v(n) by the function

$$w_0(n) = v_1(n) - v_1(n-M),$$

in the time domain is equivalent to convolution of V(k) = DFT v(n) with

$$W_0(k) = \text{DFT } w_0(n)$$

in the frequency domain. Inspection of Transform 1 of Table A.1 of Appendix A shows that the envelope of $W_0(k)$ is an

$\text{Sa } x = \frac{\sin x}{x}$ function, and the side lobes allow "leakage" of frequency components of V(k) that cause $C_V(k)$ to be different from the true spectrum of an infinite-time record. $W_o(k)$ is called a spectral window. It has been found that lag windows other than $w_o(n)$ can reduce the inaccuracies associated with the side lobes of the corresponding spectral window, and the use of these is illustrated in the example given later in this section. In general, for any lag window $w_\ell(n)$, Eq. 6.4 is replaced by

$$V_\ell(k) = \sum_{n=0}^{M-1} w_\ell(n) v(n) (\exp -j\beta_o kn) \tag{6.11}$$

$w_\ell(n)$ must be one at the center of the data interval, zero outside the interval, and symmetric about the center of the interval.

6.2.5 An Example

To illustrate the use of the DFT in spectral estimation and in realizing the digital filter designed above, a Gauss-Markov sequence was generated using a model as in Chapter 4, Section 4.5. For

$$r_V(k) = (\exp -a|k|), \qquad a > 0$$

$$= b^{|k|} \qquad \text{for} \quad \varepsilon^{-a} = b ,$$

we have

$$r_V(z) = \frac{b^2-1}{b} \frac{z}{(z-b)(z-b^{-1})}$$

$$= (1-b^2) \left(\frac{z}{z-b}\right) \left(\frac{z^{-1}}{z^{-1}-b}\right) .$$

We let

$$H(z) = \frac{1-b^2}{\sqrt{K}} \quad \frac{z}{z-b}$$

where K is the intensity (or mean square) of a zero-mean white input to H(z). The output of H(z) is {V(n)}, the input to the digital filter. The output of the digital filter is {Z(n)}. For a = 0.39, the bandwidth of {V(n)} is about two times BW of the digital filter in the previous section.

A sequence of M = 4096 values of v(n) was generated and then the following were calculated:

1. a) $$c_V(n) = \sum_{m=0}^{M-1} v(m)v(m+n)$$

   b) $C_{V1}(k) = \text{DFT } c_V(n)$ (using $M' = 64$)

2. a) $V(k) = \text{DFT } v(n)$ (using M = 4096)

   b) $C_{V2}(k) = \frac{1}{M} |V(k)|^2$

3. a) For $M' = 64$, there are 64 nonoverlapping sections of the original record; using these

   $$V^i(k) = \text{DFT } v^i(n), \quad i = 1,2,\ldots,64 .$$

   b) $C_V^i(k) = \frac{1}{M'} |V^i(k)|^2$

   c) $C_{V3}(k) = \text{Avg } C_V^i(k)$

4. Part 3. repeated using a lag window [23.].

$$w_4(n) = 1 - \left| \frac{n - \left(\frac{M'-1}{2}\right)}{\frac{M'+1}{2}} \right|$$

$$n = 0,1,2,\ldots,M'-1 \ .$$

The result is $C_{V4}(k)$.

Figure 6.16 shows the results for the computed spectra as well as the true $R_V(\exp j\theta_k)$, $\theta_k = k\Delta\theta = \frac{2\pi k}{M'}$. It is clear that C_{V1} is very close to R_V, but the computation time is larger than that for the others. C_{V2} is not satisfactory; in fact, since the figure shows only every 64th point, C_{V2} is worse than it appears here. C_{V3} and C_{V4} are both reasonably good estimates of R_V. C_{V4} would have to be rescaled to

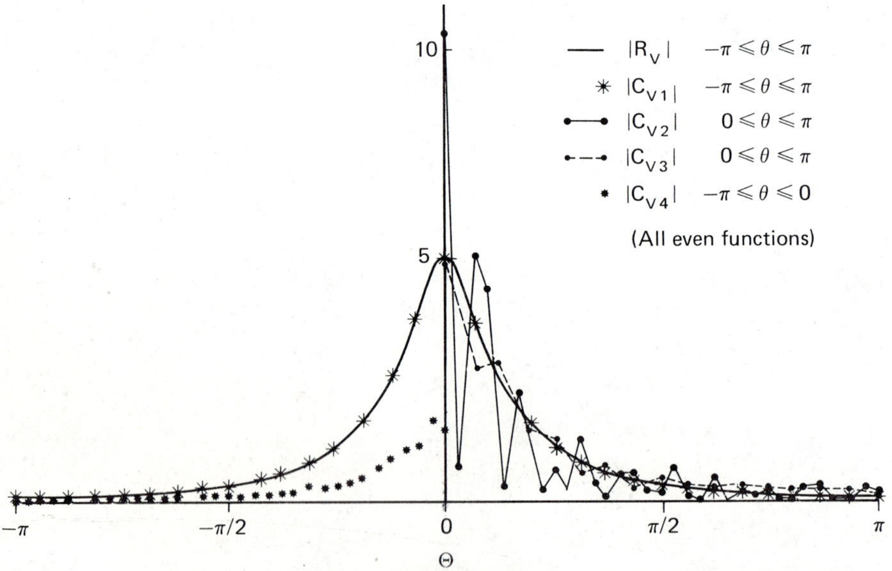

Figure 6.16

account for attenuation of the window w_4. Since the functions are even, only one side of C_{V2}, C_{V3}, and C_{V4} is plotted. We also can see similar results plotted on an oscilloscope controlled by the computer. The first figure shows C_{V1} compared to $R_V(\varepsilon^{j\theta})$ (dashed) where the v(n) were used to find c_V for 64 values. $C_V(k)$ was then found as $DFTc_V$. The second figure shows C_{V2} where the frequency increment is 1/64 of that for C_{V1}. The third figure shows C_{V3}, and the improvement over the first figure is apparent. Variances for the estimates, i.e., error variances, are given in [23.] for C_{V3} and C_{V4}. References [67.] and [68.] are excellent general sources on spectral estimation.

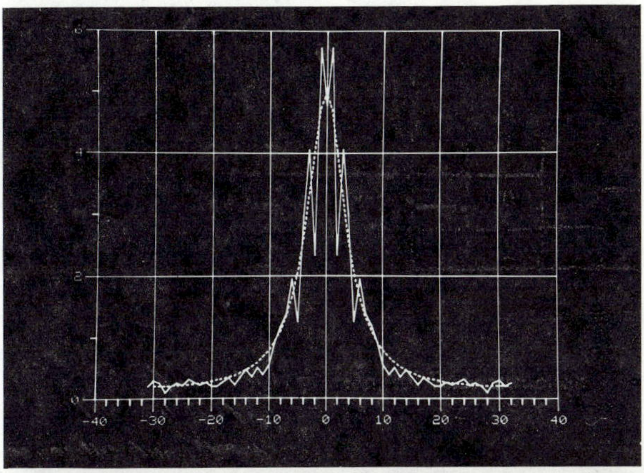

C_{V1}

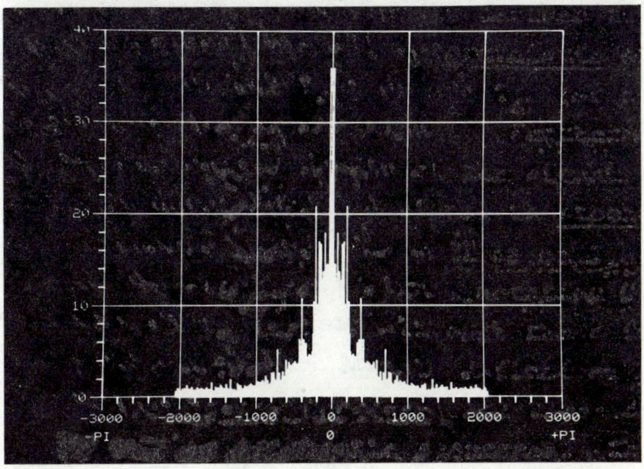

C_{V2}

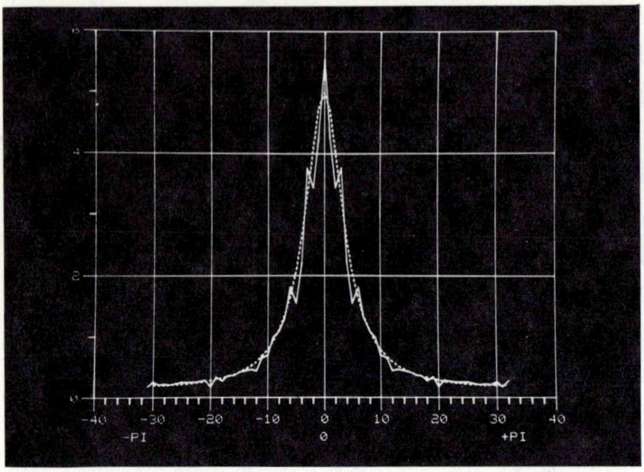

C_{V3}

To show the effect of the digital filter, the segments of 3a. were processed as

$$Z^i(k) = H_k V^i(k), \qquad k = 0,1,2,\ldots,63$$

$$i = 1,2,\ldots,64.$$

The 64 records of $Z^i(k)$ were then processed as in 3. and 4. to produce $C_{Z3}(k)$ and $C_{Z4}(k)$. The reduction in bandwidth caused by the low-pass filter is shown in Figure 6.17. The filter has been used as if the sequences of 64 data points were being received in real time, i.e., the realization of Section 6.2.3 is used. An alternative, which would be equivalent in a simulation, is to compute $C_{Z3}(k)=H_k^2 C_{V3}(k)$ and $C_{Z4}(k)=H_k^2 C_{V4}(k)$. We note too that if an output time history z(n) were of interest, we would use the segments $z^i(n)=DFT^{-1}Z^i(k)$, $i=1,2,\ldots,64$, to produce 4096 values of z(n) corresponding to the values of v(n). We also can see the effect of the low-pass filter by comparing an oscilloscope plot, a filtered C_{V2} (i.e., a C_{Z2}), with the plot of C_{V2} shown previously.

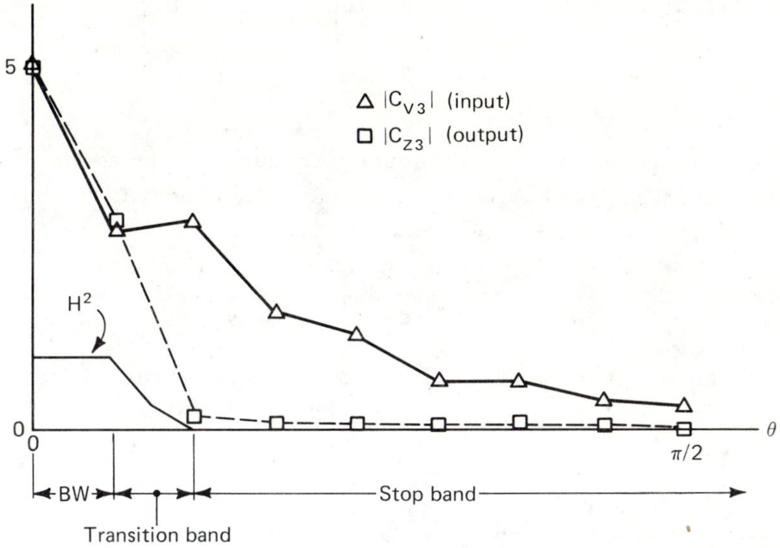

Figure 6.17 Low pass filtering

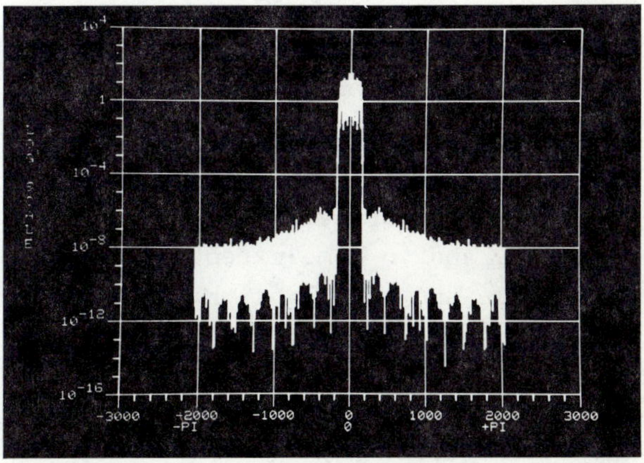

C_{Z2}

We complete the example by returning to the question of choosing T_s and M when an original signal u(t) is to be processed using the DFT. T_s is chosen such that one of the time sampling theorems of Chapter 2 is approximately satisfied. This also determines the frequency range or maximum frequency

$1/2T_s$ Hz. Thus there is an implied ω_o such that

$$\omega_s = \frac{2\pi}{T_s} \geq 2\omega_o = 4\pi f_o \ .$$

M is chosen to give a satisfactory frequency increment $1/MT_s$ Hz. (or $\Omega=2\pi/MT_s$ rad./sec.). However, in choosing M there is an implied t_o such that

$$MT_s \geq 2t_o \ .$$

We note too that combining these two inequalities gives

$$f_o t_o \leq \frac{M}{4}$$

a "time-frequency" product. Finally, we recall that for a deterministic signal, t_o is relative to u(t) and ω_o to $U(j\omega) = \mathfrak{F}\, u(t)$. For an ergodic process {U(t)}, τ_o (in place of t_o) is relative to $r_U(\tau)$ and ω_o is relative to $R_U(\omega)$. These were discussed in Chapter 2, Section 2.9.1. In a practical case, we generally do not know t_o and ω_o. However, it is sometimes advantageous to establish an ω_o by prefiltering with a sharp cutoff low-pass filter to avoid processing what may be more noise than signal. There is then an obvious T_s. The choice of M frequently is a matter of the frequency resolution required, and to some extent, the length of the data record available and the algorithm (FFT) used.

Periodic components in u(t) (or v(n)) and $r_U(\tau)$ (or $r_V(n)$) are not time limited. If the periods are known or estimated, the appropriate Fourier Series can be used to represent these components (e.g., Eq. 6.7).

A set of numerical values related to this particular simulation is as follows. Consider a process {U(t)} which has

$$r_U(\tau) = (\exp -c|\tau|),$$

$$R_U(\omega) = \frac{2c}{\omega^2 + c^2}$$

and $\omega_h = c$ is the half power frequency. If U(t) is sampled at $1/T_s$ per second,

$$r_V(n) = r_U(nT_s) = (\exp -cT_s|n|)$$
$$= (\exp -a|n|) = b^{|n|}$$

and $a = cT_s$ and $b = \varepsilon^{-a}$. Also

$$R_V(\gamma) = \frac{b^2-1}{b} \frac{\gamma}{(\gamma-b)(\gamma-b^{-1})} .$$

Now since $r_U(\tau)$ is not time-limited and $R_U(\omega)$ is not band-limited, we let

$$\omega_o = \alpha\omega_h \quad \text{and} \quad \tau_o = \beta\tau_1 ,$$

where α and β are factors that are chosen to give good approximations. In this example, $\omega_h = c$ and $\tau_1 = \frac{1}{c}$. Then we choose

$$\omega_s = 2\omega_o = 2\alpha c$$

or

$$T_s = \frac{\pi}{\alpha c}$$

and choose

$$M'T_s = 2\tau_o = \frac{2\beta}{c}$$

or

$$M' = \frac{2\alpha\beta}{\pi} .$$

In this example, $\alpha=8$ and $\beta=4\pi$ so that $M' = 64$. These choices mean that $|R_U(\omega_o)|$ is about 1.5% of $|R_U(0)|$, and $r_U(\tau_o)$ is about $6\text{x}10^{-4}\%$ of $r_U(0)$. T_s is small enough to avoid serious aliasing, i.e., overlap of sections of $R_U^*(\omega)$ as shown in Chapter 2. To further illustrate with numerical values for this example, assume $c = 200\pi$ rad./sec., i.e., the half power frequency for {U} is 100 Hz. Then T_s = 1/1600 or f_s = 1600 Hz. Also, $\Omega = 50\pi$ or Δf = 25 Hz, the frequency spacing for $C_V(k)$. The bandwidth of the digital filter is about 50 Hz and the total amount of data is contained on a record that is about 2.5 seconds (4096 T_s) long.

The design of digital filters and the use of the DFT (or FFT) in implementing them and in estimating signal spectra is still a relatively new field. References [17.] and [24.] contain some of the earlier results with extensive reference to classical time-series analysis, and References [25.] and [64.] provide more recent material for further study.

6.3 NETWORK ANALYSIS

The analysis and synthesis of electric networks is a subject of major interest in engineering. The traditional methods of writing and solving rth-order ordinary differential equations, of using Fourier and Laplace Transforms and of using phasors when the signals are sinusoidal are well known. State variable models have come into use in the last fifteen years and the advantages of these methods mentioned in Chapter 3 are especially important in the analysis of complex electric networks. The principal purposes of this section are to show how the conventional and state-variable methods of network analysis are related, to show how to obtain state-variable equations, and to illustrate the use of a digital computer in finding time domain responses. Only the restricted class of networks with constant resistors, inductors, and capacitors as well as ideal voltage and current sources is considered. Much more complete developments which include mutual inductance, time-varying elements, and controlled sources are available in [18.] and [19.].

6.3.1 Definitions

The three network elements, their symbols, and voltage-current relationships are given in Figure 6.2a and the sources are shown in Figure 6.2b.

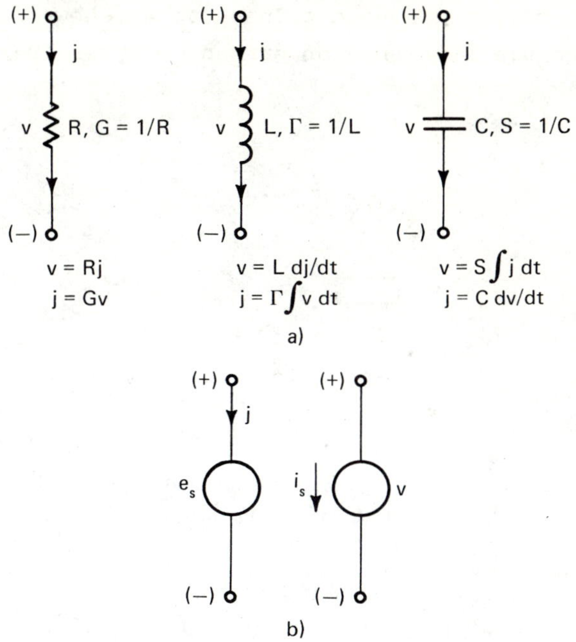

Figure 6.2 Network elements

The general branch configuration to be used is shown in Figure 6.3.

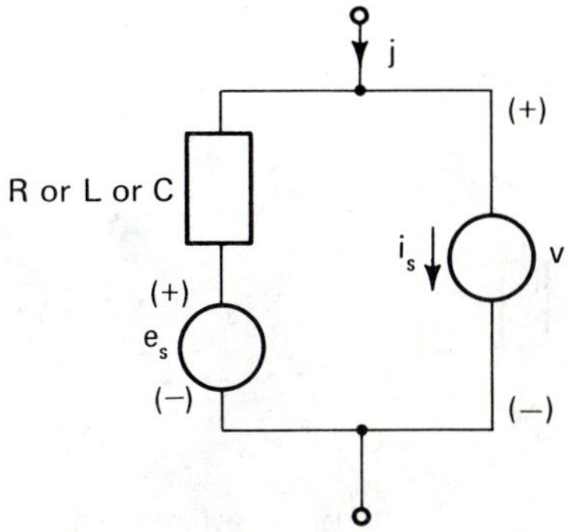

Figure 6.3 General branch

In the development of node voltage and loop current equations it is convenient to have a passive element associated with each source. If the network contains sources without the associated passive element, a preliminary modification is made by performing source transports as shown for voltage sources in Figure 6.4a and for current sources in Figure 6.4b.

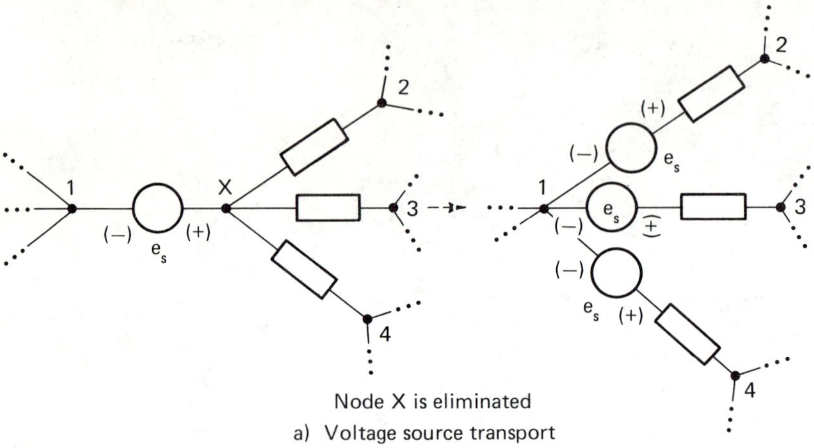

Node X is eliminated

a) Voltage source transport

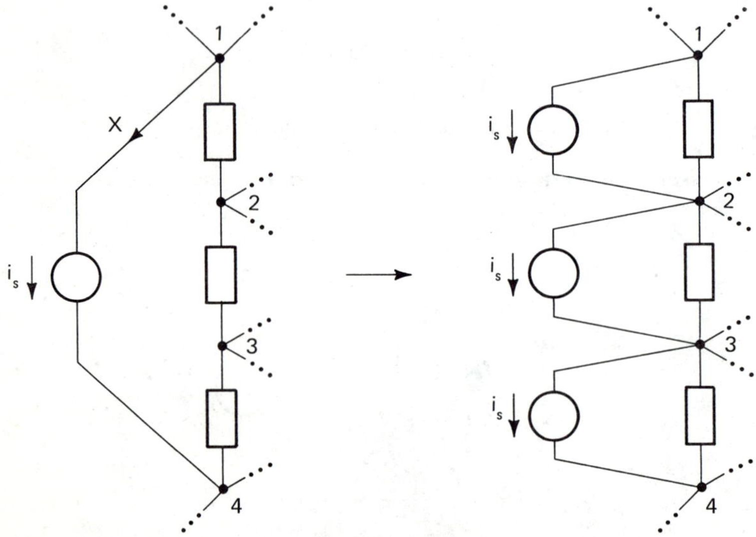

Branch X is eliminated

b) Current source transport

Figure 6.4

The common definitions of terms used in the description of networks are first summarized:

1. Branch - a) If all sources have associated passive

elements (i.e., source transport is used), a branch is a line segment representing one R, L, or C. All sources are inactive (i.e., short circuit e_s and open circuit i_s).

Branch - b) A line segment representing one R, L, C, e_s, or i_s. This will be used for the state variable forms.

2. Node - a terminal of one or more branches
3. Graph - a collection of nodes and branches
4. Oriented Graph - a graph with all branches numbered and positive directions (as in Figure 6.2) assigned.
5. Loop - two or more branches forming a closed path with each branch traversed once.
6. Tree (of a connected graph) - a collection of branches and nodes such that 1) all nodes of the graph are included, 2) all nodes are connected, and 3) there are no loops.

The collection of branches that form a tree are the tree branches, and the remaining branches are the link branches which form a cotree. Further assumptions on the class of networks are that the graphs are not hinged (i.e., not separable) and there is one separate part [20.] . Figure 6.5 illustrates some of these ideas. The self-loop A is elimin-

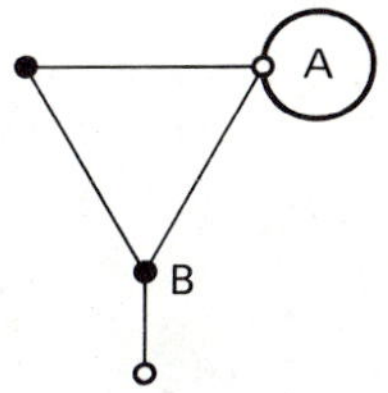

Figure 6.5

ated by requiring that a loop contain two or more branches, and branch B is eliminated since its inclusion would form a hinged graph.

Now let

N = number of nodes,

B = number of branches,

$b = N-1$ is the number of tree branches for any tree of the graph, and

$\ell = B-b$ is the number of link branches.

Also, b is the number of independent node pair voltages and ℓ is the number of independent loop currents. The following notational convention is used:

v_k = kth tree branch voltage

v_k' = kth link branch voltage

j_k = kth tree branch current

j_k' = kth link branch current

e_k = kth node to reference voltage

i_k = kth loop current

e_{sk} = kth voltage source

i_{sk} = kth current source.

6.3.2 Nodal Analysis

It is convenient to first select a tree for the given network graph (after source transport if necessary) and then to number the branches such that the branch voltage vector $\underline{v}_B$ is partitioned as

$$\underline{v}_B^T = [\underline{v}^T \vdots \underline{v}'^T] .$$

Now $\underline{v}$, which is bxl, is a branch voltage basis, and $\underline{e}$, also bxl, is a node (to reference) voltage basis. There is a non-singular transformation T_1 such that

$$\underline{e} = T_1\underline{v} \tag{6.12}$$

Now we define a <u>cut set</u> as a set of branches such that removal of these branches separates the graph into two parts (and any smaller set does not). If the convention that (for a given tree) each cut set is chosen so that it contains one tree branch and possibly several link branches, it is a <u>fundamental cut set</u>. There are then b fundamental cut sets. Then from these, we form a fundamental cut set matrix Q, which is bxB, such that the elements

q_{rs} = +1 if j_s has the same orientation as j_r

= -1 if j_s has the opposite orientation to j_r

= 0 if branch s is not in the rth cut set.

The orientation of j_r and j_s is relative to a line that cuts the graph as shown in Figure 6.18.

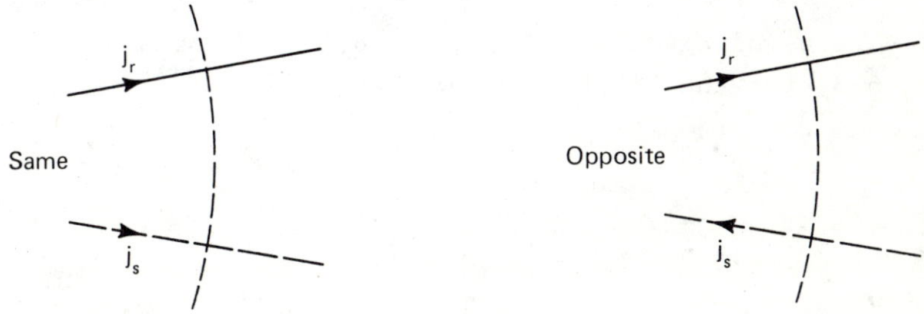

Figure 6.18

Now Kirchoff's Current Law implies that

$$Q\underline{j}_B = \underline{0} \tag{6.13}$$

where $\underline{j}_B^T = (j_1 \dots j_B)$. Also, for I_b a bxb unit matrix, we have

$$Q\underline{j}_B = Q\begin{bmatrix}\underline{j}\\ \\ \underline{j}'\end{bmatrix} = [I_b \vdots Q_o]\begin{bmatrix}\underline{j}\\ \\ \underline{j}'\end{bmatrix} = \underline{j}+Q_o\underline{j}' = \underline{0}$$

and using Q_o defined by this partition, we have

$$\underline{j} = -Q_o\underline{j}' ,$$

where Q_o is bxℓ. It also can be shown that

$$\underline{v}_B = Q^T\underline{v} . \tag{6.14}$$

This is plausible since by Tellegen's Theorem

$$\underline{v}_B^T\,\underline{j}_B = 0$$

or using 6.14,

$$\underline{v}^T(Q\underline{j}_B) = 0.$$

Since $\underline{v}$ is a voltage basis, Equation 6.13 follows. From Figure 6.3, for the kth branch, we have

$$j_k = Y_k(s)\ (v_k-e_{sk}) + i_{sk}$$

for k = 1,2,...,B and each

$$Y_k(s) = G_k \quad \text{or} \quad \frac{\Gamma_k}{s} \quad \text{or} \quad C_k s.$$

(The j_k and v_k may be primed or not according to whether they are for tree or link branches). Then, in vector notation,

$$\underline{j}_B = Y_B(s)\ (\underline{v}_B - \underline{e}_s) + \underline{i}_s \tag{6.15}$$

where the BxB diagonal matrix Y_B is the <u>branch admittance matrix</u>. Substituting Eq. 6.15 in Eq. 6.13, we get

$$Q[Y_B(s)\ (\underline{v}_B - \underline{e}_s) + \underline{i}_s] = \underline{0}$$

and using Eq. 6.14,

$$QY_B(s)\ Q^T\underline{v} = Q(Y_B(s)\underline{e}_s - \underline{i}_s)\ , \tag{6.16}$$

a set of linear equations in the tree branch voltages. Using Eq. 6.12, we obtain

$$QY_B(s)\ Q^T T_1^{-1}\underline{e} = Q(Y_B(s)\underline{e}_s - \underline{i}_s), \tag{6.17}$$

a set of linear equations in the node (to reference) voltages. Comparing this with the node voltage equations in Section 3.3.4, i.e.,

$$Y(D)\underline{e} = \underline{i}_s$$

it is clear that in terms of the notation in this section we have the equivalents:

$$Y(D) \leftrightarrow QY_B(s)\ Q^T T_1^{-1}$$

$$\underline{i}_s \leftrightarrow Q(Y_B(s)\underline{e}_s - \underline{i}_s)$$

$$\underline{e} \leftrightarrow \underline{e}\ .$$

Example 6.2a illustrates the use of Eq. 6.17.

6.3.3 Loop Analysis

Using the same numbering convention as before (after source transport if necessary),

$$\underline{j}_B^T = [\underline{j}^T \,\vdots\, \underline{j}'^T]$$

and $\underline{j}'$ is a branch current basis which is ℓx1. For $\underline{i}$ an ℓx1 loop current basis, there is a nonsingular transformation T_2 such that

$$\underline{i} = T_2\underline{j}' \; . \tag{6.18}$$

Now we define a <u>tie set</u> as a set of branches such that a loop is formed in the graph. If the convention that (for a given tree) each tie set is chosen so that it contains one link branch and possibly several tree branches, it is a <u>fundamental tie set</u>. There are ℓ fundamental tie sets. Then, we form a fundamental tie set matrix P, which is ℓxB, such that the elements

p_{rs} = +1 if j_s has the same orientation as j_r

= -1 if j_s has the opposite orientation to j_r

= 0 if j_s is not in the rth tie set.

The orientation of j_s is relative to the direction of j_r in the rth loop as shown in Figure 6.19.

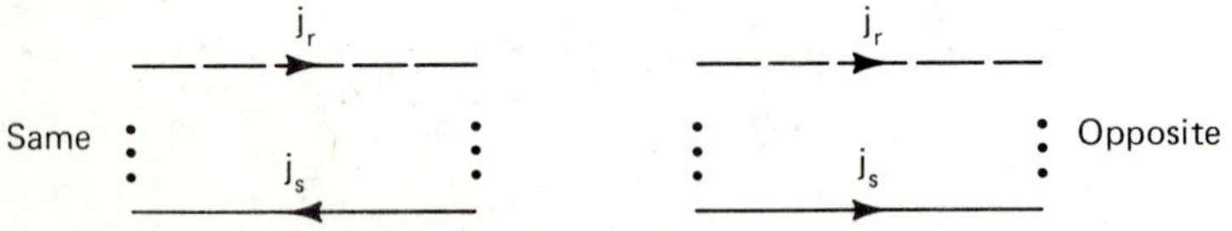

Figure 6.19

Now Kirchoff's Voltage Law implies that

$$P\underline{v}_B = \underline{0} \tag{6.19}$$

where $\underline{v}_B^T = (v_1 \ldots v_B)$. Also, we define P_o from

$$P\underline{v}_B = P \begin{bmatrix} \underline{v} \\ \\ \underline{v}' \end{bmatrix} = [P_o \vdots I_\ell] \begin{bmatrix} \underline{v} \\ \\ \underline{v}' \end{bmatrix} = P_o\underline{v}+\underline{v}' = \underline{0}$$

and

$$\underline{v}' = -P_o\underline{v} ,$$

where P_o is ℓxb and I_ℓ is an ℓxℓ unit matrix. It also can be shown that

$$\underline{j}_B = P^T\underline{j}'. \qquad (6.20)$$

This is plausible since by Tellegen's Theorem

$$\underline{j}_B^T \underline{v}_B = 0$$

or using Eq. 6.20,

$$\underline{j}'^T(P\underline{v}_B) = 0 .$$

Since $\underline{j}'$ is a current basis, Equation 6.19 follows. From Figure 6.3, for the kth branch,

$$v_k = Z_k(s)(j_k - i_{sk})+e_{sk} . \qquad (6.21)$$

for k = 1,2,...,B and

$$Z_k(s) = R_k \text{ or } L_k s \text{ or } \frac{S_k}{s} .$$

Then, in vector-matrix form,

$$\underline{v}_B = Z_B(s)\ (\underline{j}_B - \underline{i}_s)+\underline{e}_s \qquad (6.22)$$

where the BxB diagonal matrix $Z_B(s)$ in the branch impedance matrix. Substituting Eq. 6.22 in Eq. 6.19, we have

$$P[Z_B(s)\,(\underline{j}_B - \underline{i}_s)+\underline{e}_s] = \underline{0}$$

and using Eq. 6.20,

$$PZ_B(s)\,P^T\underline{j}' = P(Z_B(s)\underline{i}_s-\underline{e}_s), \qquad (6.23)$$

a set of linear equations in the link branch currents. Using Eq. 6.18, we obtain

$$PZ_B(s)\,P^T T_2^{-1}\underline{i} = P(Z_B(s)\underline{i}_s-\underline{e}_s), \qquad (6.24)$$

a set of linear equations in the loop currents. Comparing this with the loop current equations in Section 3.3.4, i.e.,

$$Z(D)\underline{i} = \underline{e}_s,$$

it is clear that we have the equivalents:

$$Z(D) \leftrightarrow PZ_B(s)P^T T_2^{-1}$$

$$\underline{e}_s \leftrightarrow P(Z_B(s)\underline{i}_s - \underline{e}_s)$$

$$\underline{i} \leftrightarrow \underline{i}\ .$$

Example 6.2b illustrates the use of Eq. 6.24.

6.3.4 State Variable Equations

It is convenient to change the method of preparing a tree before developing the state variable equations. The node voltage and loop current equations were derived using 1.a) of Section 6.3.1 as a branch definition. Ideal sources are now counted as separate branch elements as in 1.b), and source transport is not used. Then we let

N = number of nodes

B = number of branches

q = number of ideal current sources

p = number of ideal voltage sources

n_t = total number of capacitors

m_t = total number of inductors.

In the simple circuit examples in Chapter 3, the state variables were chosen as voltages across capacitors and currents through inductors. It was pointed out that the state dimension r may be less than n_t+m_t under certain conditions. Therefore, we let

n' = number of tie sets containing capacitors only or capacitors and voltage sources.

m' = number of cut sets containing inductors only or inductors and current sources.

Then, we let

$$n = n_t - n'$$

$$m = m_t - m'$$

and the state dimension is

$$r = n + m.$$

The branches are now divided into eight groups and ordered as shown below where all voltage sources are chosen as tree branches,* all current sources are chosen as link branches, n C's are tree branches, m L's are link branches, n´ C's are link branches, m´ L's are tree branches, and the resistors are assigned to tree or link branches as necessary to satisfy the other requirements. In general, this is not a unique ordering. The numerical subscripts in what follows refer to these groups:

1. Voltage-source tree branches
2. C - tree branches
3. R - tree branches
4. L - tree branches

5. C - link branches
6. R - link branches
7. L - link branches
8. Current-source link branches.

The branch current vector $\underline{j}_B$ and branch voltage vector $\underline{v}_B$ are then

$$\underline{j}_B = \begin{bmatrix} \underline{j} \\ \\ \underline{j}' \end{bmatrix} = \begin{bmatrix} \underline{j}_1 \\ \underline{j}_2 \\ \underline{j}_3 \\ \underline{j}_4 \\ \underline{j}_5 \\ \underline{j}_6 \\ \underline{j}_7 \\ \underline{j}_8 \end{bmatrix} \quad \text{and} \quad \underline{v}_B = \begin{bmatrix} \underline{v} \\ \\ \underline{v}' \end{bmatrix} = \begin{bmatrix} \underline{v}_1 \\ \underline{v}_2 \\ \underline{v}_3 \\ \underline{v}_4 \\ \underline{v}_5 \\ \underline{v}_6 \\ \underline{v}_7 \\ \underline{v}_8 \end{bmatrix}$$

*We should always be able to do this except in special cases that are not sensible, e.g., voltage sources in parallel and current sources in series.

and the dimensions are

Tree

$$\dim \underline{j}_1 = \dim \underline{v}_1 = p$$
$$\dim \underline{j}_2 = \dim \underline{v}_2 = n$$
$$\dim \underline{j}_3 = \dim \underline{v}_3 = b-p-n-m' = \beta$$
$$\dim \underline{j}_4 = \dim \underline{v}_4 = m'$$

Cotree

$$\dim \underline{j}_5 = \dim \underline{v}_5 = n'$$
$$\dim \underline{j}_6 = \dim \underline{v}_6 = \ell-q-m-n' = \alpha$$
$$\dim \underline{j}_7 = \dim \underline{v}_7 = m$$
$$\dim \underline{j}_8 = \dim \underline{v}_8 = q \ .$$

Also, $\underline{v}_1 = \underline{e}_s$ and $\underline{j}_8 = \underline{i}_s$, and $b = N-1$, $\ell = B-b$.

After forming a tree and cotree and numbering and assigning directions to all branches, a fundamental cut set is formed as

$$Q = [I_b \ \vdots \ Q_o]$$

and from Eq. 6.14,

$$\underline{v}_B = \begin{bmatrix} I_b \\ \hline Q_o^T \end{bmatrix} \underline{v}$$

or recalling the partitioning of $\underline{v}_B$, the link voltages are

$$\underline{v}' = Q_o^T \underline{v} \ . \tag{6.25}$$

Also, as shown previously,

$$\underline{j} = -Q_o \, \underline{j}' \, . \tag{6.26}$$

Now we partition Q_o, which is b×ℓ, as shown (dimensions at top and left).

$$Q_o = \begin{array}{c} \\ p \\ n \\ \beta \\ m' \end{array} \begin{array}{c} \begin{array}{cccc} n' & \alpha & m & q \end{array} \\ \left[\begin{array}{c|c|c|c} Q_{o11} & Q_{o12} & Q_{o13} & Q_{o14} \\ \hline Q_{o21} & Q_{o22} & Q_{o23} & Q_{o24} \\ \hline \Theta & Q_{o32} & Q_{o33} & Q_{o34} \\ \hline \Theta & \Theta & Q_{o43} & Q_{o44} \end{array}\right] \end{array} \tag{6.27}$$

The submatrix Q_{o31} is Θ since any capacitive tie set would contain no resistors, submatrix Q_{o42} is Θ since any inductive cut set would contain no resistors, and Q_{o41} is Θ since a capacitive tie set would contain no inductors and an inductive cut set would contain no capacitors.

In addition to the topological information contained in Eq. 6.27, the volt-ampere relationships for the six groups of branches that do not contain sources are required. Thus, we have

$$\underline{j}_2 = C_o \, \dot{\underline{v}}_2 \tag{6.28}$$

where $C_o = \text{diag}\,[C_{p+1} \cdots C_{p+n}]$, and

$$\underline{j}_3 = G_o \, \underline{v}_3 \tag{6.29}$$

where $G_o = \text{diag}\,[G_{p+n+1} \cdots G_{b-m'}]$, and

$$\underline{j}_4 = \Gamma_o \int \underline{v}_4 \, dt$$

where $\Gamma_o = \text{diag}\,[\Gamma_{b-m+1} \cdots \Gamma_b]$ or

$$\dot{\underline{j}}_4 = \Gamma_o \, \underline{v}_4 , \tag{6.30}$$

and

$$\underline{v}_5 = S_o \int \underline{j}_5 \, dt$$

where S_o = diag $[S_{b+1} \ldots S_{b+n'}]$ or

$$\dot{\underline{v}}_5 = S_o \, \underline{j}_5 \tag{6.31}$$

and

$$\underline{v}_6 = R_o \, \underline{j}_6 \tag{6.32}$$

where R_o = diag $[R_{b+n'+1} \ldots R_{B-q+m}]$, and

$$\underline{v}_7 = L_o \, \dot{\underline{j}}_7 \tag{6.33}$$

where L_o = diag $[L_{B-q+m+1} \ldots L_{B-q}]$.

Now we combine Eq. 6.28 and 6.31 as

$$\begin{bmatrix} \underline{j}_2 \\ \hline \underline{j}_5 \end{bmatrix} = \left[\begin{array}{c|c} C_o & \Theta \\ \hline \Theta & S_o^{-1} \end{array}\right] \begin{bmatrix} \dot{\underline{v}}_2 \\ \hline \dot{\underline{v}}_5 \end{bmatrix} \tag{6.34}$$

and we combine Eq. 6.33 and 6.30 as

$$\begin{bmatrix} \underline{v}_7 \\ \hline \underline{v}_4 \end{bmatrix} = \left[\begin{array}{c|c} L_o & \Theta \\ \hline \Theta & \Gamma_o^{-1} \end{array}\right] \begin{bmatrix} \dot{\underline{j}}_7 \\ \hline \dot{\underline{j}}_4 \end{bmatrix} . \tag{6.35}$$

From Eq. 6.26, using the partitioning of Eq. 6.27, we get

$$\underline{j}_2 + Q_{o21} \, \underline{j}_5 = -Q_{o22} \, \underline{j}_6 - Q_{o23} \, \underline{j}_7 - Q_{o24} \underline{i}_s . \tag{6.36}$$

Changing the left side of Eq. 6.36 and then using Eq. 6.34

we have

$$[I_n \vdots Q_{o21}] \begin{bmatrix} \underline{j}_2 \\ \hline \underline{j}_5 \end{bmatrix} = [I_n \vdots Q_{o21}] \left[\begin{array}{c|c} C_o & \Theta \\ \hline \Theta & S_o^{-1} \end{array}\right] \begin{bmatrix} \dot{\underline{v}}_2 \\ \hline \dot{\underline{v}}_5 \end{bmatrix} \tag{6.37}$$

From Eq. 6.25, again using Eq. 6.27, we obtain

$$\underline{v}_5 = Q_{o11}^T \underline{e}_s + Q_{o21}^T \underline{v}_2$$

and therefore using this in Eq. 6.37, the left side of Eq. 6.36 can be written as

$$[I_n \vdots Q_{o21}] \left[\begin{array}{c|c} C_o & \Theta \\ \hline \Theta & S_o^{-1} \end{array}\right] \begin{bmatrix} I_n \\ \hline Q_{o21}^T \end{bmatrix} \dot{\underline{v}}_2 +$$

$$+ \; [I_n \vdots Q_{o21}] \left[\begin{array}{c|c} C_o & \Theta \\ \hline \Theta & S_o^{-1} \end{array}\right] \begin{bmatrix} \Theta \\ \hline Q_{o11}^T \end{bmatrix} \dot{\underline{e}}_s$$

Now we let

$$C_a = [I_n \vdots Q_{o21}] \left[\begin{array}{c|c} C_o & \Theta \\ \hline \Theta & S_o^{-1} \end{array}\right] \begin{bmatrix} I_n \\ \hline Q_{o21}^T \end{bmatrix}$$

and

$$\hat{C}_a = -[I_n \vdots Q_{o21}] \left[\begin{array}{c|c} C_o & \Theta \\ \hline \Theta & S_o^{-1} \end{array}\right] \begin{bmatrix} \Theta \\ \hline Q_{o11}^T \end{bmatrix},$$

and Eq. 6.36 becomes

$$C_a \dot{\underline{v}}_2 = -Q_{o22}\, \underline{j}_6 - Q_{o23}\, \underline{j}_7 - Q_{o24}\, \underline{i}_s + \hat{C}_a\, \dot{\underline{e}}_s. \tag{6.38}$$

Similarly, from Eq. 6.25, using the partitioning of Eq. 6.27, we get

$$\underline{v}_7 - Q_{o43}^T \underline{v}_4 = Q_{o13}^T \underline{e}_s + Q_{o23}^T \underline{v}_2 + Q_{o33}^T \underline{v}_3 \ . \tag{6.39}$$

Changing the left side of Eq. 6.39 and then using Eq. 6.35 we have

$$[I_m \vdots -Q_{o43}^T] \begin{bmatrix} \underline{v}_7 \\ \hline \underline{v}_4 \end{bmatrix} = [I_m \vdots -Q_{o43}^T] \left[\begin{array}{c|c} L_o & \Theta \\ \hline \Theta & \Gamma_o^{-1} \end{array}\right] \begin{bmatrix} \dot{\underline{j}}_7 \\ \hline \dot{\underline{j}}_4 \end{bmatrix} \tag{6.40}$$

From Eq. 6.26, again using Eq. 6.27, we obtain

$$\underline{j}_4 = -Q_{o43}\underline{j}_7 - Q_{o44}\underline{i}_s$$

and using this in Eq. 6.40, the left side of Eq. 6.39 can be written as

$$[I_m \vdots -Q_{o43}^T] \left[\begin{array}{c|c} L_o & \Theta \\ \hline \Theta & \Gamma_o^{-1} \end{array}\right] \begin{bmatrix} I_m \\ \hline -Q_{o43} \end{bmatrix} \dot{\underline{j}}_7 + [I_m \vdots -Q_{o43}^T] \left[\begin{array}{c|c} L_o & \Theta \\ \hline \Theta & \Gamma_o^{-1} \end{array}\right] \begin{bmatrix} \Theta \\ \hline -Q_{o44} \end{bmatrix} \dot{\underline{i}}_s$$

Now we let

$$L_a = [I_m \vdots -Q_{o43}^T] \left[\begin{array}{c|c} L_o & \Theta \\ \hline \Theta & \Gamma_o^{-1} \end{array}\right] \begin{bmatrix} I_m \\ \hline -Q_{o43} \end{bmatrix}$$

and

$$\hat{L}_a = [I_m \vdots -Q_{o43}^T] \left[\begin{array}{c|c} L_o & \Theta \\ \hline \Theta & \Gamma_o^{-1} \end{array}\right] \left[\begin{array}{c} \Theta \\ \hline -Q_{o44} \end{array}\right] ,$$

Eq. 6.39 becomes

$$L_a \dot{\underline{i}}_7 = Q_{o23}^T \underline{v}_2 + Q_{o33}^T \underline{v}_3 + Q_{o13}^T \underline{e}_s + \hat{L}_a \dot{\underline{i}}_s \qquad (6.41)$$

Now Eq. 6.38 and 6.41 are almost in state variable form except that $\underline{i}_6$ and $\underline{v}_3$ are for resistive branches and their components are not state variables. From Eq. 6.26 again we have

$$\underline{i}_3 = -Q_{o32}\underline{i}_6 - Q_{o33}\underline{i}_7 - Q_{o34}\underline{i}_s \qquad (6.42)$$

and from Eq. 6.25,

$$\underline{v}_6 = Q_{o12}^T \underline{e}_s + Q_{o22}^T \underline{v}_2 + Q_{o32}^T \underline{v}_3 . \qquad (6.43)$$

Now we combine Eq. 6.38 and Eq. 6.41 to obtain

$$\left[\begin{array}{c|c} C_a & \Theta \\ \hline \Theta & L_a \end{array}\right] \left[\begin{array}{c} \dot{\underline{v}}_2 \\ \hline \dot{\underline{i}}_7 \end{array}\right] = \left[\begin{array}{c|c} \Theta & -Q_{o23} \\ \hline Q_{o23}^T & \Theta \end{array}\right] \left[\begin{array}{c} \underline{v}_2 \\ \hline \underline{i}_7 \end{array}\right] +$$

$$+ \left[\begin{array}{c|c} \Theta & -Q_{o22} \\ \hline Q_{o33}^T & \Theta \end{array}\right] \left[\begin{array}{c} \underline{v}_3 \\ \hline \underline{i}_6 \end{array}\right] + \left[\begin{array}{c|c} \hat{C}_a \frac{d}{dt} & -Q_{o24} \\ \hline Q_{o13}^T & \hat{L}_a \frac{d}{dt} \end{array}\right] \left[\begin{array}{c} \underline{e}_s \\ \hline \underline{i}_s \end{array}\right] \qquad (6.44)$$

Also, we combine Eq. 6.42 and Eq. 6.43 to obtain

$$\begin{bmatrix} \underline{j}_3 \\ \hline \underline{v}_6 \end{bmatrix} = \left[\begin{array}{c|c} \Theta & -Q_{o33} \\ \hline Q_{o22}^T & \Theta \end{array}\right] \begin{bmatrix} \underline{v}_2 \\ \hline \underline{j}_7 \end{bmatrix} + \left[\begin{array}{c|c} \Theta & -Q_{o32} \\ \hline Q_{o32}^T & \Theta \end{array}\right] \begin{bmatrix} \underline{v}_3 \\ \hline \underline{j}_6 \end{bmatrix} +$$

$$+ \left[\begin{array}{c|c} \Theta & -Q_{o34} \\ \hline Q_{o12}^T & \Theta \end{array}\right] \begin{bmatrix} \underline{e}_s \\ \hline \underline{i}_s \end{bmatrix} . \qquad (6.45)$$

Now using Eq. 6.29 and Eq. 6.32, we get

$$\begin{bmatrix} \underline{v}_3 \\ \hline \underline{j}_6 \end{bmatrix} = \left[\begin{array}{c|c} G_o^{-1} & \Theta \\ \hline \Theta & R_o^{-1} \end{array}\right] \begin{bmatrix} \underline{j}_3 \\ \hline \underline{v}_6 \end{bmatrix} . \qquad (6.46)$$

Substituting Eq. 6.45 on the right side of Eq. 6.46, we have

$$\begin{bmatrix} \underline{v}_3 \\ \hline \underline{j}_6 \end{bmatrix} = \left[\begin{array}{c|c} G_o^{-1} & \Theta \\ \hline \Theta & R_o^{-1} \end{array}\right] \left\{ \left[\begin{array}{c|c} \Theta & -Q_{o33} \\ \hline Q_{o22}^T & \Theta \end{array}\right] \begin{bmatrix} \underline{v}_2 \\ \hline \underline{j}_7 \end{bmatrix} + \right.$$

$$\left. \left[\begin{array}{c|c} \Theta & -Q_{o32} \\ \hline Q_{o32}^T & \Theta \end{array}\right] \begin{bmatrix} \underline{v}_3 \\ \hline \underline{j}_6 \end{bmatrix} + \left[\begin{array}{c|c} \Theta & -Q_{o34} \\ \hline Q_{o12}^T & \Theta \end{array}\right] \begin{bmatrix} \underline{e}_s \\ \hline \underline{i}_s \end{bmatrix} \right\}$$

Rearranging and letting

$$F = \left[\begin{array}{c|c} I_\beta & G_o^{-1} Q_{o32} \\ \hline -R_o^{-1} Q_{o32}^T & I_\alpha \end{array}\right] , \qquad (6.47)$$

we obtain

$$\begin{bmatrix} \underline{v}_3 \\ \hline \underline{j}_6 \end{bmatrix} = F^{-1} \left[\begin{array}{c|c} G_o^{-1} & \Theta \\ \hline \Theta & R_o^{-1} \end{array}\right] \left[\begin{array}{c|c} \Theta & -Q_{o33} \\ \hline Q_{o22}^T & \Theta \end{array}\right] \begin{bmatrix} \underline{v}_2 \\ \hline \underline{j}_7 \end{bmatrix} + F^{-1} \left[\begin{array}{c|c} G_o^{-1} & \Theta \\ \hline \Theta & R_o^{-1} \end{array}\right] \left[\begin{array}{c|c} \Theta & -Q_{o34} \\ \hline Q_{o12}^T & \Theta \end{array}\right] \begin{bmatrix} \underline{e}_s \\ \hline \underline{i}_s \end{bmatrix} .$$

Substituting this in the right side of Eq. 6.44, we get

$$\left[\begin{array}{c|c} C_a & \Theta \\ \hline \Theta & L_a \end{array}\right] \begin{bmatrix} \dot{\underline{v}}_2 \\ \hline \dot{\underline{j}}_7 \end{bmatrix} = \left\{ \left[\begin{array}{c|c} \Theta & -Q_{o23} \\ \hline Q_{o23}^T & \Theta \end{array}\right] + \left[\begin{array}{c|c} \Theta & -Q_{o22} \\ \hline Q_{o33}^T & \Theta \end{array}\right] \times F^{-1} \left[\begin{array}{c|c} G_o^{-1} & \Theta \\ \hline \Theta & R_o^{-1} \end{array}\right] \left[\begin{array}{c|c} \Theta & -Q_{o33} \\ \hline Q_{o22}^T & \Theta \end{array}\right] \right\} \begin{bmatrix} \underline{v}_2 \\ \hline \underline{j}_7 \end{bmatrix}$$

$$+ \left\{ \left[\begin{array}{c|c} \Theta & -Q_{o24} \\ \hline Q_{o13}^T & \Theta \end{array}\right] + \left[\begin{array}{c|c} \Theta & -Q_{o22} \\ \hline Q_{o33}^T & \Theta \end{array}\right] F^{-1} \left[\begin{array}{c|c} G_o^{-1} & \Theta \\ \hline \Theta & R_o^{-1} \end{array}\right] \times \left[\begin{array}{c|c} \Theta & -Q_{o34} \\ \hline Q_{o12}^T & \Theta \end{array}\right] \right\} \begin{bmatrix} \underline{e}_s \\ \hline \underline{i}_s \end{bmatrix} + \left\{ \left[\begin{array}{c|c} \hat{C}_a & \Theta \\ \hline \Theta & \hat{L}_a \end{array}\right] \right\} \begin{bmatrix} \dot{\underline{e}}_s \\ \hline \dot{\underline{i}}_s \end{bmatrix} \quad (6.48)$$

Now we let

$$A_1 = \left[\begin{array}{c|c} \Theta & -Q_{o23} \\ \hline Q_{o23}^T & \Theta \end{array}\right] + \left[\begin{array}{c|c} \Theta & -Q_{o22} \\ \hline Q_{o33}^T & \Theta \end{array}\right] \times$$

$$\times F^{-1} \left[\begin{array}{c|c} \Theta & -G_o^{-1}Q_{o33} \\ \hline R_o^{-1}Q_{o22}^T & \Theta \end{array}\right] \qquad (6.49)$$

and

$$B_1 = \left[\begin{array}{c|c} \Theta & -Q_{o24} \\ \hline Q_{o13}^T & \Theta \end{array}\right] + \left[\begin{array}{c|c} \Theta & -Q_{o22} \\ \hline Q_{o33}^T & \Theta \end{array}\right] \times$$

$$\times F^{-1} \left[\begin{array}{c|c} \Theta & -G_o^{-1}Q_{o34} \\ \hline R_o^{-1}Q_{o12} & \Theta \end{array}\right] \qquad (6.50)$$

and we let the state vector be

$$\underline{x} = \left[\begin{array}{c} \underline{v}_2 \\ \hline \underline{i}_7 \end{array}\right] - \left[\begin{array}{c|c} C_a^{-1}\hat{C}_a & \Theta \\ \hline \Theta & L_a^{-1}\hat{L}_a \end{array}\right] \left[\begin{array}{c} \underline{e}_s \\ \hline \underline{i}_s \end{array}\right] .$$

Then Eq. 6.48 becomes

$$\dot{\underline{x}} = A\underline{x} + B\underline{u} \qquad (6.51)$$

where

$$A = \left[\begin{array}{c|c} C_a^{-1} & \Theta \\ \hline \Theta & L_a^{-1} \end{array}\right] A_1$$

$$B = \left[\begin{array}{c|c} C_a^{-1} & \Theta \\ \hline \Theta & L_a^{-1} \end{array}\right] \left\{ B_1 + A_1 \left[\begin{array}{c|c} C_a^{-1}\hat{C}_a & \Theta \\ \hline \Theta & L_a^{-1}\hat{L}_a \end{array}\right] \right\}$$

and

$$\underline{u} = \left[\begin{array}{c} \underline{e}_s \\ \hline \underline{i}_s \end{array}\right] .$$

An output equation

$$\underline{y} = C\underline{x} + D\underline{u} \tag{6.52}$$

can be written for the specified outputs in terms of $\underline{x}$ and $\underline{u}$. It is worth noting that $\hat{C}_a$ is nonnull (i.e., $Q_{o11} \neq \Theta$) only if there are any tie sets of capacitors and voltage sources, and $\hat{L}_a$ is nonnull (i.e., $Q_{o44} \neq \Theta$) only if there are any cut sets of inductors and current sources. If $\hat{C}_a = \hat{L}_a = \theta$, the state vector $\underline{x}^T = [\underline{v}_2^T \vdots \underline{i}_7^T]$ and the second term in the brackets in B does not occur so that simple forms for A and B follow from A_1 and B_1.

An orderly procedure for obtaining the state equations is:

1. From the given network, form a tree and a cotree by arranging the 8 groups of components as suggested. Number and assign positive directions.
2. Form Q and partition as in Eq. 6.27.
3. Form the C_o, G_o, Γ_o, S_o, R_o, and L_o matrices (Eq. 6.28 through 6.33).
4. Obtain C_a, $\hat{C}_a$, and L_a, $\hat{L}_a$.

5. Obtain F (Eq. 6.47) and F^{-1}.
6. Obtain A_1 and B_1 (Eq. 6.49 and 6.50).
7. Obtain A and B (Eq. 6.51).
8. Obtain C and D (Eq. 6.52).

The following two examples illustrate our results for obtaining node voltage, loop current, and state variable equations.

Example 6.2

Consider the network shown.

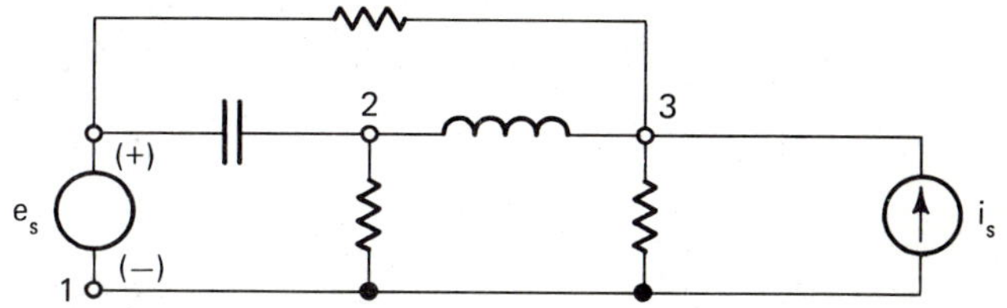

a) Find a set of node voltage equations. First use voltage source transport:

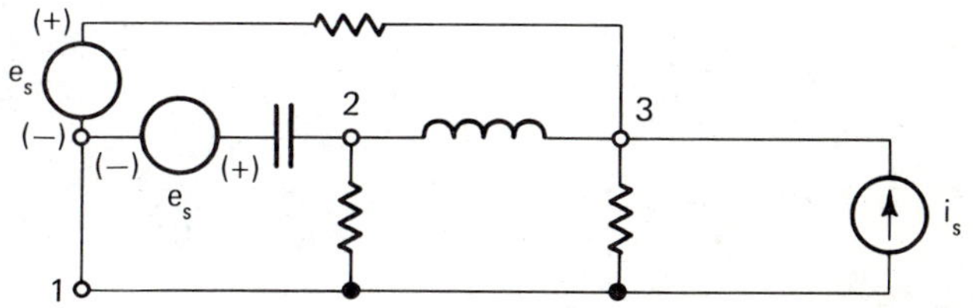

A graph is:

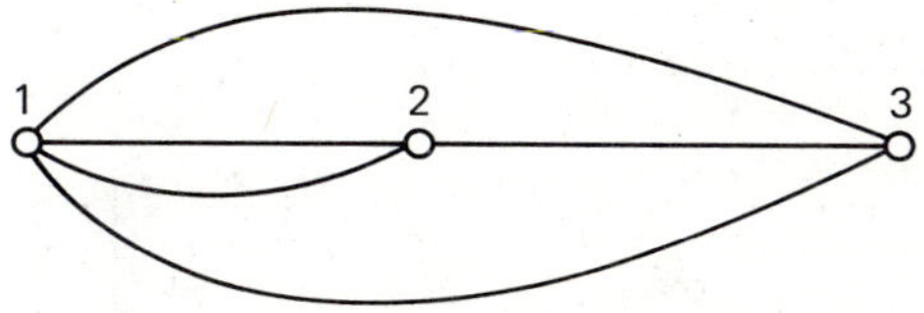

A tree and cotree from this graph are: (solid lines for tree)

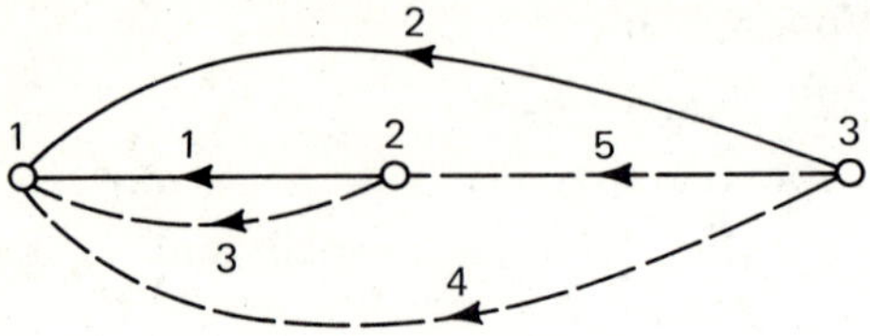

From this N=3, B=5, b=2, ℓ=3, and the branch voltages are

$$\underline{v}^T = (v_1 \quad v_2), \qquad \underline{v}'^T = (v_3 \quad v_4 \quad v_5).$$

If node 1 is the reference,

$$e_1 = v_1, \; e_2 = v_2,$$

and therefore,

$$T_1 = I.$$

To form Q, we use the cut sets formed by isolating nodes 2 and 3:

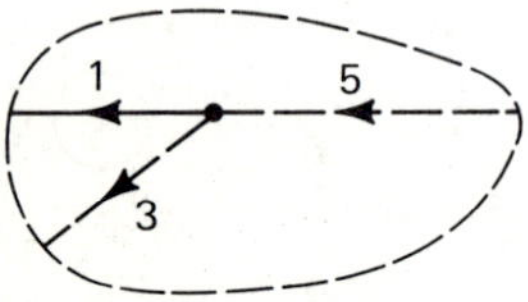

Node 2

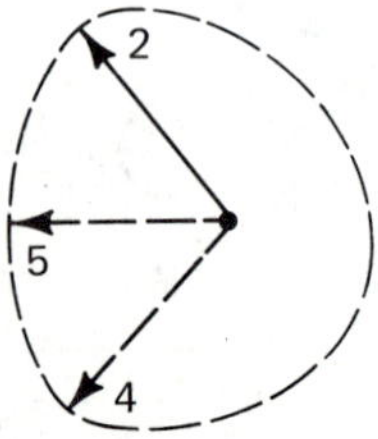

Node 3

The cut set matrix is

$$Q = \begin{bmatrix} 1 & 0 & 1 & 0 & -1 \\ 0 & 1 & 0 & 1 & 1 \end{bmatrix}$$

Also,

$$Y_B(s) = \text{diag}\ [C_1 s \quad G_2 \quad G_3 \quad G_4 \quad \frac{1}{L_5 s}] \ .$$

Therefore for

$$\underline{e}_s{}^T = (e_s \quad e_s \quad 0 \quad 0 \quad 0)$$

and

$$\underline{i}_s{}^T = (\ 0 \quad 0 \quad 0 \quad -i_s \quad 0)$$

and using Eq. 6.17, we have

$$\begin{bmatrix} C_1 s + G_3 + \frac{1}{L_5 s} & -\frac{1}{L_5 s} \\ -\frac{1}{L_5 s} & G_2 + G_4 + \frac{1}{L_5 s} \end{bmatrix} \underline{e} = \begin{bmatrix} C_1 s \ \ e_s \\ G_2 \ e_s + i_s \end{bmatrix} \ .$$

In the original diagram we get from $\underline{e}$,

e_1 = voltage of node 2 relative to node 1

e_2 = voltage of node 3 relative to node 1

b) Find a set of loop current equations. From the tree and cotree of Part a) we choose loops:

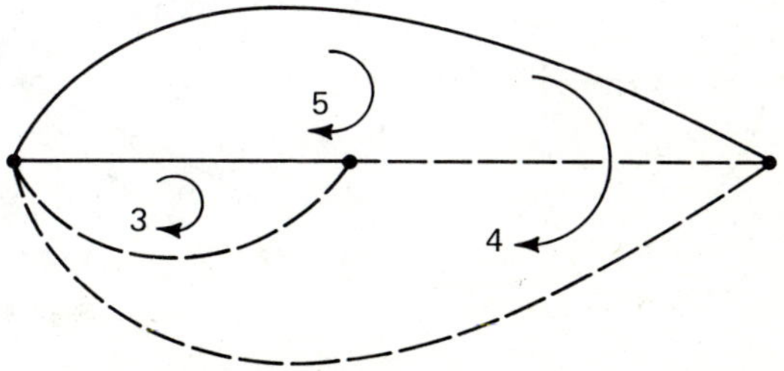

Then the branch currents are given by

$$\underline{j}'^T = (j_3 \quad j_4 \quad j_5), \quad \underline{j}^T = (j_1 \quad j_2)$$

and the loop currents by

$$\underline{i}^T = (i_3 \quad i_4 \quad i_5).$$

Now $j_3 = i_3$, $j_4 = i_4$, $j_5 = i_5$ and therefore, $T_2 = I$. For the loops chosen, we have the tie set matrix:

$$P = \begin{bmatrix} -1 & 0 & 1 & 0 & 0 \\ 0 & -1 & 0 & 1 & 0 \\ 1 & -1 & 0 & 0 & 1 \end{bmatrix}.$$

Also, $Z_B(s) = \text{diag}\,[\frac{1}{C_1 s} \quad R_2 \quad R_3 \quad R_4 \quad L_5 s]$. Using Eq. 6.24 we obtain

$$\begin{bmatrix} \frac{1}{C_1 s} + R_3 & 0 & -\frac{1}{C_1 s} \\ 0 & R_2 + R_4 & R_2 \\ -\frac{1}{C_1 s} & R_2 & \frac{1}{C_1 s} + R_2 + L_5 s \end{bmatrix} \underline{i} = \begin{bmatrix} e_s \\ e_s \; -R_4 \; i_s \\ 0 \end{bmatrix}$$

c) Find a set of state variable equations. A new tree and cotree are formed from the original diagram:

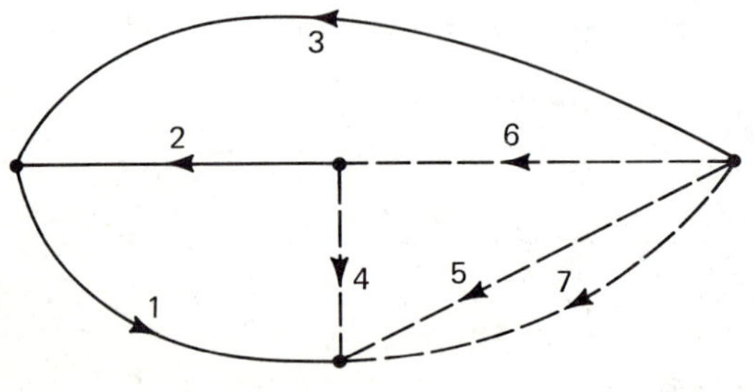

The numbering is according to the required grouping. Also,

$$N=4, \; B=7, \; b=3, \; \ell=4$$

and

$$p=1,\ n_t=n=1,\ n'=0$$

$$q=1,\ m_t=m=1,\ m'=0$$

$$\alpha=2,\ \beta=1.$$

Second, the fundamental cut set matrix is

$$Q = \begin{bmatrix} 1 & 0 & 0 & 1 & 1 & 0 & 1 \\ 0 & 1 & 0 & 1 & 0 & -1 & 0 \\ 0 & 0 & 1 & 0 & 1 & 1 & 1 \end{bmatrix}$$

and

$$Q_o = \begin{array}{c} \\ p \\ n \\ \beta \end{array} \begin{array}{c} \begin{array}{cccc} \alpha & & m & q \end{array} \\ \left[\begin{array}{cc|c|c} 1 & 1 & 0 & 1 \\ \hline 1 & 0 & -1 & 0 \\ \hline 0 & 1 & 1 & 1 \end{array}\right] \end{array} .$$

From this

$$Q_{o12} = [1 \quad 1]$$
$$Q_{o13} = 0$$
$$Q_{o14} = 1$$
$$Q_{o22} = [1 \quad 0]$$
$$Q_{o23} = -1$$
$$Q_{o24} = 0$$
$$Q_{o32} = [0 \quad 1]$$
$$Q_{o33} = 1$$
$$Q_{o34} = 1$$

Now for the parameters we have

$$C_o = C_2$$

$$G_o = \frac{1}{R_3}$$

$$R_o = \begin{bmatrix} R_4 & 0 \\ 0 & R_5 \end{bmatrix}$$

$$L_o = L_6$$

and

$$C_a = C_o, \; L_a = L_o$$

since

$$\hat{C}_a = 0, \quad \hat{L}_a = 0.$$

Using these we get

$$F = \left[\begin{array}{cc|c} 1 & 0 & R_3 \\ 0 & 1 & 0 \\ \hline -G_5 & 0 & 1 \end{array}\right]$$

and

$$F^{-1} = \frac{1}{1+R_3G_5} \begin{bmatrix} 1 & 0 & -R_3 \\ 0 & 1+R_3G_5 & 0 \\ G_5 & 0 & 1 \end{bmatrix} .$$

Substituting in Eq. 6.49,

$$A_1 = \begin{bmatrix} -G_4 & 1 \\ -1 & \dfrac{-R_3}{1+R_3G_5} \end{bmatrix}$$

and in Eq. 6.50,

$$B_1 = \frac{1}{1+R_3G_5}\begin{bmatrix} -G_4(1+R_3G_5) & 0 \\ -R_3G_5 & -R_3 \end{bmatrix}$$

Finally, for

$$\underline{x} = \begin{bmatrix} v_2 \\ j_6 \end{bmatrix} \quad \text{and} \quad \underline{u} = \begin{bmatrix} e_s \\ -i_s \end{bmatrix}$$

we have

$$\dot{\underline{x}} = A\underline{x} + B\underline{u}$$

where

$$A = \begin{bmatrix} \frac{1}{C_2} & 0 \\ 0 & \frac{1}{L_6} \end{bmatrix} A_1 = \begin{bmatrix} -\frac{G_4}{C_2} & \frac{1}{C_2} \\ -\frac{1}{L_6} & \frac{R_3}{L_6(1+R_3G_5)} \end{bmatrix} .$$

and

$$B = \begin{bmatrix} \frac{1}{C_2} & 0 \\ 0 & \frac{1}{L_2} \end{bmatrix} B_1 = \frac{1}{1+R_3G_5}\begin{bmatrix} -G_4(1+R_3G_5) & 0 \\ -\frac{R_3G_5}{L_6} & -\frac{R_3}{L_6} \end{bmatrix} .$$

If the output is the voltage across R_4, we have

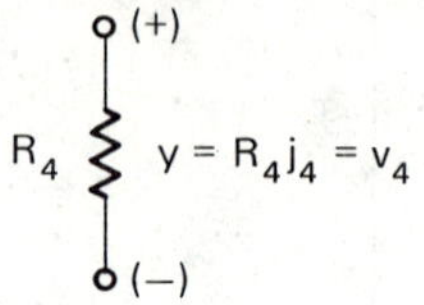

Since

$$e_s = v_2 - v_4 = x_1 - y,$$

we have

$$y = x_1 - e_s$$

or

$$y = [1 \quad 0]\underline{x} + [-1 \quad 0]\underline{u}$$

which defines $\underline{c}^T$ and $\underline{d}^T$.

Example 6.3

Consider the network shown

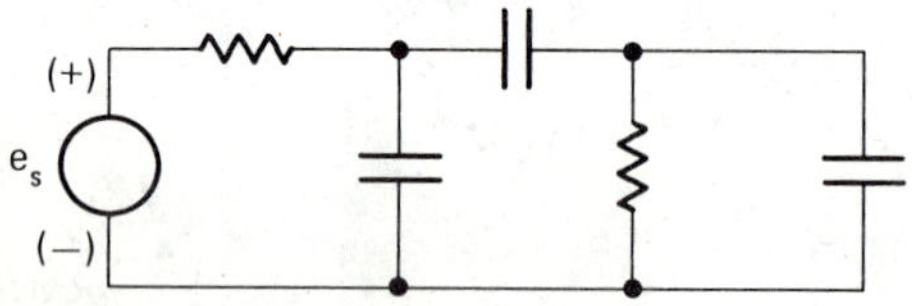

A suitable tree and cotree are chosen; note that there is a capacitive loop and C_4 is assigned to the cotree.

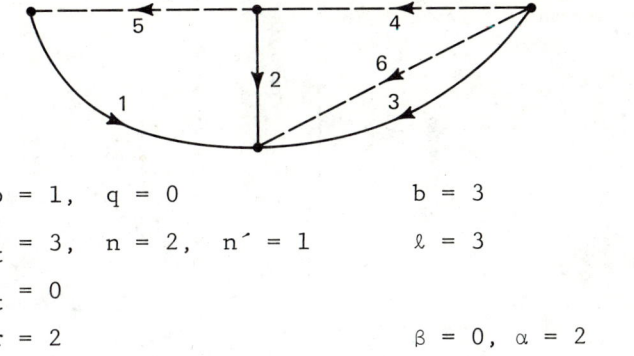

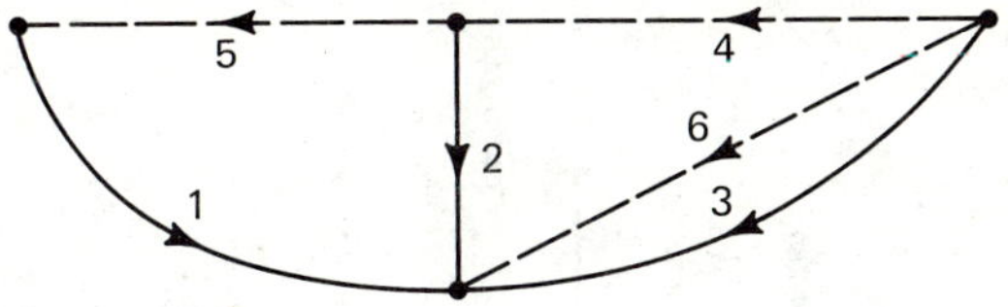

$$p = 1, \quad q = 0 \qquad\qquad b = 3$$

$$n_t = 3, \quad n = 2, \quad n' = 1 \qquad\qquad \ell = 3$$

$$m_t = 0$$

$$r = 2 \qquad\qquad \beta = 0, \ \alpha = 2$$

$$\underline{v}^T = (v_1 \quad v_2 \quad v_3), \qquad\qquad \underline{v}'^T = (v_4 \quad v_5 \quad v_6)$$

Then we have

$$Q = \begin{bmatrix} 1 & 0 & 0 & 0 & -1 & 0 \\ 0 & 1 & 0 & -1 & 1 & 0 \\ 0 & 0 & 1 & 1 & 0 & 1 \end{bmatrix}$$

and

$$Q_o = \begin{array}{c} \\ p \\ \\ n \\ \end{array} \overset{\begin{array}{cc} n' & \quad\alpha \end{array}}{\left[\begin{array}{c|cc} 0 & -1 & 0 \\ \hline -1 & 1 & 0 \\ 1 & 0 & 1 \end{array}\right]}$$

From this

$$Q_{o11} = 0$$

$$Q_{o12} = [-1 \qquad 0]$$

$$Q_{o21} = \begin{bmatrix} -1 \\ \\ 1 \end{bmatrix}$$

$$Q_{o22} = \begin{bmatrix} 1 & 0 \\ \\ 0 & 1 \end{bmatrix} = I_2 \ .$$

From the parameters, we have

$$C_o = \begin{bmatrix} C_2 & 0 \\ 0 & C_3 \end{bmatrix}$$

$$S_o = \begin{bmatrix} \frac{1}{C_4} \end{bmatrix}$$

$$R_o = \begin{bmatrix} R_5 & 0 \\ 0 & R_6 \end{bmatrix}$$

Now $F = I_2 = F^{-1}$ and

$$C_a = \begin{bmatrix} 1 & 0 & -1 \\ 0 & 1 & 1 \end{bmatrix} \begin{bmatrix} C_2 & 0 & 0 \\ 0 & C_3 & 0 \\ 0 & 0 & C_4 \end{bmatrix} \begin{bmatrix} 1 & 0 \\ 0 & 1 \\ -1 & 1 \end{bmatrix}$$

$$= \begin{bmatrix} C_2+C_4 & -C_4 \\ -C_4 & C_3+C_4 \end{bmatrix}$$

and

$$L_a = 0, \quad \hat{L}_a = 0, \quad \hat{C}_a = 0.$$

Then

$$A_1 = -Q_{o22} R_o^{-1} Q_{o22}^T = \begin{bmatrix} -\frac{1}{R_5} & 0 \\ 0 & -\frac{1}{R_6} \end{bmatrix}$$

and

$$B_1 = -Q_{o22}\, R_o^{-1}\, Q_{o12}^T = \begin{bmatrix} \frac{1}{R_5} \\ 0 \end{bmatrix}$$

Also

$$C_a^{-1} = \frac{1}{C_T} \begin{bmatrix} C_3+C_4 & C_4 \\ C_4 & C_2+C_4 \end{bmatrix}$$

where

$$C_T = C_2C_3 + C_3C_4 + C_2C_4.$$

Therefore for

$$\underline{x}^T = (v_2 \quad v_3),$$

we get

$$\dot{\underline{x}} = A\underline{x} + \underline{b}u$$

with

$$A = C_a^{-1}A_1 = \frac{1}{C_T} \begin{bmatrix} -\frac{1}{R_5}(C_3+C_4) & -\frac{C_4}{C_6} \\ -\frac{C_4}{C_5} & -\frac{1}{R_6}(C_2+C_4) \end{bmatrix}$$

$$\underline{b} = C_a^{-1}B_1 = \frac{1}{C_T} \begin{bmatrix} \frac{1}{R_5}(C_3+C_4) \\ \frac{C_4}{C_5} \end{bmatrix}$$

$$u = e_s.$$

6.3.5 Examples

To illustrate the use of state-variable models of RLC networks, two networks, for which time-domain response to a unit step function is important, are analyzed by using a digital program (see Appendix D). The first is a lumped element approximation to an ideal delay line. The important properties of the network are 1) a specified time delay between t_o and t_o+T_d, the time when the output reaches 50% of its final value when the input is $u_1(t-t_o)$, and 2) small overshoot and ringing as the response approaches its final value. The ideal transfer function is

$$H(s) = K(\exp -sT_d)$$

and many design methods for lumped element delay lines have been developed by using various rational function approximations to this H(s). One recent study [21.] has used generalized Bessel polynomials, and charts which enable the designer to choose the order of the network required and the element values have been prepared. An example of a low-pass ladder network is taken from [21.]. The specifications are:

1 μs delay

0.5% maximum overshoot

Attenuation ≤ 6 db, 0 - 55.5 kHz

2000 ohm load

voltage source input

The normalized network (normalized T_d is 0.935 sec.) is shown in Figure 6.20 with element values in ohms, henries, and farads.

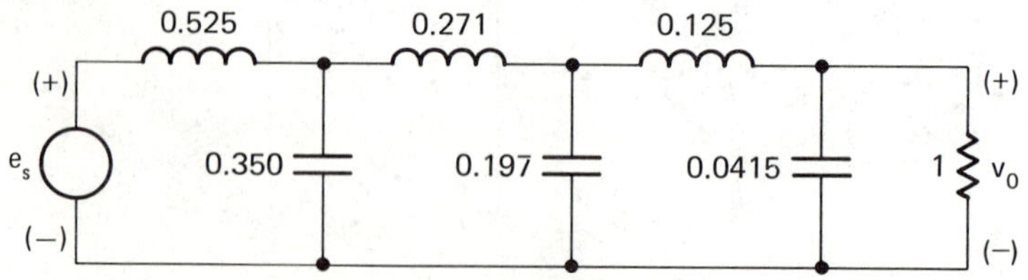

Figure 6.20

A tree and cotree are shown in Figure 6.21.

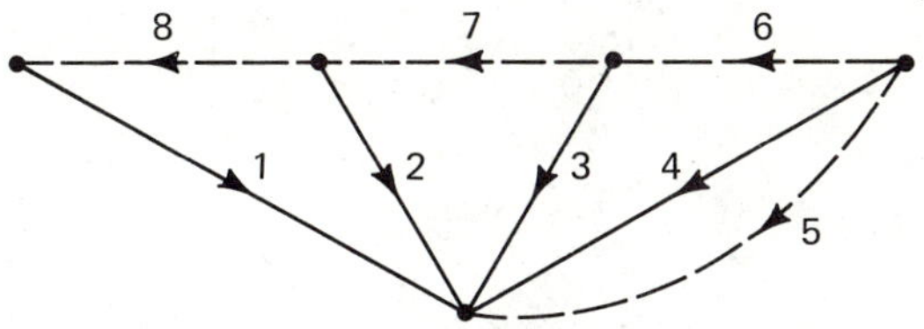

Figure 6.21

$$N = 5,\ B = 8,\ b = 4,\ \ell = 4 \qquad r = 6$$

$$p = 1,\ q = 0 \qquad \beta = 0,\quad \alpha = 1$$

$$n_t = n = 3$$

$$m_t = m = 3$$

Now

$$Q = \begin{bmatrix} 1 & 0 & 0 & 0 & 0 & 0 & 0 & -1 \\ 0 & 1 & 0 & 0 & 0 & 0 & -1 & 1 \\ 0 & 0 & 1 & 0 & 0 & -1 & 1 & 0 \\ 0 & 0 & 0 & 1 & 1 & 1 & 0 & 0 \end{bmatrix}$$

and

$$Q_{o12} = 0$$

$$Q_{o13} = [0 \quad 0 \quad -1]$$

$$Q_{o22} = \begin{bmatrix} 0 \\ 0 \\ 1 \end{bmatrix}$$

$$Q_{o23} = \begin{bmatrix} 0 & -1 & 1 \\ -1 & 1 & 0 \\ 1 & 0 & 0 \end{bmatrix}$$

$$C_o = \begin{bmatrix} C_2 & 0 & 0 \\ 0 & C_3 & 0 \\ 0 & 0 & C_4 \end{bmatrix}$$

$$R_o = R_5$$

$$L_o = \begin{bmatrix} L_6 & 0 & 0 \\ 0 & L_7 & 0 \\ 0 & 0 & L_8 \end{bmatrix}$$

Then

$$F = I_1 = 1 = F^{-1}$$

$$A_1 = \begin{bmatrix} 0 & 0 & 0 & 0 & 1 & -1 \\ 0 & 0 & 0 & 1 & -1 & 0 \\ 0 & 0 & -\frac{1}{R_5} & -1 & 0 & 0 \\ 0 & -1 & 1 & 0 & 0 & 0 \\ -1 & 1 & 0 & 0 & 0 & 0 \\ 1 & 0 & 0 & 0 & 0 & 0 \end{bmatrix}$$

$$\underline{b}_1^T = [0 \quad 0 \quad 0 \quad 0 \quad 0 \quad -1]$$

and

$$A = \begin{bmatrix} 0 & 0 & 0 & 0 & \frac{1}{C_2} & -\frac{1}{C_2} \\ 0 & 0 & 0 & \frac{1}{C_3} & -\frac{1}{C_3} & 0 \\ 0 & 0 & -\frac{1}{R_5 C_4} & -\frac{1}{C_4} & 0 & 0 \\ 0 & -\frac{1}{L_6} & \frac{1}{L_6} & 0 & 0 & 0 \\ -\frac{1}{L_7} & \frac{1}{L_7} & 0 & 0 & 0 & 0 \\ \frac{1}{L_8} & 0 & 0 & 0 & 0 & 0 \end{bmatrix}$$

$$\underline{b}^T = [0 \quad 0 \quad 0 \quad 0 \quad 0 \quad -\frac{1}{L_8}] \, .$$

Also,

$$\underline{c}^T = [0 \quad 0 \quad 1 \quad 0 \quad 0 \quad 0]$$

for

$$\underline{x}^T = [v_2 \quad v_3 \quad v_4 \quad j_6 \quad j_7 \quad j_8] \ .$$

Using the numerical values given for the normalized network, the response to $u_1(t)$ was computed using a standard digital simulation (see Appendix D), and the result is shown in Figure 6.22. The maximum overshoot is 0.3%. The analysis of this network for any set of numerical values is rather easy since all that is required is the preparation of the state variable model by the method of Section 6.3.4 and the use of a standard digital simulation program (see App. D).

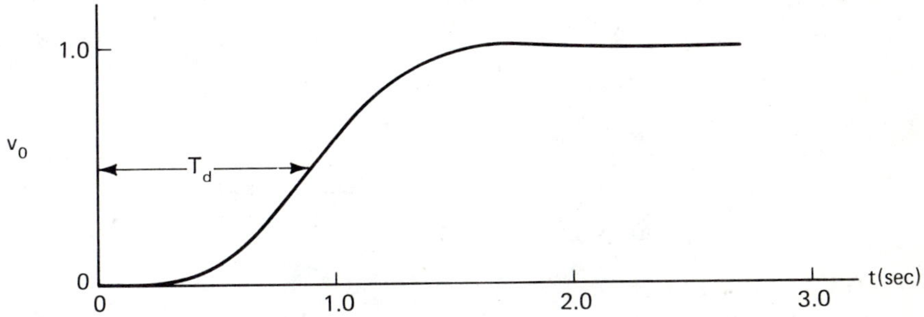

Figure 6.22 Normalized delay line response

A second example concerns the design of pulse-forming networks [22.]. Here the desirable properties are 1) a short rise time and 2) small overshoot when the step function input is applied. The design of LC ladder networks with a resistive load and a voltage source input is discussed in [22.], and tables of element values are given. For a 9-element network, the result is shown in Figure 6.23. The element values are in ohms, henries, and farads for this normalized network. The specified overshoot is 1% and the calculated rise time was 1.5 sec.

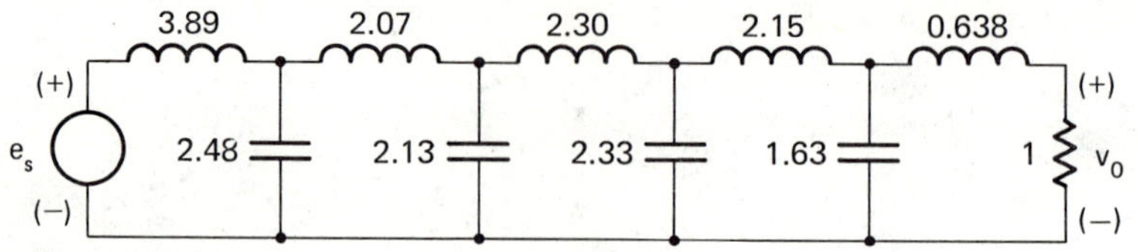

Figure 6.23

A tree and cotree are shown in Figure 6.24.

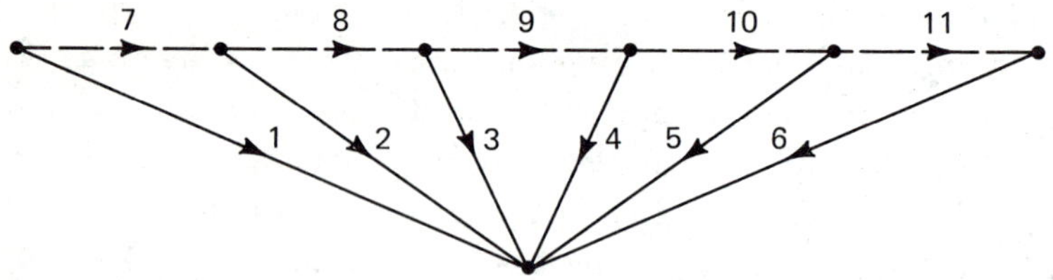

Figure 6.24

$$N = 7,\ B = 11,\ b = 6,\ \ell = 5$$

$$p = 1,\ q = 0$$

$$n_t = n = 4$$

$$m_t = m = 5$$

$$r = 9$$

$$\beta = 1,\quad \alpha = 0$$

Now

$$Q = \left[\begin{array}{c|ccccc} & 1 & 0 & 0 & 0 & 0 \\ \hline & -1 & 1 & 0 & 0 & 0 \\ & 0 & -1 & 1 & 0 & 0 \\ I_6 & 0 & 0 & -1 & 1 & 0 \\ & 0 & 0 & 0 & -1 & 1 \\ \hline & 0 & 0 & 0 & 0 & -1 \end{array}\right]$$

and Q_{o13}, Q_{o23}, and Q_{o33} are shown. Also,

$$C_o = \text{diag}\,[C_2 \quad C_3 \quad C_4 \quad C_5]$$

$$L_o = \text{diag}\,[L_7 \quad L_8 \quad L_9 \quad L_{10} \quad L_{11}]$$

$$G_o = \frac{1}{R_6}\,.$$

Then $F = I_1 = 1 = F^{-1}$ and after finding A_1 and B_1, we get

$$A = \begin{bmatrix} 0 & 0 & 0 & 0 & \frac{1}{C_2} & -\frac{1}{C_2} & 0 & 0 & 0 \\ 0 & 0 & 0 & 0 & 0 & \frac{1}{C_3} & -\frac{1}{C_3} & 0 & 0 \\ 0 & 0 & 0 & 0 & 0 & 0 & \frac{1}{C_4} & -\frac{1}{C_4} & 0 \\ 0 & 0 & 0 & 0 & 0 & 0 & 0 & \frac{1}{C_5} & -\frac{1}{C_5} \\ -\frac{1}{L_7} & 0 & 0 & 0 & 0 & 0 & 0 & 0 & 0 \\ \frac{1}{L_8} & -\frac{1}{L_8} & 0 & 0 & 0 & 0 & 0 & 0 & 0 \\ 0 & \frac{1}{L_9} & -\frac{1}{L_9} & 0 & 0 & 0 & 0 & 0 & 0 \\ 0 & 0 & \frac{1}{L_{10}} & -\frac{1}{L_{10}} & 0 & 0 & 0 & 0 & 0 \\ 0 & 0 & 0 & \frac{1}{L_{11}} & 0 & 0 & 0 & 0 & -\frac{1}{R_6 L_{11}} \end{bmatrix}$$

$$\underline{b}^T = [0 \quad 0 \quad 0 \quad 0 \quad \frac{1}{L_7} \quad 0 \quad 0 \quad 0 \quad 0]$$

$$\underline{c}^T = [0 \quad 0 \quad 0 \quad 0 \quad 0 \quad 0 \quad 0 \quad 0 \quad 1]$$

for

$$\underline{x}^T = [v_2 \quad v_3 \quad v_4 \quad v_5 \quad j_7 \quad j_8 \quad j_9 \quad j_{10} \quad j_{11}]\,.$$

Using the numerical values given, the response to $u_1(t)$ was found and is shown in Figure 6.25. The rise time (1% to 99%) was 9.4 sec. The normalized value specified was 1.5, and for $\omega_c = 1(f_c = 2\pi)$, $9.4/2\pi$ is approximately 1.5.

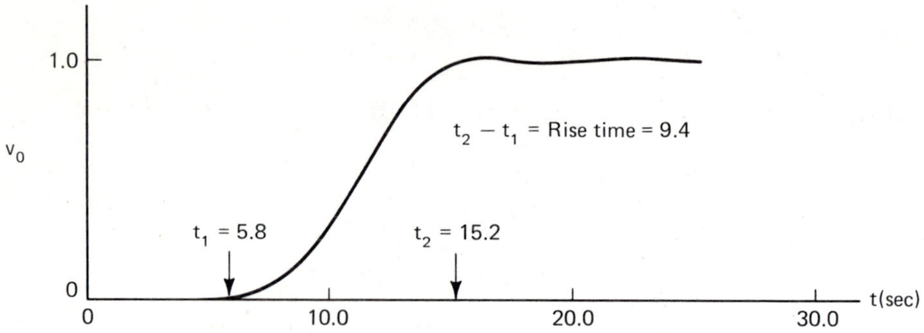

Figure 6.25 Normalized pulse-forming network response

It is worth noting that in these two networks, the ladder structure gave an obvious pattern to A (depending on the numbering used) and the state equations could have been written by inspection. State variable models of the networks made it possible to use the available digital computer program (Appendix D). Thus we have a simple method of finding the required step function responses.

6.4 AIRCRAFT CONTROL

An important first step in the design of large systems is an analysis which starts with the development of a model and then proceeds to an extensive computer simulation. The following is an example of this approach, and the emphasis here is on the details of the development of a linear model. A brief introduction to the kinds of results that can be obtained also is given.

The system to be designed is an automatic control for an aircraft during the final phase of a landing -- i.e., from an approach window to touchdown [26.]*[69.][70.]. The control of the longitudinal mode of the aircraft from an altitude of 100 feet to touchdown is to be accomplished by using instruments in the aircraft as well as elevation angle and range measurements from external equipment. The aircraft is stabilized in its lateral mode so that changes in roll and yaw are not considered; the motion is in a vertical plane. The particular aircraft used for the model is a DC-8, and the numerical values used for the linearized model are for low altitude, level flight. The automatic landing is to be accomplished satisfactorily for a variety of wind conditions including steady winds, wind sheer, and gusts. There also are inaccuracies in the measurements, and the elevation angle measurements are sampled. The trajectory which the aircraft is to follow is specified, and control is accomplished by making deviations from this desired trajectory small. Figure 6.26 shows the basic geometry. The aircraft enters the system

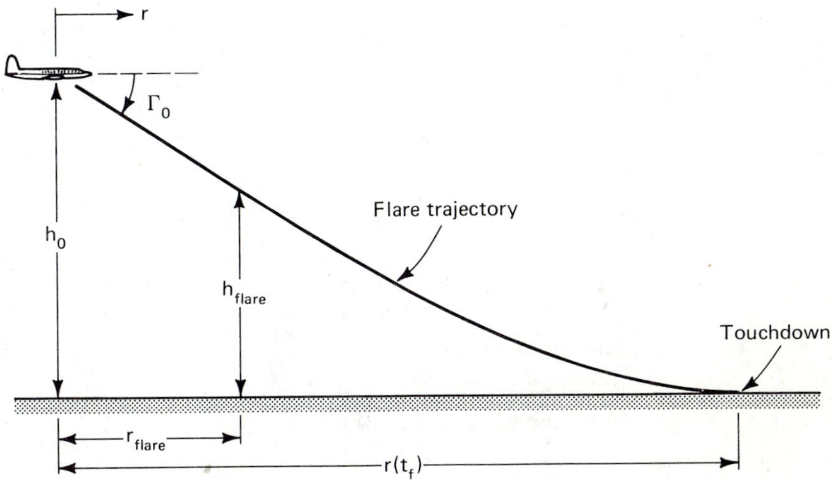

Figure 6.26

*This section is based, in part, on the Ph.D. thesis of Robert Huber, The Penn State University, 1972.

at the approach window at an altitude of h_o with a flight path angle of Γ_o and a horizontal speed of U_o. The desired trajectory is a straight line followed by a flare starting at an altitude h_{flare}. The flare path is followed to touch-down. The elevation angle and range measuring equipment is down range.

6.4.1 Linear Model

The first step is the development of a linearized model of the aircraft in the logitudinal mode. The basic configuration is shown in Figure 6.27.

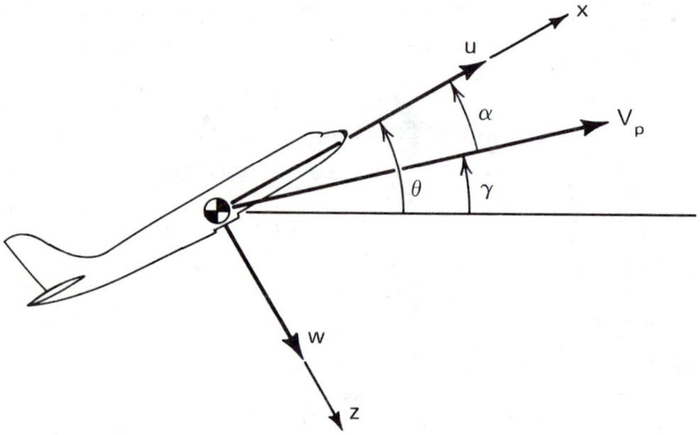

Figure 6.27

x,z	stability axes, fixed to the aircraft
u,w	perturbed inertial velocities
θ	perturbed pitch angle
α	perturbed angle of attack
γ	perturbed flight path angle
V_p	velocity along flight path.
	(y - axis is outward)

Equations in the perturbed variables are required.

To obtain a linearized model for small perturbations about a nominal flight condition, total forces are summed along the x and z-axes and moments are summed about the y-axis [27.]:

$$X - mg \sin \Theta = m[\dot{U} + QW] \qquad (6.53)$$

$$Z + mg \cos \Theta = m[\dot{W} - QU]$$

$$M_p = I_{yy}\dot{Q}$$

where

m = aircraft mass
g = gravitational acceleration
Θ = total angle about y-axis from inertial reference
M_p = pitch moment
X, Z = aerodynamic and propulsive forces in x,z directions
U, W = velocities in x,z directions
Q = angular velocity about y-axis
I_{yy} = moment of inertia about y-axis.

The earth is assumed to be an inertial reference frame for the relatively slow speeds involved. The terms QW and QU follow from the Coriolis Law because of the rotation of the aircraft axes relative to the fixed (earth) reference. Using a zero subscript for the nominal steady conditions, we let

$$X = X_o + \Delta X$$

$$Z = Z_o + \Delta Z$$

$$U = U_o + u$$

$$W = w \text{ for } W_o = 0$$

$$\Theta = \Theta_o + \theta$$

$$Q = q \text{ for } Q_o = 0$$

$$M_p = \Delta M_p \text{ for } M_{po} = 0$$

and assume $|u|$, $|w| << U_o$ and $|\theta| << 1$ radian. These expressions are substituted into Eq. 6.53 and all products of "small" terms are removed to give

$$\Delta X - \theta mg \cos \Theta_o = m\dot{u} \tag{6.54}$$

$$\Delta Z - \theta mg \sin \Theta_o = m\dot{w} - mqU_o$$

$$\Delta M_p = I_{yy}\dot{q}$$

when the steady conditions in Eq. 6.53 are subtracted. Small angle approximations for sin θ and cos θ also are used. To linearize the aerodynamic forces, we let (see Section 3.7)

$$\Delta X_1 = \frac{\partial X}{\partial u}u + \frac{\partial X}{\partial w}w + \frac{\partial X}{\partial \delta_e}\delta_e = X_u' u + X_w' w + X_{\delta_e}' \delta_e$$

and similarly,

$$\Delta Z_1 = Z_u' u + Z_w' w + Z_{\delta_e}' \delta_e$$

$$\Delta M_{p1} = M_u' u + M_w' w + M_q' q + M_{\delta_e}' \delta_e$$

where δ_e is the perturbed elevator deflection. This is an important step that requires a good understanding of the aerodynamics involved and the relative magnitudes of the many possible forces and moments acting on the aircraft for the flight conditions being used. The terms retained here were considered adequate [26.], and a more complete evaluation is given, for example, in [27.]. Now ΔX_1, ΔZ_1, and ΔM_{p1} are combined with the effects of perturbed propulsive forces and moments to give

$$\Delta X = \Delta X_1 + X_{\delta th}' \delta th$$

$$\Delta Z = \Delta Z_1 + Z_{\delta th}' \delta th$$

$$\Delta M_p = \Delta M_{p1} + M_{\delta th}' \delta th.$$

These are substituted in Eq. 6.54 and the first two equations are divided by m and the third by I_{yy} to give equations in the perturbed variables and normalized (unprimed) coefficients:

$$\dot{u} = X_u u + X_w w - (g \cos \Theta_o)\theta + X_{\delta_e}\delta_e + X_{\delta th}\delta th \qquad (6.55)$$

$$\dot{w} = Z_u u + Z_w w + U_o q - (g \sin \Theta_o)\theta + Z_{\delta_e}\delta_e + Z_{\delta th}\delta th$$

$$\dot{q} = M_u u + M_w w + M_q q + M_{\delta_e}\delta_e + M_{\delta th}\delta th$$

$$\dot{\theta} = q \ .$$

The fourth equation simply relates pitch rate and q. The dimensions are

X_u, X_w, Z_u, Z_w, M_q - $(\text{sec})^{-1}$

M_u, M_w - $(\text{ft-sec})^{-1}$

X_{δ_e}, Z_{δ_e} - (ft/sec^2)

M_{δ_e} - $(\text{sec})^{-2}$

$X_{\delta th}$, $Z_{\delta th}$ - (ft/sec^2)/% thrust

$M_{\delta th}$ - $(\text{sec})^{-2}$/% thrust

In terms of a state variable model, the state variables are u, w, q, θ, and the input variables are δ_e and δth. To relate the aircraft to an earth reference, equations in altitude, h, and range, r, are added to those of Eq. 6.55. From Figure 6.28

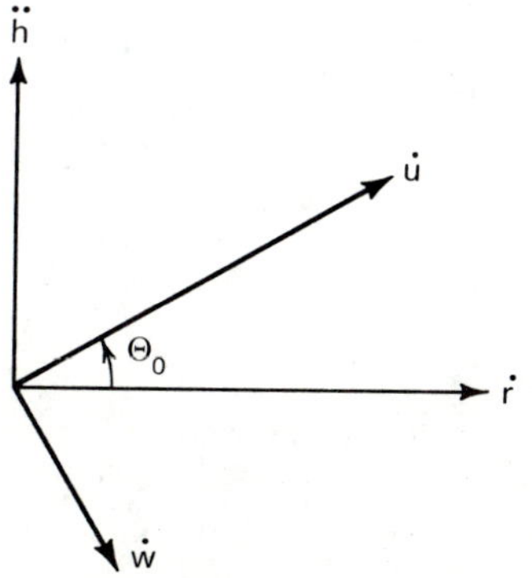

Figure 6.28

we have

$$\ddot{h} = \dot{u}\sin\Theta_o - \dot{w}\cos\Theta_o$$

$$= X_u \sin\Theta_o + X_w \sin\Theta_o w + X_{\delta_e}\sin\Theta_o\,\delta_e + X_{\delta th}\sin\Theta_o\,\delta th$$

$$-Z_u\cos\Theta_o u - Z_w\cos\Theta_o w - Z_{\delta_e}\cos\Theta_o\delta_e - Z_{\delta th}\cos\Theta_o\delta th$$

where terms in $\theta \sin\Theta_o$ and $\dot{\theta} = q$ are neglected. Also,

$$\dot{r} = (u+U_o)\cos(\Theta_o+\theta) + w\sin\Theta_o$$

$$= u\cos\Theta_o + U_o\cos\Theta_o - U_o\sin\Theta_o\theta + w\sin\Theta_o.$$

In addition to the equations in 6.55 and the two above, an equation to account for the time lag between thrust command and the change in thrust is included:

$$\dot{\delta}_{th} = -\frac{1}{T_{th}}\delta_{th} + \frac{1}{T_{th}}\delta_{thc},$$

where T_{th} is the time constant and δ_{thc} is the command. Then an augmented set of state variables is u, w, q, θ, $\dot{h}$, r, and δ_{th} and if h is added to these, the following is obtained:

$$\begin{bmatrix}\dot{u}\\ \dot{w}\\ \dot{q}\\ \dot{\theta}\\ \ddot{h}\\ \dot{h}\\ \dot{r}\\ \dot{\delta}_{th}\end{bmatrix} = \begin{bmatrix} X_u & X_w & 0 & -g\cos\Theta_o & 0 & 0 & 0 & X_{\delta th}\\ Z_u & Z_w & U_o & -g\sin\Theta_o & 0 & 0 & 0 & Z_{\delta th}\\ M_u & M_w & M_q & 0 & 0 & 0 & 0 & M_{\delta th}\\ 0 & 0 & 1 & 0 & 0 & 0 & 0 & 0\\ a_{51} & a_{52} & 0 & 0 & 0 & 0 & 0 & a_{58}\\ 0 & 0 & 0 & 0 & 1 & 0 & 0 & 0\\ \cos\Theta_o & \sin\Theta_o & 0 & -U_o\sin\Theta_o & 0 & 0 & 0 & 0\\ 0 & 0 & 0 & 0 & 0 & 0 & 0 & -\frac{1}{T_{th}}\end{bmatrix}\begin{bmatrix}u\\ w\\ q\\ \theta\\ \dot{h}\\ h\\ r\\ \delta_{th}\end{bmatrix} +$$

$$
+ \begin{bmatrix} X_{\delta_e} & 0 \\ Z_{\delta_e} & 0 \\ M_{\delta_e} & 0 \\ 0 & 0 \\ b_{51} & 0 \\ 0 & 0 \\ 0 & 0 \\ 0 & \frac{1}{T_{th}} \end{bmatrix} \begin{bmatrix} \delta_e \\ \delta_{thc} \end{bmatrix} + \begin{bmatrix} 0 \\ 0 \\ 0 \\ 0 \\ 0 \\ 0 \\ U_o \cos \Theta_o \\ 0 \end{bmatrix} \qquad (6.56)
$$

where

$$a_{51} = X_u \sin \Theta_o - Z_u \cos \Theta_o$$

$$a_{52} = X_w \sin \Theta_o - Z_w \cos \Theta_o$$

$$a_{58} = X_{\delta th} \sin \Theta_o - Z_{\delta th} \cos \Theta_o$$

$$b_{51} = X_{\delta_e} \sin \Theta_o - Z_{\delta_e} \cos \Theta_o \ .$$

This is the basic model when there are no atmospheric disturbances.

Three types of atmospheric disturbances are considered: steady winds, wind shear, and random gusts. The first two are deterministic and are assumed to occur in the direction of the x-axis. Thus we have

$$u_{det} = u_{steady} + u_{shear}$$

and

$$u_{shear} = s(h_o - h)$$

where s is the wind gradient. u_{steady} and s are measured and therefore are known disturbances. The gusts are included in the model as zero-mean, stationary Gauss-Markov stochastic processes $\{u_g\}$ and $\{w_g\}$. Therefore (see Chapter 4, Section 4.3), using lower case letters, we have

$$\dot{u}_g = -\omega_{ug}\, u_g + \xi_{ug}$$

and

$$\dot{w}_g = -\omega_{wg}\, w_g + \xi_{wg}$$

where ξ_{ug} and ξ_{wg} are uncorrelated white Gaussian inputs with intensities $\sigma^2_{\xi_{ug}}$ and $\sigma^2_{\xi_{wg}}$, and

$$\sigma^2_{\xi_{ug}} = 2\omega_{ug}\sigma_{ug}^2$$

and

$$\sigma^2_{\xi_{wg}} = 2\omega_{wg}\sigma_{wg}^2.$$

The half power frequencies ω_{ug} and ω_{wg} and the variances σ^2_{ug} and σ^2_{wg} are specified for a particular study.

The elevation angle measured by a microwave system contains errors due to "beam bends" associated with atmospheric disturbances and errors associated with the sampling process. Again assuming a zero-mean, stationary Gauss-Markov process $\{\rho_{bb}\}$ for the beam bend error, we have

$$\dot{\rho}_{bb} = -\omega_{bb}\rho_{bb} + \xi_{bb}$$

where ξ_{bb} is a zero mean white Gaussian input with intensity $\sigma^2_{\xi_{bb}} = 2\omega_{bb}\sigma^2_{bb}$. The sampling error is included in the output equation as shown below.

The 8-th order model of Eq. 6.56 is now augmented to account for the three stochastic processes $\{u_g\}$, $\{w_g\}$, and $\{\rho_{bb}\}$.

However, since the air mass in which the aircraft is flying also is in motion, some modifications are required. Thus we have

$$u = u_{as} + u_g + u_{steady} + u_{shear}$$

$$w = w_{as} + w_g$$

where u_{as} and w_{as} are airspeeds. Now u and w on the right side of the first two equations of 6.56 are first replaced by u_{as} and w_{as} since the forces are with respect to the air mass. Then, in the first equation of 6.56 on the left, we let

$$\dot{u} = \dot{u}_{as} + \dot{u}_g - s\dot{h}$$

and in the second equation on the right, we let

$$w_{as} = w - w_g$$

to use u_{as} and w as state variables. The complete result is

$$\dot{\underline{x}} = A\underline{x} + B\underline{u} + \underline{d} + \underline{\xi} \qquad (6.57)$$

with $\underline{x}(0) = \underline{x}_o$, and where

$$\underline{x}^T = [u_{as} \quad w \quad q \quad \theta \quad \dot{h} \quad h \quad r \quad \delta_{th} \quad u_g \quad w_g \quad \rho_{bb}]$$

$$\underline{d} = \underline{d}_w + \underline{d}_n$$

$$\underline{d}_w^T = [0 \quad 0 \quad 0 \quad 0 \quad 0 \quad 0 \quad \cos\Theta_o(u_{steady}+h_o s) \quad 0 \quad 0 \quad 0 \quad 0]$$

$$\underline{d}_n^T = [0 \quad 0 \quad 0 \quad 0 \quad 0 \quad 0 \quad U_o \cos\Theta_o \quad 0 \quad 0 \quad 0 \quad 0]$$

$$\underline{\xi}^T = [-\xi_{ug} \quad 0 \quad 0 \quad 0 \quad 0 \quad 0 \quad 0 \quad 0 \quad \xi_{ug} \quad \xi_{wg} \quad \xi_{bb}]$$

$$\underline{u}^T = [\delta_e \quad \delta_{thc}]$$

$$A = \begin{bmatrix} X_u & X_w & 0 & -g\cos\Theta_o & s & 0 & 0 & X_{\delta th} & \omega_{ug} & -X_w & 0 \\ Z_u & Z_w & U_o & -g\sin\Theta_o & 0 & 0 & 0 & Z_{\delta th} & 0 & -Z_w & 0 \\ M_u & M_w & M_q & 0 & 0 & 0 & 0 & M_{\delta th} & 0 & -M_w & 0 \\ 0 & 0 & 1 & 0 & 0 & 0 & 0 & 0 & 0 & 0 & 0 \\ a_{51} & a_{52} & 0 & 0 & 0 & 0 & 0 & a_{58} & 0 & -a_{52} & 0 \\ 0 & 0 & 0 & 0 & 1 & 0 & 0 & 0 & 0 & 0 & 0 \\ \cos\Theta_o & \sin\Theta_o & 0 & -U_o\sin\Theta_o & 0 & -s\cos\Theta_o & 0 & 0 & \cos\Theta_o & 0 & 0 \\ 0 & 0 & 0 & 0 & 0 & 0 & 0 & -\frac{1}{T_{th}} & 0 & 0 & 0 \\ 0 & 0 & 0 & 0 & 0 & 0 & 0 & 0 & -\omega_{ug} & 0 & 0 \\ 0 & 0 & 0 & 0 & 0 & 0 & 0 & 0 & 0 & -\omega_{wg} & 0 \\ 0 & 0 & 0 & 0 & 0 & 0 & 0 & 0 & 0 & 0 & -\omega_{bb} \end{bmatrix}$$

$$B^T = \begin{bmatrix} X_{\delta e} & Z_{\delta e} & M_{\delta e} & 0 & b_{51} & 0 & 0 & 0 & 0 & 0 & 0 \\ 0 & 0 & 0 & 0 & 0 & 0 & 0 & \frac{1}{T_{th}} & 0 & 0 & 0 \end{bmatrix}$$

Equation 6.57 is the augmented state equation for the aircraft, range and altitude, thrust control, and atmospheric disturbances. The deterministic disturbance $\underline{d}$ is known and $\underline{\xi}$ is a zero-mean white Gaussian input with intensity matrix:

$$M_{\underline{\xi}} \text{ with } m_{11\xi} = m_{99\xi} = \sigma^2_{\xi_{ug}}$$

$$m_{10,10\xi} = \sigma^2_{\xi_{wg}}$$

$$m_{11,11\xi} = \sigma^2_{\xi_{bb}}$$

$$m_{19\xi} = m_{91\xi} = -\sigma^2_{\xi_{ug}}$$

and all other elements zero.

The output or measurement equation also requires a linearization. The measured variables are q, θ, r, and ρ, where q and θ are measured in the aircraft and r and ρ are measured externally. The first three are state variables and each is assumed to include a measurement error which is a white stationary, zero-mean Gaussian stochastic process. Thus we assume

$$q_{meas.} = q + \eta_q$$

$$\theta_{meas.} = \theta + \eta_\theta$$

$$r_{meas.} = r + \eta_r$$

where $\{\eta_q\}$, $\{\eta_\theta\}$, $\{\eta_r\}$ have intensities $\sigma^2_{\eta_q}$, $\sigma^2_{\eta_\theta}$, $\sigma^2_{\eta_r}$. The geometry associated with the elevation angle measurement is shown in Figure 6.29.

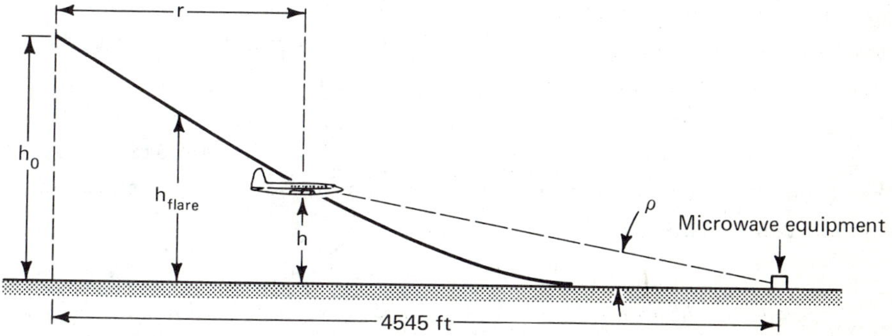

Figure 6.29

Now

$$\tan \rho = \frac{h}{4545-r}$$

and for ρ small, tan ρ ≅ ρ, therefore, we have

$$\rho(h,r) \cong \frac{h}{4545-r} .$$

To linearize this, we use a Taylor Series expansion about the desired trajectory, i.e., about $h_d(t)$ and $r_d(t)$ (which are developed later):

$$\rho(h,r) = \rho(h_d,r_d) + \left.\frac{\partial\rho}{\partial h}\right|_d (h-h_d) + \left.\frac{\partial\rho}{\partial r}\right|_d (r-r_d) + \ldots$$

where higher order terms are neglected. Using the required derivatives,

$$\rho(h,r) = \frac{h_d}{4545-r_d} + \frac{1}{4545-r_d}(h-h_d) + \frac{h_d}{(4545-r_d)^2}(r-r_d)$$

or

$$\rho(h,r) = \frac{1}{4545-r_d}h + \frac{h_d}{(4545-r_d)^2}r - \frac{h_d r_d}{(4545-r_d)^2}\,.$$

The first two terms are linear in h and r which are two state variables, and the third term is a known bias. In the actual system a measurement of $\rho(t)$ is sampled to give $\rho(nT_s)$ and is then reconstructed to give $\rho_r(nT_s)$ by using a zero-order hold (see Chapter 2, Section 2.9.3). The sampling and reconstruction error is modeled here as a random fluctuation plus a bias. Thus we let

$$\rho_{meas.} = \frac{1}{4545-r_d}h + \frac{h_d}{(4545-r_d)^2}r + \rho_{bb} + \eta_{fn} + f$$

where $\{\eta_{fn}\}$ is a white stationary, zero-mean Gaussian process with intensity $\sigma^2_{\eta_{fn}}$, and f is a known bias which includes the term $h_d r_d/(4545-r_d)^2$ as well as a bias from the sampling and reconstruction that is estimated by using the desired trajectory. The dominant terms will be a trend polynomial plus a sinusoidal component of frequency $f_s = \frac{1}{T_s}$. The output or measurement equation is now

$$\underline{y} = C\underline{x} + \underline{\eta} + \underline{f} \qquad (6.58)$$

where

$$\underline{y}^T = [q_{meas.} \quad \theta_{meas.} \quad \rho_{meas.} \quad r_{meas.}]$$

$$\underline{\eta}^T = [\eta_q \quad \eta_\theta \quad \eta_{fn} \quad \eta_r]$$

$$\underline{f}^T = [0 \quad 0 \quad f \quad 0]$$

and

$$C(t) = \begin{bmatrix} 0 & 0 & 1 & 0 & 0 & 0 & 0 & 0 & 0 & 0 & 0 \\ 0 & 0 & 0 & 1 & 0 & 0 & 0 & 0 & 0 & 0 & 0 \\ 0 & 0 & 0 & 0 & 0 & c_{36} & c_{37} & 0 & 0 & 0 & 1 \\ 0 & 0 & 0 & 0 & 0 & 0 & 1 & 0 & 0 & 0 & 0 \end{bmatrix}$$

with

$$c_{36}(t) = \frac{1}{4545 - r_d(t)}$$

$$c_{37}(t) = \frac{h_d(t)}{(4545 - r_d(t)^2} \quad .$$

The process $\{\underline{\eta}\}$ is white, Gaussian zero-mean with intensity matrix:

$$M_{\underline{\eta}} = \text{diag}\ [\sigma^2_{\eta_q} \quad \sigma^2_{\eta_\theta} \quad \sigma^2_{\eta_{fn}} \quad \sigma^2_{\eta_r}]$$

Equations 6.57 and 6.58 are the final linear state variable model of the system for which a control is required.

6.4.2 Desired Trajectory

The equations for the desired trajectory are developed as follows. First, we consider the case of zero winds and the constants:

$$U_o = 228 \text{ ft/sec.}$$

$$h_o = 100 \text{ ft}$$

$$h_{flare} = 60 \text{ ft}$$

$$\Gamma_o = -2.8^o.$$

Now from the geometry we have

$$\dot{h}_d(t) = U_o \sin \Gamma_o = -11.14 \text{ ft/sec}, \quad 100 \geq h_d \geq 60$$

and therefore

$$h_d(t) = -11.14t + 100 \qquad 0 \leq t \leq t_{flare}$$

and

$$t_{flare} = \frac{100 - 60}{11.14} = 3.6 \text{ sec.}$$

Also, from the geometry,

$$r_d(t) = (U_o \cos \Gamma_o)t \qquad \begin{matrix} 100 \geq h_d \geq 0 \\ 0 \leq t \leq t_f \end{matrix}$$

and

$$r_d(t_{flare}) = 820 \text{ ft.}$$

An exponential flare path is used, thus we let

$$h_d(t) = c_1(\exp -c_2t)+c_3 \qquad 60 \geq h_d \geq 0$$

and

$$\dot{h}_d(t) = -c_1c_2(\exp -c_2t).$$

Now using t_{flare}, we have

$$h_d(3.6) = 60 = c_1(\exp -c_2(3.6)) + c_3$$

and at the final time,

$$h_d(t_f) = 0 = c_1(\exp -c_2(t_f)) + c_3$$

and differentiating we get

$$\dot{h}_d(3.6) = -11.14 = -c_1c_2(\exp -c_2(3.6))$$

$$\dot{h}_d(t_f) = -2 = -c_1c_2(\exp -c_2t_f).$$

These four equations are solved to find

$$t_f = 14.9 \text{ sec}$$
$$c_1 = 126.5$$
$$c_2 = 0.152$$
$$c_3 = -13.13 \ .$$

Therefore, the desired trajectory for <u>zero winds</u> is

$$h_d(t) = -11.14t + 100 \qquad 0 \leq t \leq 3.6$$
$$= 126.5(\exp -0.152t)-13.13 \qquad 3.6 \leq t \leq 14.9$$

and

$$r_d(t) = 228t \qquad 0 \leq t \leq 14.9.$$

In particular, $r_d(t_f) = 3390$ ft. Now when steady winds and wind shear are included, the average ground speed is

$$U_{gs} = U_o \cos \Gamma_o + u_{steady} + \frac{s}{2} h_o,$$

then we let

$$t_f' = \frac{3390}{U_{gs}}$$

and

$$t_{flare}' = \frac{820}{U_{gs}} \ .$$

Now we let

$$r_d(t) = U_{gs}t \ . \qquad 0 \leq t \leq t_f' \ . \qquad (6.59)$$

Also, we substitute $t = r_d/U_o$ in $h_d(t)$ for the zero wind conditions and this gives

$$h_d(r_d) = \frac{-11.14}{228} r_d + 100 \qquad 100 \geq h_d \geq 60$$

$$= 126.5(\exp - \frac{0.152}{228} r_d) - 13.13 \qquad 60 \geq h_d \geq 0.$$

Then for $r_d(t) = U_{gs}t$, the desired altitude for nonzero winds is

$$h_d(t) = -0.049\ U_{gs}t + 100 \qquad 0 \leq t \leq t'_{flare}$$

$$= 126.5(\exp -0.00067\ U_{gs}t)$$

$$- 13.13 \qquad t'_{flare} \leq t \leq t'_f \ . \qquad (6.60)$$

Thus we have $r_d(t)$ and $h_d(t)$ given in terms of U_{gs} to account for the effect of the winds on changing the time interval of control.

6.4.3 Control System

The control system provides an input $\underline{u}$ in Eq. 6.57 that is a function of the output $\underline{y}$ of Eq. 6.58, i.e., it is a feedback control system. The design of such controls is a major task in itself, but the special case of interest here has a well-known solution, and these results are outlined here. First, the model of Eq. 6.57 and 6.58 has the following features:

1. It is linear
2. $\underline{d}$ and $\underline{f}$ are known time functions
3. $\{\underline{\xi}\}$ and $\{\underline{\eta}\}$ are uncorrelated zero mean, white Gaussian stochastic processes.

The control is designed by specifying a performance measure (Chapter 5, Section 5.2) and then finding the $\underline{u}$ for which it is minimal. Thus, we let

$$J_c = E\ \{\tfrac{1}{2}(\underline{x}(t_f') - \underline{x}_d(t_f'))^T Q_1 (\underline{x}(t_f') - \underline{x}_d(t_f')) + $$
$$+ \tfrac{1}{2}\int_0^{t_f'} [(\underline{x}(t)-x_d(t))^T Q(\underline{x}(t)-\underline{x}_d(t)) +$$
$$+ \underline{u}^T(t)R\underline{u}(t)]dt\} \qquad (6.61)$$

where Q_1 and Q are symmetric, positive semidefinite weighting matrices and R is symmetric and positive definite. This is a general form of Eq. 5.42. $\underline{x}_d(t)$ is the desired trajectory for $\underline{x}(t)$. Now an important result in control theory is that the control $\underline{u}$ for this problem has the same form as that when $\underline{\xi} = \underline{0}$ and the expectation of Eq. 6.61 is removed [28.]. For this case, where Eq. 6.61 is a quadratic form and t_f' is finite, a form of the optimal control is [29.].

$$\underline{u}(t) = F_{cc}(t)\underline{x}(t) + \underline{f}_{cc}(t) \qquad (6.62)$$

where

$$F_{cc}(t) = -R^{-1}B^T K_{cc}(t)$$

$$\underline{f}_{cc}(t) = -R^{-1}B^T \underline{k}_{cc}(t)$$

and $K_{cc}(t)$ and $\underline{k}_{cc}(t)$ satisfy the differential equations:

$$\dot{K}_{cc} + K_{cc}A + A^T K_{cc} - K_{cc}BR^{-1}B^T K_{cc} + Q = \Theta$$

$$\dot{k}_{cc} - K_{cc}BR^{-1}B^T \underline{k}_{cc} + A^T \underline{k}_{cc} - Q\underline{x}_d + K_{cc}\underline{d} = \underline{0}$$

with boundary conditions $K_{cc}(t_f') = Q_1$ and $\underline{k}_{cc}(t_f') = -Q_1\underline{x}_d(t_f')$. Eq. 6.62 is a linear, time-varying feedback control law, but it is not a satisfactory result since $\underline{x}(t)$ is not measured.

A second component of the control system is a filter which uses $\underline{y}$ as an input and produces an estimate of $\underline{x}$, $\hat{\underline{x}}$, as an output. A second important result is control theory is that if $\underline{\hat{x}}$ is chosen such that

$$J_f = E[\underline{e}^T(t)\underline{e}(t)] \tag{6.63}$$

is minimal, $\underline{x}$ in Eq. 6.62 can be replaced by $\underline{\hat{x}}$, and the combined system of filter and controller is optimal [28.]. This is a separation principle. Eq. 6.63 is a general form like Eq. 5.27 and J_f is the mean square error where the error is

$$\underline{e}(t) = \underline{x}(t) - \underline{\hat{x}}(t) .$$

Also, as shown in Chapter 5, we can write

$$J_f = \text{tr } C_{\underline{E}}(t)$$

where

$$C_{\underline{E}}(t) = E\ [\underline{e}(t)\underline{e}^T(t)] ,$$

the error second moment matrix; the process $\{\underline{e}(t)\}$ is not stationary. The optimal filter is a dynamic linear system [28.]:

$$\dot{\underline{\hat{x}}} = A\underline{\hat{x}} + B\underline{u} + F_{cf}(t)\ [\underline{y}-\underline{f}-C(t)\underline{\hat{x}}] + \underline{d} \tag{6.64}$$

with $\underline{\hat{x}}(0) = \underline{\hat{x}}_o$ and where

$$F_{cf}(t) = S_{cf}(t)C^T(t)M_{\underline{\eta}}^{-1}$$

and a $S_{cf}(t)$ satisfies the differential equation:

$$\dot{S}_{cf} = M_{\underline{\eta}} - S_{cf}C^TM_{\underline{\eta}}^{-1}C\ S_{cf} + S_{cf}A^T + AS_{cf}$$

with boundary condition $S_{cf}(0)$. The intensity matrix $M_{\underline{\eta}}$ was given in Section 6.4.1. The initial state of Eq. 6.57 is assumed to be a Gaussian random variable with known

$$E[\underline{x}_o] = \underline{m}_{\underline{x}}(0)$$

and covariance

$$E[(\underline{x}_o - \underline{m}_{\underline{x}}(0))\,(\underline{x}_o - \underline{m}_{\underline{x}}(0))^T] = P_o \ .$$

Also, we require in Eq. 6.64,

$$\hat{\underline{x}}_o = \underline{m}_{\underline{x}}(0) \ .$$

The matrix S_{cf} also is the matrix $C_{\underline{E}}(t)$ and the filter gives an unbiased estimate, i.e., $E[\hat{\underline{x}}]$ or $E[\underline{x}]$ or $E[\underline{e}] = \underline{0}$, so that $C_{\underline{E}}(t)$ is error covariance. The complete system is now described by

$$\begin{bmatrix} \dot{\underline{x}} \\ \cdots\cdots \\ \dot{\hat{\underline{x}}} \end{bmatrix} = \left[\begin{array}{c|c} A & BF_{cc} \\ \hline F_{cf}C & A+BF_{cc}-F_{cf}C \end{array}\right] \begin{bmatrix} \underline{x} \\ \\ \hat{\underline{x}} \end{bmatrix} + \left[\begin{array}{c} B\underline{f}_{cc} + \underline{d} + \underline{\xi} \\ \hline B\underline{f}_{cc} + \underline{d} + F_{cf}\underline{n} \end{array}\right] \ . \qquad (6.65)$$

This is a linear, time-varying system with deterministic and stochastic inputs. Figure 6.6 is a description of the system. The time-varying matrices $F_{cc}(t)$ and $F_{cf}(t)$ and the vector $\underline{f}_{cc}(t)$ could be precomputed and stored or computed on line depending on the available computing capability.

The complete continuous-time system is now converted to a discrete-time system for two reasons: it is of interest to consider a digital rather than an analog implementation, and digital computations are to be used in an analysis of the system. Starting with Eq. 6.57, Conversion VII of Chapter 3 is used. Thus, for a time increment Δt, we get

$$\underline{x}(n+1) = A_d\underline{x}(n) + B_d\underline{v}(n) + \underline{d}_d(n) + \underline{\xi}_d(n) \qquad (6.66)$$

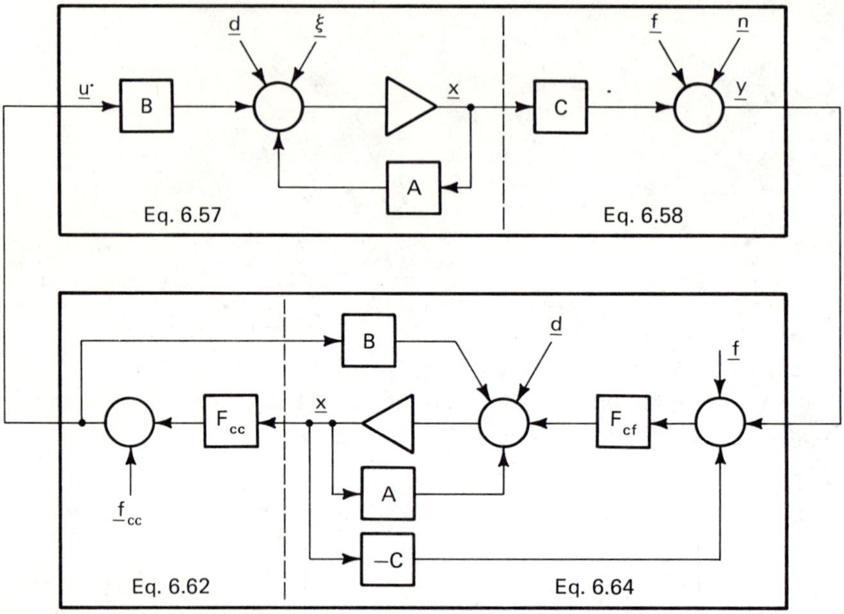

Figure 6.6 Complete system

where $\underline{x}(n)$ replaces $\underline{x}(t_n) = \underline{x}(n\Delta t)$, $\underline{x}(0) = \underline{x}_o$, and

$$A_d = (\exp A\Delta t)$$

$$B_d = [\int_0^{\Delta t} \varepsilon^{A\lambda}\, d\lambda]B = HB$$

$\underline{v}(n) = \underline{u}(n\Delta t)$ ($\underline{u}(t)$ approximated as piecewise constant)

$\underline{d}_d(n) = H\underline{d}(n\Delta t)$ ($\underline{d}(t)$ approximated as piecewise constant)

$$\underline{\xi}_d(n) = \int_{n\Delta t}^{(n+1)\Delta t} (\exp A[(n+1)\Delta t - \lambda)]\ \underline{\xi}(\lambda)d\lambda$$

Also,

$$E[\underline{\xi}_d] = \underline{0} \quad \text{since} \quad E[\underline{\xi}] = \underline{0}\ ,$$

and

$$E[\underline{\xi}_d(n)\ \underline{\xi}_d^T(n+k)] = M_{\underline{\xi}_d}\ v_o(k)$$

with

$$M_{\underline{\xi}_d} = \int_0^{\Delta t} (\exp A(\Delta t-\lambda))\ M_{\underline{\xi}}(\exp A^T(\Delta t-\lambda))d\lambda$$

$$\cong \Delta t\ M_{\underline{\xi}} \quad \text{for} \quad \Delta t \text{ small.}$$

From Eq. 6.58 we have

$$\underline{y}(n) = C(n)\underline{x}(n) + \underline{n}_d(n) + \underline{f}_d(n) \tag{6.67}$$

where

$$\underline{y}(n) \text{ replaces } \underline{y}(n\Delta t)$$

$$C(n) \text{ replaces } C(n\Delta t)$$

$$\underline{n}_d(n) = \underline{n}(n\Delta t)$$

$$\underline{f}_d(n) = \underline{f}(n\Delta t)$$

and

$$M_{\underline{n}_d} = M_{\underline{n}}.$$

The control for the discrete-time system of Eq. 6.66 and 6.68 is formed by minimizing the performance measure

$$J_c = \tfrac{1}{2}(\underline{x}(n_f)-\underline{x}_d(n_f))^T Q_1 (\underline{x}(n_f)-\underline{x}_d(n_f)) + \\ + \tfrac{1}{2}\sum_{n=0}^{n_f-1} [(\underline{x}(n)-\underline{x}_d(n))^T Q(\underline{x}(n)-\underline{x}_d(n)) + \\ + v^T(n)R\underline{v}(n)] \tag{6.68}$$

where Q_1, Q, and R are as in Eq. 6.61 and $n_f = t_f'/\Delta t$ (integer). $\underline{x}_d(n)$ has replaced $\underline{x}_d(n\Delta t)$. A form of control analogous to Eq. 6.62 can be found [29.] as:

$$\underline{v}(n) = F_{dc}(n)\underline{x}(n) + \underline{f}_{dc}(n) \tag{6.69}$$

where

$$F_{dc}(n) = -R^{-1}B_d^T[A_d^T]^{-1}\ [K_{dc}(n)-Q]$$

$$\underline{f}_{dc}(n) = -R^{-1}B_d^T[A_d^T]^{-1}\ [\underline{k}_{dc}(n) - Q\underline{x}_d(n)]$$

and $K_{dc}(n)$ and $\underline{k}_{dc}(n)$ satisfy the difference equations:

$$K_{dc}(n) = A_d^T K_{dc}(n+1)[I+B_d R^{-1}B_d^T\ K_{dc}(n+1)]^{-1}A_d+Q$$

and

$$\{I+A_d^T K_{dc}(n+1)[I+B_d R^{-1}B_d K_{dc}(n+1)]^{-1}B_d R^{-1}B_d^T\}\underline{k}_{dc}(n) =$$

$$= A_d^T\underline{k}_{dc}(n+1)+A_d^T K_{dc}(n+1)[I+B_d R^{-1}B_d^T\ K_{dc}(n+1)]^{-1}\underline{d}_d(n)$$

$$- Q\underline{x}_d(n)$$

with boundary conditions $K_{dc}(n_f) = Q_1$ and $\underline{k}_{dc}(n_f) = -Q_1\underline{x}_d(n_f)$. These are rather awkward forms for computation; and alternative is to use $F_{dc}(n) = F_{cc}(n\Delta t)$ and $\underline{f}_{dc}(n) = \underline{f}_{cc}(n)$ for Δt small with F_{cc} and $\underline{f}_{cc}$ computed from a discrete version of Eq. 6.22.

Since $\underline{x}(n)$ is not measured, a filter to find $\hat{\underline{x}}(n)$ is required. For

$$J_f = E[\underline{e}^T(n)\underline{e}(n)] \tag{6.70}$$

with

$$\underline{e}(n) = \underline{x}(n) - \hat{\underline{x}}(n)$$

the optimal filter is [28.]:

$$\underline{\hat{x}}(n+1) = A_d\underline{\hat{x}}(n)+B_d\underline{v}(n)+F_{df}(n)[\underline{y}(n)-\underline{f}_d(n)-C(n)\underline{\hat{x}}(n)] + \underline{d}_d(n) \tag{6.71}$$

with $\underline{\hat{x}}(0) = \underline{m}_{\underline{x}}(0)$ and where

$$F'_{df}(n) = A_d S_{df}(n) C^T(n) \; [M_{\underline{n}_d} + C(n) S_{df}(n) C^T(n)]^{-1}$$

and $S_{df}(n)$ satisfies the difference equation:

$$S_{df}(n+1) = [A_d - F_{df}(n)C(n)] \; S_{df}(n) A_d^T + M_{\underline{n}_d}$$

with boundary condition $S_{df}(0)$. Equations 6.66, 6.67, 6.69, and 6.71 can now be used to obtain a figure similar to Figure 6.6 for the discretized system and an equation similar to Eq. 6.65 can be written.

6.4.4 System Evaluation

The evaluation of the complete system design can now be done. The two performance measures of interest are the mean value of $\underline{x}(n)$ and the covariance of $\underline{x}(n)$. This problem was discussed in Chapter 4, Section 4.3. Taking the expected value of Eq. 6.68 after substituting

$$\underline{v}(n) = F_{dc}(n) \; \underline{\hat{x}}(n) + \underline{f}_{dc}(n)$$

gives

$$\underline{m}_{\underline{x}}(n+1) = [A_d + B_d F_{dc}(n)] \; \underline{m}_{\underline{x}}(n) + B_d \underline{f}_{dc}(n) + \underline{d}_d(n) \tag{6.72}$$

with $\underline{m}_{\underline{x}}(0) = E[\underline{x}(0)]$. A measure of the dispersion of a trajectory $\underline{x}(n)$ about the mean $\underline{m}_{\underline{x}}(n)$ is given by the covariance

$$P(n) = E[(\underline{x}(n) - \underline{m}_{\underline{x}}(n))(\underline{x}(n) - \underline{m}_{\underline{x}}(n))^T]$$

in the notation of Chapter 4. By using

$$\underline{x}(n+1) = A_d\underline{x}(n)+B_dF_{dc}(n)\hat{\underline{x}}(n)+B_d\underline{f}_{dc}(n)+d_d(n)+\underline{\xi}_d(n)$$

and subtracting Eq. 6.72, we obtain

$$\underline{x}(n+1)-\underline{m}_x(n+1) = A_d\underline{e}(n)+[A_d+B_dF_{dc}(n)][\hat{\underline{x}}(n)-\underline{m}_x(n)]+\underline{\xi}_d$$

since $E[\underline{x}] = E[\hat{\underline{x}}] = \underline{m}_x = \underline{m}_{\hat{x}}$ Then proceeding as in the derivation of Eq. 4.69,

$$P(n+1) = A_dS_{df}(n)A_d^T+M_{\underline{\xi}d} +$$

$$+ [A_d+B_dF_c(n)]\ N\ [A_d+B_dF_{dc}(n)]^T$$

where $N = E[(\hat{\underline{x}}(n)-\underline{m}_{\hat{x}}(n))(\hat{\underline{x}}(n)-\underline{m}_{\hat{x}}(n))^T]$. It can be shown that

$$P(n) = N+S_{df}(n)$$

by using $\underline{x}-\underline{m}_x = \underline{e}+(\hat{\underline{x}}-\underline{m}_{\hat{x}})$ and the fact that $E[\underline{e}\ \hat{\underline{x}}^T] = \Theta$ and $E[\underline{e}] = \underline{0}$. Finally we obtain

$$P(n+1) = A_dS_{df}(n)A_d^T+M_{\underline{\xi}d} +$$

$$+ [A_d+B_dF_{dc}(n)][P(n)-S_{df}(n)][A_d+B_dF_{dc}(n)]^T \quad (6.73)$$

with $P(0) = P_o$. Equations 6.72 and 6.73 are the basic equations for evaluating the system performance. In addition to these two equations, $F_{dc}(n)$ and $\underline{f}_{dc}(n)$ in Eq. 6.69 and $F_{df}(n)$ and $S_{df}(n)$ in Eq. 6.71 must be programmed. It is worth noting that since measurements on the aircraft are being made prior to the entry into the approach window at n=0, the initial covariance matrices $S_{df}(0)$ (for $\underline{e}$) and P_o (for $\underline{x}$) can be estimated from these conditions [26.]; the initial filter performance is therefore better than that of the usual case where $S_{df}(0) = P_o$.

Reference [26.] gives both mean values and 2σ dispersions found from P(n) for a great variety of conditions. Other variables such as control deflections and normal acceleration also are given. Two typical results are shown in Figures 6.7 and 6.8.* It can be seen that the trajectory given by mean altitude and range is very close to h_d and r_d. The stochastic inputs, as well as initial conditions, were assumed Gaussian, and the 2σ dispersions therefore indicate that any variable of interest will be within those limits with probability = 0.95. The results in [26.] are based on a modification of the control in Eq. 6.69 in that no feedback from the last three state variables, i.e., u_g, w_g, and ρ_{bb}, is used.

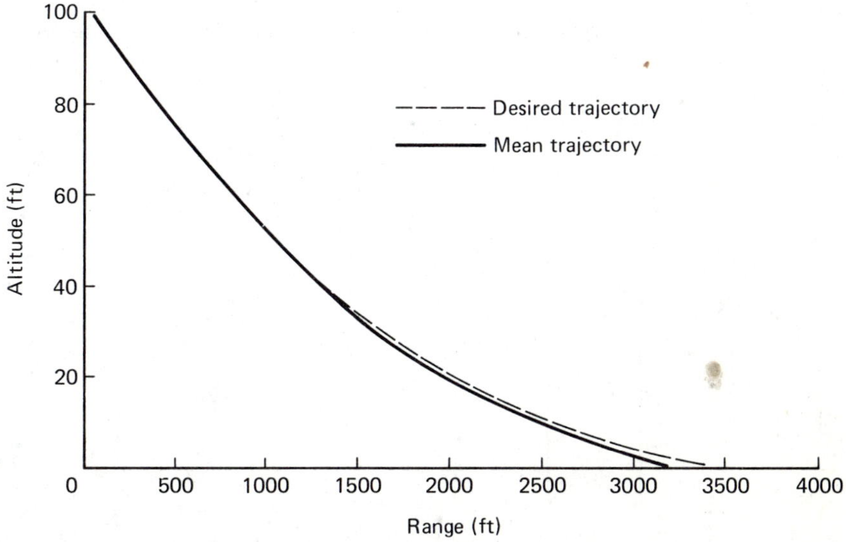

Figure 6.7a Mean and desired trajectory

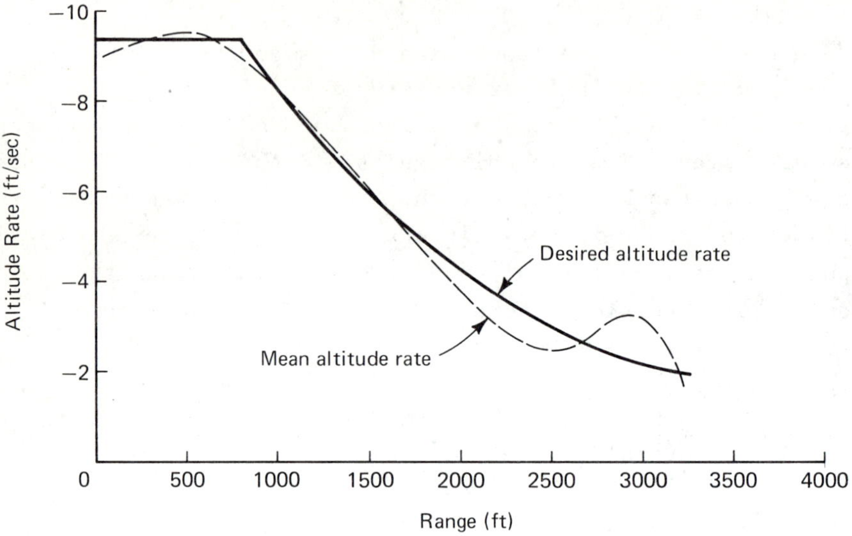

Figure 6.7b Mean and desired altitude rate

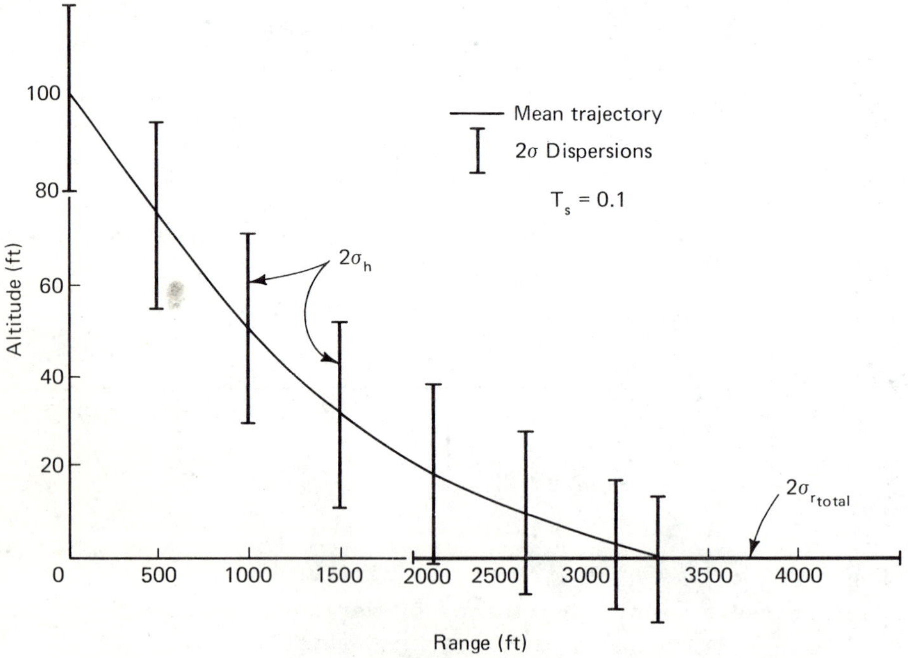

Figure 6.8 2σ trajectory dispersions

The parameters for which numerical values are required are tabulated below.

Aircraft

$X_u = -0.0373$	$M_u = 0$	$X_{\delta e} = 0$
$X_w = 0.136$	$M_w = -0.000461$	$Z_{\delta e} = -9.25$
$Z_u = -0.283$	$M_q = -0.594$	$M_{\delta e} = -0.923$
$Z_w = -0.750$	$M_{\delta th} = 0.000623$	
$X_{\delta th} = 0.106$		
$Z_{\delta th} = -0.00097$		

Flight Path

$\Theta_o = -2.8^o$

$U_o = 228$ ft/sec

$h_o = 100$ ft

Wind

$s = 8$ Knots/100 ft

$u_{steady} = -25$ Knots

$\omega_{ug} = 0.34$

$\omega_{wg} = 3.95$

$\sigma_{ug} = 10$ ft/sec

$\sigma_{wg} = 6.25$ ft/sec

Measurements

$\sigma_{\eta_r} = 20$ ft	range
$\omega_{bb} = 0.0775$	elevation angle
$\sigma_{bb} = 0.00017$ radians	elevation angle
$\sigma_{\eta_{fn}}$, T_s, f	elevation angle
σ_{η_q}, σ_{η_θ}	pitch rate, pitch

Control

$$Q = \text{diag}\,[0 \quad 0 \quad 0 \quad 0 \quad 0 \quad 5 \quad 0.1 \quad 0 \quad 0 \quad 0 \quad 0]$$

$$R = \text{diag}\,[1.5\text{x}10^6 \qquad 10^2]$$

$$Q_1 = \text{diag}\,[0 \quad 0 \quad 100 \quad 10^7 \quad 500 \quad 100 \quad 0.1 \quad 0 \quad 0 \quad 0 \quad 0]$$

$$\underline{x}_d^T(t) = [- \; - \; - \; - \; - \; h_d(t) \quad r_d(t) \; - \; - \; - \; -], \quad 0 \leq t < t_f'$$

$$\underline{x}_d^T(t_f) = [- \; - \quad 0 \quad 0.093 \quad -2 \quad 0 \quad 3387 \quad - \quad - \quad - \quad -]$$

(The components marked - are not specified since Q and Q_1 do not weight them.)

Initial Conditions

P_o

$S_{df}(0)$

$\underline{m}_{\underline{x}}(0)$

Other

$$\Delta t = 1/50 \text{ sec}$$

$$T_{th} = 1 \text{ sec.}$$

$$g = 32.2 \text{ ft/sec}^2$$

Δt is chosen small in [26.] since a simulation of a continuous-time implementation was required.

It is clear that if the system to be analyzed can be modeled as approximately linear with deterministic and Gauss-Markov stochastic inputs, a large amount of information on its performance can be obtained. It also is obvious that a rather large computer capability is required.

6.5 ELECTRIC POWER SYSTEMS

The analysis of complex electric power systems is a difficult problem, and many techniques for solving it have been used over the past several decades. There has been an increased reliance on computers as their capabilities have evolved from the early special-purpose network analyzers to the more versatile general-purpose analog computers and large general-purpose digital computers. As in the preceding section, the first step is the development of a system model, and that is the principal concern of this discussion.

Generally, a power system is a complex interconnection of generating plants through a network of transmission lines and distribution systems. The essentials of interest here are the synchronous generators and prime movers with their associated controls (voltage regulators and speed governors) in the plants and the network of transmission lines with connected loads. The large amount of switching equipment, protective devices, and communication and instrumentation systems are not included. The major part of this section is the development of a state variable model of a single synchronous generator being driven by a prime mover. This machine is then connected to a system of several generators with interconnecting network impedances -- the transmission lines.

6.5.1 Synchronous Generator Model

Figure 6.9 shows a schematic of a three-phase machine. The salient-pole rotor contains the main field winding as well as damper windings, and the stator contains a three-phase, Y-connected armature winding (ungrounded). The angular displacement of the rotor from a fixed reference is $\theta(t)$ (mechanical radians) and the corresponding electrical displacement is $\rho(t) = n\theta(t)$ where n is the number of <u>pole pairs</u>. It is convenient to use a two-axis representation of the machine, and Park's transformation converts the stator vari-

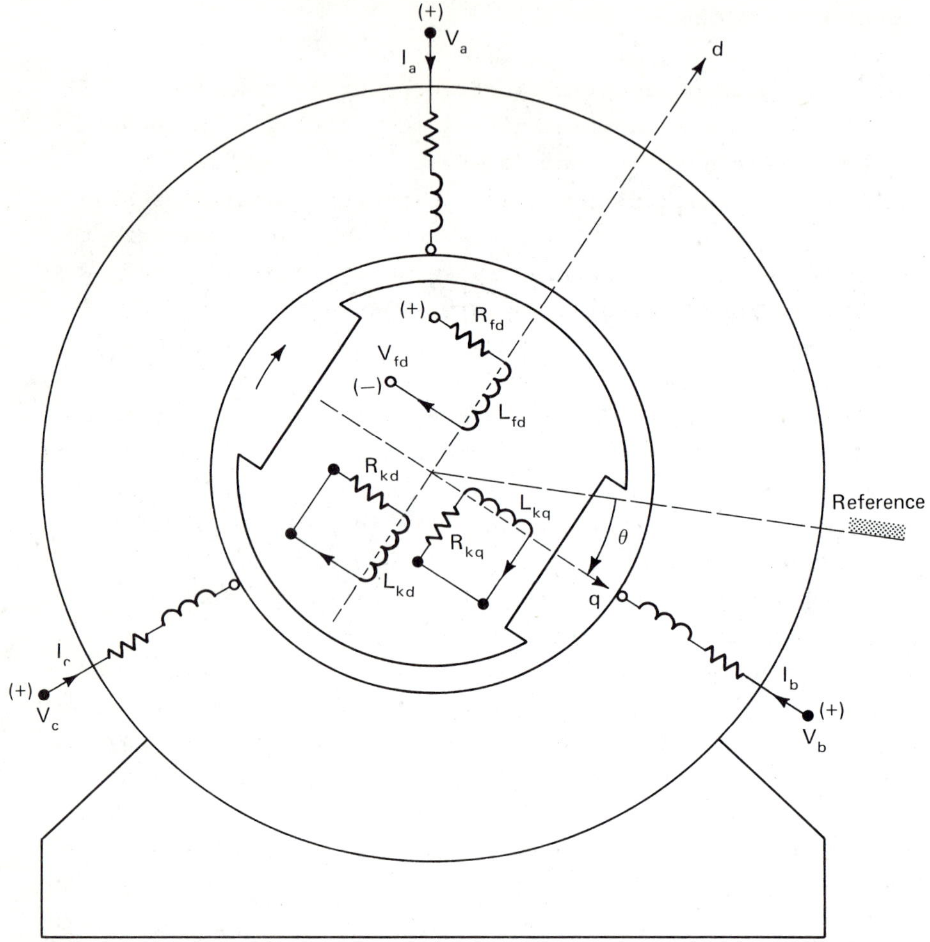

Figure 6.9

ables of interest to a direct and quadrature axis rotating reference frame. In this way stator and rotor variables are viewed in a common frame; the result is a demodulation of the stator variables as viewed from the rotor. The transformation is

$$T_p = \frac{2}{3}\begin{bmatrix} \cos\rho & \cos(\rho - 120^\circ) & \cos(\rho + 120^\circ) \\ -\sin\rho & -\sin(\rho - 120^\circ) & -\sin(\rho + 120^\circ) \\ \frac{1}{2} & \frac{1}{2} & \frac{1}{2} \end{bmatrix} \quad (6.74)$$

and

$$\begin{bmatrix} I_d \\ I_q \\ I_z \end{bmatrix} = T_p \begin{bmatrix} I_a \\ I_b \\ I_c \end{bmatrix} , \quad \begin{bmatrix} E_d \\ E_q \\ E_z \end{bmatrix} = T_p \begin{bmatrix} V_a \\ V_b \\ V_c \end{bmatrix}$$

where I_z and E_z are the zero sequence current and voltage. These are zero in this discussion since the Y-connected winding of the stator is ungrounded; furthermore, balanced conditions will be assumed. The result of this transformation is shown schematically in Figure 6.10. (Note that the capital letters used for currents and voltages here symbolize instantaneous quantities. This is to distinguish them from the per unit variables used later.)

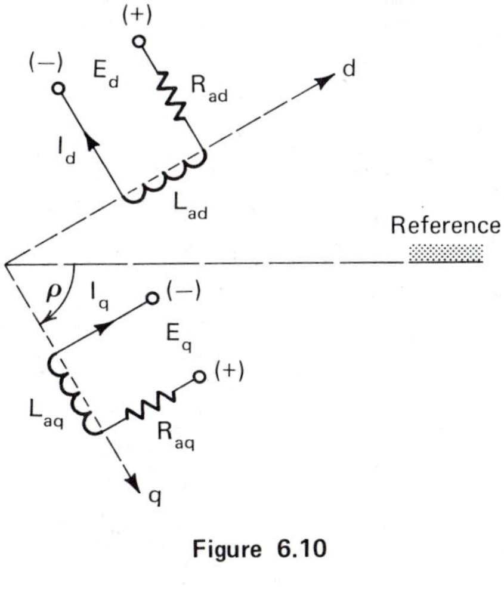

Figure 6.10

Differential equations can now be written for the three field circuits of Figure 6.9 and the two transformed armature circuits of Figure 6.10. These are [30.]:

<u>Field Circuit</u>

$$V_{fd} = (R_{fd}+pL_{fd})I_{fd}+p\ L_{fkd}I_{kd}-pL_{fad}I_d \quad . \qquad (6.75)$$

Direct-Axis Damper

$$V_{kd} = 0 = pL_{fkd}I_{fd}+(R_{kd}+pL_{kd})I_{kd}-pL_{kad}I_d \;. \tag{6.76}$$

Quadrature-Axis Damper

$$V_{kq} = 0 = (R_{kq}+pL_{kq})I_{kq}-pL_{kaq}I_q \;. \tag{6.77}$$

Direct-Axis Armature

$$E_d = pL_{fad}I_{fd}+pL_{kad}I_{kd}-n\dot{\theta}\, L_{kaq}I_{kq} - (R_{ad}+pL_{ad})I_d+n\dot{\theta}L_{aq}I_q \;. \tag{6.78}$$

Quadrature-Axis Armature

$$E_q = n\dot{\theta}L_{fad}I_{fd}+n\dot{\theta}L_{kad}I_{kd}+pL_{kaq}I_{kq} - n\dot{\theta}L_{ad}I_d-(R_{aq}+pL_{aq})I_q \;. \tag{6.79}$$

In these equations, $p \equiv \frac{d}{dt}$. The four inductances with three subscripts are mutual inductances, and those associated with $\dot{\theta}$ produce "speed" voltage terms. If the inductances are constant (e.g., no iron saturation), then p and any inductance may be interchanged. Finally, a torque equation can be written:

$$T = J\ddot{\theta}+K\dot{\theta}+n[L_{fad}I_{fd}+L_{kad}I_{kd}I_q - L_{kaq}I_{kq}I_d+(L_{aq}-L_{ad})I_dI_q], \tag{6.80}$$

where T is the mechanical torque developed by the prime mover, J is the moment of inertia, and K is a friction coefficient.

The usual phasor diagram for a synchronous machine operating under steady-state conditions is shown in Figure 6.11. This shows the power angle δ and the phase angle ϕ, as well as the excitation voltage E_f.

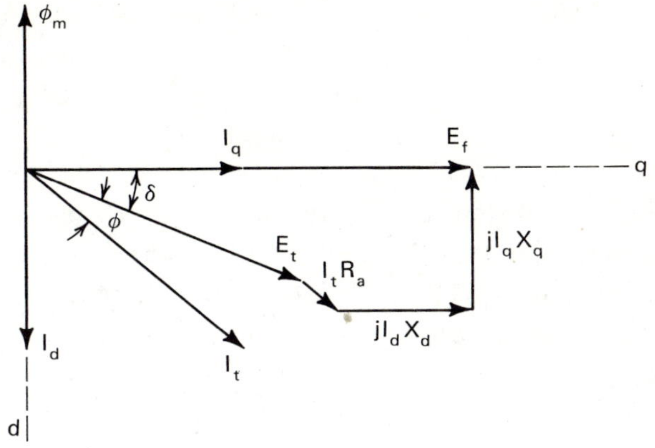

Figure 6.11

It is now convenient to convert the variables to per unit quantities [30.]. For any variable Q, its per unit value q is

$$q = \frac{Q}{Q_o}$$

where Q_o is the base (usually rated) value. In addition, an impedance ratio that refers rotor variables to the stator is required. This has the effect of producing equivalent one-one turn ratios in the coupled circuits as is common in transformer studies, and the result is that all per unit mutual inductances and reactances in the d-axis are equal, and all those in the q-axis are equal. Using lower case for per unit variables,

$$I \rightarrow i \quad \text{using } I_o \text{ (rms per phase)}$$

$$E,V \rightarrow e,v \quad \text{using } E_o, V_o \text{ (rms per phase)}$$

$$X \rightarrow x \quad \text{using } X_o(V_o/I_o)$$

$$R \rightarrow r \quad \text{using } X_o$$

$$L \rightarrow \ell \quad L/X_o$$

in Equations 6.75 to 6.79. The per unit flux linkage is ψ, and is defined as

$$\psi = xi = \omega_o \ell i \ .$$

The instantaneous electrical radian frequency is $\omega = n\dot{\theta} = \dot{\rho}$ and the rated value is $\omega_o = 2\pi f_o$. Reactances x are assumed to be constant for small deviations of ω from ω_o.

The flux linkage in the field circuit is

$$\psi_{fd} = \psi_{fm} + \psi_{f\ell}$$

in terms of mutual (ψ_{fm}) and leakage ($\psi_{f\ell}$) values. Therefore, we have

$$\begin{aligned}
\psi_{fd} &= [x_{ad}(i_{fd}+i_{kd}-i_d)+x_{f\ell}\ i_{fd}] \\
&= [(x_{ad}+x_{f\ell})i_{fd}-x_{ad}\ i_d + x_{ad}\ i_{kd}] \\
&= [x_{ffd}\ i_{fd}-x_{ad}\ i_d+x_{ad}\ i_{kd}] \ , \qquad (6.81)
\end{aligned}$$

where x_{ad} is the direct-axis mutual reactance and x_{ffd} is the self reactance of the main field referred to the stator. Similarly, we get

$$\psi_d = [x_{ad}\ i_{fd}-x_d\ i_d+x_{ad}\ i_{kd}] \qquad (6.82)$$

$$\psi_{kd} = [x_{ad}\ i_{fd}-x_{ad}\ i_d+x_{kkd}\ i_{kd}] \qquad (6.83)$$

$$\psi_q = [-x_q\ i_q+x_{aq}\ i_{kq}] \qquad (6.84)$$

$$\psi_{kq} = [-x_{aq}\ i_q+x_{kkq}\ i_{kq}] \qquad (6.85)$$

where x_d, x_q, x_{kkd}, and x_{kkq} are self reactances of the direct and quadrature armature and damper windings, and x_{aq} is the quadrature-axis mutual reactance.

Using the per unit form of Eq. 6.75, we obtain

$$v_{fd} = r_{fd}\, i_{fd} + p[\ell_{fd}\, i_{fd} + \ell_{fkd}\, i_{kd} - \ell_{fad}\, i_d] \ .$$

For $\ell_{fd} = \ell_{fdm} + \ell_{fd\ell}$, the sum of mutual and leakage per unit inductance, we have, in terms of per unit reactances,

$$v_{fd} = r_{fd}\, i_{fd} + \frac{1}{\omega_o}\, p\, [x_{ad}(i_{fd} + i_{kd} - i_d) + x_{f\ell}\, i_{fd}]$$

since the per unit mutual inductances are equal. Then from Eq. 6.81, we get

$$v_{fd} = r_{fd}\, i_{fd} + \frac{1}{\omega_o}\, p\, \psi_{fd} \ . \qquad (6.86)$$

Similarly using Eq. 6.78, 6.82, and 6.84,

$$e_d = -r_{ad}\, i_d + \frac{1}{\omega_o}\, p\, \psi_d - \frac{\omega}{\omega_o}\, \psi_q \qquad (6.87)$$

and from Eq. 6.76 and 6.83,

$$0 = r_{kd}\, i_{kq} + \frac{1}{\omega_o}\, p\, \psi_{kd} \qquad (6.88)$$

and from Eq. 6.79, 6.82, and 6.84,

$$e_q = -r_{aq}\, i_q + \frac{1}{\omega_o}\, p\, \psi_q + \frac{\omega}{\omega_o}\, \psi_d \qquad (6.89)$$

and from Eq. 6.77 and 6.85,

$$0 = r_{kq}\, i_{kq} + \frac{1}{\omega_o}\, p\, \psi_{kq} \ . \qquad (6.90)$$

In Eq. 6.87 and 6.89, we let $r = r_{ad} = r_{aq}$.

The torque equation, Eq. 6.80, also is converted to per unit form. Thus for T_o a base or rated torque at ω_m where ω_m is the rated (mechanical) speed, we have

$$\omega_m T_o = \frac{\omega_o T_o}{n} = P_o = 3 V_o I_o$$

where P_o is the corresponding base power. Eq. 6.80 is divided by T_o, but the bracketed term on the right is divided by the equivalent nP_o/ω_o. Then we get

$$\frac{T}{T_o} = \frac{J}{T_o}\ddot{\theta} + \frac{K}{T_o}\dot{\theta} + \frac{\omega_o}{P_o}[L_{fad} I_{fd} I_q + L_{kad} I_{kd} I_q$$
$$- L_{kaq} I_{kq} I_d + (L_{aq}-L_{ad}) I_d I_q].$$

Using previous definitions we have the general equality of the forms on the right as

$$\frac{1}{P_o} L I_1 I_2 = \frac{1}{3} \ell i_1 i_2$$

and therefore, we get

$$\frac{T}{T_o} = \frac{J}{T_o}\ddot{\theta} + \frac{K}{T_o}\dot{\theta} + \frac{1}{3}\omega_o [\ell_{fad} i_{fd} i_q + \ell_{kad} i_{kd} i_q$$
$$- \ell_{kaq} i_{kq} i_d + (\ell_{aq}-\ell_{ad}) i_d i_q].$$

Now using Eq. 6.82 and 6.84, we have

$$[\psi_d i_q - \psi_q i_d] = [x_{ad} i_{fd} - x_d i_d + x_{ad} i_{kd}] i_q$$
$$- [-x_q i_q + x_{aq} i_{kq}] i_d .$$

Also, we have

$$\frac{1}{\omega_o} x_{ad} = \ell_{fad} = \ell_{kad}$$

$$\frac{1}{\omega_o} x_{aq} = \ell_{kaq}$$

$$\frac{1}{\omega_o} x_d = \ell_{ad}$$

$$\frac{1}{\omega_o} x_q = \ell_{aq}$$

and therefore,

$$\frac{T}{T_o} = \frac{J}{T_o}\,\ddot{\theta} + \frac{K}{T_o}\,\dot{\theta} + \frac{1}{3}[\psi_d i_q - \psi_q i_d].$$

Since $\rho = n\theta$, this becomes

$$\frac{T}{T_o} = \frac{J}{nT_o}\,\ddot{\rho} + \frac{K}{nT_o}\,\dot{\rho} + \frac{1}{3}[\psi_d i_q - \psi_q i_d].$$

Now we let

$$\frac{T}{T_o} = T_m \text{ (dimensionless)}$$

$$\frac{J}{nT_o} = J_n \text{ (sec}^2\text{)}$$

$$\frac{K}{nT_o} = K_n \text{ (sec)}$$

$$\frac{1}{3}[\psi_d i_q - \psi_q i_d] = T_g \text{ (dimensionless)}$$

and the final normalized torque equation is

$$T_m = J_n\ddot{\rho} + K_n\dot{\rho} + T_g. \tag{6.91}$$

J_n is usually given in terms of H, an inertia constant, and

$$H = \frac{\text{stored energy at } \omega_o}{\text{rated power}} = \frac{W_o}{P_o}\,.$$

The stored energy at $\dot{\theta} = \omega_o/n$ is

$$\begin{aligned} W_o &= \tfrac{1}{2}\, J\, \dot{\theta}^2 \\ &= \tfrac{1}{2}\, nT_o\, J_n \left(\frac{\omega_o}{n}\right)^2 \\ &= \tfrac{1}{2}\left(\frac{\omega_o T_o}{n}\right) J_n\, \omega_o = \tfrac{1}{2} P_o\, J_n \omega_o \end{aligned}$$

and we have then

$$H = \tfrac{1}{2}J_n\omega_o$$

or

$$J_n = \frac{2H}{\omega_o}\ .$$

The model given by Eq. 6.86 through 6.91 has been widely used to study synchronous machine dynamics on analog computers [30.][31.][32.][33.][34.], hybrid computers [35.], and digital computers [36.].

In anticipation of extending the model to a system containing several machines, Eq. 6.91 is modified by introducing another angle [36.]. Let $\omega-\omega_o$ be the deviation from ω_o, then

$$\int_o^t (\omega-\omega_o)dt + \rho_o = \rho-\omega_o t + \rho_o,$$

and if a system reference frame of D and Q-axis is chosen such that $\rho_o = 0$,

$$\alpha = \rho-\omega_o t$$

is the deviation of the d and q-axes of the machine from the system reference frame. This is shown for steady-state conditions in the per unit phasor diagram of Figure 6.12. The angle β_o must be known and could be made zero by choosing the Q-axis the same as e_t. Differentiating α,

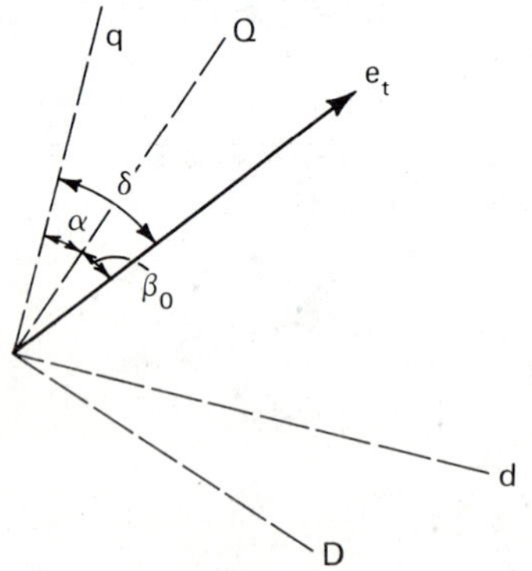

Figure 6.12

$$\dot{\alpha} = \dot{\rho} - \omega_o = \omega - \omega_o = \dot{\delta}$$

$$\ddot{\alpha} = \ddot{\rho} = \ddot{\delta}$$

and Eq. 6.91 becomes

$$T_m = J_n\ddot{\alpha} + K_n\dot{\alpha} + T_g + K_n\omega_o. \tag{6.92}$$

6.5.2 Governor and Voltage Regulator

A simple model of governor action is included as follows. Basically, for a decrease in load, T_g will decrease and there will be an accelerating torque $T_m - T_g$ to produce an increase in α. As a result of governor action, a decrease in T_m is required so that ω tends to ω_o, i.e., $\dot{\alpha}$ tends to zero. T_m is decreased by reducing the throttle valve (or penstock gate) opening. Thus, a simple model (in per unit torque) with one time constant is [33.]:

$$T_m = \frac{G_r}{(1+\tau_g p)} g_r - \frac{G_s}{(1+\tau_g p)}\dot{\alpha} + K_g \tag{6.93}$$

where g_r is a reference (speed gear) setting, G_r and G_s are gains, K_g is a constant. A more complete discussion of a governor is given in [37.]. Since $p\alpha = \dot{\alpha} = \omega - \omega_o$, inspection of Eq. 6.92 shows that changes in T_g are balanced by small changes in ω if the machine is operating alone to supply a load. Where the steady-state ω must equal ω_o exactly, an integral or reset term, i.e., a term proportional to α, is needed in Eq. 6.93. If the machine is operating in a system with a frequency reference, torque balance is maintained at ω_o by supplying excess power to (or drawing needed power from) the "infinite bus" to which the individual machine is connected. It is the latter case that is of primary interest here. In this mode of operation, g_r is the basic control variable on the real power supplied by the machine.

A model for the voltage regulator is

$$e_f = \frac{G_e}{(1+\tau_e p)}(v_r - e_t) \tag{6.94}$$

where $e_t^2 = e_d^2 + e_d^2$ and e_f is the per unit value of E_f, the excitation voltage. v_r is the regulator reference voltage and is the basic control on reactive power. It also can be shown that

$$v_{fd} = \frac{r_{fd}}{x_{ad}} e_f \qquad (6.95)$$

which is used in Eq. 6.86 later.

6.5.3 State Variable Model*

A state variable model of the synchronous machine can now be developed from Eq. 6.86 through 6.90 and 6.92. The choice of state variables is not unique although two common sets are the flux linkages ψ_{fd}, ψ_d, ψ_{kd}, ψ_q, and ψ_{kq} and the currents i_{fd}, i_d, i_{kd}, i_q, and i_{kq}. These are related in Eq. 6.81 through 6.85. The choice of flux linkages has some advantage in developing an accurate nonlinear model, e.g., in modelling iron saturation [35.]. Currents have been used in the PQR method [38.] where digital computer manipulation of the matrices is emphasized. The five flux linkages are chosen as state variables here, and in addition, α and $\dot{\alpha}$ are included from Eq. 6.92 so that the basic machine is modelled as a seventh-order system.

First we let

$$\underline{\psi}^T = [\psi_{fd} \quad \psi_d \quad \psi_{kd} \quad \psi_q \quad \psi_{kq}] ,$$

then Eq. 6.86 through 6.90 can be written as

---

*The following three sections are based, in part, on the Ph.D. thesis of William L. Miller, The Penn State University, 1970.

$$\dot{\underline{\psi}} = \begin{bmatrix} \omega_o (v_{fd} - r_{fd} i_{fd}) \\ \omega_o (e_d + r i_d) + \omega \psi_q \\ \omega_o (-r_{kd} i_{kd}) \\ \omega_o (e_q + r i_q) - \omega \psi_d \\ \omega_o (-r_{kq} i_{kq}) \end{bmatrix}$$

or

$$\dot{\underline{\psi}} = \Omega\underline{\psi} + \omega_o\underline{e} + \omega_o R_o \underline{i} \qquad (6.96)$$

where the elements of Ω, ω_{ij}, are

$$\omega_{ij} = 0$$

except

$$\omega_{24} = \omega \quad \text{and} \quad \omega_{42} = -\omega,$$

and

$$R_o = \text{diag}\,[-r_{fd} \quad r \quad -r_{kd} \quad r \quad -r_{kq}],$$

and

$$\underline{e}^T = [v_{fd} \quad e_d \quad 0 \quad e_q \quad 0],$$

and

$$\underline{i}^T = [i_{fd} \quad i_d \quad i_{kd} \quad i_q \quad i_{kq}] .$$

From Eq. 6.81 through 6.85, we obtain

$$\underline{\psi} = X\underline{i}$$

where

$$X = \left[\begin{array}{ccc|cc} x_{ffd} & -x_{ad} & x_{ad} & & \\ x_{ad} & -x_{d} & x_{ad} & \multicolumn{2}{c}{\Theta} \\ x_{ad} & -x_{ad} & x_{kkd} & & \\ \hline & & & -x_{q} & x_{aq} \\ & \Theta & & -x_{aq} & x_{kkq} \end{array}\right]$$

From this, we have

$$\underline{i} = X^{-1}\,\underline{\psi},$$

and substituting in Eq. 6.96, we get

$$\dot{\underline{\psi}} = \Omega\underline{\psi} + \omega_{o}\,R_{o}X^{-1}\,\underline{\psi} + \omega_{o}\underline{e}\ .$$

If X^{-1} is written as

$$X^{-1} = \left[\begin{array}{c|c} M & \Theta \\ \hline \Theta & N \end{array}\right],$$

this equation becomes

$$\dot{\underline{\psi}} = \omega_{o}\left[\begin{array}{ccc|cc} -r_{fd}m_{11} & -r_{fd}m_{12} & -r_{fd}m_{13} & & \\ r\ m_{21} & r\ m_{22} & r\ m_{23} & \multicolumn{2}{c}{\Theta} \\ -r_{kd}m_{31} & -r_{kd}m_{32} & -r_{kq}m_{33} & & \\ \hline & & & r\ n_{11} & r\ n_{12} \\ & \Theta & & -r_{kq}n_{21} & -r_{kq}n_{22} \end{array}\right]\underline{\psi}$$

$$+\ \omega_{o}\underline{e} + \Omega\underline{\psi}\ . \qquad (6.97)$$

In practice, X^{-1} would be precomputed from the per unit reactances.

To include the torque equation as well as the governor in the model, it is convenient to use deviations from a steady-state condition. If it is assumed that $\omega-\omega_o = \dot{\alpha} = 0$ and g_r is a constant g_{ro} for steady-state conditions, Eq. 6.92 and 6.93 give for the steady-state torque:

$$T_{mo} = T_{go} + K_n\omega_o = G_r g_{ro} + K_g$$

which relates T_{go}, the steady-state load with g_{ro}, the governor setting. Now we subtract T_{mo} from Eq. 6.92 to obtain

$$\delta T_m = T_m - T_{mo} = J_n\ddot{\alpha} + K_n\dot{\alpha} + T_g - T_{go} \tag{6.98}$$

and from Eq. 6.93 to obtain

$$\delta T_m = T_m - T_{mo} = \frac{G_r\ \delta g_r - G_s\dot{\alpha}}{(1+\tau_g p)} \tag{6.99}$$

where $\delta g_r = g_r - g_{ro}$. In Eq. 6.94, we let $v_r = v_{ro} + \delta v_r$, then we have

$$e_f = \frac{G_e\ \delta v_r}{(1+\tau_e p)} + \frac{G_e\ v_{ro}}{(1+\tau_e p)} - \frac{G_e\ e_t}{(1+\tau_e p)} \tag{6.100}$$

A state vector $\underline{x}$ is now chosen as

$$\underline{x}^T = [\psi_{fd}\quad \psi_d\quad \psi_{kd}\quad \psi_q\quad \psi_{kq}\quad \alpha\quad \dot{\alpha}\quad \delta T_m\quad e_f]$$

and the nine first-order differential equations are as follows:

$$\dot{x}_1 = -\omega_o r_{fd} m_{11} x_1 - \omega_o r_{fd} m_{12} x_2 - \omega_o r_{fd} m_{13} x_3 + \omega_o \frac{r_{fd}}{x_{ad}} x_9$$

where Eq. 6.95 is used;

$$\dot{x}_2 = \omega_o r m_{21} x_1 + \omega_o r m_{22} x_2 + \omega_o r m_{23} x_3 + \omega_o e_d + x_4 x_7 + \omega_o x_4$$

where $\omega = \dot{\alpha} + \omega_o$ is used;

$$\dot{x}_3 = -\omega_o r_{kd} m_{31} x_1 - \omega_o r_{kd} m_{32} x_2 - \omega_o r_{kd} m_{33} x_3;$$

$$\dot{x}_4 = \omega_o r n_{11} x_4 + \omega_o r n_{12} x_5 + \omega_o e_q - x_2 x_7 - \omega_o x_2$$

where $\omega = \dot{\alpha} + \omega_o$ is used;

$$\dot{x}_5 = -\omega_o r_{kq} n_{21} x_4 - \omega_o r_{kq} n_{22} x_5;$$

$$\dot{x}_6 = x_7;$$

$$\dot{x}_7 = -\frac{K_n}{J_n} x_7 + \frac{1}{J_n} x_8 - \frac{1}{3J_n} i_q x_2 + \frac{1}{3J_n} i_d x_4 + \frac{T_{go}}{J_n}$$

from Eq. 6.98;

$$\dot{x}_8 = -\frac{1}{\tau_g} x_8 - \frac{G_s}{\tau_g} x_7 + \frac{G_r}{\tau_g} \delta g_r$$

from Eq. 6.99;

$$\dot{x}_9 = -\frac{1}{\tau_e} x_9 + \frac{G_e}{\tau_e} \delta v_r + \frac{G_e}{\tau_e} v_{ro} - \frac{G_e}{\tau_e} (e_d^2 + e_q^2)^{\frac{1}{2}}$$

from Eq. 6.100. Inspection of the right hand sides of these nine equations shows that there are linear and nonlinear terms in the state variables, linear and nonlinear terms in e_d, e_q, i_d, i_q and the state variables, terms δg_r and δv_r, and constant terms. It is possible to eliminate i_d and i_q since

$$\underline{i} = X^{-1} \underline{\psi},$$

and we get

$$i_d = m_{21}\psi_{fd} + m_{22}\psi_d + m_{23}\psi_{kd}$$

and

$$i_q = n_{11}\psi_q + n_{12}\psi_{kq}.$$

Using these,

$$\dot{x}_7 = -\frac{K_n}{J_n}x_7 + \frac{1}{J_n}x_8 - \frac{1}{3J_n}(n_{11}x_2x_4 + n_{12}x_2x_5)$$

$$+ \frac{1}{3J_n}(m_{21}x_1x_4 + m_{22}x_2x_4 + m_{23}x_3x_4) + \frac{T_{go}}{J_n}.$$

The basic machine state equation is now written as

$$\underline{\dot{x}} = A\underline{x}+B\underline{u}+\underline{f}(\underline{x})+\underline{g}(e_d,\ e_q)+\underline{d} \qquad (6.101)$$

where $\underline{u}^T = [\delta g_r \quad \delta v_r]$,

and the nonzero elements of A are

$$a_{11} = -\omega_o\ r_{fd}m_{11}$$

$$a_{12} = -\omega_o\ r_{fd}m_{12}$$

$$a_{13} = -\omega_o\ r_{fd}m_{13}$$

$$a_{19} = \omega_o\ \frac{r_{fd}}{x_{ad}}$$

$$a_{21} = \omega_o\ rm_{21}$$

$$a_{22} = \omega_o\ rm_{22}$$

$$a_{23} = \omega_o\ rm_{23}$$

$$a_{24} = \omega_o$$

$$a_{31} = -\omega_o \, r_{kd} m_{31}$$

$$a_{32} = -\omega_o \, r_{kd} m_{32}$$

$$a_{33} = -\omega_o \, r_{kd} m_{33}$$

$$a_{42} = -\omega_o$$

$$a_{44} = \omega_o \, rn_{11}$$

$$a_{45} = \omega_o \, rn_{12}$$

$$a_{54} = -\omega_o \, r_{kq} n_{21}$$

$$a_{55} = -\omega_o \, r_{kq} n_{22}$$

$$a_{67} = 1$$

$$a_{77} = -\frac{K_n}{J_n} = -\frac{K}{J}$$

$$a_{78} = \frac{1}{J_n}$$

$$a_{87} = -\frac{G_s}{\tau_s}$$

$$a_{88} = -\frac{1}{\tau_g}$$

$$a_{99} = -\frac{1}{\tau_e} ,$$

and the nonzero elements of B are

$$b_{81} = \frac{G_r}{\tau_g}$$

$$b_{92} = \frac{G_e}{\tau_e}$$

and

$$\underline{f}^T = [0 \quad x_4x_7 \quad 0 \quad -x_2x_7 \quad 0 \quad 0 \quad f_7 \quad 0 \quad 0]$$

with

$$f_7 = \frac{1}{3J_n} [(m_{22}-n_{11})x_2x_4 + m_{21}x_1x_4 + m_{23}x_3x_4 - n_{12}x_2x_5],$$

and

$$\underline{g}^T = [0 \quad \omega_o e_d \quad 0 \quad \omega_o e_q \quad 0 \quad 0 \quad 0 \quad 0 \quad g_9]$$

with

$$g_9 = -\frac{G_e}{\tau_e}(e_d^2 + e_q^2)^{\frac{1}{2}},$$

and

$$\underline{d}^T = [0 \quad 0 \quad 0 \quad 0 \quad 0 \quad 0 \quad \frac{T_{go}}{J_n} \quad 0 \quad \frac{G_e}{\tau_e} v_{ro}].$$

In addition to the state equation, Eq. 6.101, an output equation is needed. It is assumed that the terminal voltages (V_a, V_b, V_c), the terminal currents (I_a, I_b, I_c), and field voltage and current (V_{fd}, I_{fd}), and the power angle and power angle rate (δ and $\dot{\delta}$) are measured in the actual machine. The latter two are not usually measured although it can be done provided that a suitable phase reference is maintained as the machine is loaded starting from zero load. These measurements are equivalent to measuring i_d, i_q, e_d, e_q, v_{fd}, i_{fd}, α, and $\dot{\alpha}$ in the model above. (In an actual machine, T_p would be used to transform the three-phase voltages and currents to the demodulated direct and quadrature axis voltages and currents for use in a filter and controller.) Also, e_f is related to v_{fd} by Eq. 6.95, and from Figure 6.12, we have $\alpha = \delta - \beta_o$ and $\dot{\alpha} = \dot{\delta}$. Thus we have x_6, x_7, and x_9 measured as well as i_d, i_q, e_d, e_q, and i_{fd}. Since e_d and e_q appear in

$\underline{g}$ in Eq. 6.101, they are regarded as input variables to the machine and i_{fd}, i_d, i_q, α, $\dot{\alpha}$, and e_f are then selected as output variables. Therefore, we let

$$\underline{y}^T = [i_{fd} \quad i_d \quad i_q \quad \alpha \quad \dot{\alpha} \quad e_f],$$

and

$$\underline{y} = C\underline{x} \tag{6.102}$$

where

$$C = \begin{bmatrix} m_{11} & m_{12} & m_{13} & 0 & 0 & 0 & 0 & 0 & 0 \\ m_{21} & m_{22} & m_{23} & 0 & 0 & 0 & 0 & 0 & 0 \\ 0 & 0 & 0 & n_{11} & n_{12} & 0 & 0 & 0 & 0 \\ 0 & 0 & 0 & 0 & 0 & 1 & 0 & 0 & 0 \\ 0 & 0 & 0 & 0 & 0 & 0 & 1 & 0 & 0 \\ 0 & 0 & 0 & 0 & 0 & 0 & 0 & 0 & 1 \end{bmatrix}.$$

A typical set of data for a synchronous machine is as follows:

$$\begin{aligned} x_d &= 2.0 & r_{fd} &= 0.001 \\ x_q &= 1.9 & r &= 0.002 \\ x_{kkd} &= 1.85 & r_{kd} &= r_{kq} = 0.003 \\ x_{ffd} &= 2.0 & \omega_o &= 337 \\ x_{ad} &= 1.8 & H &= 5.3 \\ x_{aq} &= 1.7 & K & \quad \text{negligible} \\ x_{kkq} &= 1.75 & & \end{aligned}$$

For the governor and voltage regulator,

$$\tau_g = 0.5, \qquad \tau_e = 0.1$$

and G_r and G_e are gain settings. Also, G_s and K_g depend on the governor, and G_s is basically a damping coefficient setting (see a_{87} of the A-matrix).

6.5.4 Connected Network

Although Eq. 6.101 and 6.102 are the state model, there is a constraint relation between i_d, i_q in the output and e_d, e_q in the input since the machine is connected to an electrical network which may include impedances at the machine terminals and impedances of transmission lines that are connected to other machines in the system. First we consider a balanced three-phase load on the machine; then the per unit equations are

$$v_a = i_a r_L + \ell_L p i_a$$

$$v_b = i_b r_L + \ell_L p i_b$$

$$v_c = i_c r_L + \ell_L p i_c$$

where r_L and ℓ_L are per unit constants. For $\underline{v}_t^T = [v_a \quad v_b \quad v_c]$, $\underline{i}_t^T = [i_a \quad i_b \quad i_c]$, we write

$$\underline{v}_t = (R + Lp)\underline{i}_t$$

with $R = r_L I_3$ and $L = \ell_L I_3$ (I_3 is the 3x3 unit matrix.) Then, using Eq. 6.74, we have

$$\begin{bmatrix} e_d \\ e_q \\ e_z \end{bmatrix} = T_p \begin{bmatrix} v_a \\ v_b \\ v_c \end{bmatrix}, \qquad \begin{bmatrix} i_d \\ i_q \\ i_z \end{bmatrix} = T_p \begin{bmatrix} i_a \\ i_b \\ i_c \end{bmatrix}$$

or

$$\begin{bmatrix} e_d \\ e_q \\ e_z \end{bmatrix} = T_p\ (R+Lp)\underline{i}_t$$

$$= R \begin{bmatrix} i_d \\ i_q \\ i_z \end{bmatrix} + L\ T_p p\ \underline{i}_t\ .$$

As an approximation we assume $\omega \cong \omega_o$ and $\dot{\omega} \cong 0$. Then $\rho \cong \omega_o t$ and we can show by computing pi_d, pi_q, and pi_z (which is zero) that we get

$$LT_p p\ \underline{i}_t \cong Lp \begin{bmatrix} i_d \\ i_q \\ 0 \end{bmatrix} + L\omega_o \begin{bmatrix} -i_q \\ i_d \\ 0 \end{bmatrix} .$$

Using this we have

$$\begin{bmatrix} e_d \\ e_q \\ 0 \end{bmatrix} \cong R \begin{bmatrix} i_d \\ i_q \\ 0 \end{bmatrix} + Lp \begin{bmatrix} i_d \\ i_q \\ 0 \end{bmatrix} + \omega_o L \begin{bmatrix} -i_q \\ i_d \\ 0 \end{bmatrix} .$$

Finally, since i_d and i_q are demodulated (i.e., low frequency variables, we use the inequalities

$$|Lp\ i_d| << |\omega_o L\ i_q|$$

and

$$|Lp\ i_q| << |\omega_o L\ i_d|$$

to obtain

$$\begin{bmatrix} e_d \\ \\ e_q \end{bmatrix} = \begin{bmatrix} r_L & 0 \\ \\ 0 & r_L \end{bmatrix} \begin{bmatrix} i_d \\ \\ i_q \end{bmatrix} + \begin{bmatrix} x_L & 0 \\ \\ 0 & x_L \end{bmatrix} \begin{bmatrix} -i_q \\ \\ i_d \end{bmatrix} \qquad (6.103)$$

The last assumption amounts to neglecting electrical transients associated with the network relative to the mechanical and magnetic flux transients in the machine. This usually is justifiable, although there are exceptions, e.g., in the study of machines in parallel [31.]. Equation 6.103 means that e_d, e_q, and i_d, i_q can be related by using the network per unit resistances and reactances (at ω_o).

From Figure 6.12,

$$\begin{bmatrix} e_D \\ \\ e_Q \end{bmatrix} = \begin{bmatrix} \cos\alpha & -\sin\alpha \\ \\ \sin\alpha & \cos\alpha \end{bmatrix} \begin{bmatrix} e_d \\ \\ e_q \end{bmatrix} = T_\alpha \begin{bmatrix} e_d \\ \\ e_q \end{bmatrix}$$

which transforms the machine direct and quadrature axis voltages to the system reference frame, and a similar relationship for currents can be written. Relative to the D and Q-axes, the network is then represented in terms of the per unit network admittance matrix Y_n (see Section 6.3):

$$\underline{i}_N = Y_n \underline{e}_N \qquad (6.104)$$

where each machine is treated as a current source. For the i-th machine with transformed currents i_{Di} and i_{Qi}, we have

$$i_i = i_{Di} + j i_{Qi}, \qquad (j = \sqrt{-1}).$$

Then the components of $\underline{i}_N$ are i_i, i=1,2,...,m for m machines, and similarly for $\underline{e}_N$. Since there may be nodes of the network at which there is no source, Eq. 6.104 is partitioned so that the first m rows are equations for nodes with machines (current sources), and the remainder for nodes without machines. Then we get

$$\begin{bmatrix} \underline{i}_{NA} \\ \underline{0} \end{bmatrix} = \begin{bmatrix} Y_{11} & Y_{12} \\ Y_{21} & Y_{22} \end{bmatrix} \begin{bmatrix} \underline{e}_{NA} \\ \underline{e}_{NB} \end{bmatrix} \tag{6.105}$$

and from this,

$$\underline{0} = Y_{21}\underline{e}_{NA} + Y_{22}\underline{e}_{NB}$$

or

$$\underline{e}_{NB} = -Y_{22}^{-1}Y_{21}\ \underline{e}_{NA} .$$

Then we obtain

$$\underline{i}_{NA} = Y_{11}\underline{e}_{NA} - Y_{12}Y_{22}^{-1}Y_{21}\underline{e}_{NA}$$

or

$$\underline{i}_{NA} = Y_R\ \underline{e}_{NA} \tag{6.106}$$

with $Y_R = Y_{11}-Y_{12}Y_{22}^{-1}Y_{21}$, the reduced admittance matrix of the system.

Solving Eq. 6.106, we get

$$\underline{e}_{NA} = Y_R^{-1}\ \underline{i}_{NA} = Z_R\ \underline{i}_{NA}$$

and

$$Z_R = R_R + jX_R$$

where R_R and X_R are mxm resistance and reactance matrices. Now

$$\underline{e}_{NA} = \underline{e}_{DA} + j\underline{e}_{QA} \quad \text{and} \quad \underline{i}_{NA} = i_{DA} + j\ i_{QA} ,$$

therefore, we have

$$\underline{e}_{DA} + j\ \underline{e}_{QA} = (R_R + j\ X_R)\ (\underline{i}_{DA} + j\ i_{QA})$$

and

$$\underline{e}_{DA} = R_R\ \underline{i}_{DA} - X_R\ \underline{i}_{QA} \qquad (6.107a)$$

$$\underline{e}_{QA} = X_R\ \underline{i}_{DA} + R_R\ \underline{i}_{QA} \qquad (6.107b)$$

which are the vector equivalents of Eq. 6.103 in the D and Q axes. This model has been used in [36.] and [35.]. A schematic representation of the entire model of a single machine is given in Figure 6.13.

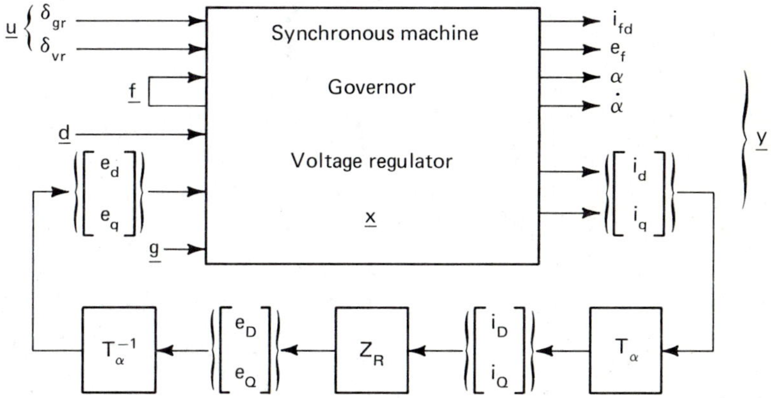

Figure 6.13

6.5.5 Extensions

Extensions from the model of Eq. 6.101, 6.102, and 6.107 can now be carried out. To model a system of m machines, we let the model of the ith machine be

$$\dot{\underline{x}}_i = A_i \underline{x}_i + B_i \underline{u}_i + \underline{f}_i + \underline{g}_i + \underline{d}_i$$

$$\underline{y}_i = C_i \underline{x}_i, \qquad i = 1,2,\ldots,m,$$

and we let the <u>system</u> state be

$$\underline{x}_s^T = [\underline{x}_1^T \quad \underline{x}_2^T \quad \cdots \quad \underline{x}_m^T] .$$

Then the system model becomes

$$\dot{\underline{x}}_s = A_s\underline{x}_s + B_s\underline{u}_s + \underline{f}_s + \underline{g}_s + \underline{d}_s \tag{6.108a}$$

$$\underline{y}_s = C_s \underline{x}_s \tag{6.108b}$$

with

$$A_s = \begin{bmatrix} A_1 & & 0 \\ & \ddots & \\ 0 & & A_m \end{bmatrix}$$

$$B_s = \begin{bmatrix} B_1 & & 0 \\ & \ddots & \\ 0 & & B_m \end{bmatrix} , \qquad C_s = \begin{bmatrix} C_1 & & 0 \\ & \ddots & \\ 0 & & C_m \end{bmatrix}$$

$$\underline{f}_s^T = [\underline{f}_1^T \vdots \cdots \vdots \underline{f}_m^T]$$

$$\underline{g}_s^T = [\underline{g}_1^T \vdots \cdots \vdots \underline{g}_m^T]$$

$$\underline{d}_s^T = [\underline{d}_1^T \vdots \cdots \vdots \underline{d}_m^T]$$

$$\underline{u}_s^T = [\underline{u}_1^T \vdots \cdots \vdots \underline{u}_m^T]$$

and the machines are interconnected through Eq. 6.107. Equation 6.108 can be discretized (e.g., by using a modification of Conversion VII of Chapter 3) to study the system dynamics on a digital computer.

A very simple example of a two-machine system is shown in Figure 6.14 where the voltages and currents are in the D-Q axes, and the per unit resistances and reactances are shown.

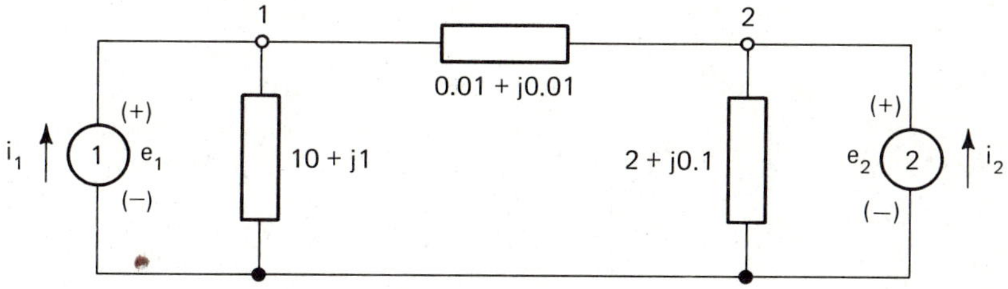

Figure 6.14

From this we have

$$i_1 = \left(\frac{1}{10+j1} + \frac{1}{0.01+j0.01}\right)e_1 - \left(\frac{1}{0.01+j0.01}\right)e_2$$

$$i_2 = \left(\frac{1}{0.01+j0.01}\right)e_1 + \left(\frac{1}{2+j0.1} + \frac{1}{0.01+j0.01}\right)e_2$$

and these define the variables in

$$\underline{i}_{NA} = Y_n\, \underline{e}_{NA} = Y_{11}\, \underline{e}_{NA} = Y_R\, \underline{e}_{NA}$$

since nodes 1 and 2 both have machines connected to them. Then we get

$$\underline{e}_{NA} = Y_R^{-1}\, \underline{i}_{NA} = Z_R\, \underline{i}_{NA} = (R_R + jX_R)\underline{i}_{NA}$$

from which e_{D1}, e_{Q1}, e_{D2}, e_{Q2} can be found by using Eq. 6.107 when the corresponding currents are known. The currents are the transformed outputs of the machines that are modelled as in Eq. 6.101 and 6.102. Transforming the D and Q-axis voltages to the d and q-axes of each machine can then be used for the next computational cycle in a digital simulation. Using the numerical values, we get

$$R_R = \begin{bmatrix} 1.674 & 1.666 \\ 1.666 & 1.668 \end{bmatrix}, \quad X_R = \begin{bmatrix} 0.1042 & 0.0958 \\ 0.0958 & 0.0974 \end{bmatrix}$$

and

$$\underline{e}_{DA} = \begin{bmatrix} e_{D1} \\ \\ e_{D2} \end{bmatrix} = R_R \begin{bmatrix} i_{D1} \\ \\ i_{D2} \end{bmatrix} - X_R \begin{bmatrix} i_{Q1} \\ \\ i_{Q2} \end{bmatrix}$$

$$\underline{e}_{QA} = \begin{bmatrix} e_{Q1} \\ \\ e_{Q2} \end{bmatrix} = X_R \begin{bmatrix} i_{D1} \\ \\ i_{D2} \end{bmatrix} + R_R \begin{bmatrix} i_{Q1} \\ \\ i_{Q2} \end{bmatrix} .$$

Thus we have

$$\begin{bmatrix} e_{d1} \\ \\ e_{q1} \end{bmatrix} = T_{\alpha_1}^{-1} \begin{bmatrix} e_{D1} \\ \\ e_{Q1} \end{bmatrix}, \begin{bmatrix} e_{d2} \\ \\ e_{q2} \end{bmatrix} = T_{\alpha_2}^{-1} \begin{bmatrix} e_{D2} \\ \\ e_{Q2} \end{bmatrix}$$

where $T_{\alpha_i}^{-1}$, $i = 1,2$, is

$$T_{\alpha_i}^{-1} = \begin{bmatrix} \cos \alpha_i & \sin \alpha_i \\ \\ -\sin \alpha_i & \cos \alpha_i \end{bmatrix} .$$

If we let $e_2 = 1+j0$ to simulate an infinite bus, we have

$$\begin{bmatrix} e_1 \\ \\ 1 \end{bmatrix} = Z_R \begin{bmatrix} i_1 \\ \\ i_2 \end{bmatrix} = \begin{bmatrix} z_{11} & z_{12} \\ \\ z_{21} & z_{22} \end{bmatrix} \begin{bmatrix} i_1 \\ \\ i_2 \end{bmatrix}$$

and

$$i_2 = \frac{1}{z_{22}} (1 - z_{21} i_1).$$

Then we get

$$e_1 = \left[z_{11} - \frac{z_{12}\, z_{21}}{z_{22}} \right] i_1 + \frac{z_{12}}{z_{22}}$$

and if i_1 is observed, i_2, the current exchanged with the infinite bus (machine 2) can be found. Only machine 1 must be modelled as a dynamic element.

A further extension, similar to the development in Section 6.4, concerns the problem of finding a suitable control input $\underline{u}$ when for each machine we know the output $\underline{y}$, i.e., we want a feedback control. Since the basic model here is nonlinear, the methods of the previous section cannot be used immediately. However, if an approximate linear model is obtained from Eq. 6.101 and 6.102, both the optimal control and optimal filtering methods shown previously can be used. For example, a linear optimal filter was designed in [35.][39.] where the linear machine state vector and output vector were

$$\underline{x}^T = [\psi_{fd} \quad \psi_d \quad \psi_{kd} \quad \psi_q \quad \psi_{kq}]$$

and

$$\underline{y}^T = [i_{fd} \quad i_d \quad i_q]$$

so that an estimate $\hat{\underline{x}}$ is obtained. An optimal control $\underline{u}$ for a linear model was studied in [40.] where it was assumed that the entire state vector was measured. An optimal control for a linear model was found in [41.] in terms of an output vector $\underline{y}$, and an optimal control obtained from the PQR model was found in [42.]. It should again be obvious that a great variety of power system stability and control problems can be studied by using state variable models. Discrete-time approximations easily can be obtained for analysis on large digital computers.

6.6 OTHER APPLICATIONS

The four previous sections considered applications of many of the ideas of signal and system modelling, response properties, and signal and system analysis. There also was a substantial reliance on the use of computers for simulation; in fact, most of the practical problems of interest are so complex (the state dimension is large) that the use of a computer is the only feasible way to do the required analysis. The combination of signal and system modelling and computer simulation has enabled engineers to analyze and design a great variety of complex systems. The purpose of this section is to indicate some of the many applications that have been made of these techniques. The list is far from complete, but some of the engineering journals in which such results are published are listed first and then some specific references are given.

The following journals are of particular interest:

- IEEE Proceedings
- IEEE Transactions
  - Aerospace and Electronic Systems
  - Circuits and Systems
  - Control Systems
  - Biomedical Engineering
  - Industry Applications
  - Information Theory
  - Instrumentation and Measurement
  - Power Engineering
  - Reliability
  - Systems, Man, and Cybernetics
- Automatica
- ASME Transactions
- ISA Journal
- AIAA Journal
- A I Ch E Journal
- Nuclear Technology
- Proceedings IEE (London)
- Automation and Remote Control (Moscow)

There are, of course, many others that could be listed, and there are several points of view from what are largely applied mathematics journals to publications that emphasize the important aspects of system implementation with available components and subsystems. In addition, there are annual conferences held by the various professional engineering societies and the published proceedings of these conferences describe many new applications each year.

Eight areas of engineering interest in which many of the ideas of signals and systems are used are as follows:

1. Communications, Signal Theory, Information Theory

Section 6.2 discussed digital signal processing or digital filtering and spectral estimation. Another example of the use of state variable models is given in [43.], where signal detection, i.e., the detection of a known or stochastic signal in the presence of stochastic noise, is discussed.

2. Electric Power Systems

Section 6.5 developed a dynamic model for a single synchronous machine as well as for a system of machines interconnected through a network. An example of a steady-state model of a power system where the state variables are the node voltages is given in [44.] where an algorithm for the estimation of the state is developed.

Steady-state conditions are of great importance for operating systems, e.g., in load dispatching. The modelling and control of power plant boilers also have been investigated, and [45.][46.][47.][48.] are examples of how these complex systems can be handled.

3. Transportation Systems

Automotive, railroad, aircraft, and ship transport all have been studied as transportation systems. Reference 49. is an entire special issue describing some of this work. Air traffic control is considered in [50.], and automated highway systems are considered in [51.].

4. Aircraft Control

One example of an aircraft control system was given in Section 6.4. An example of the difficulties in modelling and controlling a flexible air frame is discussed in [52.]. Some interesting instrumentation problems are covered, and these illustrate the need for a careful coordination of system modelling and system instrumentation.

5. Process Control

Process control includes the control of chemical processes (distillation, absorption, extraction, etc.), nuclear processes, manufacturing processes (steel making, paper making, cement making, etc.), and a great variety other processes for refining and manufacturing materials. An example of a chemical process control is given in [53.] -- the manufacture of ethylene. Two aspects of steel making are given in [54.], which describes the modelling and control of a reheat furnace, and [55.], which describes a five-stand cold rolling mill. The manufacture of paper is covered in [56.] and [57.], and an example of a problem in nuclear reactor control is given in [58.]. It is important to note that linearization (see Chapter 3) is almost always required for such systems, and in addition, approximations of pure time delays and distributed parameter models are required if the ideas of linear, lumped systems of Chapter 3 are to be used.

6. Navigation Systems

The design of automatic navigation systems for aircraft, ships, and space vehicles has developed over the past two decades. The use of several sources of navigational information such as radio/radar equipment, star trackers, inertial acceleration and velocity sensors as well as the conventional air speed and water speed indicators and the use of modern, high-speed digital computers have made very accurate navigation possible. Reference [59.] is an example of a system of this type used on the C-5 aircraft.

7. Economic Systems

Although the previous applications have been largely associated with engineering, many of the methods of signal and system analysis can be used in other areas too. The modelling of economic systems is one activity when there has been much interest in state-variable models, and Reference [60.] is a broad survey of recent trends in using system theories for studying economic systems. This reference contains an extensive bibliography on a wide range of topics related to the study of economic systems.

8. Biomedical Systems

Although physiological systems generally involve properties that make the use of the system models of this book questionable -- i.e., properties such as nonlinearity and nonstationarity, nonuniform elasticity, complex modes of fluid flow, hysteresis -- there are some cases where useful approximations can be made. References [61.], [62.], and [63.] are examples of some of these instances.

6.7 POSTSCRIPT

A review of Chapters 2, 3, 4, and 5 shows a considerable range of ideas and techniques concerning signal and system modelling as well as methods of analysis. Chapter 6 has suggested some of the ways in which these can be applied. Some words of caution as well as words of encouragement to go on to more advanced studies are in order here. First, in using almost all of the material in this book, it is most important that the requirements of linearity and frequently stationarity be kept in mind. Extensions to nonlinear and nonstationary cases can sometimes be made if the approximation methods shown earlier are not satisfactory. However, these extensions always involve a loss in generality. Extensions to cover systems with pure time delays and distributed parameters also are available, but again there is a compromise between the accuracy of the model and its complexity. A third extension to systems or signals where the parameters vary in time -- but the time variation is described stochastically -- has been attempted, and some results are available.

Finally, there are many complex systems of interest today where the analyst must work with considerably less information than is implied in all of the models in this book. Recall that the signal and system identification methods introduced in Chapters 2 and 4 required some very strong assumptions in order that they work. A challenging area for further work lies in what is generally called the study of complex uncertain systems, and techniques other than the probabilistic methods suggested by this book are needed. There are opportunities for systems analysts to make new applications of present methods and to develop new theories and methods.

REFERENCES

1. Ruston, H. and Bordogna, "Electric Networks: Functions, Filters, Analysis," New York: McGraw-Hill Book Co., 1966.
2. Van Valkenburg, M., "Introduction to Modern Network Synthesis," New York: John Wiley & Sons, 1960.
3. Weinburg, L., "Network Analysis and Synthesis," New York: McGraw-Hill Book Co., 1962.
4. Mitra, S., "Analysis and Synthesis of Linear Active Networks," New York: John Wiley & Sons, 1969.
5. Huelsman, L., "Theory and Design of Active RC Circuits," New York: McGraw-Hill Book Co., 1968.
6. Tow, J., "A Step-by-Step Active-Filter Design," IEEE Spectrum, Dec. 1969, pp. 64-68.
7. Geffe, P., "Toward High Stability in Active Filters," IEEE Spectrum, May 1970, pp. 63-66.
8. Rader, C. and Gold, B., "Digital Filter Design Techniques in the Frequency Domain," Proc. IEEE, v. 55, Feb. 1967, pp. 149-171.
9. Kaiser, J. F., "Digital Filters," in "System Analysis by Digital Computer," (Kuo, F. and Kaiser, J., Ed.) New York: John Wiley & Sons, 1966.

10. Steiglitz, K., "Computer-Aided Design of Recursive Digital Filters," IEEE Trans. Audio Electroacoust., v. AU-18, June 1970, pp. 123-129.
11. Hamming, R., "Numerical Methods for Scientists and Engineers," New York: McGraw-Hill Book Co., 1962.
12. Fleischer, P., "Digital Realization of Complex Transfer Functions," Simulation, March 1966, pp. 171-180.
13. Rabiner, R., Gold, B., and McGonigal, C., "An Approach to the Approximation Problem for Nonrecursive Digital Filters," IEEE Trans. Audio Electroacoust, v. AU-18, June 1970, pp. 83-106.
14. Hermann, O., "Design of Nonrecursive Digital Filters with Linear Phase,"Electronics Letters, v. 6, n. 11, May 28, 1970, p. 328.
15. Hofstetter, E., Oppenheim, A., and Siegel, J., "A New Technique for the Design of Non-Recursive Digital Filters," Proc. 5th Annual Princeton Conf., 1971, pp. 64-72.
16. Bergland, G., "A Guided Tour of the Fast Fourier Transform," IEEE Spectrum, July, 1969, pp. 41-51.
17. Blackman, R. and Tukey, J., "The Measurement of Power Spectra," New York: Dover Publications, 1959.
18. Balabanian, N. and Bickart, T., "Electrical Network Theory," New York: John Wiley & Sons, 1969.
19. Anderson, B. D. and Vongpanitlerd, S., "Network Analysis and Synthesis," Englewood Cliffs, N.J., Prentice-Hall Book Co., 1973.
20. Weinberg, L., "Network Analysis and Synthesis," New York: McGraw-Hill Book Co., 1962.
21. Bloom, M. P., "The Generalized Bessel Polynomials as Applied to the Approximation Problems of Network Synthesis," Ph.D. Thesis, The Penn State University, 1971.
22. Jess, J. and Schuessler, H., "A Class of Pulse Forming Networks," IEEE Trans. Circuit Theory, v. CT-12, n. 2, June, 1965, pp. 296-299.
23. Welch, P., "The Use of Fast Fourier Transform for the

Estimation of Power Spectra: A Method Based on Time Averaging Over Short, Modified Periodograms," IEEE Trans. Audio Electrocoust., v. AU-15, June, 1967, pp. 70-73.

24. Blackman, R., "Linear Data Smoothing and Prediction in Theory and Practice," Reading, Mass: Addison-Wesley, 1965.
25. Brigham, E., "The Fast Fourier Transform," Englewood Cliffs, N.J.: Prentice-Hall Book Co., 1974.
26. Huber, R., "Optimal Control Aircraft Landing Analysis," Ph.D. Thesis, The Penn State University, August, 1973.
27. Ashley, H., "Engineering Analysis of Flight Vehicles," Reading, Mass.: Addison-Wesley, 1974.
28. Kwakernaak, H. and Sivan, R., "Linear Optimal Control Systems," New York: Wiley, 1972.
29. Sage, A., "Optimum Systems Control," Englewood Cliffs, N.J.: Prentice-Hall, 1968.
30. White, D. and Woodson, H., "Electromechanical Energy Conversion," New York: Wiley, 1959.
31. Riaz, M., "Analogue Computer Representations of Synchronous Generators in Voltage-Regulation Studies," AIEE Trans., v. 75, Dec. 1956, pp. 1178-1182.
32. Aldred, A., "Electronic Analogue Computer Simulation of Multi-Machine Power-System Networks," IEE Proc., v. 109A, June 1962, pp. 195-202.
33. Shackshaft, G., "General-Purpose Turbo-Alternator Model," IEE Proc., v. 110, April, 1963, pp. 703-713.
34. Laughton, M., "Matrix Analysis of Dynamic Stability in Synchronous Multimachine Systems," IEE Proc., v. 113, Feb. 1966, pp. 325-336.
35. Miller, W. L., "Dynamic State Estimation in Nonlinear Multimachine Power Systems," Ph.D. Thesis, The Penn State University, Dec. 1970.
36. Prabhashankar, K. and Janischewsyj, W., "Digital Simulation of Multimachine Power Systems for Stability Studies," IEEE Trans., PAS, v. PAS-87, Jan. 1968, pp. 73-80.

37. Dineley, J. and Powner, E., "Power - System Governor Simulation," IEE Proc., v. 111, Jan. 1964, pp. 115-124.
38. Anderson, J., "Matrix Methods for the Study of a Regulated Synchronous Machine," IEEE Proc., v. 57, n. 12, Dec. 1969, pp. 2122-2136.
39. Miller, W. and Lewis, J., "Dynamic State Estimation in Power Systems," IEEE Trans., v. AC-16, n. 6, Dec. 1971, pp. 841-846.
40. Yu, Y.-N., et al., "Application of an Optimal Control Theory to a Power System," IEEE Trans., PAS, v. PAS-89, Jan. 1970, pp. 55-62.
41. Davison, E. and Rau, N., "The Optimal Output Feedback Control of a Synchronous Machine," IEEE Trans., v. PAS-90, n. 5, September/October, 1971, pp. 2123-2135.
42. Anderson, J., "The Control of a Synchronous Machine Using Optimal Control Theory," IEEE Trans., v. 59, n. 1, Jan. 1971, pp. 25-35.
43. Van Trees, H. L., "Applications of State-Variable Techniques in Detection Theory," IEEE Proc., v. 58, n. 5, May 1970, pp. 653-669.
44. Debs, A., "Estimation of Steady-State Power System Parameters," IEEE Trans., v. PAS-93, n. 5, September/October 1974, pp. 1260-1266.
45. Enns, M., et al., "Practical Aspects of State-Space Methods, Part I - System Formulation and Reduction," IEEE Trans., v. MIL-8, n. 2, April 1964, pp. 81-93.
46. McDonald, J., et al.,"A Nonlinear Model for Reheat Boiler-Turbine-Generator Systems, Part I - General Description and Evaluation," Preprints J.A.C.C. (IEEE 71C 36-AC) 1971, pp. 219-226.
47. Kwatny, H., et al., "A Nonlinear Model for Reheat Boiler-Turbine-Generator Systems, Part II - Development," Preprints J.A.C.C. (IEEE 71C 36-AC) 1971, pp. 227-236.
48. Bengtsson, G. and Lindahl, S., "A Design Scheme for Incomplete State or Output Feedback with Applications to Boiler and Power System Control," Automatica, v. 10,

n. 1, January 1974, pp. 15-30.
49. ____________ Special Issue on Transportation, IEEE Proc., v. 56, n. 4, April 1968.
50. Athans, M., "System Aspects of Air Traffic Control," Proc. 4th Hawaii International Conference on Systems Sciences, January 1971.
51. Levine, W. and Athans, M., "On the Optimal Error Regulation of a String of Moving Vehicles," IEEE Trans., v. AC-11, n. 3, July 1966, pp. 355-361.
52. Rynaski, E., "Case Study on the Optimal Control of a Flexible Launch Vehicle," 1968 Case Studies in System Control (IEEE 68 C 40-AC), pp. 193-219.
53. Wilson, J. and Belanger, P., "Application of Dynamic Optimization Techniques to Reactor Control in the Manufacture of Ethylene," IEEE Trans., v. AC-17, n. 6, December 1971, pp. 756-762.
54. Pike, H., "Case Study on the Control of a Slab Reheating Furnace," 1969 Case Studies in System Control (IEEE 69 C 41-AC), pp. 63-88.
55. Bryant, G. and Higham, J., "A Method for Realizable Non-Interactive Control Design for a Five Stand Cold Rolling Mill," Automatica, v. 9, n. 4, July 1973, pp. 453-466.
56. Bakke, R., "Case Study in the Paper Making Industry," 1968 Case Studies in System Control (IEEE 68 C 40-AC), pp. 111-192.
57. Astrom, K., "Application of Linear Stochastic Control Theory to Paper Machine Control," 1970 Case Studies in System Control (IEEE 70 C 50-AC).
58. Tsai, T. and Grigsby, L., "An Approach to Spatial Power Regulation in Nuclear Reactors," Preprints J.A.C.C. (IEEE 71C 36-AC) 1971, pp. 255-260.
59. Schmidt, S., et al., Case Study of Kalman Filtering in the C-5 Aircraft Navigation Systems," 1968 Case Studies in System Control (IEEE 68 C 40-AC), pp. 57-110.
60. Athans, M. and Kendrick, D., "Control Theory and

Economics," IEEE Trans., v. AC-19, n. 5, October 1974, pp. 518-523.

61. Golden, J., et al., "Mathematical Modeling of Pulmonary Airway Dynamics," IEEE Trans., v. BME-20, n. 6, November 1973, pp. 397-403.

62. Yao, H. and Diana, J., "Computer Simulation Model for Transient Transcapillary Fluid Exchange," v. BME-20, n. 6, November 1973, pp. 427-433.

63. Fan, L. T., et al., "A Review on Mathematical Models of the Human Thermal System," IEEE Trans., v. BME-18, n. 3, May 1971, pp. 218-234.

64. Schwartz, M. and Shaw, L., "Signal Processing: Discrete Spectral Analysis, Detection, and Estimation," New York: McGraw-Hill Book Co., 1975.

65. Oppenheimer, A. and Schafer, R., "Digital Signal Processing," Englewood Cliffs, N.J.: Prentice-Hall, 1975.

66. Rabiner, L. and Gold, B., "Theory and Application of Digital Signal Processing," Englewood Cliffs, N.J.: Prentice-Hall, 1974.

67. Jenkins, G. and Watts, D., "Spectral Analysis and Its Applications," San Francisco, Ca.: Holden-Day, 1968.

68. Koopmanns, L. "Spectral Analysis of Time Series," New York: Academic Press, 1974.

69. Huber, R., "Application of Modern Control Theory to the Analysis of Aircraft Autoland Performance Using a Scanning Beam Guidance System," Proc. JACC, June 1974.

70. Huber, R., "Optimal Control Aircraft Landing Analysis," Air Force Flight Dynamics Lab., TR-73-141, Dec. 1973.

Appendix A

FOURIER, LAPLACE, AND Z- TRANSFORMS

Transform methods play an important role in the study of continuous-time and discrete-time linear systems. Summaries of Fourier, Laplace, and Z-Transforms introduced in Chapter 1 are given here to provide a convenient reference in a form which relates these transforms within a common framework of notation and terminology. The books cited at the end of this appendix should be consulted when more complete explanations are required. In most cases, the stronger conditions in common use have been given whenever there are differences among the references.

A.1 THE FOURIER TRANSFORM

A.1.1 Fourier Integral Theorem

The Fourier Integral Theorem provides the basis for the Fourier Transform.

<u>Theorem A.1</u>

1. f(t) piecewise continuous and real, $-\infty < t < \infty$, and

2. $\int_{-\infty}^{\infty} |f(t)|dt < \infty$ (the integral of $|f(t)|$ is convergent or f(t) is absolutely integrable)

$\Rightarrow$

1. $$F(j\omega) = \int_{-\infty}^{\infty} f(t)(\exp -j\omega t)dt \qquad (A.1)$$

is defined for real ω, is continuous for all ω, and

2. $$I(t) = \frac{1}{2\pi}\int_{-\infty}^{\infty} F(j\omega)(\exp j\omega t)d\omega \text{ (principal value)} \qquad (A.2)$$

converges for all t, and

3. $I(t) = f(t)$ where f(t) is continuous or

$I(t) = \frac{f(t+)+f(t-)}{2}$ where f(t) has jumps.

A.1.2 Transform Pair

The Fourier Integral A.1 and the Inversion Integral A.2 can be shown to relate the functions $F(j\omega)$ and f(t) uniquely under the hypothesis of Theorem A.1. The Fourier Transform is, therefore, given by the pair:

$$F(j\omega) = \int_{-\infty}^{\infty} f(t)(\exp -j\omega t)dt = \mathfrak{F}\, f(t) \qquad (A.3)$$

$$f(t) = \frac{1}{2\pi}\int_{-\infty}^{\infty} F(j\omega)(\exp jt\omega)d\omega = \mathfrak{F}^{-1}F(j\omega) \qquad (A.4)$$

and the conditions of Theorem A.1 are understood.

A.1.3 Properties

It is assumed that the indicated transforms exist.

1. Linearity

$$\mathcal{F}\,[af(t) + bg(t)] = aF(j\omega) + bG(j\omega)$$

2. Differentiation of f(t) $(D \equiv \frac{d}{dt})$

   If the $D^k f$ are continuous and have absolutely convergent integrals for $k = 0,1,2,...$ and $f(t) \to 0$ for $t \to \pm\infty$,

$$\mathcal{F}\,[D^k f(t)] = (j\omega)^k F(j\omega) \ .$$

3. Integration of f(t)

   If the $D^{-k} f$ have absolutely convergent integrals for $k = 1,2,...$ and f(t) is piecewise continuous,

$$\mathcal{F}\,[D^{-k} f(t)] = \frac{1}{(j\omega)^k} F(j\omega), \qquad \omega \neq 0 \ .$$

4. Time Shift

$$\mathcal{F}\, f(t-\tau) = (\exp -j\omega\tau)F(j\omega) \ .$$

5. Frequency Shift

$$\mathcal{F}\,[(\exp j\omega_o t)f(t)] = F(j\omega - j\omega_o) = F(j(\omega-\omega_o)).$$

6. Time Multiplication

$$\mathcal{F}\,[t^n f(t)] = j^n \frac{d^n F(j\omega)}{d\omega^n}$$

7. Time Scale

$$\mathcal{F}\,[f(at)] = \frac{1}{a} F(\frac{j\omega}{a}) \ , \qquad a > 0 \ .$$

8. Symmetry

$$F(jt) \leftrightarrow 2\pi f(-\omega) \ .$$

A.1.4 Theorems

1. Parseval
   If f(t) satisfies the hypothesis of Theorem A.1,

$$\int_{-\infty}^{\infty} f(t)^2 dt = \frac{1}{2\pi} \int_{-\infty}^{\infty} |F(j\omega)|^2 d\omega \; .$$

2. Real Convolution (Complex Multiplication)
   If f(t) and g(t) satisfy the hypothesis of Theorem A.1 and are bounded, let

$$f*g = \int_{-\infty}^{\infty} f(\lambda)g(t-\lambda)d\lambda \; ,$$

   then

$$\mathfrak{F}\,[f*g] = F(j\omega)G(j\omega) \; .$$

   Furthermore, $f*g = g*f$.

3. Complex Convolution (Real Multiplication)
   For

$$F*G = \frac{1}{2\pi}\int_{-\infty}^{\infty} F(\beta)G(\omega-\beta)d\beta \; ,$$

   the transform of the product fg is

$$\mathfrak{F}\,[fg] = F*G \; .$$

A.1.5 Tables

Two classes of functions are considered. Those in the first group satisfy the hypothesis of Theorem A.1 (Table A.1) and those in the second do not (Table A.2). Included in the second class are several forms of interest in signal and system analysis such as periodic functions and functions containing impulses. Signals containing impulses are properly treated as generalized functions [10.].

A.1.6 Fourier Transform of f(n) — Definition

It is possible to define transforms of functions of integers, f(n), which are analogous to the previous transforms. Thus there are

$$\mathfrak{F}_d f(n) = F(\gamma)$$

and

$$\mathfrak{F}_d^{-1} F(\gamma) = f(n)$$

where $\gamma = \varepsilon^{j\theta}$, $-\pi < \theta \leq \pi$. These are defined and a table is given after the summary of the Z-Transforms in Section A.3.6.

TABLE A.1

f(t)	F(jω)
1. $u_1(t+T)-u_1(t-T)$	$2T \dfrac{\sin \omega T}{\omega T}$
2. $\varepsilon^{-at}u_1(t)$	$\dfrac{1}{a+j\omega}$
3. $\varepsilon^{-a\|t\|}$	$\dfrac{2a}{a^2+\omega^2}$
4. $1-2(\|t\|/T), \quad \|t\| \leq T/2$ $0 \quad , \quad \|t\| > T/2$	$\dfrac{T}{2}\left[\dfrac{\sin \frac{\omega T}{4}}{\frac{\omega T}{4}}\right]^2$
5. $2\omega_0 \dfrac{\sin \omega_0 t}{\omega_0 t}$	$2\pi[u_1(\omega+\omega_0)-u_1(\omega-\omega_0)]$

Note: A sufficient condition that F(jω) be the $\mathfrak{F}$ f(t) is that

$$\int_{-\infty}^{\infty} |F(j\omega)|\, d\omega < \infty \quad .$$

Note: u_1 is a unit step function (see Chapter 2).

TABLE A.2

	f(t)	F(jω)
1.	$u_o(t)$	1
2.	$u_1(t)$	$\pi u_o(\omega) + \frac{1}{j\omega}$
3.	K	$2\pi K u_o(\omega)$
4.	$\cos \omega_o t$	$\pi[u_o(\omega-\omega_o)+u_o(\omega+\omega_o)]$
5.	$\sin \omega_o t$	$-j\pi[u_o(\omega-\omega_o)-u_o(\omega+\omega_o)]$
6.	$(\cos \omega_o t)u_1(t)$	$\pi/2[u_o(\omega-\omega_o)+u_o(\omega+\omega_o)] + \frac{j\omega}{\omega_o^2-\omega^2}$
7.	$(\sin \omega_o t)u_1(t)$	$-j\pi/2[u_o(\omega-\omega_o)-u_o(\omega+\omega_o)]+ \frac{\omega_o}{\omega_o^2-\omega^2}$
8.	$\operatorname{sgn} t$	$\frac{2}{j\omega}$
9.	$(\exp j\omega_o t)$	$2\pi u_o(\omega-\omega_o)$

Note: u_o is a unit impulse (see Chapter 2)

A.2 THE LAPLACE TRANSFORM

A.2.1 Two-sided Laplace Transform

The Two-Sided (or Bilateral) Laplace Transform is defined as

$$F_2(s) = \mathcal{L}_2 f(t) = \int_{-\infty}^{\infty} f(t)(\exp -st)dt \qquad (A.5)$$

where

$$s = \sigma + j\omega$$

and the integral converges for

$$c_1 < \sigma < c_2 \quad .$$

The numbers c_1 and c_2 depend on f(t), and define R, the region of convergence. The Inverse Laplace Transform is

$$f(t) = \mathcal{L}_2^{-1} F_2(s) = \frac{1}{2\pi j} \int_{\sigma - j\infty}^{\sigma + j\infty} F_2(s)\varepsilon^{ts} ds \ . \qquad (A.6)$$

It is necessary to consider the region of convergence of $F_2(s)$ to determine the correct f(t). For the usual cases of signals which are bounded for large $|t|$, poles of $F_2(s)$ in the left half s-plane (LHP) are associated with $t>0$ and poles of $F_2(s)$ in the RHP are associated with $t<0$. This is discussed below with regard to the use of Jordan's Lemma.

Theorem A.2

1. f(t) piecewise continuous and real, $-\infty < t < \infty$, and
2. $\int_{-\infty}^{\infty} |f(t)\varepsilon^{-\sigma t}| dt < \infty$ for some real σ

$\Rightarrow$

1. $F_2(s) = \mathcal{L}_2 f(t)$ exists, and
2. $f(t) = \mathcal{L}_2^{-1} F_2(s)$ where f(t) is continuous.

TABLE A.3

$f(t)$	$F_2(s)$
1. $\varepsilon^{-a\|t\|}$, $a > 0$	$\frac{2a}{a^2-s^2}$ $-a < \sigma < a$
2. $\frac{2a}{(t^2+a^2)}$, $a > 0$	$2\pi\varepsilon^{-a\|s\|}$ $\sigma = 0$
3. $u_1(t+T)-u_1(t-T)$ $= u_1(-t^2+T^2)$	$(2T)\ \frac{\sinh(Ts)}{(Ts)}$ $\|\sigma\| < \infty$
4. $(2\omega_o)\ \frac{\sin(\omega_o t)}{(\omega_o t)}$	$2\pi u_1(s^2+\omega_o^2)$ $\sigma = 0$
5. $\cos \omega_o t$	$\pi u_o(s^2+\omega_o^2)$

Note the symmetry property illustrated by the pair 1. and 2. and the pair 3. and 4. Note that 5. does not satisfy the hypothesis of Theorem A.2; it is simply another form of 4. of Table A.2.

A.2.2 One-sided Laplace Transform

The One-Sided (or Unilateral) Laplace Transform is used for functions $f(t)$ where $f(t) = 0$, $t<0$. This can be emphasized by using the notation $f(t)u_1(t)$ and

$$F(s) = \mathcal{L}\, f(t)u_1(t) = \int_0^\infty f(t)\varepsilon^{-st}dt \qquad (A.7)$$

where $s = \sigma + j\omega$ and the integral converges for $\sigma > c$. The Inverse Laplace Transform is

$$f(t)u_1(t) = \mathcal{L}^{-1}F(s) = \frac{1}{2\pi j}\int_{\sigma-j\infty}^{\sigma+j\infty} F(s)\,\varepsilon^{ts}ds \qquad \text{(A.8)}$$

Theorem A.3

1. $$\int_0^T |f(t)|dt \leq K < \infty$$

   for some $T>0$ and $K>0$, and

2. $|f(t)| \leq M\varepsilon^{ct}$ for real $M>0$ and c, $t>T$.

$\Rightarrow$

1. $F(s) = \int_0^\infty f(t)\,\varepsilon^{-st}dt$ converges absolutely for $\sigma > c$.

   The region R of the s-plane such that $\sigma > c$ is the region of convergence of the integral.

It also can be shown that if the integral converges absolutely for some σ_o, the integral converges uniformly and absolutely in the region given by $\sigma \geq \sigma_o$. Furthermore, the transform pair A.7 and A.8 can be shown to relate $f(t)u_1(t)$ and $F(s)$ uniquely [6.].

There are two conventions in use regarding the lower limit in A.7. In cases where $f(t)$ or its derivatives are not continuous at $t=0$, it is important to specify whether the limit 0+ or 0- is used and the notation $\mathcal{L}_{o+}$ and $\mathcal{L}_{o-}$ is common. Either convention gives correct results provided it is used consistently. As a simple illustration, consider the transform

$$\mathcal{L}_{o+} \cos \omega_o t\, u_1(t) = \frac{s}{s^2+\omega_o^2} = F(s) = \mathcal{L}_{o+} f(t).$$

and

$$\mathcal{L}_{o-} \cos \omega_o t\, u_1(t) = \frac{s}{s^2+\omega_o^2} .$$

However,

$$\mathcal{L}_{o+}\frac{d}{dt}(\cos \omega_o t\, u_1(t)) = \mathcal{L}_{o+}[-\omega_o \sin \omega_o t\, u_1(t) + \cos \omega_o t\, u_o(t)]$$

$$= \frac{-\omega_o^2}{s^2+\omega_o^2}$$

$$= sF(s) - f(0+)$$

but

$$\mathcal{L}_{o-}\frac{d}{dt}(\cos \omega_o t\, u_1(t)) = \frac{-\omega_o^2}{s^2+\omega_o^2} + 1$$

$$= \frac{s^2}{s^2+\omega_o^2}$$

$$= sF(s) - f(0-).$$

Theorem A.4

1. $$I(t) = \frac{1}{2\pi j}\int_{a-jb}^{a+jb} F(s)\varepsilon^{ts}ds, \quad \text{and}$$

2. $$F(s) = \int_0^{\infty} f(t)\varepsilon^{-st}dt, \quad \sigma > c, \text{ and}$$

3. $a > c$, and

4. $f(t)$ of bounded variation [6.]

$\Rightarrow$

1. $$\lim_{b\to\infty} I(t) = 0, \qquad t < 0$$

$$= f(0+)/2, \quad t = 0$$

$$= \frac{f(t+)+f(t-)}{2}, \qquad t > 0 .$$

Corrolary A.4

1. $f(t)$ continuous, $t > 0$

2. $\sigma > c$

$\Rightarrow$

1. $$f(t) = \frac{1}{2\pi j} \int_{\sigma-j\infty}^{\sigma+j\infty} F(s)\varepsilon^{ts}ds, \qquad t > 0 .$$

The Inversion Theorem A.4 provides the general method for finding inverse Laplace Transforms. When $F(s)$ is meromorphic, the integral is evaluated by using the calculus of residues. Inverse transforms also can be found by the use of a table (see Section A.2.5) computed by the use of the direct transform integral; this follows from the uniqueness property. Also, in cases where $F(s)$ is a rational function, a partial fraction expansion of $F(s)$ reduces it to simpler forms which can be found in the table.

Partial Fraction Method

For $F(s) = N(s)/D(s)$ with $N(s)$ and $D(s)$ polynomials such that n, the degree of $D >$ the degree of N,

1. $$F(s) = \sum_{j=1}^{n} \frac{A_j}{s-\lambda_j}$$

where the n λ_j are simple zeroes of $D(s)$ and

$$A_j = \left.\frac{N(s)(s-\lambda_j)}{D(s)}\right|_{s=\lambda_j}$$

or

2. $$F(s) = \sum_{j=1}^{m} \sum_{i=1}^{j=1} \frac{A_{ij}}{(s-\lambda_j)^i}$$

where m_j is the order of the zero of $D(s)$ at $s=\lambda_j$,

and

$$n = \sum_{j=1}^{m} m_j,$$

$$A_{ij} = \left\{ \frac{1}{(m_j-i)!} \frac{d^{m_j-i}}{ds^{m_j-i}} \left[\frac{N(s)(s-\lambda_j)^{m_j}}{D(s)} \right] \right\} \Bigg|_{s=\lambda_j}.$$

The inverse transforms are then

1. $$f(t) = \sum_{j=1}^{n} A_j (\exp \lambda_j t), \quad t>0$$

or

2. $$f(t) = \sum_{j=1}^{m} (\exp \lambda_j t) \sum_{i=1}^{m_j} A_{ij} \frac{t^{i-1}}{(i-1)!}, \quad t>0.$$

Use of the Calculus of Residues

Jordan's Lemma - A

1. $F(s)$ meromorphic for $\sigma \le c$

2. $|F(s)| \to 0$ as $\rho \to \infty$, $s = \rho\varepsilon^{j\theta}$

$\Rightarrow$

1. $$I(t) = \int_{C_\ell} F(s)\varepsilon^{ts} ds \to 0 \text{ as } \rho \to \infty \quad t>0$$

where C_ℓ is a semicircular path to the left with radius ρ and center at the origin of the s-plane.

Also,

Jordan's Lemma - B

1. $F(s)$ meromorphic for $\sigma \ge c$

2. $|F(s)| \to 0$ as $\rho \to \infty$, $s = \rho\varepsilon^{j\theta}$

1. $$I(t) = \int_{C_r} F(s)\varepsilon^{ts} ds \to 0 \text{ as } \rho \to \infty, \quad t<0$$

where C_r is a semicircular path to the right with radius ρ and center at the origin of the s-plane.

For

$$\frac{1}{2\pi j}\int_C F(s)\varepsilon^{ts}ds = \frac{1}{2\pi j}\int_{\sigma-j\infty}^{\sigma+j\infty} F(s)\varepsilon^{ts}ds + \frac{1}{2\pi j}\int_{C_\ell} F(s)\varepsilon^{ts}ds$$

where C is a closed contour as indicated, the Cauchy Residue Theorem implies that the left side is

$$\sum \text{Residues of } F(s)\varepsilon^{ts}\Big|_{\text{poles inside C}}$$

and Jordan's Lemma - A implies the second term on the right is zero (for suitable F(s)). Then, since $\sigma = c$ is to the right of the poles of F(s),

$$\sum \text{Residues of } F(s)\varepsilon^{ts}\Bigg|_{\text{poles of } F(s)} = \frac{1}{2\pi}\int_{\sigma-j\infty}^{\sigma+j\infty} F(s)\varepsilon^{ts}ds, \quad t>0$$

$$= f(t), \qquad t>0 .$$

By similar argument using Jordan's Lemma - B, $f(t) = 0$, $t<0$, since there are no poles to the right of $\sigma = c$.

The Inverse Two-Sided Laplace Transform of Section A.2 can be evaluated by similar procedures when the convergence strip is given by $c_1 < \sigma < c_2$. Thus, using Jordan's Lemma - A with $c = c_1$,

$$\sum \text{Residues of } F_2(s)\varepsilon^{ts}\Bigg|_{\substack{\text{poles of } F_2(s)\\ \text{left of } \sigma = c_1}} = f(t), \quad t > 0$$

and using Jordan's Lemma - B with $c = c_2$,

$$-\sum \text{Residues of } F_2(s)\varepsilon^{ts}\Big|_{\substack{\text{poles of } F_2(s)\\ \text{right of } \sigma = c_2}} = f(t), \quad t > 0.$$

The Partial Fraction Method also can be used with analogous interpretation of the poles of $F_2(s)$ and associated components of $f(t)$ for $t>0$ and $f(t)$ for $t<0$.

Finally, the required residues are found from

1. Residue of $F(s)\varepsilon^{ts}$

$$= \lim_{s \to \lambda_j} [(s-\lambda_j)F(s)\varepsilon^{ts}]$$

for a simple pole of $F(s)$ at $s = \lambda_j$ or

2. Residue of $F(s)\varepsilon^{ts}$

$$= \frac{1}{(m-1)!} \lim_{s \to \lambda_j} \left[\frac{d^{m-1}}{ds^{m-1}} (s-\lambda_j)^m F(s)\varepsilon^{ts} \right]$$

for an m-th order pole of $F(s)$ at $s = \lambda_j$. Note that if the singularities of $F(s)$ are not all poles, i.e., $F(s)$ is not meromorphic, a more general method of evaluating the inverse transform is required [6.].

A.2.3 Properties of the One-sided Transform

1. Linearity

$$\mathcal{L}\,[af(t)+bg(t)] = aF(s) + bG(s)$$

2. Differentiation of f(t)

$$\mathcal{L}\,[Df(t)] = sF(s) - f(0)$$

and

$$\mathcal{L}\ [D^k f(t)] = s^k F(s) - s^{k-1} f(0) - \ldots - f^{(k-1)}(0)$$

where $f^{(m)} = D^m f$

3. Integration of f(t)

$$\mathcal{L}\ [D^{-1} f(t)] = \frac{F(s)}{s} + \frac{1}{s} \int f dt \Big|_{t=0} .$$

4. Time Shift

$$\mathcal{L}\ f(t-\tau) = \varepsilon^{-\tau s} F(s), \quad \tau > 0$$

or

$$\mathcal{L}\ f(t+\tau) = \varepsilon^{\tau s} F(s) - \varepsilon^{\tau s} \int_0^{\tau} f(t) \varepsilon^{-st} dt, \quad \tau > 0$$

5. Frequency Shift (Exponential Multiply)

$$\mathcal{L}\ f(t) \varepsilon^{at} = F(s-a) .$$

6. Time Multiplication

$$\mathcal{L}\ t f(t) = - \frac{dF(s)}{ds} .$$

7. Time Scale

$$\mathcal{L}\ f(at) = \frac{1}{a} F(s/a), \quad a > 0.$$

A.2.4 Theorems

1. Integral Square
   If f(t) satisfies the hypothesis of Theorem A.2 for $\sigma = 0$,

$$\int_{-\infty}^{\infty} f(t)^2 dt = \frac{1}{2\pi j} \int_{-j\infty}^{j\infty} F_2(s) F_2(-s) ds .$$

2. Real Convolution (Complex Multiplication)

$$\mathcal{L}_2 f*g = F_2(s)G_2(s)$$

where

$$f*g = \int_{-\infty}^{\infty} f(\lambda)g(t-\lambda)d\lambda$$

and F_2 and G_2 have regions of convergence specified by c_{f1}, c_{f2} and c_{g1}, c_{g2}.

3. Complex Convolution (Time Multiplication)
For

$$F_2*G_2 = \frac{1}{2\pi j}\int_{\sigma-j\infty}^{\sigma+j\infty} F_2(\mu)G_2(s-\mu)d\mu \ ,$$

$$\mathcal{L}_2 fg = F_2*G_2 \text{ with } c_{f1}<\sigma<c_{f2}$$

and

$$c_{f1}+c_{g1} < \text{Re } s < c_{f2}+c_{g2}.$$

Similar forms for one-sided transforms can be written with appropriate changes in the limits on the time-domain integrals.

4. Initial Value

$$f(0+) = \lim_{t\to 0+} f(t) = \lim_{s\to\infty} sF(s)$$

5. Final Value

$$f(\infty) = \lim_{t\to\infty} f(t) = \lim_{s\to 0} sF(s) \ ,$$

where sF(s) has poles only in LHP.

A.2.5 Table

TABLE A.4

	f(t)	F(s)	
1.	$u_o(t)$	1	$\lvert\sigma\rvert<\infty$
2.	$u_1(t)$	$1/s$	$\sigma>0$
3.	$\varepsilon^{-at}u_1(t)$	$\frac{1}{s+a}$	$\sigma>-a$
4.	$\sin \omega_o t\ u_1(t)$	$\frac{\omega_o}{s^2+\omega_o^2}$	$\sigma>0$
5.	$\cos \omega_o t\ u_1(t)$	$\frac{\omega_o}{s^2+\omega_o^2}$	$\sigma>0$

Note that in 1., F(s) is found using the sifting property of $u_o(t)$. Transforms of the singularity functions are discussed in Chapter 2.

A.3 THE Z-TRANSFORM

The Fourier Transform of Section A.1 and the Laplace Transform of Section A.2 are defined on functions of the real variable t. Similar transforms can be defined on functions of integers. In some cases, the functions of integers originate in that form, for example, in synchronous sequential circuits. In other cases, a uniform sampling process selects a set of times $t_n = nT_s$. In either case, the notation f(n) is used in this section. The relation between the Laplace Transform of uniform impulse-sampled f(t) and the Z-Transform of f(n) is discussed in Section A.4.

A.3.1 Two-sided Z-Transform

The Two-Sided Z-Transform is defined as

$$F_2(z) = \mathfrak{Z}_2 f(n) = \sum_{n=-\infty}^{\infty} f(n) z^{-n} \qquad \text{(A.9)}$$

where

$$z = \alpha + j\beta = |z|\varepsilon^{j\theta} = \rho\varepsilon^{j\theta}$$

and the sum converges in an annulus about z=0, i.e., for $r_1 < \rho < r_2$. The Inverse Z-Transform is

$$f(n) = \mathfrak{Z}_2^{-1} F_2(z) = \frac{1}{2\pi j} \oint_C F_2(z) z^{n-1} dz \qquad \text{(A.10)}$$

where C is a closed contour about z=0 lying in the region of convergence of $F_2(z)$. As in the case of Eq. A.6, the region of convergence determines f(n). For the usual cases of signals which are bounded for large |n|, poles of $F_2(z)$ inside the unit circle Γ are associated with n>0 and poles of $F_2(z)$ outside Γ are associated with n<0. This may be seen by observing that A.9 is a Laurent series expansion of $F_2(z)$ about z=0 which converges uniformly in the annulus. The integral A.10 may be evaluated using the calculus of residues, and is discussed in Section A.3.2.

Theorem A.5

1. f(n) real, $-\infty < n < \infty$, and

2. $$\sum_{n=-\infty}^{\infty} |f(n)\rho^{-n}| < \infty$$

   for some real $\rho > 0$

$\Rightarrow$

1. $F_2(z) = \mathfrak{Z}_2 f(n)$ exists, and

2. $f(n) = \mathfrak{Z}_2^{-1} F_2(z)$.

TABLE A.5

f(n)	$F_2(z)$	
1. $b^{\|n\|}$, $0 < \|b\| < 1$	$\frac{b^2-1}{b} \frac{z}{(z-b)(z-b^{-1})}$	$\|b\|<\rho<\|b^{-1}\|$
2. $v_1(n+k) - v_1(n-k)$	$\frac{z(z^k - z^{-k})}{(z-1)}$	$\rho<\infty$
3. $v_1(n+k) - v_1(n-k-1)$ (symmetric)	$\frac{z(z^k - z^{-(k+1)})}{z-1}$	$\rho<\infty$

A.3.2 One-sided Z-Transform

The One-Sided Z-Transform is used for functions f(n) where $f(n)=0$, $n<0$. This is emphasized by using the notation $f(n)v_1(n)$ and

$$F(z) = \mathfrak{z}\, f(n)v_1(n) = \sum_{n=0}^{\infty} f(n)z^{-n} \tag{A.11}$$

where $z=\rho\varepsilon^{j\theta}$ and the sum converges for $\rho>r>0$, the region of convergence. The Inverse Z-Transform is

$$f(n)v_1(n) = \mathfrak{z}^{-1}F(z) = \frac{1}{2\pi j}\oint_C F(z)z^{n-1}dz \tag{A.12}$$

where C is a closed contour about $z=0$ which is entirely in the region of convergence.

<u>Theorem A.6</u>

1. $$\sum_{n=0}^{N} |f(n)| \leq K < \infty$$

for some $N>0$ and $K>0$, and

2. $|f(n)| \leq Mr^n$, for some real M>0 and r>0, n>N.

$\Rightarrow$

1. $$F(z) = \sum_{n=0}^{\infty} f(n) z^{-n}$$

converges absolutely for $\rho > r$. The power series also converges uniformly in the region of the z-plane $\rho \geq \rho_o > r$.

Theorem A.7

1. F(z) converges uniformly, $\rho \geq \rho_o > r$

$\Rightarrow$

1. $$f(n) = \frac{1}{2\pi j} \oint_C F(z) z^{n-1} dz, \quad n \geq 0$$

where C is as in A.12.

The Inversion Theorem A.7 provides the general method for finding inverse Z-Transforms. When F(z) is meromorphic, the integral is evaluated using the calculus of residues. Inverse transforms also can be found by the use of a table (see Section A.3.5) computed by the use of the direct transform summation. Also, in cases where F(z) is a rational function, a partial fraction expansion of F(z)/z reduces it to simpler forms which can be found in the table. Finally, for F(z) rational, a series expansion of F(z) about z=0 in terms of z^{-1}, which converges for $\rho > r$, can be used. The coefficient of z^{-n} is f(n), and the expansion is obtained easily by the use of long division.

Partial Fraction Method

For F(z) = zN(z)/D(z) with N(z) and D(z) polynomials such that the degree p of D > the degree of N,

1. $$F(z)/z = \sum_{j=1}^{p} \frac{A_j}{z-\lambda_j}$$

where the p λ_j are simple zeroes of D(z) and

$$A_j = \left. \frac{N(z)(z-\lambda_j)}{D(z)} \right|_{z=\lambda_j}$$

or

2. $$F(z)/z = \sum_{j=1}^{k} \sum_{i=1}^{m_j} \frac{A_{ij}}{(z-\lambda_j)^i}$$

where m_j is the order of the zero of D(z) at $z=\lambda_j$, and

$$p = \sum_{j=1}^{k} m_j ,$$

$$A_{ij} = \frac{1}{(m_j-i)!} \frac{d^{m_j-i}}{dz^{m_j-i}} \left[\frac{N(z)(z-\lambda_j)^{m_j}}{D(z)} \right] \Bigg|_{z=\lambda_j}$$

The inverse transforms are then*

1. $$f(n) = \sum_{j=1}^{p} A_j (\lambda_j)^n , \quad n \geq 0$$

or

2. $$f(n) = \sum_{j=1}^{k} \sum_{i=1}^{m_j} A_{ij} (\lambda_j)^{n-i+1} \frac{n!}{(i-1)!(n-i+1)!} , \quad n \geq 0$$

Use of the Calculus of Residues

For F(z) meromorphic, the evaluation of A.12 is by direct

*Note that $\lambda_j=0$ requires the definition $0^0=1$.

application of the Cauchy Residue Theorem:

$$\sum \text{Residues of } F(z)z^{n-1}\Big|_{\text{poles of } F(z)z^{n-1}} =$$

$$= \frac{1}{2\pi j} \oint_C F(z)z^{n-1}dz = f(n), \qquad n \geq 0,$$

where C is a closed contour about z=0 which is entirely in the region of convergence. Special care should be taken to include the residue of poles of finite order at the origin.

Since the integrand contains an essential singularity at the origin for $n \to -\infty$, the application of the Cauchy Residue Theorem for n<0 is carried out by altering the contour as shown in Figure A.1.

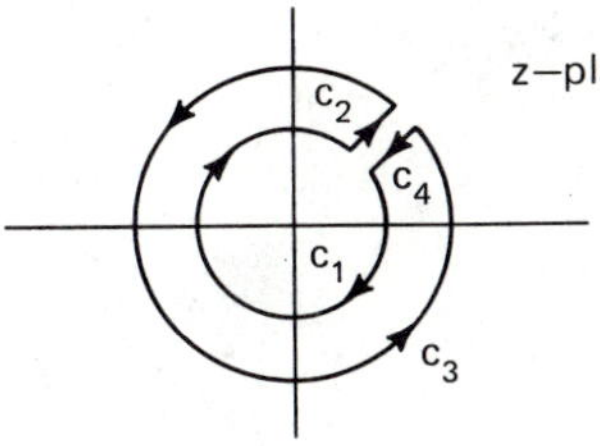

Figure A.1

For C_1 in the region of convergence, the residue within the closed contour $C_1+C_2+C_3+C_4$ is zero and

$$-\frac{1}{2\pi j} \oint_{C_1} F(z)z^{n-1}dz = \frac{1}{2\pi j} \oint_C F(z)z^{n-1}dz = f(n) = 0, \quad n<0,$$

for

$$\lim_{\rho\to\infty} F(z) < \infty \quad .$$

The Inverse Two-Sided Transform of Section A.3.1 can be evaluated by similar procedures when the region of convergence is given by $r_1<\rho<r_2$. Since the integrand of A.10 contains an essential singularity at infinity for $n \to \infty$ and an

essential singularity at z=0 for $n \to -\infty$, the integral is evaluated as:

$$\sum \text{Residues of } F_2(z)z^{n-1}\Bigg|_{\substack{\text{poles of } F_2(z)z^{n-1} \\ \text{inside of } \rho=r_1}} = f(n), \quad n \geq 0,$$

and

$$-\sum \text{Residues of } F_2(z)z^{n-1}\Bigg|_{\substack{\text{poles of } F_2(z)z^{n-1} \\ \text{outside of } \rho=r_2}} = f(n), \quad n<0.$$

The residues are found as in Section A.2.2.

A.3.3 Properties of the One-sided Z-Transform

1. Linearity

$$\mathcal{Z}\,[af(n) + bg(n)] = aF(z) + bG(z)$$

2. Difference of f(n)

$$\mathcal{Z}\,[\Delta_b f(n)] = (1-z^{-1})F(z) - f(-1)$$

and

$$\mathcal{Z}\,[\Delta_b^k f(n)] = (1-z^{-1})^k F(z) - (1-z^{-1})^{k-1} f(-1) - (1-z^{-1})^{k-2}\Delta_b f(-1) - \cdots - \Delta_b^{k-1} f(-1),$$

where Δ_b is the first backward difference operator.

3. Sum of f(n)

$$\mathcal{Z}[\Delta_b^{-1} f(n)] = \mathcal{Z} \sum^{n} f(k)$$

$$= \frac{1}{1-z^{-1}} F(z) + \frac{1}{1-z^{-1}} \sum^{-1} f(k) .$$

4. "Time" Shift

$$\mathcal{Z}\, f(n-k) = z^{-k} F(z), \quad k > 0$$

or

$$\mathcal{Z}\, f(n+k) = z^{k} F(z) - z^{k} \sum_{j=0}^{k-1} f(j) z^{-j}, \qquad k > 0$$

5. z-Change (a^n Multiply)

$$\mathcal{Z}\, f(n) a^{n} = F(a^{-1} z) .$$

6. "Time" Multiplication

$$\mathcal{Z}\, n f(n) = z^{-1} \frac{d}{dz^{-1}} F(z) .$$

A.3.4 Theorems

1. Sum Square
   If f(n) satisfies the hypothesis of Theorem A.5 for $\rho=1$,

$$\sum_{n=-\infty}^{\infty} f(n)^2 = \frac{1}{2\pi j} \oint_{\Gamma} F_2(z) F_2(z^{-1}) z^{-1} dz$$

   where Γ is the unit circle.

2. Real Convolution

$$\mathcal{Z}_2 f{*}g = F_2(z) G_2(z)$$

where

$$f*g = \sum_{k=-\infty}^{\infty} f(k)g(n-k) \quad .$$

3. Complex Convolution

For

$$F_2*G_2 = \frac{1}{2\pi j} \oint_C F_2(\eta)G_2(z\eta^{-1})\eta^{-1}d\eta$$

$\mathfrak{z}_2 fg = F_2*G_2$ where C is a closed contour about z=0 lying in the region of convergence.

4. Initial Value

$$f(0) = \lim_{z\to\infty} (1-z^{-1})F(z) \quad .$$

5. Final Value

$$f(\infty) = \lim_{z\to 1} (1-z^{-1})F(z)$$

where $(1-z^{-1})F(z)$ has poles only inside the unit circle.

A.3.5 Table

TABLE A.6

	f(n)	F(z)	
1.	$v_o(n)$	1	$\rho<\infty$
2.	$v_1(n)$	$\frac{z}{z-1}$	$\rho>1$
3.	$a^n v_1(n)$	$\frac{z}{z-a}$	$\rho>\|a\|$
4.	$(\sin \lambda_o n)v_1(n)$	$\frac{z \sin\lambda_o}{z^2-2z \cos \lambda_o+1}$	$\rho>1$
5.	$(\cos \lambda_o n)v_1(n)$	$\frac{z(z-\cos \lambda_o)}{z^2-2z \cos \lambda_o+1}$	$\rho>1$

Note: $v_o(n)$ is a unit pulse and $v_1(n)$ a unit step (see Chapter 2)

A.3.6 Fourier Transform of f(n)

For the transform

$$\mathfrak{F}_d f(n) = F(\gamma) = \sum_{n=-\infty}^{\infty} f(n)\gamma^{-n} \qquad \text{(A.13)}$$

and the corresponding inverse

$$\mathfrak{F}_d^{-1} F(\gamma) = \frac{1}{2\pi j} \oint_\Gamma F(\gamma)\gamma^{n-1} d\gamma = f(n) \qquad \text{(A.14)}$$

where $\gamma=\varepsilon^{j\theta}$, $-\pi<\theta\leq\pi$, a connection between the Z-Transform of f(n) (with a convergence region $r_1<\rho<r_2$) and the Fourier Transform of f(n) (which converges for $\rho=1$) is established. The analogy to the relationship between the Laplace Transform of Section A.2 and the Fourier Transform of Section A.1 is obvious. These relationships are summarized in Table A.9 in Section A.4. A sufficient condition on f(n) is

$$\sum_{n=-\infty}^{\infty} |f(n)| < \infty .$$

Table A.7 shows some of these transforms.

A.4 RELATIONSHIPS AMONG TRANSFORMS

There are several situations where it is convenient to change from one transform of a function f(t) or f(n) to another transform. For example, this is done in finding amplitude spectra from Laplace or Z-Transforms by converting to Fourier Transforms. Second, the computation of two-sided transforms from known one-sided transforms is sometimes easy. Third, operations on signals in the time domain, for example, sampling and truncating, may be done equivalently by operations on the signal transforms. Table A.9 summarizes the forms discussed here.

TABLE A.7

$f(n)$	$F(\gamma) = F(\epsilon^{j\theta})$
1. $v_o(n)$	1
2. $v_1(n)$	$\pi u_o(\gamma-1) + \dfrac{\gamma}{\gamma-1}$
3. K	$2\pi K u_o(\gamma-1)$
4. $(\exp j\lambda_o n)$	$2\pi u_o(\gamma-\gamma_o),\ \gamma_o = (\exp j\lambda_o)$
5. $\cos \lambda_o n$	$\pi[u_o(\gamma-\gamma_o) + u_o(\gamma-\gamma_o^*)]$
6. $a^n v_1(n),\ 0<|a|<1$	$\dfrac{\gamma}{\gamma-a}$
7. $b^{|n|},\ 0<|b|<1$	$\dfrac{(b^2-1)}{b}\ \dfrac{\gamma}{(\gamma-b)(\gamma-b^{-1})}$

A.4.1 Fourier, Laplace, and Z-Transform

1. Two of the common relationships are those between
$F(s)(= \mathcal{L} f(t))$ and $F(j\omega)(= \mathcal{F} f(t))$ and between $F(z)(= \mathfrak{z} f(n))$
and $F(\gamma)(= \mathcal{F}_d f(n))$. The entries marked A1 in Table A.9 are
illustrated first. For the transforms of $f(t)$,

a. If the poles of $F(s)$ lie in the LHP, the
substitution $\delta=j\omega$ for s or vice versa relates
the Fourier and Laplace Transforms.
For the transforms of $f(n)$,

b. If the poles of $F(z)$ lie inside Γ, the substitution $\gamma=\epsilon^{j\theta}$ for z or vice versa relates the
Fourier and Z-Transforms.
As examples, compare Table A.1 (entry 2) with Table A.4
(entry 3) and compare Table A.6 (entry 3) with Table A.7
(entry 6).

2. Similar to the above relationships, those between $F_2(s)(= \mathcal{L}_2 f(t))$ and $F(j\omega)(= \mathcal{F} f(t))$ and between $F_2(z)(= \mathfrak{z}_2 f(n))$ and $F(\gamma)(= \mathcal{F}_d f(n))$ are shown as the entries marked A2 in Table A.9. For the transforms of f(t),

   a. If the poles of $F_2(s)$ lie in LHP or RHP, the substitution $\delta=j\omega$ for s or vice versa, relates the Fourier and Laplace Transforms.

   b. If the poles of $F_2(z)$ lie inside and outside Γ, the substitution $\gamma=\varepsilon^{j\theta}$ for z or vice versa, relates the Fourier and Z-Transforms.

As examples, compare Table A.1 (entry 3) with Table A.3 (entry 1) and compare Table A.5 (entry 1) with Table A.7 (entry 7). The above substitutions also occur as trivial cases when $F_2(s)$ converges only for $\sigma=0$ or $F_2(z)$ converges only for $\rho=1$. See Table A.1 (entry 5) and Table A.3 (entry 4) as an example.

3. The special cases of simple poles of F(s) on $\Delta(s=j\omega)$ or simple poles of F(z) on Γ also can relate transforms for the entries marked A3 in Table A.9. For transforms of f(t),

   a. If the poles of $F(s)(= \mathcal{L} f(t))$ are simple poles at $s=s_m$, $s_m \varepsilon \Delta$, the imaginary axis,

$$\mathcal{F} f(t) = F(s)\Big|_{s=j\omega} + \pi \sum_m c_m u_o(\omega-\omega_m)$$

   where c_m is the residue of F(s) at $s = s_m = j\omega_m$. For the transform of f(n),

   b. If the poles of $F(z)(= \mathfrak{z} f(n))$ are simple poles at $z = z_m$, $z_m \varepsilon \Gamma$, the unit circle,

$$\mathcal{F}_d f(n) = F(z)\Big|_{z=\gamma} + \pi \sum_m c_m u_o(\gamma-\varepsilon^{+j\theta_m})$$

   where c_m is the residue of F(z) at $z = z_m = \varepsilon^{j\theta_m}$.

As examples, compare Table A.4 (entry 5) with Table A.2

(entry 6) and compare Table A.6 (entry 2) with Table A.7 (entry 2).

4. Another special case occurs where f(t) or f(n) is periodic; see A4 in Table A.9. For f(t) periodic, $T_o=2\pi/\omega_o$,

a.

$$\mathcal{F}\ f(t) = 2\pi \sum_{k=-\infty}^{\infty} C_k u_o(\omega - k\omega_o)$$

where the C_k are coefficients of a Fourier Series (exponential form) expansion of f(t). For f(n) periodic, $N_o=2\pi/\lambda_o$, N_o odd,

b.

$$\mathcal{F}_d f(n) = 2\pi \sum_{k=-K}^{K} C_k u_o(\gamma - \gamma_o^{\,k}), \quad \gamma_o = (\exp j\lambda_o)$$

where the C_k are coefficients of a Fourier Series (exponential form) expansion of f(n). A similar form for N_o even follows, see Chapter 2.

As examples see Table A.2 (entry 4) and Table A.7 (entry 5).

A.4.2 One- and Two-sided Transforms

The Two-Sided Laplace Transform is

$$F_2(s) = \int_{-\infty}^{\infty} f(t)\varepsilon^{-st}dt = \int_{-\infty}^{0} f(t)\varepsilon^{-st}dt + \int_{0}^{\infty} f(t)\varepsilon^{-st}dt$$

$$= \int_{0}^{\infty} f(-\tau)\varepsilon^{s\tau}d\tau + F(s) ,$$

thus

$$F_2(s) = \mathcal{L}\ f(-\tau)\Big|_{s\to -s} + F(s) .$$

The section of f(t), $t\geq 0$ is transformed to F(s) in the normal one-sided computation, and the section of f(t), $t<0$ is converted to $f(-\tau)$, transformed, and then changed by the substitution $s \to -s$. The transformations can be done using a table of one-sided transforms. In general, f(t) must satisfy the hypothesis of Theorem A.2, and special care must be taken when singularity functions $u_j(t)$, $j \leq 0$, occur. In the special case where <u>f(t) is an even function</u>,

$$F_2(s) = F(-s) + F(s).$$

The two-sided Z-Transform is treated similarly to give

$$F_2(z) = \mathcal{Z}\, f(-k)\Big|_{z \to z^{-1}} + F(z) - f(0).$$

In the special case where <u>f(n) is an even function</u>,

$$F_2(z) = F(z^{-1}) + F(z) - f(0).$$

See B in Table A.9.

A.4.3 Truncation of f(t) or f(n)

There are cases, for example, in assuring the realizability of a unit impulse response h(t) or a unit pulse response h(n), where the elimination of the negative argument section of the function is required. If the transforms $F_2(s)$ or $F_2(z)$ are known, the corresponding F(s) and F(z) are required. It is assumed that the regions of convergence of $F_2(s)$ and $F_2(z)$ include the imaginary axis of the s-plane or the unit circle in the z-plane, respectively.

The operation on $F_2(s)$ is commonly denoted as

$$F(s) = [F_2(s)]_+$$

where the symbol on the right means to expand $F_2(s)$ in a partial fraction expansion and then retain only those terms with poles in the LHP, i.e., with a time domain response for $t > 0$.

The operation on $F_2(z)$ is denoted similarly as

$$F(z) = [F_2(z)]_+$$

where the symbol on the right means to expand $F_2(z)/z$ in a partial fraction expansion and then retain only those terms with poles inside the unit circle. The terms that are retained are then multiplied by z and these have a time domain response for n>0. See C in Table A.9.

As examples, consider Table A.3 (entry 1) with

$$F_2(s) = \frac{2a}{a^2-s^2} = \frac{-2a}{(s+a)(s-a)}, \quad a > 0$$

$$= \frac{1}{s+a} - \frac{1}{s-a}$$

Then,

$$[F_2(s)]_+ = \frac{1}{s+a}$$

and this checks Table A.4 (entry 3). Similarly, consider Table A.5 (entry 1) with

$$F_2(z) = \frac{b^2-1}{b} \quad \frac{z}{(z-b)(z-b^{-1})}$$

and

$$F_2(z)/z = \frac{1}{z-b} - \frac{1}{z-b^{-1}} .$$

Then

$$[F_2(z)]_+ = \frac{z}{z-b}$$

and this checks Table A.6 (entry 3).

A.4.4 Uniform Inpulse Sampling

All of the cases in Appendix A that have been considered up to this point were considered either with an f(t) and its transforms or with an f(n) and its transforms, i.e., only with the left or right columns of Table A.9. By uniformly impulse sampling (see Chapter 2) a g(t) to obtain $g(nT_s)$, it is possible to relate their corresponding transforms. Let

$$u_{T_s}(t) = \sum_{n=0}^{\infty} u_o(t-nT_s), \quad t \geq 0$$

be a sequence of unit impulses spaced T_s seconds apart. Also, let

$$g^*(t) = g(t)u_{T_s}(t) = \sum_{n=0}^{\infty} g(nT_s)u_o(t-nT_s)$$

be the impulse-modulated g(t). For g(t) = 0, t<0,

$$\mathcal{L}\, g^*(t) = G^*(s) = \sum_{n=0}^{\infty} g(nT_s)(\exp -nT_s s)$$

If the substitution $z = (\exp T_s s)$ is used,

$$G^*(s)\Big|_{z=(\exp sT_s)} = \sum_{n=0}^{\infty} g(nT_s)z^{-n} = \mathcal{Z}\, g(nT_s).$$

It is common to shorten the left side to G(z) and previous Z-Transforms are related when T_s is included appropriately. For example, if for b>0,

$$g(t) = \varepsilon^{-bt}u_1(t) ,$$

the corresponding

$$g(nT_s) = (\exp -bnT_s)u_1(nT_s),$$

and

$$G(z) = \sum_{n=0}^{\infty} (\exp -bnT_s)z^{-n} = \frac{z}{z-(\exp -bT_s)}.$$

Then consider the discrete-time function

$$f(n) = a^n v_1(n), \qquad 0 < a < 1 ,$$

and

$$F(z) = \frac{z}{z-a} .$$

Thus, in this example,

$$g(nT_s) = f(n) \quad \text{and} \quad G(z) = F(z) \quad \text{for} \quad (\exp -bT_s) = a.$$

In this way, the Z-Transform tables of Section A.3 can also be used for finding G(z). See D of Table A.9.

It is also possible to find $G^*(s)$ using the Complex Convolution Theorem of Section A.2.4. Thus

$$G^*(s) = \mathcal{L}\, g(t)u_{T_s}(t) = \frac{1}{2\pi j}\int_{c-j\infty}^{c+j\infty} U_{T_s}(s-\mu)G(\mu)d\mu$$

where

$$U_{T_s} = \mathcal{L}\, u_{T_s}(t) = \frac{1}{1-(\exp -T_s s)}$$

converges for $\sigma>0$, and if $\mathcal{L}\, g(t)$ converges for $\sigma>\sigma_o$, require $\sigma_o<c<\sigma_1$ with $\sigma_1>\max(\sigma_o, 0)$. Therefore, using $z = (\exp T_s s)$,

$$G(z) = \frac{1}{2\pi j}\int_{c-j\infty}^{c+j\infty} \frac{G(\mu)d\mu}{1-z^{-1}(\exp T_s\mu)}$$

If the magnitude of the integrand approaches zero for μ on C_ℓ as $p\to\infty$, Jordan's Lemma A and the Cauchy Residue Theorem lead to the result

$$G(z) = \sum \text{Residues}\left[\frac{G(\mu)}{1-z^{-1}(\exp T_s\mu)}\right]_{\text{poles of } G(\mu)} \qquad \text{(A.15)}$$

This is a useful form for finding the Z-Transform of a uniformly impulse sampled function g(t). Extensive tables of $\mathcal{Z}\, g(nT_s) = G(z)$ have been published [9.], and may be used as additions to Table A.6 with proper definition of constants. See Table A.8.

A.4.5 The Modified Z-Transform

Another relationship between a g(t) and $g(nT_s)$ and their transforms can be established through the use of the Modified Z-Transform [9.]:

$$\mathcal{Z}_m g(t) = G(z,m) = z^{-1} \sum_{n=0}^{\infty} g(nT_s+mT_s) z^{-n}$$

$$= z^{-1} \sum \text{Residues} \left. \frac{G(\mu)(\exp mT_s\mu)}{1-(\exp \mu T_s) z^{-1}} \right|_{\substack{\text{poles} \\ \text{of } G(\mu)}}$$

where $z = (\exp sT_s)$ and $0 \leq m \leq 1$. Also,

$$g(nT,m) = \mathcal{Z}_m^{-1} G(z,m)$$

$$= \frac{1}{2\pi j} \oint_C G(z,m) z^{n-1} dz$$

where C encloses the poles of $G(z,m)z^{n-1}$.

Both G(z) and G(s) can be obtained from G(z,m) or $g(nT_s,m) = \mathcal{Z}_m^{-1} G(z,m)$ as follows:

$$G(z) = \lim_{m \to 0} zG(z,m) = \mathcal{Z}\, g(nT_s)$$

and

$$G(s) = \mathcal{L}\, g(t) = \mathcal{L} \left\{ g(nT_s,m) \Big|_{n=0,1,2\ldots} \right\}, \qquad 0 \leq m \leq 1.$$

TABLE A.8

	$g(nT_s)$	$F(z)$	Compare Table A.6
1.	$u_1(nT_s)$	$\frac{z}{z-1}$	---
2.	$(\exp -bnT_s)u_1(nT_s)$	$\frac{z}{z-(\exp -bT_s)}$	$(\exp -bT_s) = a$
3.	$(\sin \omega_o nT_s)u_1(nT_s)$	$\frac{z \sin \omega_o T_s}{z^2-2z \cos \omega_o T_s + 1}$	$\omega_o T_s = \lambda_o$
4.	$(\cos \omega_o nT_s)u_1(nT_s)$	$\frac{z(z-\cos \omega_o T_s)}{z^2-2z \cos \omega_o T_s + 1}$	$\omega_o T_s = \lambda_o$

This establishes a two-way link between the two columns in Table A.9 (see E). However, the latter operation is not convenient, and one alternative is through the Zeta Transform [11.]. For

$$\xi = \frac{z-1}{T_s} = \frac{(\exp sT_s)-1}{T_s} ,$$

$$G_1(\xi) = G(z)\Big|_{z=\xi T_s+1}$$

and then

$$G(s) = \lim_{T_s \to 0} T_s G_1(\xi)$$

with

$$\lim_{T_s \to 0} \xi = s .$$

There is one restriction of this operation. Since only the principal strip of the s-plane should be mapped into the z-plane for a single-valued result, the sampling period T_s should be such that the poles of G(s) are in the strip $-\pi/T_s < \omega \leq \pi/T_s$.

A.4.6 Transform Relationships

TABLE A.9

1. $\mathcal{F}$ Series
    (periodic $f(t)$)
    (See Chapter 2)

    $\mathcal{F}_d$ Series
    (periodic $f(n)$)
    (See Chapter 2)

    (A4)   (A4)

2. $\mathcal{F}$ Transform
    $\mathcal{F} f(t) = F(\delta)$

    $\mathcal{F}_d$ Transform
    $\mathcal{F}_d f(n) = F(\gamma)$

    (A3)   (A1)   (A3)   (A1)

3. $\mathcal{L}$ Transform
    $\mathcal{L} f(t) = F(s)$

    (A2)

    $\mathcal{Z}$ Transform
    $\mathcal{Z} f(n) = F(z)$

    (C)   (B) — (D) —→ (C)   (B)   (A2)

            — (E) —→

4. $\mathcal{L}_2$ Transform
    $\mathcal{L}_2 f(t) = F_2(s)$

    $\mathcal{Z}_2$ Transform
    $\mathcal{Z}_2 f(n) = F_2(z)$

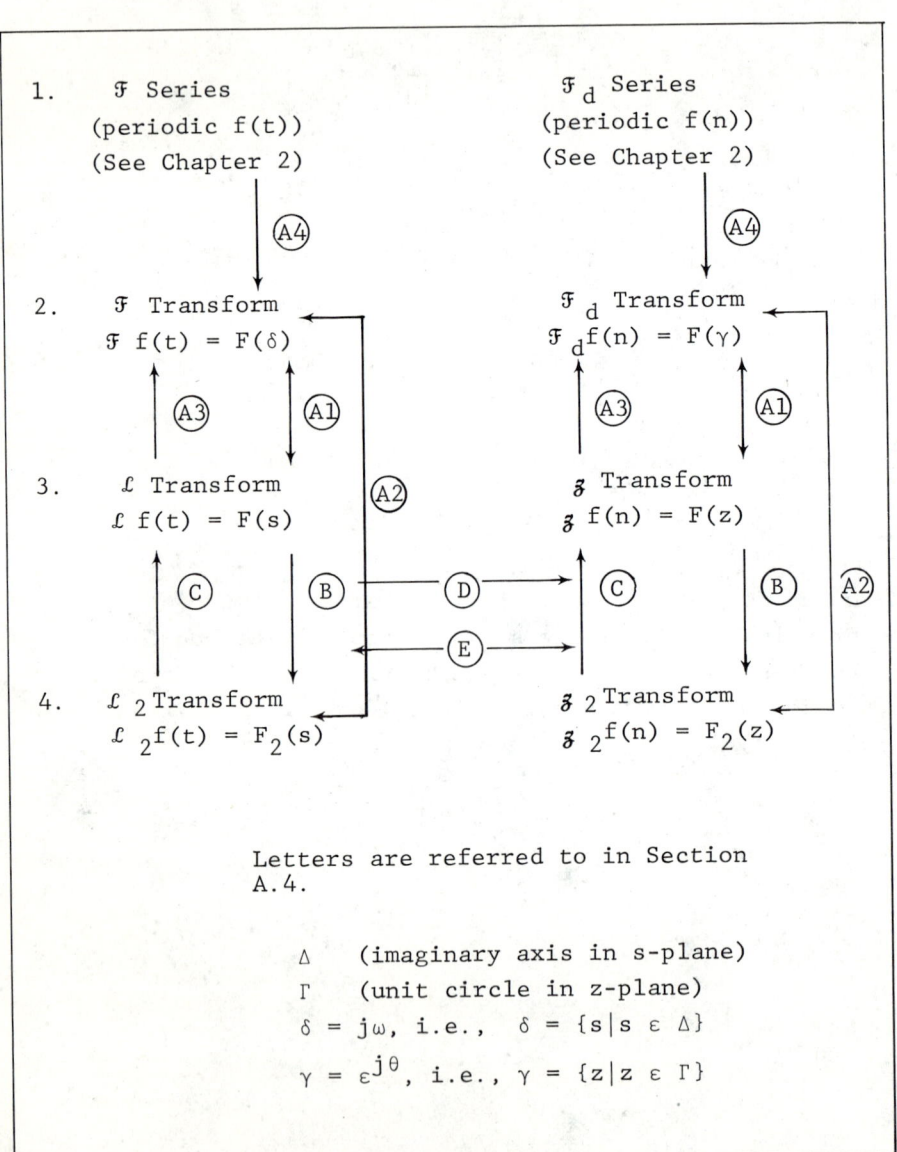

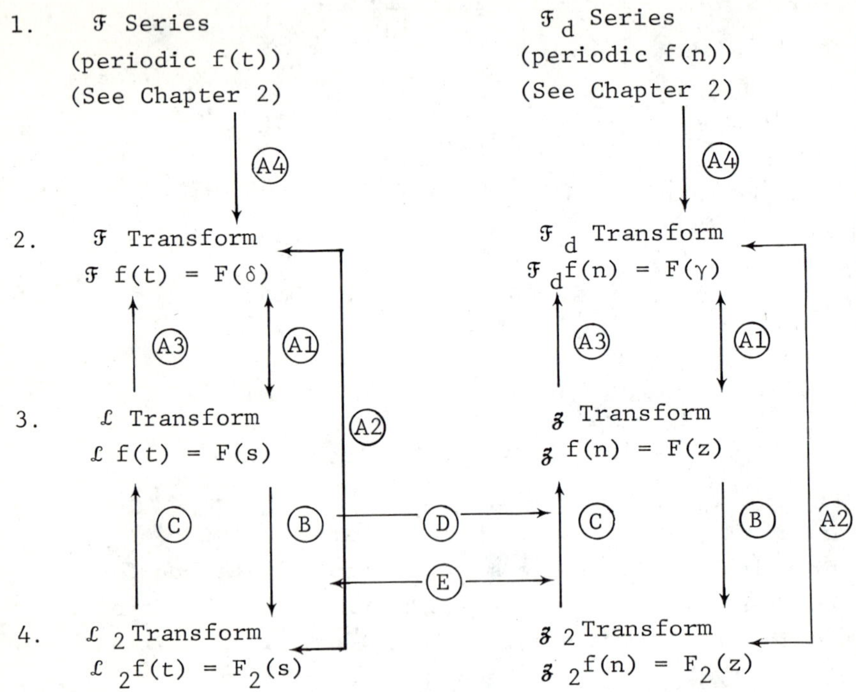

Letters are referred to in Section A.4.

Δ (imaginary axis in s-plane)
Γ (unit circle in z-plane)
$\delta = j\omega$, i.e., $\delta = \{s \mid s \in \Delta\}$
$\gamma = \varepsilon^{j\theta}$, i.e., $\gamma = \{z \mid z \in \Gamma\}$

REFERENCES

1. Cooper, G. and McGillem, C., "Methods of Signal and System Analysis," New York: Holt, Rinehart, Winston, Inc., 1967.
2. Lathi, B., "An Introduction to Random Signals and Communication Theory," Scranton, Pa.: Int'l. Textbook Co., 1968.
3. Papoulis, A., "The Fourier Integral and Its Applications," New York: McGraw-Hill Book Co., 1962.
4. LePage, W., "Complex Variables and the Laplace Transform for Engineers," New York: McGraw-Hill Book Co., 1961.
5. Van der Pol, B. and Bremmer, H., "Operational Calculus Based on the Two-Sided Laplace Integral," 2nd Ed., New York: Cambridge Univ. Press, 1955.
6. Schwarz, R. and Friedland, B., "Linear Systems," New York: McGraw-Hill Book Co., 1965.
7. DeRusso, P., et al., "State Variables for Engineers," New York: John Wiley and Sons, Inc., 1965.
8. Freeman, H., "Discrete-Time Systems," New York: John Wiley and Sons, Inc., 1965.
9. Jury, E., "Theory and Application of the Z-Transform Method," New York: John Wiley and Sons, Inc., 1964.
10. Bracewell, R., "The Fourier Transform and Its Applications," New York: McGraw-Hill Book Co., 1965.
11. Gupta, S. C., "Transform and State Variable Methods in Linear Systems," New York: John Wiley and Sons, Inc., 1966.

Appendix B

MATRICES AND VECTOR SPACES

The notation of vectors and matrices and associated operations in linear vector spaces simplify much of the description and analysis of signals and linear systems. Also, many fundamental concepts are readily explained in these terms, and the calculations required in the study of signals and systems are easily programmed on computers when standard algebraic operations are used. This appendix is a very brief summary of material that is used in this book. The references at the end of the appendix give more complete developments.

B.1 VECTOR SPACES

The n-ordered real numbers $v_1, v_2, \ldots, v_n$ are the coordinates of the <u>column vector</u> $\underline{v}$:

$$\underline{v} = \begin{bmatrix} v_1 \\ v_2 \\ \vdots \\ v_n \end{bmatrix}$$

A linear vector space S is a set of vectors satisfying the following axioms for a sum of vectors and multiplication by a scalar:

1. $\underline{x} + \underline{y} = \underline{y} + \underline{x}$
2. $(\underline{x}+\underline{y}) + \underline{z} = \underline{x} + (\underline{y}+\underline{z})$
3. $\exists\ \underline{0} \ni \underline{x} + \underline{0} = \underline{x}$
4. For every $\underline{x}$, $\exists$ unique $\underline{y} \ni \underline{x} + \underline{y} = \underline{0}$
5. $\exists\ 1 \ni 1\underline{x} = \underline{x}$
6. $a(b\underline{x}) = (ab)\underline{x}$
7. $(a+b)\underline{x} = a\underline{x} + b\underline{x}$
8. $a(\underline{x}+\underline{y}) = a\underline{x} + a\underline{y}$.

For the m vectors $\underline{v}_1, \underline{v}_2, \ldots, \underline{v}_m$, a linear combination is

$$\underline{x} = \sum_{i=1}^{m} a_i \underline{v}_i ,$$

and the space S of all such linear combinations is spanned by the vectors $\underline{v}_i$, i = 1,..., m. If $\underline{x} = \underline{0}$ only for all $a_i = 0$, the vectors $\underline{v}_1, \ldots, \underline{v}_m$ are linearly independent. In this case, $\underline{v}_1, \ldots, \underline{v}_m$ form a basis is S, and every $\underline{x}$ is S can be written as a unique linear combination of these basis vectors. The dimension of S is the maximum number of linearly independent vectors in S.

A particular basis of an n-dimensional vector space which

is frequently used in this book is the set of n-component unit vectors:

$$\underline{e}_1 = \begin{bmatrix} 1 \\ 0 \\ \vdots \\ 0 \end{bmatrix}, \ldots, \underline{e}_i = \begin{bmatrix} 0 \\ \vdots \\ 1 \\ \vdots \\ 0 \end{bmatrix}, \ldots, \underline{e}_n = \begin{bmatrix} 0 \\ \vdots \\ 0 \\ 1 \end{bmatrix}. \qquad (B.1)$$

In a finite-dimensional, linear vector space, the operations of addition and multiplication by a scalar then follow. For $\underline{x}$ and $\underline{y}$ in S and using the basis B.1:

$$\underline{x} + \underline{y} = \sum_i x_i \underline{e}_i + \sum_i y_i \underline{e}_i$$

$$= \sum_i (x_i + y_i)\underline{e}_i \equiv \sum_i z_i \underline{e}_i = \underline{z}$$

and

$$a\underline{x} = a \sum_i x_i \underline{e}_i$$

$$= \sum_i (ax_i)\underline{e}_i = \sum_i p_i \underline{e}_i = \underline{p}$$

and $\underline{z}$ and $\underline{p}$ belong to S, i.e., the set of vectors in S is closed under addition and multiplication by a scalar. The x_i are the coordinates of $\underline{x}$ relative to the given basis, and a similar comment applies to $\underline{y}$, $\underline{z}$, and $\underline{p}$.

For the basis B.1 in S, an inner product of two vectors in S is defined as:

$$\langle \underline{x}, \underline{y} \rangle = \sum_i x_i y_i \quad . \qquad (B.2)$$

This simple form occurs because the basis B.1 is orthonormal, i.e.,

$$\langle \underline{e}_i, \underline{e}_j \rangle = 1 \qquad j = i$$
$$= 0 \qquad j \neq i$$

The Gram-Schmidt [4.] procedure can be used to derive an orthonormal basis in S from any basis in S. The inner product B.2 has the properties of a norm. The length or norm of any $\underline{x}$ in S is defined here as

$$\text{norm } \underline{x} = ||\underline{x}|| = \langle \underline{x}, \underline{x} \rangle^{1/2}. \qquad \text{(B.3)}$$

The symbol R_n is used for a real, linear, finite-dimensional, inner product vector space or Euclidean space. The operations of addition and multiplication by a scalar for vectors in R_n are written in the equivalent form:

$$\underline{x}+\underline{y} = \begin{bmatrix} x_1 + y_1 \\ \vdots \\ x_n + y_n \end{bmatrix} = \begin{bmatrix} z_1 \\ \vdots \\ z_n \end{bmatrix} = \underline{z}$$

and

$$a\underline{x} = \begin{bmatrix} ax_1 \\ \vdots \\ ax_n \end{bmatrix} = \begin{bmatrix} p_1 \\ \vdots \\ p_n \end{bmatrix} = \underline{p} \ .$$

Since $\underline{z}$ and $\underline{p}$ are also in R_n, the space is closed under addition and scalar multiplication. The space R_n (which includes B.1, B.2, and B.3) is assumed in the following sections except where noted (as in Section B.3) when complex numbers occur and a unitary space C_n is introduced.

B.2 MATRICES

B.2.1 Definitions

The ordered array of real numbers

$$A = [a_{ij}] = \begin{bmatrix} a_{11} & \cdots & a_{1n} \\ \vdots & & \vdots \\ a_{m1} & \cdots & a_{mn} \end{bmatrix}$$

is an mxn matrix, and for n=1, a column matrix or vector is denoted

$$\underline{a} = \begin{bmatrix} a_1 \\ \vdots \\ a_m \end{bmatrix} .$$

The transpose of A, A^T, is

$$A^T = [a_{ji}] = \begin{bmatrix} a_{11} & \cdots & a_{m1} \\ \vdots & & \vdots \\ a_{1n} & \cdots & a_{mn} \end{bmatrix}$$

obtained by interchanging rows and columns, and

$$\underline{a}^T = [a_1 \ \cdots \ a_m] ,$$

is a row vector. For A square, it is a symmetric matrix if $A = A^T$. A diagonal matrix A is such that for A square,

$$a_{ij} = 0, \qquad i \neq j$$

and the elements a_{ii} form the principal diagonal. If all $a_{ii} = 0$, the diagonal matrix is the null matrix Θ. If all $a_{ii} = 1$, the diagonal matrix is the unit or identity matrix I. The cofactor of any a_{ij} of A (for A square) is

$$\alpha_{ij} = (-1)^{i+j} |A_r| = (-1)^{i+j} \det A_r$$

where the ith row and jth column are deleted from A to form the reduced matrix A_r. The symbol det A_r is the determinant of A_r, and is evaluated using the Laplace expansion [2.]. The adjoint of a square matrix A is

$$\text{adj } A = [\alpha_{ij}]^T = [\alpha_{ji}] .$$

B.2.2 Algebra

The algebraic operations associated with matrices are defined as follows:

1. Equality: $A = B$ if $a_{ij} = b_{ij}, \quad \forall\, i,j$
   where A and B are both mxn.

2. Addition: $C = A+B$ for $c_{ij} = a_{ij} + b_{ij}, \quad \forall\, i,j$
   where A and B are both mxn and addition is commutative and associative.

3. Multiplication:
   a. $$C = AB \text{ for } c_{ij} = \sum_{k=1}^{r} a_{ik} b_{kj}, \quad \forall\, i,j$$
   where A is mxr and B is rxn. In general, $AB \neq BA$, but matrix multiplication is associative and distributive.
   b. A special case occurs for n=1,
   $$\underline{c} = A\underline{b}$$
   and $\underline{c}$ is an m-dimension column vector.
   c. Multiplication by a scalar is
   $$C = aA \text{ for } c_{ij} = aa_{ij}, \quad \forall\, i,j .$$

4. Inverse
   The inverse of a square matrix A, denoted A^{-1}, is defined such that the products
   $$A\,A^{-1} = A^{-1}A = I$$

and if A^{-1} exists, A is nonsingular. Also,

$$A^{-1} = \frac{\text{adj } A}{\det A}, \qquad \det A \neq 0.$$

5. Rank
   The rank of a matrix A is the dimension of the largest square array in A with a nonzero determinant. If A is an nxn nonsingular matrix, its rank is n. Conversely, if its rank is less than n, A is singular.

6. Trace:
   For A square, the trace of A is

$$\text{tr } A = \sum_i a_{ii}$$

7. Norm:
   The norm of a matrix A, denoted $||A||$, is

$$||A|| = \max ||A\underline{u}||$$

   where $||\underline{u}||=1$. Thus the norm of a matrix is given in terms of the norm of a vector. Consistent with B.3, for A real,

$$||A|| = \lambda_{max}^{\frac{1}{2}}$$

   where λ_{max} is the maximum characteristic value (see Section B.3.2) of A^TA.

Useful identities for common forms in matrix algebra are given in Table B.1. The inverses are assumed to exist and the dimensions are assumed consistent for the required calculations.

TABLE B.1

1.	$(AB)^T = B^T A^T$
2.	$\det AB = (\det A)(\det B)$
3.	$(AB)^{-1} = B^{-1}A^{-1}$
4.	$(A^T)^{-1} = (A^{-1})^T$
5.	$(A+B)^T = A^T + B^T$
6.	$\langle \underline{x}, \underline{y} \rangle = \underline{x}^T \underline{y}$
7.	$\underline{x}\,\underline{y}^T = Z$ where $z_{ij} = x_i y_j$
8.	$\mathrm{tr}(A+B) = \mathrm{tr}A + \mathrm{tr}B$
9.	$\mathrm{tr}(AB) = \mathrm{tr}(BA)$
10.	$\langle \underline{x}, A\underline{y} \rangle = \langle \underline{y}, A^T\underline{x} \rangle$
11.	$\langle A, B \rangle = \sum_i \sum_j a_{ij} b_{ij} = \mathrm{tr}(AB^T)$

B.2.3 Calculus

The matrices and vectors of the book frequently are functions of real scalar arguments t (continuous-time) or n (discrete-time). The operations of differentiation and integration (or differencing and summing) with respect to t (or with respect to n) are required. Furthermore, scalar, vector, and matrix functions of vectors and matrices occur in system optimization where derivatives with respect to the (continuous) components of the vectors or matrices are required. Thus (using the convention of lower case letters for scalars, underlined lower case letters for vectors, and capital letters for matrices) the nine forms of interest are:

$$\begin{array}{ccc} f(t) & f(\underline{x}) & f(X) \\ \underline{f}(t) & \underline{f}(\underline{x}) & \underline{f}(X) \\ F(t) & F(\underline{x}) & F(X) \end{array}$$

For the entries in the first column, the appropriate transforms are also listed below to show the notation. No confusion should occur if the combination of lower case, underlines, and capital letters, as well as the arguments, are used.

1. For f(t) or f(n)

$$\frac{d}{dt} f(t) = \dot{f}(t),$$

and

$$\int^{t} f(\lambda)d\lambda = g(t) + c$$

or

$$\Delta_b f(n) = f(n) - f(n-1) \quad ,$$

and

$$\sum^{n} f(j) = g(n) + c \;.$$

Using the transforms of Appendix A,

$$\mathcal{L}\, \dot{f}(t) = sF(s) - f(0) \;,$$

and

$$\mathcal{L} \int^{t} f(\lambda)d\lambda = \frac{F(s)}{s} + \frac{c}{s} = G(s) + \frac{c}{s}$$

or

$$\mathcal{Z}\,\Delta_b f(n) = (1-z^{-1})F(z) - f(-1) \ ,$$

and

$$\mathcal{Z} \sum^{n} f(j) = \frac{1}{1-z^{-1}} F(z) + \frac{1}{1-z^{-1}} c = G(z) + \frac{1}{1-z^{-1}} c$$

2. For $\underline{f}(t)$ or $\underline{f}(n)$

$$\frac{d}{dt}\underline{f}(t) = \begin{bmatrix} \dot{f}_1(t) \\ \vdots \\ \dot{f}_p(t) \end{bmatrix} = \underline{\dot{f}}(t) \ ,$$

and

$$\int^t \underline{f}(\lambda)d\lambda = \underline{g}(t) + \underline{c}$$

with similar analogous forms for $\underline{f}(n)$.

Again using the transforms of Appendix A,

$$\mathcal{L}\ \underline{\dot{f}}(t) = s\underline{F}(s) - \underline{f}(0),$$

and

$$\mathcal{L} \int^t \underline{f}(\lambda)d\lambda = \frac{1}{s}\underline{F}(s) + \frac{1}{s}\underline{c} = \underline{G}(s) + \frac{1}{s}\underline{c}$$

with similar forms for $\mathcal{Z}\ \underline{f}(n)$.

3. For $F(t)$ or $F(n)$, F_{pxq}

$$\frac{d}{dt}F(t) = \left[\frac{d}{dt}f_{ij}(t)\right] = \dot{F}(t),$$

and

$$\int^t F(\lambda)d\lambda = \left[g_{ij}(t)+c_{ij}\right] = G(t) + C$$

with similar forms for F(n). The transforms are

$$\mathcal{L}\ \dot{F}(t) = sF(s) - F(0),$$

and

$$\mathcal{L}\int^t F(\lambda)d\lambda = \frac{1}{s}F(s) + \frac{1}{s}C = G(s) + \frac{1}{s}C$$

with similar forms for F(n).

4. For $f(\underline{x})$

$$\frac{df(\underline{x})}{d\underline{x}} = \nabla_{\underline{x}} f = \text{grad}_{\underline{x}} f = \left[\frac{\partial f}{\partial x_1}, \ldots, \frac{\partial f}{\partial x_j}, \ldots, \frac{\partial f}{\partial x_n}\right]^T,$$

the gradient of $f(\underline{x})$. (The gradient is also sometimes defined as a row vector, and in 5., the extension is the standard Jacobian matrix).

5. Fox $\underline{f}(\underline{x})$

$$\frac{d\underline{f}(\underline{x})}{d\underline{x}} = \left[\frac{\partial f_j}{\partial x_i}\right] = \begin{bmatrix} \frac{\partial f_1}{\partial x_1} & \cdots & \frac{\partial f_p}{\partial x_1} \\ \vdots & & \vdots \\ \frac{\partial f_1}{\partial x_n} & & \frac{\partial f_p}{\partial x_n} \end{bmatrix} = J^T_{\underline{x}}(\underline{f})$$

where $J_{\underline{x}}(f)$ is the Jacobian matrix. This definition is an extension of 4. and

$$\left(\frac{d\underline{f}(\underline{x})}{d\underline{x}}\right)^T = J_{\underline{x}}(\underline{f}),$$

or

$$\frac{d}{d\underline{x}}\underline{f}(\underline{x}) = [\nabla_{\underline{x}} f_1 \vdots \cdots \vdots \nabla_{\underline{x}} f_p]$$

6. For $F(\underline{x})$, F_{pxq}

$$\frac{dF(\underline{x})}{d\underline{x}} = \left[\left(\frac{\partial \underline{f}_1}{\partial \underline{x}}\right)^T \vdots \cdots \vdots \left(\frac{\partial \underline{f}_q}{\partial \underline{x}}\right)^T \right]^T .$$

This is an extension of 5. where each submatrix for each column of F is nxp. Using the Jacobian notation,

$$\frac{dF(\underline{x})}{d\underline{x}} = \left[J_{\underline{x}}(\underline{f}_1) \vdots \cdots \vdots J_{\underline{x}}(\underline{f}_q) \right]^T .$$

7. For $f(X)$, X_{nxm}

$$\frac{df(X)}{dX} = \left[\frac{\partial f}{\partial x_{ij}}\right] = \begin{bmatrix} \frac{\partial f}{\partial x_{11}} & \cdots & \frac{\partial f}{\partial x_{1m}} \\ \vdots & & \vdots \\ \frac{\partial f}{\partial x_{n1}} & & \frac{\partial f}{\partial x_{nm}} \end{bmatrix} = Gr_X(f)$$

the <u>gradient</u> <u>matrix</u> of $f(X)$. This is an extension of 4.

8. For $\underline{f}(X)$,

$$\frac{d\underline{f}(X)}{dX} = \left[\left[\frac{\partial f_1}{\partial x_{ij}}\right] \vdots \cdots \vdots \left[\frac{\partial f_p}{\partial x_{ij}}\right] \right]$$

$$= \left[Gr_X(f_1) \vdots \cdots \vdots Gr_X(f_p) \right] .$$

9. For $F(X)$, F_{pxq}, X_{nxm}

$$\frac{dF}{dX} = \left[Gr_X(f_{ji}) \right]_{nqxmp}$$

$$= \begin{bmatrix} Gr_X(f_{11}) & \cdots & Gr_X(f_{p1}) \\ \vdots & & \vdots \\ Gr_X(f_{1q}) & \cdots & Gr_X(f_{pq}) \end{bmatrix} .$$

Useful identities for common forms in matrix calculus are given in Table B.2. These forms follow from the previous definitions and are consistent within this appendix. However, there are differences of definition among various references.

TABLE B.2

---

1. $\frac{d}{d\mathbf{x}}(\mathbf{a}^T\mathbf{x}) = \mathbf{a} = \frac{d}{d\mathbf{x}}(\mathbf{x}^T\mathbf{a})$

2. $\frac{d}{d\mathbf{x}}(A\mathbf{x}) = A^T, \quad \frac{d}{d\mathbf{x}}(\mathbf{x}^T A^T) = A$

3. $\frac{d}{d\mathbf{x}}(\mathbf{a}^T\mathbf{b}) = \frac{d\mathbf{a}}{d\mathbf{x}}\mathbf{b} + \frac{d\mathbf{b}}{d\mathbf{x}}\mathbf{a}$

4. $\frac{d}{d\mathbf{x}}(\mathbf{x}^T A\mathbf{x}) = A\mathbf{x} + A^T\mathbf{x}$

5. $\frac{\partial}{\partial \mathbf{u}}(\mathbf{u}^T A\mathbf{x}) = A\mathbf{x}$

6. $\frac{\partial}{\partial \mathbf{u}}(\mathbf{x}^T A\mathbf{u}) = A^T\mathbf{x}$

7. For $f(AX) = \text{tr } AX$, [7.]

   $\frac{df}{dX} = \frac{d}{dX}(\text{tr } AX) = A^T$, also

   for $f(AX^T) = \text{tr } AX^T$

   $\frac{df}{dX} = \frac{d}{dX}(\text{tr } AX^T) = A$

8. For $\mathbf{y} = \mathbf{y}(\mathbf{x})$ and $f = f(\mathbf{x}, \mathbf{y})$,

   $\frac{df}{d\mathbf{x}} = \frac{\partial f}{\partial \mathbf{x}} + \frac{d\mathbf{y}}{d\mathbf{x}} \frac{\partial f}{\partial \mathbf{y}}$

9. For $\mathbf{y} = \mathbf{y}(\mathbf{x})$ and $\mathbf{f} = \mathbf{f}(\mathbf{x}, \mathbf{y})$,

   $\frac{d\mathbf{f}}{d\mathbf{x}} = \frac{\partial \mathbf{f}}{\partial \mathbf{x}} + \frac{d\mathbf{y}}{d\mathbf{x}} \frac{\partial \mathbf{f}}{\partial \mathbf{y}}$

TABLE B.2 *(Continued)*

10. For $f = f(\underline{x},t)$ and $\underline{x} = \underline{x}(t)$,

$$\frac{df}{dt} = \frac{\partial f}{\partial t} + \left(\frac{\partial f}{\partial \underline{x}}\right)^T \frac{d\underline{x}}{dt}$$

11. For $\underline{f} = \underline{f}(\underline{x},t)$ and $\underline{x} = \underline{x}(t)$,

$$\frac{d\underline{f}}{dt} = \frac{\partial \underline{f}}{\partial t} + \left(\frac{\partial \underline{f}}{\partial \underline{x}}\right)^T \frac{d\underline{x}}{dt}$$

12. For $F = F(\underline{x},t)$ and $\underline{x} = \underline{x}(t)$

$$\frac{dF}{dt} = \frac{\partial F}{\partial t} + \left(\frac{\partial F}{\partial \underline{x}}\right)^T \frac{d\underline{x}}{dt}$$

13. $\frac{d}{dt}(A^{-1}) = -A^{-1}\left(\frac{dA}{dt}\right)A^{-1}$

14. $\frac{d^k}{dt^k} \varepsilon^{At} = A^k \varepsilon^{At}$

for $\varepsilon^{At} = \left(I+At + \ldots + \frac{A^j t^j}{j!} + \ldots\right)$

15. $\int_0^t \varepsilon^{A\alpha} d\alpha = A^{-1}(\varepsilon^{At} - I)$

16. For $\underline{x}(t)$,

$$\frac{d}{dt}(\underline{x}^T Q \underline{x}) = \underline{x}^T (Q^T + Q)\dot{\underline{x}}$$

17. For $\underline{x}(t)$ and $\dot{\underline{x}} = A\underline{x}$,

$$\int_0^t \underline{x}^T Q \underline{x}\, d\alpha = \underline{x}^T P \underline{x} - \underline{x}(0)^T P \underline{x}(0)$$

where $Q = PA + A^T P$

B.2.4 Partial Fraction Expansion

The partial fraction expansion of scalar rational functions F(s) and F(z) given in Appendix A can be extended to the partial fraction expansion of matrix functions of s or z where the elements are rational functions. The elements of F(s) are assumed to have numerator degree less than the denominator degree, and the elements of F(z) to have numerator degree less than or equal to denominator degree.

1. For the pxq matrix F(s) with elements containing only simple poles at $s=\lambda_j$

$$F(s) = \sum_{j=1}^{n} \frac{A_j}{s-\lambda_j}$$

where the pxq residue matrices are

$$A_j = \lim_{s\to\lambda_j} [(s-\lambda_j)F(s)]$$

and the set of finite poles $s=\lambda_j$, $j=1, \ldots, n$, is found from the finite poles of the elements of F(s). Each λ_j must occur in at least one element.

2. For the pxq matrix F(s) with elements containing poles at $s=\lambda_j$ of maximum order m_j,

$$F(s) = \sum_{j=1}^{m} \sum_{i=1}^{m_j} \frac{A_{ij}}{(s-\lambda_j)^i}$$

where the pxq matrices are

$$A_{ij} = \left\{ \frac{1}{(m_j-i)!} \frac{d^{m_j-i}}{ds^{m_j-i}} \left[(s-\lambda_j)^{m_j} F(s) \right] \right\}_{s=\lambda_j}$$

3. For the pxq matrix F(z) with elements containing only simple poles at $z=\lambda_j$

$$F(z)/z = \sum_{j=1} \frac{A_j}{z-\lambda_j}$$

where

$$A_j = \lim_{z\to\lambda_j} [(z-\lambda_j)\, F(z)/z]$$

where the λ_j, j=1,...,n are as in 1.

4. For the pxq matrix F(z) with elements containing poles at $z=\lambda_j$ of maximum order m_j,

$$F(z)/z = \sum_{j=1}^{k} \sum_{i=1}^{m_j} \frac{A_{ij}}{(z-\lambda_j)^i}$$

where the pxq matrices are

$$A_{ij} = \left\{ \frac{1}{(m_j-i)!} \frac{d^{m_j-i}}{dz^{m_j-i}} \left[(z-\lambda_j)^{m_j} F(z)/z \right] \right\}_{z=\lambda_j}$$

B.3 CHARACTERISTIC VALUES

B.3.1 Linear Algebraic Equations

The vector-matrix notation of Sections B.1 and B.2 can be used to write a set of m linear algebraic equations in n unknowns as

$$A\underline{x} = \underline{y} \tag{B.4}$$

where A is mxn. When $\underline{y}=\underline{0}$, it is a homogeneous set of equaations, otherwise a nonhomogeneous set. A complete theory exists for the solution of such equations [2.], but the case of m=n is of particular interest here.

If r, the rank of A, equals n, A is nonsingular and the solution is

$$\underline{x} = A^{-1}\underline{y} .$$

If the rank r is less than n, there are n-r arbitrary components of the solution. The remaining r components can be found by first reducing the original set of n equations (using elementary operations) to a set of r equations with a nonsingular coefficient matrix. It is clear then that for the $\underline{y}=\underline{0}$, only the solution $\underline{x}=\underline{0}$ exists if r=n and nonzero solutions may exist if r<n, i.e., det A=0.

B.3.2 Characteristic Values and Vectors

A case of special interest in the study of linear systems is the linear equation B.4 where the solution $\underline{x}$ is such that for $\underline{x}\neq 0$

$$A\underline{x} = \lambda\underline{x}$$

where λ is a scalar and A is nxn. Thus, $\underline{x}$ must be a solution of the homogeneous equation

$$(\lambda I-A)\underline{x} = \underline{0} \tag{B.5}$$

and for nontrivial solutions it is necessary that

$$\det(\lambda I-A) = p(\lambda) = 0. \tag{B.6}$$

Values of λ which satisfy B.6 are characteristic values of A and corresponding solutions $\underline{x}$ of B.5 are characteristic vectors of A. The nth-degree polynomial $p(\lambda)$ is the characteristic polynomial of A, and B.6 is the characteristic equation of A. There are n characteristic values $\lambda_1, \lambda_2, \ldots, \lambda_n$ and the associated characteristic vectors are $\underline{x}_1, \underline{x}_2, \ldots, \underline{x}_n$ or more generally, $\underline{k}_1 = b_1\underline{x}_1, \ldots, \underline{k}_n = b_n\underline{x}_n$, where each b_i is any scalar. It is sometimes convenient to choose each b_i such that $||\underline{k}_i||=1$, thus constructing a normalized set of characteristic vectors. It is possible for some of the characteristic values of the matrix A to be complex, and the corresponding characteristic vectors also would contain complex components. For this, the definitions of inner product B.2 and norm B.3 must be changed to

$$<\underline{x},\underline{y}> = \sum_i x_i y_i^* , \tag{B.7}$$

where y_i^* means complex conjugate, and with this,

$$\text{norm } \underline{x} = ||\underline{x}|| = \langle \underline{x}, \underline{x} \rangle^{1/2} \tag{B.8}$$

where the basis B.1 is used. The symbol C_n designates this complex, inner product space in which the eight properties listed for R_n are true for complex scalars and vectors with complex coordinates.

If the characteristic polynomial is written as

$$p(\lambda) = \lambda^n + a_{n-1}\lambda^{n-1} + \ldots + a_1\lambda + a_o \tag{B.9}$$

the coefficients $a_{n-1}, \ldots, a_o$ may be found by expanding the determinant in B.6, by using Bocher's formula [8.], or by the method of Fadeev [6.]. The characteristic values are then found by solving

$$p(\lambda) = \lambda^n + a_{n-1}\lambda^{n-1} + \ldots + a_1\lambda + a_o = 0$$

for the λ_i, i=1, 2,...,n.

There are several methods for constructing a set of characteristic vectors. Basically, it is necessary to solve

$$A\underline{x}_i = \lambda_i \underline{x}_i$$

for $\underline{x}_i$ using λ_i, i=1, 2,...,n. However, in some cases, the number of characteristic vectors is less than n. Therefore, it is necessary to consider the cases of <u>distinct</u> characteristic values and <u>multiple</u> characteristic values separately. Furthermore, for multiple characteristic values, the <u>degeneracy</u> (see below) of the matrix $(\lambda_i I-A)$ for each λ_i that is multiple must be checked.

One systematic method [6.] is developed as follows. Using $p(\lambda)$, a function $f(\lambda,\alpha)$ is defined as

$$f(\lambda,\alpha) = \frac{p(\alpha) - p(\lambda)}{(\lambda-\alpha)}$$

or

$$(\lambda-\alpha)f(\lambda,\alpha) = p(\lambda) - p(\alpha) . \tag{B.10}$$

Using B.9, $f(\lambda,\alpha)$ can be found as

$$f(\lambda,\alpha) = \frac{(\lambda^n-\alpha^n)}{(\lambda-\alpha)} + a_{n-1}\frac{\lambda^{n-1}-\alpha^{n-1}}{(\lambda-\alpha)} + \ldots + a_1 . \qquad \text{(B.11)}$$

Now a matrix equation corresponding to B.10 is formed (see Section B.5 on functions of a matrix), where λI replaces λ, A replaces α, and a_1 is multiplied by I. The result is

$$(\lambda I-A)F(\lambda I,A) = P(\lambda I) - P(A) .$$

The Cayley-Hamilton Theorem (Section B.5.1) implies $P(A) = \Theta$, and since $P(\lambda I) = p(\lambda)I$,

$$(\lambda I-A)C(\lambda) = p(\lambda)I \qquad \text{(B.12)}$$

where $C(\lambda) = F(\lambda I,A)$. For any λ_i,

$$(\lambda_i I-A)C(\lambda_i) = \Theta$$

and comparing this to B.5, each column of $C(\lambda_i)$ <u>may</u> be a characteristic vector of A. Some columns may be zero and not all columns are independent. An alternative to using B.11 to find $C(\lambda) = F(\lambda I,A)$ is

$$C(\lambda) = \text{adj}[\lambda I-A]$$

which follows from B.12.* A special case occurs if all of the elements of $C(\lambda)$ have a common divisor $d(\lambda)$. In this instance let

$$C_r(\lambda) = \frac{1}{d(\lambda)} C(\lambda).$$

---

*This also can be shown by first considering the case of $\lambda \neq \lambda_i$, and then using the fact that the elements of both sides are polynomials of maximum degree n. Since they are equal for all $\lambda \neq \lambda_i$, i.e., for more than n+1 values, they are equal for all λ.

$C_r(\lambda)$ is called the reduced adjoint matrix of $(\lambda I-A)$. If $d(\lambda)$ is the greatest common divisor, $p(\lambda)=m(\lambda)d(\lambda)$, and $m(\lambda)$ is the <u>minimal polynomial</u> of A.

To distinguish between the several cases for constructing characteristic vectors, the following definitions are used:

$$n = \text{dimension of } (\lambda I-A)$$
$$m_i = \text{multiplicity of } \lambda_i$$
$$r_i = \text{rank of } (\lambda_i I-A)$$
$$d_i = \text{degeneracy of } (\lambda_i I-A) = n-r_i$$

and

$$1 \le d_i \le m_i.$$

1. Distinct λ_i: $m_i = 1$, $r_i = n-1$, $d_i = 1$

A set of characteristic vectors $\underline{u}_1, \underline{u}_2, \ldots, \underline{u}_n$ of the nxn matrix A can be found by using the nonzero columns of $C(\lambda_1)$, $C(\lambda_2), \ldots, C(\lambda_n)$ where $\underline{x}_i$ is taken from $C(\lambda_i)$ and the b_i are chosen for convenience to give the $\underline{u}_i$. The characteristic vectors are linearly independent and form a basis in R_n or C_n. When there are complex conjugate pairs λ_i, λ_i^*, the corresponding characteristic vectors must be chosen as complex conjugates $\underline{u}_i$, $\underline{u}_i^*$.

2. Multiple λ_i: $1 < m_i \le n$

In order to simplify the discussion, assume that only one characteristic value is multiple. The the characteristic values are λ_i, with multiplicity m_i, and λ_j, with multiplicity $m_j = 1$, $j \neq i$.

(The extension for more than one multiple characteristic value follows similarly.) Three possibilities occur:

a) $m_i>1$, $r_i=n-1$, $d_i=1$.

There is only one characteristic vector associated with λ_i in this case of <u>simple degeneracy</u>. The single $\underline{u}_i$ can be chosen proportional to any nonzero column of $C(\lambda_i)$. A subspace of R_n or C_n is spanned by the $(1 + \sum_j m_j)$ characteristic vectors of A, and these are then a basis of the subspace.

b) $m_i>1,\ n-m_i<r_i<n-1,\ 1<d_i<m_i$

There are d_i characteristic vectors associated with λ_i in this case of *intermediate degeneracy*. $C(\lambda_i)$ is a null matrix, and characteristic vectors can be chosen proportional to the nonzero independent columns of $C_r(\lambda_i)$. The method in Section B.4.3 can be used to find d_i-m_i generalized characteristic vectors. A subspace of R_n or C_n is spanned by the $(d_i + \sum_j m_j)$ characteristic vectors of A.

c) $m_i>1,\ r_i=n-m_i,\ d_i=m_i$

There are m_i characteristic vectors associated with λ_i in this case of *full degeneracy*. Again $C(\lambda_i)$ is a null matrix, but m_i characteristic vectors can be chosen from the columns of $C_r(\lambda_i)$ or from the columns of

$$\left[\frac{d^{m_i-1}}{d\lambda^{m_i-1}} C(\lambda)\right]_{\lambda=\lambda_i} = C^{(m_i-1)}(\lambda_i) .$$

Thus, the $(m_i + \sum_j m_j)$ characteristic vectors of A span the space R_n or C_n and form a basis of the space.

One use of the characteristic vectors of a matrix A in the book is in the construction of an nxn matrix of a similarity transformation. The columns are characteristic vectors or a combination of characteristic vectors and generalized characteristic vectors if there are fewer than n characteristic vectors. These transformations are discussed in Section B.4 along with a discussion of Jordan blocks and Jordan forms.

B.4 LINEAR TRANSFORMATIONS

B.4.1 Definition

A *linear transformation* L, or linear mapping, is a function whose domain and range are linear vector spaces. Furthermore, for $\underline{x}$ and $\underline{y}$ represented relative to a basis in R_n or C_n (e.g., B.1), the properties

$$L(\underline{x}+\underline{y}) = L(\underline{x}) + L(\underline{y})$$

$$L(a\underline{x}) = aL(\underline{x})$$

must be true. A linear transformation can be represented by a matrix equation. Thus the equation (see Section B.3)

$$A\underline{x} = \underline{y}$$

has the dual interpretation of a set of linear algebraic equations or a linear transformation of $\underline{x}$ into $\underline{y}$ in R_n or C_n. The same notation is used in both cases, and A is the matrix representation of the linear transformation L. In this way many of the concepts of Sections B.1, B.2, and B.3 are brought together. In particular, the matrix algebra of Section B.2 carries over to operations with linear transformations. The study of linear transformations is much more general then suggested here, but the case of an nxn matrix representation is of special importance.

B.4.2 Change of Basis and Similarity

For the vector space R_n, it is frequently convenient to <u>change the basis</u> with a consequent change of the coordinates of vectors in the space. Thus, for a basis $\underline{v}_1, \underline{v}_2, \ldots, \underline{v}_n$, any vector in the space can be written uniquely as

$$\sum_i x_i \underline{v}_i = M\underline{x}$$

where M is formed using $\underline{v}_i$ as the ith column, i = 1,2,...,n. Similarly, for a basis $\underline{v}_1', \underline{v}_2', \ldots, \underline{v}_n'$, the result is

$$\sum x_i' \underline{v}_i' = M'\underline{x}' \quad .$$

If the x_i and x_i' are the coordinates of the same vector relative to the two bases, we have

$$M\underline{x} = M'\underline{x}'$$

or

$$\underline{x} = M^{-1}M'\underline{x}' = T\underline{x}'$$

where T represents a linear transformation of the coordinates relative to the change of basis.

For a matrix A representing a transformation of $\underline{x}$ into $\underline{y}$, where $\underline{x}$ and $\underline{y}$ are coordinate vectors in the $\underline{v}$-basis,

$$\underline{y} = A\underline{x}$$

and for a matrix B representing a transformation of $\underline{x}'$ into $\underline{y}'$, where $\underline{x}'$ and $\underline{y}'$ are coordinate vectors in the $\underline{v}'$-basis,

$$y' = B\underline{x}' \quad .$$

The transformation of coordinates T can be used to relate A and B. Thus,

$$\underline{y} = A\underline{x} = AT\underline{x}'$$

and since

$$\underline{y}' = T^{-1}\underline{y}$$

it follows that

$$\underline{y}' = T^{-1}AT\underline{x}' = B\underline{x}'$$

and

$$B = T^{-1}AT. \qquad (B.13)$$

This is a <u>similarity</u> transformation, and A and B are similar matrices. Other related forms are <u>orthogonal</u> <u>similarity</u>, where

$$B = T^{-1}AT \qquad (B.14)$$

with $T^{-1} = T^T$ and the old and new bases are orthogonal; and <u>congruence</u>, where

$$B = S^TAS \quad .$$

An important property of the similarity transformation B.13 is that the characteristic values of B are the same as those of A. A property of the orthogonal transformation B.14 is that the lengths and relative angles of vectors are not changed in the transformation. Several other useful forms can be developed from the general linear transformation $B = PAQ$ by appropriate choice of P and Q [3.]. Also, any real, symmetric matrix A is similar to a diagonal matrix.

B.4.3 Jordan Forms

The rxr matrix

$$J_r = \begin{bmatrix} \lambda_r & 1 & 0 & \cdots & & & 0 \\ 0 & \lambda_r & 1 & 0 & \cdots & & 0 \\ 0 & 0 & \lambda_r & 1 & 0 & \cdots & 0 \\ & & & \cdots & 0 & \lambda_r & 1 \\ & & & & \cdots & 0 & \lambda_r \end{bmatrix}$$

is a Jordan block. Any nxn matrix A with characteristic values λ_i, can be transformed by means of a similarity transformation to an nxn matrix J, or Jordan form. The matrix J is a block diagonal form where one or more Jordan blocks J_i, $i \leq n$, associated with the λ_i, form the blocks on the diagonal. The cases concerning characteristic values in Section B.3.2 are used here to classify the possible forms of J.

1. Distinct characteristic values

There is one Jordan block for each λ_i and the block is the single element λ_i. J is, therefore, a diagonal matrix:

$$J = \text{diag}[\lambda_1, \lambda_2, \ldots, \lambda_n] .$$

2. Multiple characteristic values

a) For simple degeneracy (d_i=1), there is one Jordan block for each λ_i of multiplicity m_i and the blocks for the multiple characteristic values are $m_i x m_i$. For example, for λ_1 with $m_1 = 3$,

$$J = \left[\begin{array}{ccc|ccc} \lambda_1 & 1 & 0 & & & \\ 0 & \lambda_1 & 1 & & \Theta & \\ 0 & 0 & \lambda_1 & & & \\ \hline & & & \lambda_2 & & 0 \\ & \Theta & & & \lambda_3 & \\ & & & 0 & & \ddots \end{array}\right]$$

b) For intermediate degeneracy ($1 < d_i < m_i$) there are d_i Jordan blocks for each multiple characteristic value λ_i, and the sum of the block dimensions is m_i. However, the dimensions of the individual blocks are determined by the elementary divisors of A [6.].

c) For full degeneracy ($d_i = m_i$) there are m_i Jordan blocks associated with the multiple characteristic value and each block is the single element λ_i. J is, therefore, a diagonal matrix with m_i elements λ_i and $n-m_i$ elements λ_j, $j \neq i$.

A transformation matrix in the similarity transformation

$$J = T^{-1}AT$$

can be constructed in several ways. For the two cases of distinct λ_i or multiple λ_i with full degeneracy ($d_i = m_i$), J is diagonal and T can be constructed by using n characteristic vectors as columns. The methods for finding the characteristic vectors were given in Section B.3.2. In these two cases where A is similar to a diagonal matrix, A is called a matrix of <u>simple structure</u>.

When $d_i=1$, simple degeneracy, m_i columns of T can be chosen from the set of matrices [6.]

$$C(\lambda_i),\ \left.\frac{dC(\lambda)}{d\lambda}\right|_{\lambda_i} = C^{(1)}(\lambda_i), C^{(2)}(\lambda_i), \ldots, C^{(m_j-1)}(\lambda_i), \quad \text{(B.15)}$$

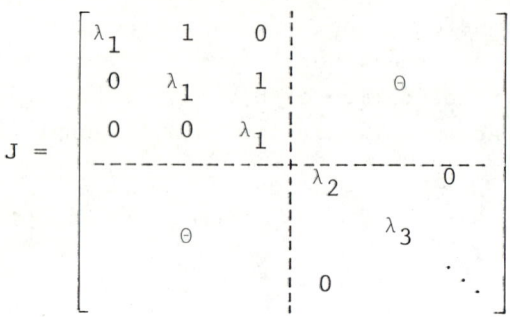

where for any nonzero column $\underline{c}_k$ of $C(\lambda_i)$, the k-th columns of the remaining set are used. These form a set of m_i independent vectors for m_i columns of T, but only one is a characteristic vector. The remaining $n-m_i$ columns of T are chosen from the nonzero columns of the matrices $C(\lambda_j)$, $j \neq i$.

Finally, when $1 < d_i < m_i$, the reduced adjoint matrix $C_r(\lambda)$ must be used in B.15. Generally, the above procedure will not produce the required set of m_i independent vectors and additional elementary operations on the columns of the matrices $C_r(\lambda_i),\ldots,C_r^{(m_i-1)}(\lambda_i)$ must be performed to find the necessary vectors [6.]. An alternative [9.] is to determine the form of J (e.g., using the elementary divisors of A or by "trial and error") and then for each Jordan block of dimension p, a set of p independent vectors $\underline{t}_1, \underline{t}_2,\ldots,\underline{t}_p$ satisfies the equations

$$\begin{aligned} A\underline{t}_1 &= \lambda_i\underline{t}_1 \\ A\underline{t}_2 &= \lambda_i\underline{t}_2 + \underline{t}_1 \\ &\vdots \\ A\underline{t}_p &= \lambda_i\underline{t}_p + \underline{t}_{p-1} \end{aligned} \tag{B.16}$$

which follow from

$$AT = TJ \ .$$

The m_i vectors $\underline{t}_i$ must also be chosen such that those associated with each Jordan block are independent of those in other blocks. The remaining $n-m_i$ vectors $\underline{t}_j$ can be chosen from the nonzero columns of $C(\lambda_j)$, $j \neq i$, or by further use of B.15.

B.5 FUNCTIONS OF A MATRIX

The notation of functions of scalars, vectors and matrices was introduced in Section B.2. The meanings of scalar and vector functions of scalars and vectors is well known, and the cases of matrix functions and matrix arguments were extended by analogy when differentiation was discussed. A situation of particular interest in this book is as follows. For a given nxn matrix A and a scalar function of a scalar argument $f(\alpha)$, properties of the corresponding matrix function $F(A)$, obtained by replacing α by A in $f(\alpha)$, are required.

B.5.1 Cayley-Hamilton Theorem

If $f(\alpha)$ is a polynomial

$$f(\alpha) = \alpha^m + b_{m-1}\alpha^{m-1} + \ldots + b_1\alpha + b_o$$

the corresponding matrix polynomial for A is

$$F(A) = A^m + b_{m-1}A^{m-1} + \ldots + b_1A + b_oI \; .$$

A theorem concerning a particular matrix polynomial is the Cayley-Hamilton Theorem.

Theorem B.1

If A is an nxn matrix and $p(\lambda) = 0$ is the characteristic equation of A, then

$$P(A) = \Theta .$$

This is true for A with distinct or multiple characteristic values [6.].

B.5.2 Matrix Function Expansions

If $f(\alpha)$ is defined by a power series whose region of convergence includes all the characteristic values of A, then the corresponding matrix series may be used to define $F(A)$. Thus, for

$$f(\alpha) = \sum_{k=0}^{\infty} b_k\alpha^k \; ,$$

the matrix series is

$$F(A) = \sum_{k=0}^{\infty} b_kA^k \; ,$$

and examples are

$$f(\alpha) = \varepsilon^{\alpha} = 1 + \alpha + \frac{\alpha^2}{2!} + \ldots + \frac{\alpha^k}{k!} + \ldots$$

with

$$F(A) = \varepsilon^{A} = I + A + \frac{A^2}{2!} + \ldots + \frac{A^k}{k!} + \ldots$$

or

$$f(\alpha) = \sin \alpha = \alpha - \frac{\alpha^3}{3!} + \ldots = \frac{1}{2j} [\varepsilon^{j\alpha} - \varepsilon^{-j\alpha}]$$

with

$$F(A) = \sin A = A - \frac{A^3}{3!} + \ldots = \frac{1}{2j} [\varepsilon^{jA} - \varepsilon^{-jA}] .$$

Two cases are now considered according to whether A is of simple structure (i.e., similar to a diagonal matrix) or not.

1. When A is of simple structure it is similar to a diagonal matrix

$$\Lambda = T^{-1}AT = \mathrm{diag}[\lambda_1, \ldots, \lambda_j, \ldots, \lambda_n]$$

where some of the λ_j may be repeated. Assume there are $s \leq n$ different characteristic values. For a given $f(\alpha)$, $F(A)$ may be found using the following theorem [10.].

<u>Theorem B.2</u>

For an nxn matrix A of simple structure with different characteristic values λ_j, $j=1, 2, \ldots, s$, and analytic function $f(\alpha)$,

$$F(A) = \sum_{j=1}^{s} f(\lambda_j) A_j \tag{B.17}$$

where

$$A_j = \frac{\prod_{\substack{k=1 \\ k \neq j}}^{s} (A - \lambda_k I)}{\prod_{\substack{k=1 \\ k \neq j}}^{s} (\lambda_j - \lambda_k)}$$

The matrices A_j are the constituent idempotents of A and have the properties

$$A_j^2 = A_j$$

$$\sum A_j = I$$

$$A_j A_k = \Theta, \qquad j \neq k.$$

Equation B.17 is a useful way of computing F(A) and is sometimes called Sylvester's Theorem [9.].

2. When A is not of simple structure, a similar theorem [10.] to the above is used.

Theorem B.3

For an nxn matrix A with different characteristic values λ_j, j = 1,2,...,s with multiplicities m_j and lying in the domain of an analytic function $f(\alpha)$,

$$F(A) = \sum_{j=1}^{s} \sum_{k=1}^{m_j} \frac{f^{(k-1)}(\lambda_j)}{(k-1)!} A_{jk} \qquad (B.18)$$

where $f^{(k-1)} = \frac{d^{k-1}}{d\lambda^{k-1}} f(\lambda)$ and where the matrices A_{jk} of A are

$$A_{jk} = (A-\lambda_j I)^{k-1} A_j \qquad k = 1,2,\ldots,m_j, \qquad j = 1,\ldots,s$$

and the A_j are the constituent idempotents of A given by

$$A_j = \sum_{k=1}^{m_j} \alpha_{jk} (A-\lambda_j I)^{k-1} P_j$$

The coefficients α_{jk} are obtained from a partial fraction expansion of $1/p(\lambda)$

$$\frac{1}{p(\lambda)} = \sum_{j=1}^{s} \sum_{k=1}^{m_j} \frac{\gamma_{jk}}{(\lambda-\lambda_j)^k}$$

with

$$\gamma_{jk} = \frac{1}{(m_j-k)!} \left\{ \frac{d^{m_j-k}}{d\lambda^{m_j-k}} \left[\frac{(\lambda-\lambda_j)^{m_j}}{p(\lambda)} \right] \right\}_{\lambda=\lambda_j}$$

Then let $\alpha_{jk} = \gamma_{ji}$, $i = m_j-k+1$, and P_j is defined as

$$P_j = \prod_{\substack{i=1 \\ i\neq j}}^{s} (A-\lambda_i I)^{m_i}$$

Alternative computational methods are given in [10.].

B.6 QUADRATIC FORMS

B.6.1 Definitions

For a given real symmetric nxn matrix Q of rank n and vector $\underline{x}$ of R_n, the scalar function of $\underline{x}$

$$q = \underline{x}^T Q \underline{x}$$

or

$$q = \langle \underline{x}, Q\underline{x} \rangle$$

is a quadratic form in the components of $\underline{x}$. Q is the matrix of the form and the rank of Q is the rank of the form. Since every real symmetric matrix is orthogonally similar to a diagonal matrix, the form q can be reduced to a sum of squares. Thus there is an orthogonal linear transformation T such that

$$\underline{x} = T\underline{y}$$

and

$$q = \underline{y}^T T^T Q T \underline{y} = \underline{y}^T T^{-1} Q T \underline{y}$$

or

$$q = \underline{y}^T D \underline{y}$$

where the diagonal matrix $D = T^{-1}QT$. A transformation T can be found by choosing the columns as a set of n orthonormal characteristic vectors (see Section B.3). Furthermore,

$$D = \text{diag}[\lambda_1, \lambda_2, \ldots, \lambda_j, \ldots \lambda_n]$$

where the λ_j are the characteristic values of Q.

In the more general case where Q is rank $r \leq n$, q can be written in terms of normalized variables as

$$q = v_1^2 + \ldots + v_a^2 - v_{a+1}^2 - \ldots - v_b^2 \qquad \text{(B.19)}$$

where a is the number of positive characteristic values of Q and b is the number of negative characteristic values of Q and a+b = r. The number a is the <u>index</u> of Q.

B.6.2 Definite Forms

Quadratic forms which are positive or negative for all $\underline{x}$ in R_n (or some subspace of R_n) are of particular interest. The following definitions are used:

1. q is <u>positive (negative) definite</u> if for all $\underline{x} \neq \underline{0}$, $q > 0$ (<0) and $q = 0$ for $\underline{x} = \underline{0}$.

2. q is <u>positive (negative) semidefinite</u> if for all $\underline{x}$, $q \geq 0$ (≤ 0) and $q = 0$ for $\underline{x} = \underline{0}$.

Furthermore, if the quadratic form q is any of these, the matrix Q is also said to have the same property. If a and b in the form B.19 are found, the properties of q are:

1. $(a = n, b = 0)$ $\Rightarrow$ positive definite q
2. $(a < n, b = 0)$ $\Rightarrow$ positive semidefinite q
3. $(a = 0, b = n)$ $\Rightarrow$ negative definite q
4. $(a = 0, b < n)$ $\Rightarrow$ negative semidefinite q
5. $(ab \neq 0)$ $\Rightarrow$ indefinite q.

The following theorem (sometimes called Sylvester's) provides an alternative test for a positive definite quadratic form.

Theorem B.4

The form $q = \underline{x}^T Q \underline{x}$, where Q is an nxn real symmetric matrix, is positive definite

$\Leftrightarrow$

det Q and all principal minors of Q are positive. Thus require

$$q_{11} > 0, \quad \begin{vmatrix} q_{11} & q_{12} \\ q_{21} & q_{22} \end{vmatrix} > 0, \ldots, \det Q > 0.$$

REFERENCES

1. Davis, P. J., "The Mathematics of Matrices," Waltham, Mass.: Blaisdell Publ. Co., 1965.
2. Hildebrand, F. B., "Methods of Applied Mathematics," New York: Prentice-Hall, Inc., 1952.
3. Perlis, S., "Theory of Matrices," Cambridge, Mass.: Addison-Wesley Publ. Co., Inc., 1952.
4. Halmos, P. R., "Finite-Dimensional Vector Spaces," Princeton, N.J.: D. Van Nostrand Co., Inc., 1958.
5. Bodewig, E., "Matrix Calculus," New York: Interscience Publishers, 1959.
6. Gantmacher, F. R., "The Theory of Matrices," v. 1, New York: Chelsea Publ. Co., 1959.
7. Athans, M. and Schweppe, F., "Gradient Matrices and Matrix Calculations," MIT Lincoln Laboratory TN 1965-53.
8. Bocher, M., "Introduction to Higher Algebra," New York: Macmillan Co., 1936.
9. De Russo, et al., "State Variables for Engineers," New York: J. Wiley & Sons, 1965.
10. Frame, J. S., "Matrix Functions and Applications," Parts II and IV, IEEE Spectrum, April 1964, pp. 102-108 and June, 1964, pp. 123-131.

Appendix C

PROBABILITY AND RANDOM VARIABLES

Many phenomena of interest in systems analysis require a probabilistic description because the occurrence of some events cannot be predicted with certainty. In Chapter 2, deterministic signal models and stochastic signal models are developed. In the first case, common continuous-time or discrete-time functions are adequate, but in the second, stochastic processes or sequences are the mathematical models of the signals. An elementary review of the fundamentals of probability theory is presented here as a background for that chapter. As in all studies of physical systems, there is the dual interest in an ideal or mathematical model and in an actual system. Probability theory is concerned with the former and the latter, where the assignment of probabilities from experimental data and the verification of predictions are of interest, is the subject of statistics.

C.1 PROBABILITY

Historically, probability theory has been developed in several ways. The classical ratio of favorable outcomes to the total possible number of outcomes for "equally likely" events, the limit of the frequency ratio, the subjective or intuitive approach, and the development from a set of axioms have all been used. The axiomatic method is outlined here, but the other points of view are useful in assigning probabilities in many of the examples used.

C.1.1 Sample Space

The symbol $\mathcal{S}$ denotes a sample space, the set of all possible outcomes of an experiment. The symbol E denotes a subset of $\mathcal{S}$ which is associated with E, an event or favorable outcome. The symbol σ denotes elements or points of $\mathcal{S}$, and there may be a finite number or an infinity of such elements depending upon the particular experiment and the sample space. On a particular trial of the experiment, the event E occurs if the element σ_i on the ith trial belongs to E. Thus

$$\text{E occurs} \quad \Leftrightarrow \quad \sigma_i \in E \quad .$$

For example, using a Venn diagram (Figure C.1), it can be seen that event E occurred on trial 1 since $\sigma_1 \in E$ and event E did not occur on trial 2 since $\sigma_2 \notin E$.

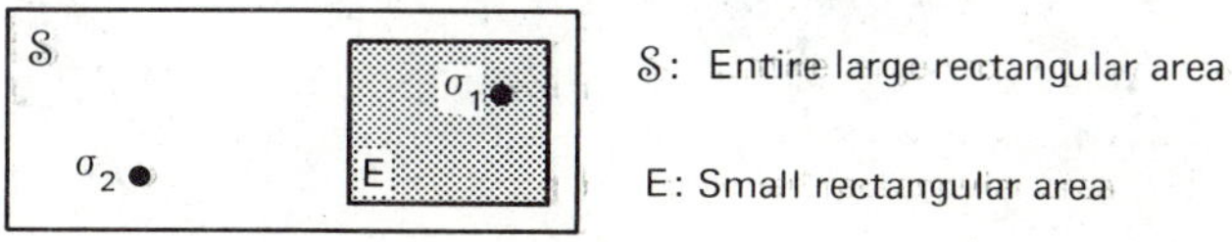

Figure C.1

C.1.2 Axioms

The axiomatic development starts with the assumption of the existence of a number Pr(E), called the probability of the event E, and the following three conditions are satisfied:

Axioms

1. For any event E,

$$0 \leq Pr(E) \leq 1$$

2. The event E_c such that $\sigma_i \in \mathcal{S}$ (a certain event) has

$$Pr(E_c) = 1$$

3. a. For two mutually exclusive events E_1 and E_2,

$$Pr(E_1 \text{ or } E_2) = Pr(E_1) + Pr(E_2)$$

or

$$Pr(E_1 \oplus E_2) = Pr(E_1) + Pr(E_2)$$

where mutually exclusive means the intersection of ε_1 and ε_2 is empty, i.e., the two events have no common elements in .

3. b. If the number of points of the sample space is infinite, for mutually exclusive events,

$$Pr(E_1 \oplus E_2 \oplus \ldots) = Pr(E_1) + Pr(E_2) + \ldots$$

4. Using 3.a. with E_2 the event $\bar{E}_1$ (not E_1),

$$Pr(E_1 \oplus \bar{E}_1) = Pr(E_1) + Pr(\bar{E}_1) = 1$$

or

$$Pr(\bar{E}_1) = 1 - Pr(E_1).$$

5. The joint event E_1 and E_2 has a probability

$$Pr(E_1 \text{ and } E_2) = Pr(E_1 \otimes E_2) = Pr(E_1, E_2) \quad .$$

6. For the joint probabilities $Pr(E_i, E_j)$,

$$i = 1,2,\ldots,n; \quad j = 1,2,\ldots,m \; ,$$

where the E_i are mutually exclusive and E_j are mutually exclusive, the marginal probabilities are

$$\Pr(E_i) = \sum_{j=1}^{m} \Pr(E_i, E_j)$$

and

$$\Pr(E_j) = \sum_{i=1}^{n} \Pr(E_i, E_j) \ .$$

7. The conditional event E_2, given that E_1 occurred, has a probability

$$\Pr(E_2 | E_1) = \frac{\Pr(E_1, E_2)}{\Pr(E_1)} \ , \quad \Pr(E_1) \neq 0$$

and interchanging E_1 and E_2

$$\Pr(E_1 | E_2) = \frac{\Pr(E_1, E_2)}{\Pr(E_2)} \ , \quad \Pr(E_2) \neq 0$$

thus

$$\Pr(E_1, E_2) = \Pr(E_2 | E_1)\Pr(E_1) = \Pr(E_1 | E_2)\Pr(E_2) \ .$$

8. a. Two events are independent if

$$\Pr(E_1, E_2) = \Pr(E_1)\Pr(E_2)$$

b. For n events, the events are independent if

$$\Pr(E_1, \ldots, E_n) = \Pr(E_1)\Pr(E_2)\ldots\Pr(E_n)$$

and if all joint events formed with n-1, n-2, ..., 3, 2 events satisfy the product rule, i.e., the probability of the joint event is the product of the individual event probabilities [5.].

The operations $\oplus$ and $\otimes$ correspond to the set operations of union (∪) and intersection (∩), and Figure C.2 illustrates some of the above properties:

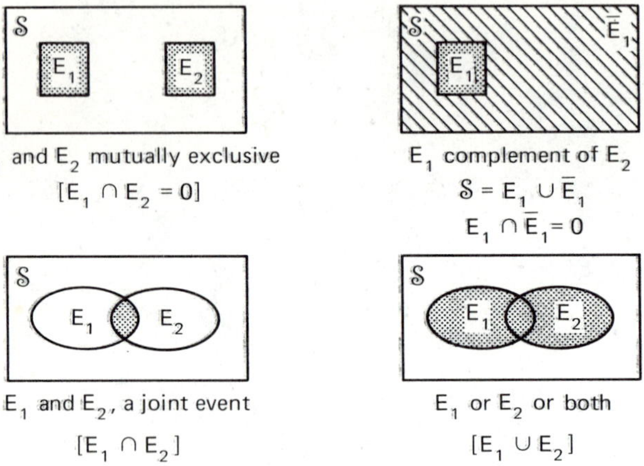

Figure C.2

C.2 RANDOM VARIABLES

The outcomes of the trials of an experiment are assigned numerical values by the introduction of a random variable:

Definition: A random variable X is a real-valued function of the elements of $\mathcal{S}$, thus $X = X(\sigma)$.

Therefore, $X(\sigma)$ is a function whose domain is $\mathcal{S}$ and whose range is a set of real numbers. It is common practice to use capital letters for random variables and corresponding lower case for real numbers.

When the range of X is a set of denumerable real numbers x_k, $k = 1,2,3,\ldots$, X is a <u>discrete</u> random variable. When the range of X is a set of nondenumerable numbers x, e.g., $a \le x \le b$, X is a <u>continuous</u> random variable.

The set of all outcomes such that a discrete random variable is less than or equal to a specified number x_k is $\{X(\sigma) \le x_k\} \subseteq \mathcal{S}$ and the set of all outcomes such that a continuous random variable is less than or equal to a number x is $\{X(\sigma) \leq x\} \subseteq \mathcal{S}$. Similarly, the sets $\{x_p < X(\sigma) \leq x_q\} \subseteq \mathcal{S}$ and $\{a < X(\sigma) \leq b\} \subseteq \mathcal{S}$ indicate all outcomes for the random variables in specified intervals.

The unique probabilities of the events given by the sets $\{X(\sigma) \leq x_k\} \subseteq \mathcal{S}$ and $\{X(\sigma) \leq x\} \subseteq \mathcal{S}$ are $\Pr(X \leq x_k)$ and $\Pr(X \leq x)$, respectively, and these must satisfy the axioms of Section C.1.

C.3 PROBABILITY DISTRIBUTION FUNCTIONS

C.3.1 Discrete Random Variables

The distribution function of a discrete random variable X is

$$P_X(x_k) = \Pr(X \leq x_k),$$

defined for all real x_k. For two numbers x_p and x_q,

$$\Pr(x_p < X \leq x_q) = P_X(x_q) - P_X(x_p).$$

The function $P_X(x_k)$ has the following properties:

1. It is a nondecreasing function of x_k.
2. $P_X(\infty) = 1$
3. $P_X(-\infty) = 0$

The density function of a discrete random variable X is

$$p_X(x_k) = \Pr(X = x_k).$$

The function $p_X(x_k)$ has the properties:

1. $p_X(x_k) \geq 0, \quad \forall k$
2. $\sum_{\forall k} p_X(x_k) = 1$

which follows from Axiom 2. Using $p_X(x_k)$,

$$\Pr(X \leq x_k) = \sum_{x_j \leq x_k} p_X(x_j)$$

and $P_X(x_k)$ and $p_X(x_k)$ are therefore related by

$$P_X(x_k) = \sum_{x_j \leq x_k} p_X(x_j) \quad .$$

C.3.2 Continuous Random Variables

The distribution function of a continuous random variable X is

$$P_X(x) = \Pr(X \leq x)$$

defined for all real x. For two numbers a and b,

$$\Pr(a < X \leq b) = P_X(b) - P_X(a) \quad .$$

The function $P_X(x)$ has the same three properties given for $P_X(x_k)$.

The density function of a continuous random variable X is

$$p_X(x)$$

and this function has the properties:

1. $p_X(x) \geq 0, \quad \forall\, x$
2. $\int_{-\infty}^{\infty} p_X(\lambda)\, d\lambda = 1$

These two properties are consistent with the three properties of $P_X(x)$ if, in addition,

$$P_X(x) = \int_{-\infty}^{x} p_X(\lambda)\, d\lambda \quad , \quad \forall\, x$$

and

$$p_X(x) = \frac{dP_X(x)}{dx}$$

for all x where $P_X(x)$ is differentiable.

C.3.3 Examples

Examples of density functions for both discrete and continuous random variables are listed below.

Discrete X

1. Binomial

$$p_X(x_k) = \binom{n}{x_k} p^{x_k} (1-p)^{n-x_k}, \quad 0 \le x_k \le n$$

p = probability of single event

n = number of trials

2. Poisson

$$p_X(x_k) = \frac{\lambda^{x_k}}{x_k!} \varepsilon^{-\lambda}, \quad x_k \ge 0, \quad \lambda > 0$$

λ = average rate (events per unit).

Other common density functions are the Hypergeometric and the Discrete Uniform.

Continuous X

1. Uniform

$$p_X(x) = \frac{1}{b-a}, \quad a < x < b$$

$$= 0, \quad \text{otherwise.}$$

2. Normal (Gaussian)

$$p_X(x) = \frac{1}{\sqrt{2\pi}\sigma_X} (\exp \frac{-(x-m_X)^2}{2\sigma_X^2})$$

$$\sigma_X^2 = \text{variance of } X$$

$$m_X = \text{mean of } X.$$

Other common density functions are the Beta, Laplace, Rayleigh, Maxwell, Cauchy, and Gamma [1.].

C.3.4 Multiple Random Variables

The idea of joint events introduced in Section 6.1 leads to functions of several random variables $X_1, X_2, \ldots, X_n$. Using the vector notation of Appendix B, the random vector $\underline{X}^T = (X_1 \;\; X_2 \;\; \ldots \;\; X_n)$ is defined, and the related discrete and continuous joint distribution and density functions are as follows:

1. Discrete

$$P_{\underline{X}}(x_{k1}, x_{k2}, \ldots, x_{kn}) = Pr(X_1 \leq x_{k1}, \ldots, X_n \leq x_{kn})$$

for all real $x_{k1}, \ldots, x_{kn}$, and

$$p_{\underline{X}}(x_{k1}, \ldots, x_{kn}) = Pr(X_1 = x_{k1}, \ldots, X_n = x_{kn})$$

and

$$P_{\underline{X}}(x_{k1}, \ldots, x_{kn}) = \sum_{x_{jn} \leq x_{kn}} \cdots \sum_{x_{j1} \leq x_{k1}} p_{\underline{X}}(x_{j1}, \ldots, x_{jn})$$

Note that k, as before, is the discrete integer variable over the sample space and the second subscript now indicates the number of the vector component.

2. Continuous

$$P_{\underline{X}}(x_1, \ldots, x_n) = Pr(X_1 \leq x_1, \ldots, X_n \leq x_n)$$

for all real $x_1, \ldots, x_n$, and

$$P_{\underline{X}}(x_1, \ldots, x_n) = \int_{-\infty}^{x_n} \cdots \int_{-\infty}^{x_1} p_{\underline{X}}(\alpha_1, \ldots, \alpha_n)\, d\alpha_1 \ldots d\alpha_n .$$

The Joint Normal density function is of great interest; it is

$$p_{\underline{X}}(\underline{x}) = p_{\underline{X}}(x_1, \ldots, x_n)$$

$$= \frac{1}{(2\pi)^{n/2} \sqrt{|C_{\underline{X}}|}} \left(\exp \tfrac{1}{2}(\underline{x}-\underline{m}_{\underline{X}})^T C_{\underline{X}}^{-1} (\underline{x}-\underline{m}_{\underline{X}})\right) \qquad (C.1)$$

where $C_{\underline{X}}$ is the covariance matrix of $\underline{X}$ and $\underline{m}_{\underline{X}}$ is the mean of $\underline{X}$ (see Section C.4.2), and $|C_{\underline{X}}| = \det C_{\underline{X}}$.

C.3.5 Marginal Distributions

Associated with marginal probability are marginal distribution and density functions. Thus, for two discrete random

$$\Pr(X_1 \le x_{k1}, X_2 \le x_{k2}) = P_{\underline{X}}(x_{k1}, x_{k2})$$

$$= \sum_{x_{j2} \le x_{k2}} \sum_{x_{j1} \le x_{k1}} p_{\underline{X}}(x_{j1}, x_{j2})$$

and

$$\Pr(X_1 \le x_{k1}) = P_{\underline{X}}(x_{k1}, \infty)$$

$$= P_{X_1}(x_{k1})$$

or

$$\Pr(X_2 \le x_{k2}) = P_{\underline{X}}(\infty, x_{k2})$$

$$= P_{X_2}(x_{k2})$$

are the distribution functions. For the density functions,

$$\Pr(X_1 = x_{k1}) = \sum_{x_{k2}} p_{\underline{X}}(x_{k1}, x_{k2})$$

$$= p_{X_1}(x_{k1})$$

or

$$\Pr(X_2 = x_{k2}) = \sum_{x_{k1}} p_{\underline{X}}(x_{k1}, x_{k2})$$

$$= p_{X_2}(x_{k2}) \ .$$

Similarly, for two continuous random variables,

$$\Pr(X_1 \le x_1, X_2 \le x_2) = P_{\underline{X}}(x_1, x_2)$$

$$= \int_{-\infty}^{x_2}\int_{-\infty}^{x_1} p_{\underline{X}}(\alpha_1, \alpha_2)\, d\alpha_1 d\alpha_2$$

and

$$\Pr(X_1 \le x_1) = P_{\underline{X}}(x_1, \infty)$$

$$= P_{X_1}(x_1)$$

or

$$\Pr(X_2 \le x_2) = P_{\underline{X}}(\infty, x_2)$$

$$= P_{X_2}(x_2).$$

For the density functions,

$$p_{X_1}(x_1) = \int_{-\infty}^{\infty} p_{\underline{X}}(x_1, \alpha)\, d\alpha$$

or

$$p_{X_2}(x_2) = \int_{-\infty}^{\infty} p_{\underline{X}}(\alpha, x_2)\, d\alpha \ .$$

Similar forms for more than two random variables can be written.

Independence of random variables is defined in terms of their joint distribution and density functions and their marginal distribution and density functions. Thus, two continuous random variables are independent if

$$P_{\underline{X}}(x_1,x_2) = P_{X_1}(x_1)P_{X_2}(x_2)$$

and it follows that

$$p_{\underline{X}}(x_1,x_2) = p_{X_1}(x_1)p_{X_2}(x_2)$$

with similar forms for discrete random variables. The converse is also true. For n random variables, they are independent if

$$P_{\underline{X}}(x_1,\ldots,x_n) = P_{X_1}(x_1)\ \cdots\ P_{X_n}(x_n)$$

and again

$$p_{\underline{X}}(x_1,\ldots,x_n) = p_{X_1}(x_1)\ \cdots\ p_{X_n}(x_n)$$

with similar forms of discrete random variables. Finally, two random vectors $\underline{X} \varepsilon R_n$ and $\underline{Y} \varepsilon R_m$ are independent if

$$P_{\underline{Z}}(x_1\ldots x_n y_1\ldots y_m) = P_{\underline{X}}(x_1,\ldots,x_n)P_{\underline{Y}}(y_1,\ldots,y_m)$$

where $P_{\underline{Z}}$ is the distribution function of the random vector $\underline{Z}^T = (X_1,\ldots,X_n,Y_1,\ldots,Y_m)$ [5.].

C.3.6 Conditional Distributions

Associated with conditional probability are <u>conditional distribution</u> and <u>density</u> functions. For two discrete random variables, and given $X_1 = x_{k1}$,

$$\Pr(X_2 \leq x_{k2}|X_1 = x_{k1}) = \frac{\Pr(X_1=x_{k1},X_2\leq x_{k2})}{\Pr(X_1=x_{k1})}$$

and the conditional distribution function is

$$P_{X_2}(x_{k2}|X_1=x_{k1}) = \frac{P_{\underline{X}}(x_{k1},x_{k2})-P_{\underline{X}}(x_{k-1,1},x_{k2})}{p_{X_1}(x_{k1})} .$$

The corresponding conditional density function is

$$p_{X_2}(x_{k2}|X_1 = x_{k1}) = \frac{p_{\underline{X}}(x_{k1},x_{k2})}{p_{X_1}(x_{k1})} .$$

Similarly for two continuous random variables and given $X_1=x_1$,

$$P_{X_2}(x_2|X_1 = x_1) = \frac{\partial P_{\underline{X}}(x_1,x_2)/\partial x_1}{p_{X_1}(x_1)}$$

and the conditional density function is

$$p_{X_2}(x_2|X_1 = x_1) = \frac{p_{\underline{X}}(x_1,x_2)}{p_{X_1}(x_1)} .$$

For other given conditions such as $X_1 \leq x_{k1}$ (or $X_1 \leq x_1$) and $x_{p1} < X_1 \leq x_{q1}$ (or $a < X_1 \leq b$), similar forms can be derived, and for more than two random variables, many conditional forms can be found [5.].

C.4 EXPECTED VALUE

C.4.1 Single Random Variable

The expected value of a function of a single discrete random variable X is

$$E[g(X)] = \sum_{k=-\infty}^{\infty} g(x_k)p_X(x_k) ,$$

and for X a continuous random variable,

$$E[h(X)] = \int_{-\infty}^{\infty} h(x)p_X(x)\,dx .$$

1. By choosing $g(X)=X^n$, where n is a positive integer, the moments of X are

$$E[X^n] = \sum_{k=-\infty}^{\infty} x_k^n p_X(x_k) = \alpha_{Xn}$$

for a discrete random variable and for $h(X)=X^n$,

$$E[X^n] = \int_{-\infty}^{\infty} x^n p_X(x)\,dx = \alpha_{Xn}$$

for a continuous random variable. Cases of particular interest are the <u>first moment</u> or <u>mean</u> value of X when n=1;

$$E[X] = \alpha_{X1} = \sum_{k=-\infty}^{\infty} x_k p_X(x_k)$$

or

$$E[X] = \alpha_{X1} = \int_{-\infty}^{\infty} x p_X(x)\; dx.$$

Other common symbols for α_{X1} are $\overline{X}$ and m_X. Similarly, the <u>second moment</u> or <u>mean square</u> value of X when n=2;

$$E[X^2] = \alpha_{X2} = \sum_{k=-\infty}^{\infty} x_k^2 p_X(x_k)$$

or

$$E[X^2] = \alpha_{X2} = \int_{-\infty}^{\infty} x^2 p_X(x)\,dx\ .$$

Other common symbols for α_{X2} are $\overline{X^2}$ and S_X^2.

2. By choosing $g(X)=(X-\alpha_{X1})^n$, the <u>central moments</u> of X are

$$E[(X-\alpha_{X1})^n] = \sum_{k=-\infty}^{\infty} (x_k-\alpha_{X1})^n p_X(x_k) = \beta_{Xn}$$

of for $h(X)=(X-\alpha_{X1})^n$,

$$E[(X-\alpha_{X1})^n] = \int_{-\infty}^{\infty} (x-\alpha_{X1})^n p_X(x)\,dx = \beta_{Xn}\ .$$

For n=1, the first central moment of X, β_{X1}, is zero, but for n=2, the second central moment of X or variance of X is

$$E[(X-\alpha_{X1})^2] = \beta_{X2} = \sum_{k=-\infty}^{\infty}(x_k-\alpha_{X1})^2 p_X(x_k)$$

or

$$E[(X-\alpha_{X1})^2] = \beta_{X2} = \int_{-\infty}^{\infty}(x-\alpha_{X1})^2 p_X(x)\,dx \ .$$

Furthermore, $\beta_{X2}=\alpha_{X2}-\alpha_{X1}^2$. Another common symbol for β_{X2} is σ_X^2, and the square root of β_{X2} is the standard deviation of X or σ_X.

Higher order moments of X or central moments of X are obtained for n=3,4,... .

C.4.2 Multiple Random Variables

For two or more random variables, the vector $\underline{X}^T=(X_1\ X_2\ \dots\ X_n)$ is used; thus the first moment vector or mean value vector of $\underline{X}$ is

$$E[\underline{X}] = \underline{\alpha}_{\underline{X}1} = (\alpha_{\underline{X}11}\ \alpha_{\underline{X}12}\ \cdots\ \alpha_{\underline{X}1i}\ \cdots\ \alpha_{\underline{X}1n})^T$$

where

$$\alpha_{\underline{X}1i} = \sum_{x_{kn}}\cdots\sum_{x_{k1}} x_{ki}\,p_{\underline{X}}(x_{k1},\dots,x_{ki},\dots,x_{kn})$$

$$i = 1,2,\dots,n \ ,$$

where $\underline{X}$ is a discrete random variable vector, or

$$\alpha_{\underline{X}1i} = \int_{-\infty}^{\infty}\cdots\int_{-\infty}^{\infty} x_i\,p_{\underline{X}}(x_1,\dots,x_n)\,dx_1\dots dx_n \ ,$$

where $\underline{X}$ is a continuous random variable vector. Observe that the multiple sum and integral are over the entire sample space for each component of $\underline{\alpha}_{\underline{X}1}$. A simpler notation for $\underline{\alpha}_{\underline{X}1}$ is $\overline{\underline{X}}$ or $\underline{m}_{\underline{X}}$.

The second moment matrix of $\underline{X}$ is

$$E[\underline{X}\ \underline{X}^T] = S_{\underline{X}} ,$$

a symmetric nxn matrix, where each element

$$s_{\underline{X}pq} = \sum_{x_{kn}} \cdots \sum_{x_{k1}} x_{kp} x_{kq} p_{\underline{X}}(x_{k1}, \ldots, x_{kn})$$

$$p,q=1,2,\ldots,n,$$

or

$$s_{\underline{X}pq} = \int_{-\infty}^{\infty} \cdots \int_{-\infty}^{\infty} x_p x_q p_{\underline{X}}(x_1, \ldots, x_n) dx_1 \ldots dx_n .$$

Similarly, the second central moment matrix of $\underline{X}$ or covariance matrix of $\underline{X}$ is

$$E[(\underline{X}-\overline{\underline{X}})(\underline{X}-\overline{\underline{X}})^T] = C_{\underline{X}}$$

where the multiple sum or integral may be inferred from above. Another common symbol for $C_{\underline{X}}$ is $Cov(\underline{X},\underline{X})$. Finally, in the case of two random vectors $\underline{X} \in R_n$ and $\underline{Y} \in R_m$, the covariance (or cross covariance) matrix of $\underline{X}$ and $\underline{Y}$ is

$$E[(\underline{X}-\overline{\underline{X}})(\underline{Y}-\overline{\underline{Y}})^T] = Cov(\underline{X},\underline{Y})$$

an nxm matrix.

C. 4.3 Conditional Expectation

The conditional density functions of Section C.3.6 can be used to find conditional expected values. Using the condition $X_1 = x_{k1}$ and the function $g(X_1,X_2)$,

$$E[g(X_1,X_2)|X_1=x_{k1}] = \sum_{x_{k2}} g(X_1,X_2) p_{X_2}(x_{k2}|X_1=x_{k1})$$

for discrete random variables, and for $X_1 = x_1$ and $h(X_1,X_2)$

$$E[h(X_1,X_2)|X_1=x_1] = \int_{-\infty}^{\infty} h(x_1,x_2)p_{X_2}(x_2|X_1=x_1)dx_2 \ .$$

Similar forms can be found for other conditional density functions.

C.4.4 Uncorrelated Random Variables

The operation of taking the expected value of a function of a random variable meets the homogeneity and additivity requirements of a linear operator. Thus, for a single random variable

$$E[a_1g_1(X) +\ldots+ a_ng_n(X)] = a_1E[g_1(X)] +\ldots+ a_nE[g_n(X)]$$

and similar forms can be written for functions of more than one random variable.

If the random variables $X_1,\ldots,X_n$ are <u>independent</u>,

$$E[g_1(X_1)g_2(X_2)\ldots g_n(X_n)] = E[g_1(X_1)]\times E[g_2(X_2)]\times\ldots\times E[g_n(X_n)].$$

Two random variables X_1 and X_2 are <u>uncorrelated</u> if their covariance is zero, i.e.,

$$\mathrm{Cov}(X_1,X_2) = E[X_1-\bar{X}_1)(X_2-\bar{X}_2)] = 0$$

from which

$$E[X_1X_2] = E[X_1]E[X_2] = \bar{X}_1\bar{X}_2 \ .$$

The components of the random vector $\underline{X}$ are therefore uncorrelated if the covariance matrix of $\underline{X}$, $C_{\underline{X}}$, is diagonal. Two random vectors $\underline{X}$ and $\underline{Y}$ are uncorrelated if $\mathrm{Cov}(\underline{X},\underline{Y})$ is a zero matrix.

In general, if the random variables $X_1, X_2, \ldots, X_n$ are independent, they are also uncorrelated. The converse generally is not true. However, one important exception is the Joint Normal density function C.1 where if $C_{\underline{X}}$ is diagonal, $p_{\underline{X}}$ can be factored into a product of n scalar Normal density functions which means that the $X_1, X_2, \ldots, X_n$ are independent. Also, since $C_{\underline{X}}$ is always a real, symmetric matrix, it is orthogonally similar to a diagonal matrix and an orthogonal transformation T exists such that for $\underline{X}' = T^T \underline{X}$, $C_{\underline{X}}$ is diagonalized to give

$$C_{\underline{X}'} = T^T C_{\underline{X}} T ,$$

and the $X_1', X_2', \ldots, X_n'$ are independent. The general linear transformation of a normal random vector is discussed in Section C.6.3.

C.5 CHARACTERISTIC FUNCTIONS

C.5.1 Definitions

The <u>characteristic function</u> of a discrete random variable X with density function $p_X(x_k)$ is

$$\psi_{Xd}(\nu) = E[\varepsilon^{j\nu X}] = \sum_k p_X(x_k)\varepsilon^{j\nu x_k}.$$

This can be written in terms of the $\mathfrak{F}_d$ transform of Appendix A with $\gamma = \varepsilon^{jQ\nu}$, $x_k = kQ$, and $Q = x_{k+1} - x_k$, $\forall\ k$:

$$\psi_{Xd}(-\nu) = \mathfrak{F}_d p_X(x_k) = \sum_k p_X(x_k)\gamma^{-k}$$

and $p_X(x_k)$ can be found from $\psi_{Xd}(\nu)$ as

$$p_X(x_k) = \mathfrak{F}_d^{-1}\ \psi_{Xd}(-\nu)$$

or

$$p_X(x_k) = \frac{Q}{2\pi}\int_{-\pi/Q}^{\pi/Q} \psi_{Xd}(\nu)(\exp -jx_k\nu)d\nu$$

If x_k is an integer variable, Q=1.

Similarly, the characteristic function of a continuous random variable X with density function $p_X(x)$ is

$$\psi_X(\nu) = E[\varepsilon^{j\nu X}] = \int_{-\infty}^{\infty} p_X(x)\varepsilon^{j\nu x}dx \quad .$$

This can be written as

$$\psi_X(-\nu) = \mathfrak{F}\ p_X(x)$$

and $p_X(x)$ can be found from

$$p_X(x) = \mathfrak{F}^{-1}\psi_X(-\nu)$$

or

$$p_X(x) = \frac{1}{2\pi}\int_{-\infty}^{\infty} \psi_X(\nu)\varepsilon^{-jx\nu}d\nu \quad .$$

One of the uses of a characteristic function is the computation of the moments of a random variable. We can show that

$$E(X^n) = \alpha_{Xn} = (-j)^n\ \psi_X^{(n)}(0), \qquad j = \sqrt{-1}$$

where the nth derivative of ψ_X is taken with respect to ν and then ν is set to zero. This is used for both continuous and discrete random variables, i.e., for both ψ_X and ψ_{Xd}, provided the derivatives exist.

C.5.2 Multiple Random Variables

For two or more continuous random variables $X_1, X_2, \ldots, X_n$, the random vector $\underline{X}^T = (X_1X_2\ldots X_n)$ has the characteristic function

$$\psi_{\underline{X}}(\nu_1,\nu_2,\ldots,\nu_n) = E[(\exp j(\nu_1X_1+\ldots+\nu_nX_n))].$$

The joint density functions also can be found from $\psi_{\underline{X}}(\nu_1,\ldots,\nu_n)$ by using multivariable forms similar to the single variable forms above.

The following theorem is useful when the components of $\underline{X}$ are independent:

Theorem C.1

1. $X_1,X_2,\ldots,X_n$ independent random variables in $\underline{X}^T = (X_1X_2\ldots X_n)$
2. $\psi_{\underline{X}}(\nu_1,\nu_2,\ldots,\nu_n)$, the characteristic function of $\underline{X}$,

$\Rightarrow$

1. $\psi_{\underline{X}}(\nu_1,\nu_2,\ldots,\nu_n) = \psi_{X_1}(\nu_1)\psi_{X_2}(\nu_2)\ldots\psi_{X_n}(\nu_n)$ where each ψ_{X_i}, $i=1,2,\ldots,n$, is found using $p_{X_i}(x_i)$.

Two useful forms using characteristic functions are:

1. $Y = aX + b$

   $\Rightarrow$

   $\psi_Y(\nu) = \varepsilon^{jb\nu}\psi_X(a\nu)$.

2. $Z = X+Y$, X and Y independent

   $\Rightarrow$

   $\psi_Z(\nu) = \psi_X(\nu)\psi_Y(\nu)$. (The converse is not true [5.])

Application of the real convolution theorem of Appendix A, Section A.2.4 to the second form gives

$$\mathcal{F}^{-1}\psi_Z(-\nu) = p_Z(z) = p_X*p_Y$$

or

$$p_Z(z) = \int_{-\infty}^{\infty} p_X(z-\lambda)p_Y(\lambda)\,d\lambda \quad .$$

Analogous results are true for discrete random variables.

C.6 FUNCTIONS OF RANDOM VARIABLES

A common problem in probability theory is the need for a density function $p_X(\cdot)$, where $X = X_1+X_2+\ldots+X_n$, when the density functions $p_{X_1}(\cdot),\ldots,p_{X_n}(\cdot)$ are known. Other general cases of interest occur when one or more new random variables are specified as known functions of other random variables.

C.6.1 Sum of Poisson Variables

If $X_1,X_2,\ldots,X_n$ are <u>independent</u> random variables, each with a Poisson density function, i.e., for each X_i, $i=1,\ldots,n$,

$$p_{X_i}(x_k) = \frac{\lambda_i^{x_k}}{x_k!}\,\varepsilon^{-\lambda_i} ,$$

and

$$X = \sum_{i=1}^{n} X_i ,$$

the density function for X is

$$p_X(x_k) = \frac{\lambda^{x_k}}{x_k!}\,\varepsilon^{-\lambda}$$

with

$$\lambda = \sum_{i=1}^{n} \lambda_i .$$

C.6.2 Sum of Normal Variables

If $X_1,X_2,\ldots,X_n$ are <u>independent</u> random variables each with a Normal density function

$$p_{X_i}(x) = \frac{1}{\sqrt{2\pi}\,\sigma_i}\left(\exp - \frac{(x-m_i)^2}{2\sigma_i^2}\right) \qquad i = 1,2,\ldots,n,$$

and

$$X = \sum_{i=1}^{n} X_i ,$$

the density function for X is

$$p_X(x) = \frac{1}{\sqrt{2\pi}\ \sigma} \left(\exp - \frac{(x-m)^2}{2\sigma^2}\right)$$

with

$$m = \sum_{i=1}^{n} m_i$$

and

$$\sigma^2 = \sum_{i=1}^{n} \sigma_i^2$$

If $X_1, X_2, \ldots, X_n$ are also identically distributed, i.e.,

$$m_o = m_1 = m_2 = \ldots = m_n$$

and

$$\sigma_o^2 = \sigma_1^2 = \ldots = \sigma_n^2 ,$$

let

$$X_o = \frac{1}{n} \sum_{i=1}^{n} X_i$$

and

$$Y = \sqrt{n}\ \frac{X_o - m_o}{\sigma_o} ,$$

then

$$p_Y(y) = \frac{1}{\sqrt{2\pi}\ \sigma_Y} \left(\exp - \frac{(y-m_Y)^2}{2\sigma_Y^2}\right)$$

$$= \frac{1}{\sqrt{2\pi}} \left(\exp - \frac{y^2}{2}\right)$$

since $m_Y=0$ and $\sigma_Y=1$.

C.6.3 Linear Transformation of Joint Normal Variables

If $X_1,\ldots,X_n$ have zero means and a Joint Normal density function, C.1 becomes

$$p_{\underline{X}}(x_1,\ldots,x_n) = \frac{1}{(2\pi)^{n/2}\sqrt{|S_{\underline{X}}|}} \left(\exp - \frac{1}{2}\underline{x}^T S_{\underline{X}}^{-1} \underline{x}\right)$$

where $S_{\underline{X}}$ is the second moment matrix ($|S_{\underline{X}}| = \det S_{\underline{X}}$), and $S_{\underline{X}} = C_{\underline{X}}$ when $\overline{\underline{X}} = \underline{0}$. Now consider the linear transformation matrix A which transforms $\underline{X}$ into a new random vector $\underline{Y}$. Thus

$$\underline{Y} = A\,\underline{X}$$

and A is rxn, $r \leq n$. It can be proved [4.][5.] that $Y_1,\ldots,Y_r$ also have zero means and a Joint Normal density function. Furthermore, the elements of $S_{\underline{Y}}$ and $S_{\underline{X}}$ are related by

$$\mu_{ij} = \sum_{p=1}^{n} \sum_{q=1}^{n} a_{ip}\lambda_{pq}a_{jq} \qquad i,j = 1,2,\ldots,r \quad ,$$

where λ_{pq} are the elements of $S_{\underline{X}}$ and μ_{ij} the elements of $S_{\underline{Y}}$. In matrix notation,

$$S_{\underline{Y}} = AS_{\underline{X}}A^T \ .$$

This is also true for $C_{\underline{X}}$ and $C_{\underline{Y}}$; thus taking the covariance of Y,

$$C_{\underline{Y}} = E[\underline{Y}-\overline{\underline{Y}})(\underline{Y}-\overline{\underline{Y}})^T] = AC_{\underline{X}}A^T \quad \text{for} \quad \underline{Y} = A\underline{X} \ .$$

C.6.4 Central Limit Theorems

An important property of a sequence of sums of random variables is the central limit property. Under certain conditions the associated distribution function of a standardized random variable approaches the Normal distribution function. Several theorems related to the central limit property have been proved [3.][5.], and only the commonest is given here.

Theorem C.2

1. $X_1, X_2, \ldots, X_n$ independent random variables.

2. $X_1, X_2, \ldots, X_n$ identically distributed with mean m and variance $\sigma^2 \neq 0$.

3. $$Y_n = \sum_{i=1}^{n} X_i$$

4. $$Z_n = \frac{Y_n - nm}{\sigma\sqrt{n}}$$

with distribution function $P_{Z_n}(z)$

$\Rightarrow$

1. The sequence $\{P_{Z_n}(z)\}$ satisfies

$$\lim_{n\to\infty} \{P_{Z_n}(z)\} = \frac{1}{\sqrt{2\pi}} \int_{-\infty}^{z} (\exp - \frac{\alpha^2}{2}) d\alpha \quad .$$

Thus, for large n, Z_n is a random variable whose distribution function approaches the Normal function. It is possible to remove Condition 2. if the moments of $X_1, X_2, \ldots, X_n$ of third order exist and a similar result to the above follows [5.].

C.6.5 Change of Random Variables

The functions discussed above involve linear combinations of random variables. The more general case is one of mapping a set of random variables $X_1, X_2, \ldots, X_n$ to a set $Y_1, Y_2, \ldots, Y_m$ with a known function

$$\underline{Y} = \underline{g}(\underline{X})$$

and it is required to find $P_{\underline{Y}}$ with $P_{\underline{X}}$ known, and the inverse mapping must exist.

The general approach is to find $P_{\underline{Y}}(Y_1,\ldots,Y_m) = Pr(Y_1 \le y_1,\ldots,Y_m \le y_m)$ by computing, using $P_{\underline{X}}$, the equal probability

$$Pr(\underline{X} \varepsilon R_{\underline{y}})$$

where $R_{\underline{y}}$ is a set of $x_1, x_2, \ldots, x_n$ such that for given $\underline{y}^T = (y_1 y_2 \ldots y_m)$,

$$g_1(\underline{x}) \le y_1,\ g_2(\underline{x}) \le y_2,\ \ldots,\ g_m(\underline{x}) \le y_m\ .$$

Three cases which are discussed in detail in [1.] are for

1. $Y_1 = g_1(X_1)$
2. $Y_1 = g_1(X_1, X_2)$
3. $Y_1 = g_1(X_1, X_2)$ and $Y_2 = g_2(X_1, X_2)$.

Once the distribution function $P_{\underline{Y}}$ is known, the density function $p_{\underline{Y}}$ can be found by differentiation (for continuous random variables) or differencing (for discrete random variables).

When n=m, it also is possible to find $p_{\underline{Y}}$ directly from $p_{\underline{X}}$ by using

$$p_{\underline{Y}}(\underline{y}) = \sum_i \frac{p_{\underline{X}}[\underline{h}_i(\underline{y})]}{|J[\underline{h}_i(\underline{y})]|}$$

where $\underline{h}(\underline{y}) = \underline{x}$ is the inverse of $\underline{y} = \underline{g}(\underline{x})$ and for a single-valued inverse, i=1. J is the Jacobian determinant of $\underline{g}$ and the vertical bars indicate magnitude of J, i.e.,

$$J = \det \begin{bmatrix} \frac{\partial g_1}{\partial x_1} & \cdots & \frac{\partial g_1}{\partial x_n} \\ \vdots & & \vdots \\ \frac{\partial g_n}{\partial x_1} & & \frac{\partial g_n}{\partial x_n} \end{bmatrix}$$

and the solutions $\underline{h}_i(\underline{y}) = \underline{x}_i$ are substituted and the magnitude of J for each solution is used.

REFERENCES

1. Papoulis, A., "Probability, Random Variables, and Stochastic Processes," New York: McGraw-Hill Book Company, 1965.
2. Beckmann, P., "Elements of Applied Probability Theory," New York: Harcourt, Brace, and World, Inc., 1967.
3. Cramér, H., "The Elements of Probability Theory," New York: John Wiley & Sons, Inc., 1955.
4. Laning, H. and Battin, R., "Random Processes in Automatic Control," New York: McGraw-Hill Book Company, 1956.
5. Fisz, M., "Probability Theory and Mathematical Statistics," New York: John Wiley & Sons, Inc., 1963.

Appendix D

A DIGITAL COMPUTER PROGRAM

Since many of the state models in the book are:

$$\dot{\underline{x}}(t) = A\underline{x}(t) + B\underline{u}(t) \tag{D.1}$$

$$\underline{y}(t) = C\underline{x}(t) + D\underline{u}(t)$$

or

$$\underline{x}(n+1) = A\underline{x}(n) + B\underline{v}(n) \tag{D.2}$$

$$\underline{z}(n) = C\underline{x}(n) + D\underline{v}(n)$$

it is convenient to have a program to solve these on a digital computer. Many programs of this type are available, and only the essential features are pointed out here. Actual program listings are available in the references.

If the given system is continuous-time as in Eq. D.1, Conversion VII of Chapter 3 can be used to obtain a discrete-time approximation. The essential steps are:

1. Choose T_s

2. Specify $\underline{u}(nT_s) = \underline{v}(n)$ from $\underline{u}(t)$

3. Program

$$\Phi(\lambda) = \varepsilon^{A\lambda}$$

4. Program

$$H = \int_{0}^{T_s} \Phi(\lambda)d\lambda$$

5. Compute

$$\Phi(T_s) = A_d$$

$$HB = B_d$$

The rest of the analysis program is concerned with an iterative solution of the form Eq. D.2, and a sequence of questions for a typical interactive program is as follows:

1. What are the dimensions?

   Input q =

   State r =

   Output p =

2. What are the matrices?

   A =

   B =

   C =

   D =

3. Continuous-time or discrete-time?

   If discrete-time, go to 4.
   If continuous-time,

   $T_s =$

   (Program computes A_d and B_d).

4. What is the initial state?

   $\underline{x}(0) =$

5. What is the input?

   a) For specified functions
      $v(n) =$
      (v_1 or v_2 or v_e or v_a or v_G where v_G is a sample function of a Gauss-Markov sequence)
      or

   b) From an input device
      (any $v(n)$ available from cards, tape, etc.)

6. What outputs are required?

   a) Final value of n
      $N =$

   b) Components of $\underline{y}$ required

   c) Components of $\underline{x}$ required

   d) Output format
      (line printer, plotter, etc.)

Discussions on the programming for $\Phi(\lambda)$ and H in 3) and 4) above can be found in [1.], [2.], and [3.]; flow diagrams and program listings are given. In addition, estimates of error bounds are included. The additional programs required are primarily standard matrix algebra routines.

REFERENCES

1. Kalman, R. and Englar, "A User's Manual for the Automatic Synthesis Program," N.A.S.A. Document N66-30527, June, 1966.
2. Melsa, J., "Computer Programs for Computational Assistance," New York: McGraw-Hill Book Co., 1970.
3. Cadzow, J. and Martens, H., "Discrete-Time and Computer Control Systems," Englewood Cliffs, N.J.: Prentice-Hall Book Co., 1970.

INDEX

G

H

I

V

W

X Y Z